SEDIMENTOLOGY

SEDIMENTOLOGY

Michael McLane

New York   Oxford
OXFORD UNIVERSITY PRESS
1995

Oxford University Press

Oxford   New York   Toronto
Delhi   Bombay   Calcutta   Madras   Karachi
Kuala Lumpur   Singapore   Hong Kong   Tokyo
Nairobi   Dar es Salaam   Cape Town
Melbourne   Auckland   Madrid

and associated companies in
Berlin   Ibadan

Published by Oxford University Press, Inc.,
200 Madison Avenue, New York, New York 10016

Oxford is a registered trademark of Oxford University Press

Library of Congress Cataloging-in-Publication Data
McLane, Michael.
  Sedimentology / Michael McLane.
    p. cm.
  Includes bibliographical references and index.
  ISBN 0-19-507868-3 (acid free)
  1. Sedimentology. I. Title.
  QE471.M3143   1995   552'.5—dc20   94-42952

95.00

9 8 7 6 5 4 3 2 1

Printed in the United States of America
on acid-free paper

In Memory of My Dad

Preface

Rapid progress in sedimentology seems to have begun in the 1960s and continues apace. To write a book about the present state of knowledge in such a field is like describing the current location of a pesky fly. I hope that I have presented a reasonably complete picture of current thinking, taking care not to ignore the good work done in earlier decades. Some of the more recent ideas and results presented here will succumb to the test of time and will be revised or abandoned, just as some old ideas and results have been. (Some "new" ideas, it turns out, are but rediscoveries or confirmations of ideas hatched out long ago.) Not only has knowledge increased rapidly, but technology has too. Small and inexpensive but powerful computers are changing the ways in which we gather, record, and analyze our data or test our ideas. Special tools have been developed that can "see" and identify single atoms or that can sort atoms and molecules according to small differences in their mass, size, or shape. But many of today's instrumental procedures are expensive and require special expertise, beyond the reach of some geology departments, or are considered too arcane by some professional geologists or their clients. Thus, in addition to describing the newer procedures and their results, I also describe the standard petrographic methods that, more often than not, achieve the same or similar results, though perhaps without the certainty, quantification, or precision that electron microprobe or isotope mass spectrometry might provide. Henry Clifton Sorby proved long ago that it is possible to do good, valid sedimentology without benefit of a million-dollar laboratory.

I programmed all of the graphs of (non-empirical) mathematical functions in this book, including statis-tical, fluid dynamics, and chemical equilibrium functions, for desktop automatic digital computation and plotting to help ensure their accuracy. Plotting-machine graphs were then turned over to Mr. Rachmat Atma (Bandung), a very fine draftsman, who reproduced them with wonderful precision.

I hope, dear student, that my book proves worthy of your time and attention, and that your study of it will help you to become a good sedimentologist. No doubt you will find errors in this book, or some topic not treated with the clarity or detail that you should hope for, or some topic neglected; your comments and suggestions would be most welcome.

I am grateful to anonymous reviewers for their observations about style, content, and organization. In general, I have taken their advice, and the book is much better (not to mention shorter) than it might have been otherwise.

My long-time friend, mentor, and correspondent Dr. Raymond C. Gutschick offered encouragement and solace during some difficult times, and, by virtue of his long and stubborn pursuit of knowledge, has been an inspiration to me. Also, I thank my friend and sometime colleague Dr. C. Kent Chamberlain, who got me off the truck. Dr. Rachman Gunawan (Bandung) was most helpful with last-minute library details and in interceding between my draftsman and me. Terima kasih juga kepada Dr. Jan Ong dan Dr. Ling Ong dari P.T. Geoservices (Ltd.), Jakarta dan Bandung, for their understanding and many favors.

Thanks also to certain oil company clients of mine, who permitted me to use various materials that I had prepared for them over the last few years provided that I not mention their names. I used the ample libraries of the Colorado School of Mines in Golden, the University of Colorado in Boulder, and the United States Geological Survey in Lakewood, Colorado. I used WordPerfect to prepare the manuscript; without this marvelous device I might never have taken on this project.

Jo Harden, Athletic Club Executive Suites, Denver, and my friend Elaine Carlson of Denver were most helpful with certain transworld logistical matters toward the end. To Joyce Berry, my editor at Oxford University Press, thank you for your faith in this project and your patience.

Jakarta   M.M.
June 1994

Contents

8. Terrigenous Depositional Environments, 179

9. Composition and Classification of the Carbonate Rocks; Depositional Environments, 227

SEDIMENTOLOGY

1

Introduction

O sweet spontaneous
earth how often have
the
doting

          fingers of
prurient philosophers pinched
and
poked

thee
,has the naughty thumb
of science prodded
thy

    beauty . . .

—"La Guerre"; e. e. cummings, 1933

This is a book about sediments and sedimentary rocks—what they are made of, what they look like, their origins and fates, how they are studied, and how they have contributed to our understanding of earth's mechanisms, history, and place in the universe.

The Origin and Nature of Sediments and Sedimentary Rocks

Sediments are earth materials accumulated at or near the surface of the lithosphere under atmospheric or hydrospheric conditions. They include particulate mineral matter, materials precipitated from solution, and biological products. There is also a minor contribution from space. Coherent, lithified masses of sediment are called *sedimentary rocks*. Processes that lead to the formation of sedimentary rocks include physical and chemical breakdown of earth materials, fluid entrainment and transport of sedimentary particles and solutions, gravitational settling or sedimentation, chemical and biological precipitation, and various chemical and physical processes operating at depths up to 10 or 20 kilometers below the surface of the earth that modify sedimentary materials.

Most sediments and sedimentary rocks are derived from breakdown of older rocks and are aggregates of independently formed grains, commonly from widely scattered sources. Some sediments are composed of particles formed de novo within the general site of deposition, and undergo minimal transport before being deposited. Other sediments are chemical precipitates—aggregates of crystals formed in situ from dissolved materials, and undergoing no transport whatsoever.

Many sedimentary particles are the products of biological processes; such products include the skeletal or other structural elements of animals and plants and plant or animal excretions, being the by-products of organic assimilation and respiration. Organisms not only contribute materials to sediments, but many of them also reorganize or redistribute existing sedimentary materials and alter in many ways the physical and chemical conditions under which deposition occurs.

3

Organisms also play a very important role in the destruction of all kinds of rocks and in the creation of detritus and dissolved mineral matter.

Composition and texture are the bases for classification of sedimentary rocks, as for all rocks. *Composition* refers to the mineralogy of grains derived from other rocks or to the types of grains formed at the depositional site and the mineralogy of matrix and chemical cements. *Texture* refers to absolute and relative sizes of grains and their shapes and to the quantity of matrix. Closely related to texture and commonly considered as a part of texture is fabric. *Fabric* is concerned with spatial relations among grains, such as nature of grain contacts; the manner in which grains are packed together; relationships between grains and the matrix, cement, or pore space that envelops them; and the spatial orientations of elongate or platy grains.

Sedimentary structure refers to still larger aspects of grain aggregates, features that are imposed by sedimentary processes acting collectively upon many grains at once. Many types of sedimentary structure, such as cross-bedding, ripples, and graded bedding, are imprints left in the sediment by the fluids that transported the detritus. Other imprints—a great variety of complicated markings, galleries, and perforations—are made by animals or plants that made their living on the detritus, foraging and seeking shelter, chasing or being chased. Still other sedimentary structures, such as convoluted bedding, dish structure, and loading structures, result from various mechanical adjustments that a sediment must sometimes make shortly after it is deposited.

Composition, texture, and structure are extremely varied in sedimentary rocks. Some sedimentary rocks are virtually monomineralic; others contain as many as twenty or thirty distinguishable types of components. In some, grains are all very nearly the same size; in others, grain size spans eight or ten orders of magnitude. The shapes of the grains are determined by many factors, mineralogical, hydrodynamic, biological. Some sedimentary rocks are intricately structured, showing hierarchies of sedimentary structure that range in scale from mappable to microscopic. Other sedimentary rocks or rock bodies seem to have very little structure at all. Composition, texture, and structure are controlled by the origin of the materials of which the sediments are made, the nature of the transporting media and duration and distance of transport; the physical, chemical, and biological conditions under which deposition occurs; and the nature and extent

of post-depositional changes. The complexity and variability of sedimentary materials gives them a certain robustness of response to the conditions that shape them, and it is this robustness that makes sedimentary rocks potentially so informative in our search for the past.

There are four stages in the formation of a sedimentary rock: (1) *weathering* of some other rocks, forming chemical solutions and particulate detritus conditioned for transport; (2) *transport* of the products of weathering; (3) *sedimentation* or chemical or biological *precipitation* of transported materials; and (4) lithification and other post-depositional processes, referred to as *diagenesis*. Each stage controls in some way the characteristics of the resulting rock. The composition of the source rock, and the nature and intensity of weathering in detrital source terrains ultimately control the composition of the final product, but transportational and depositional processes modify the composition of this material and distribute various components into separate venues. The nature of the transporting media and the distance and duration of their influence control texture of the final product, though texture is also inherited to some degree from the source rock. The physical, chemical, and biological conditions at the depositional site impose structure, or large-scale organization, on the material deposited, and commonly contribute additional solid materials precipitated chemically or biologically from solution; also fluids present at the depositional site become entrapped in spaces between the sedimentary particles. Diagenesis modifies the texture, structure, thickness, and shape of the deposit, adds or subtracts mineral matter, replaces or modifies interstitial fluids, and alters the derived properties of the material, such as mechanical strength and porosity. By consolidating a deposit into a monolithic mass, diagenesis also helps to preserve and protect the deposit, after some upheaval and exhumation, against weathering and erosion.

How Sedimentary Rocks Are Studied and Why

Some of the major goals of the sedimentologist are to estimate the contributions made to the final product at each stage and to describe the processes operating and the environmental conditions obtaining at each stage. To achieve these goals, the sedimentologist resorts to a wide range of observational and analytical procedures, including stratigraphy, paleontology, petrography, and chemical analysis.

The study of sedimentary materials benefits from observations made at scales from the atomic to the planetary level. At one end of the scale, the scanning electron microscope (SEM) and its various modifications and augmentations, such as the electron probe microanalyzer and energy-dispersive x-ray analyzer (EDX), have been indispensable in the examination of grain surfaces, pores and pore throats, clay particles, chemical cements, and nannofossils. The x-ray diffractometer (XRD) is in wide service in the study of clay minerals. The gas chromatograph and mass spectrometer, commonly working in concert, are used in the absolute dating of rocks; tracing the provenance of detritus; determining certain aspects of the depositional environment, such as paleotemperature; and analyzing the organic matter contained in sediments and sedimentary rocks.

At a higher scale, the light microscope, especially the petrographic microscope and its various adjuncts, such as the cathodoluminescence apparatus (CLM), is a standard tool of the sedimentologist. It is used in the routine identification of mineral grains and organic detritus in rocks, examination of the spaces between grains and the chemical cements that fill these spaces, and the study of texture. The observations made by the sedimentary petrologist lead to conclusions about source of detritus, mode of transport, depositional environment, and progress of diagenesis.

At a more human scale—where demands on instrumental aids to the senses are much gentler but demands on the human powers of observation (and sometimes on human endurance) are nonetheless great—is the examination of vertically and areally extensive bodies of rock. This is mainly the realm of the geologists working in the field—the stratigraphers, paleontologists, or geochemists who make preliminary estimates of rock type, measure sedimentary structures, identify fossils and describe their distribution, determine the local thicknesses of rock bodies and the vertical trends in composition and texture of deposits, and take samples for detailed work in the laboratory. Their tools are the hammer and hand lens, jacob staff and tape, or, for some, the core bit, "shale-shaker," and wire-line sonde.

Results obtained at the scale of the atom, mineral grain, and outcrop or well head might then be assimilated into large-scale syntheses—the sizes and shapes of entire deposits and depositional basins, lateral and vertical relationships among bodies of rock, regional patterns of composition and texture, and patterns of vector properties and fossil occurrences. At this level, the tools of long-range correlation are important: the stratigraphic cross-section, isopach and facies maps, and other large-scale maps, and, for some, the images of the earth's surface obtained by orbiting satellites. Much work on the shapes and organization of large-scale sedimentary bodies is now being done in collaboration with the exploration seismologist.

Finally, at the global or planetary scale, sedimentologists attempt to place their results into the context of continental or global tectonics or other planet-embracing processes or phenomena, such as the astronomical tides, catastrophic volcanism and the impacts of large cosmic bodies, the geomagnetic field, fluctuating solar output, eustatic sea-level excursions, secular changes in composition of the atmosphere and oceans, and organic evolution.

Typically, the sedimentologist begins in the middle, at the scale of the outcrop or drill core. Skillful stratigraphers are needed here; commonly they approach the outcrop or core not knowing what secrets and wonders it might hold. It is the source of first impressions, the director of the research, and the final arbiter of disputes and uncertainties. Stratigraphers come away from an outcrop or core with a document called a measured stratigraphic section, which is a description—commonly partly verbal, partly graphical—of the succession of strata that the outcrop or core has laid bare. The stratigraphic section is a medium by which the pertinent characteristics of the succession of strata are communicated. Insofar as a stratigraphic description falls short of our analytical goals, we take samples, little pieces of the outcrop, home with us to the laboratory, where further observation, analysis, and description can be accomplished.

At all stages of sedimentological inquiry, computing machines have come into wide use—to store, retrieve, display, organize, sort, and compare the data gathered from the far-flung outposts; to search for the patterns and trends; and to help the human researcher know what matters and what does not. Moreover, in its capacity as a "universal machine," the computer is a means of performing experiments, not by setting up complicated and expensive apparatuses but by the simple expedient of manipulating symbols. Without the computer, many such experiments, referred to as (numerical) simulations, would be impossible.

Physicists and chemists, in their pursuit of understanding of physical and chemical processes, like to examine closed systems, and will go to great lengths to isolate their systems from external influence and to control the conditions under which they perform their experiments. Sedimentologists and other geologists ordinarily cannot do this; indeed, in doing so, a geologist *becomes* a physicist or chemist. Geologists must deal with systems in which complexity is an essential characteristic, a complexity that exists at several different levels. Sedimentologists must understand not only the substances that constitute the grains, matrix, cements, and interstitial fluids of sedimentary rocks but also the ways in which these materials are made, organized, and assembled into bodies of rock and how these bodies of sediment interact with their surroundings and with one another. Thus the sedimentologist takes the fundamental processes and reactions discovered and analyzed, prodded and poked at by the chemist and the physiologist and the physicist and examines their consequences and products in complex natural systems in which a host of chemical, biological, and physical processes are interacting in the long term.

Many geological processes are very slow, and natural objects are the non-equilibrium products of incomplete or interrupted processes. Thus, when we examine a body of sedimentary rock so as to understand its origin, we are typically dealing with the results of a natural experiment that was never finished and was performed under a host of conditions always changing. From time to time we will find ourselves ignoring some sedimentological problem because it seems too messy and beyond our understanding. Commonly it is necessary to compensate for some extraneous effect, to peer beneath some overprint, or to reconstruct the mangled and scattered remnants of some sedimentological object before its true properties and proportions can be discerned and its origin diagnosed.

Nonetheless, Nature's sedimentological kitchen does occasionally produce some object that suggests a task diligently and carefully completed. There are quartz sandstones that have attained great purity and textural simplicity through long-continued or oft-renewed sedimentary processes and chemical precipitates that the fastidious chemist would be proud to claim as a product of his or her own laboratory, limestones so homogeneous that they can serve as the medium for printers' lithographic plates, and bedding planes so flat that one could shoot billiards on them. Explanations of these objects may be every bit as elusive as explanations of the complicated, adulterated,

"messy" things that dissembling Nature is more apt to present to us.

That Nature commonly leaves some task unfinished permits us to see ancient sedimentological results, in various stages of development, as a series of embryos. Thus we find sandstones, said to be immature, that have progressed only a little way through the sedimentary cycle and the supermature sandstones whose grains have sustained the hard knocks for extended periods. Or plant matter in various stages of conversion to coal, and coal on its way to graphite. Examining the youthful and mature sandstones, the low-rank and the high-rank coals, we learn about the processes that make sandstones and coals; the "finished" products by themselves are never quite so informative.

Sedimentary rocks record, however imperfectly, earth's history, including the history of organic evolution. These rocks contain much of the evidence of continental drift and the history of orogenesis. The records of paleogeography, ancient climates and glaciation, even of the changing rate of earth's rotation and variations in earth's magnetic field are stored chronologically in the sedimentary rocks. Some of the oldest rocks known on earth are sedimentary. Probably every instant of geologic time of the last half-billion years or so is represented somewhere on earth by a sediment. Earlier times are less completely represented, but records of some events that occurred as long as 3 or 4 billion years ago have been recovered from sedimentary rocks. This attests not so much to the durability or immutability of these rocks but merely to chance survival of some of them. In fact, Nature has been rather casual about keeping records, and the sedimentary record that she *has* made is corrupted and incomplete. Sedimentation is unsteady and sporadic; a deposit now is later debited. Of all the sedimentary rock ever made, probably less than 20 percent is preserved today, due in large part to Nature's habit of recycling the old stuff.

In reconstructing a history of the earth, we sedimentologists must be content with a fragmentary record of what have been rather local, special, and perhaps ephemeral environments where sedimentation happens to have been going on. We might delight in the opportunity to examine a large delta somewhere, but the existence of this delta implies some larger expanse of earth's surface where, though many wonderful and significant events might be taking place, little or no rock record is being kept; it is a place, in fact, where rock records are being destroyed. Thus the body of historical information that we sedimentologists have assembled is biased toward natural systems that are capable of keeping and have kept records.

Sediments and sedimentary rocks are of supreme importance to the well-being of the biosphere and the human condition. They play a vital role as a regulator of atmospheric and oceanic composition, acting as a reservoir for excess carbon dioxide, chloride, sulfate, and other chemical species and as a pH buffer. They store solar energy in chemical form. They store water and nutrients and are the substrate for many living things.

Sediments and sedimentary rocks are the raw material for many of humankind's important products, commonly fulfilling needs available from no other source. Glass and ceramics, including Portland cement, are made from sandstone, limestone, and shale. Limestone is a flux in steelmaking; quartz sand is an abrasive and a metal-casting medium; clays are used in filters and ion-exchange reactors by the chemical industry, as fillers and glazes by the paper industry, as a drilling medium for boring deep wells, and as a catalyst in petroleum refining. Some dimension stones, crushed rock, and concrete aggregate are products of sedimentary materials. The fossil fuels—coal, oil, and natural gas—are formed by sedimentary processes and are obtained from sedimentary rocks; these are also the raw materials of our plastics and petrochemicals industries. Much of our primary nuclear fuel also is extracted from sedimentary rocks. Sediments are a major source of iron, aluminum, salt, gypsum and other sulfates, potash, phosphate, nitrate, fluoride, boron, barium, and selenium. Gold, silver, tin, titanium, and diamonds are commonly beneficiated from placer deposits. Underground space—as a receptacle of high-value hydrothermal deposits; a container for ground water, oil, and gas; and a place for storing human waste materials—is hosted mainly by sediments and sedimentary rocks.

Study of sedimentary processes is vital to an understanding of how Nature disposes of the residues of human enterprise and how natural surficial processes respond to human alterations and manipulations. What becomes of the solid, liquid, and gaseous wastes that we pour into the ground, disperse to the wind, or dump into the rivers and oceans? How and where might nuclear wastes be stored for the long term? What are the effects of forestry, agriculture, mining, and the production of crude oil and water from wells, construction of levees and dams, groins and breakwaters, and dredging and draining of wetlands? These are questions that sedimentology and sedimentologists can help to answer.

The Relationship of Sedimentary Rocks to Other Rocks and of Sedimentology to Other Endeavors

Sedimentary rocks are part of a continuum that embraces all rocks (Fig. 1-1). There is wide latitude in the physical conditions, most importantly temperature and pressure, obtaining in earth's crust, where rocks are formed, destroyed, and reformed. Sedimentary rocks lie in the region of "moderate" temperatures and pressures that are, by and large, the conditions experienced and enjoyed by living things; a broad range of higher temperatures and pressures is the realm of the metamorphic rocks; at still higher temperatures, igneous melts arise. The different rocks are not isolated from one another; rather, the same raw materials drift from field to field, pausing here and now as part of some great mass of sediment and then moving passively on to pursue a temporary existence inside some large pluton or metamorphic aureole, destined to return in some unknown time and place to the sedimentary realm. Many of the materials of sediments were wrought and forged under conditions very different from the conditions under which they exist as sediments.

Sedimentary materials envelop the lithosphere, covering almost all of it, though in some places only very thinly. Some of these materials, such as those in soils or regoliths, have only recently been released from older rocks (Fig. 1-2); some of these materials are in transport in rivers or on beaches; some are participating in life processes; some are in temporary storage on flood plains or aeolian dunes; some are in long-term storage, buried beneath thousands of meters of younger sediment; and some are departing this realm of the *sedimentary cycle*, suffering death by fire and pressure in the roots of mountains and the nether regions of subduction zones.

The transformations in mineral composition and texture that occur with increasing temperatures and pressures are generally irreversible, but weathering ultimately brings refractory igneous and metamorphic rocks back into the sedimentary field. Weathering is driven by the energy of the sun; igneous and metamorphic processes are driven by energy of earth's in-

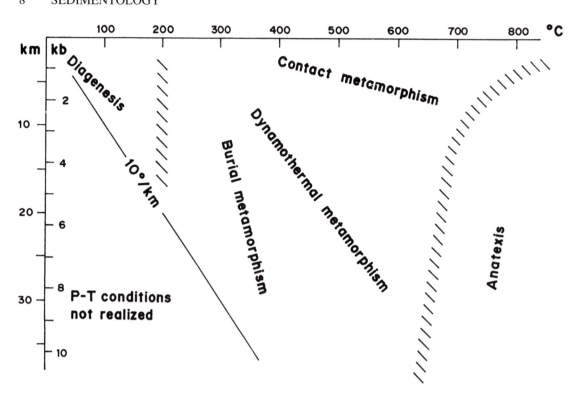

Fig. 1-1. Temperature and pressure conditions of formation of rocks. (After Winkler, 1967, fig. 1.)

terior. These two energy sources have been independent almost from the beginning of the solar system, and the volume of sediment on the earth reflects the relative importance of these energy sources and some sort of compromise or equilibrium between them. The total volume of sediments in earth's crust and the rate at which sediments are driven through the cycle represent a balance between the processes that construct sediments and the processes that destroy them.

Many volcanic rocks and plutonic rocks display textures and structures similar to some of those encountered in sedimentary rocks, and these features result from processes similar to sedimentary processes. Ignimbrites are products of hot air- or gas-suspension volcanic turbidity currents; in some ways these rocks are similar and in some ways different from oceanic and lacustrine sedimentary turbidites. Lahars are volcanic mudflows; they behave much like the ordinary alluvial-fan mudflows familiar to the sedimentologist and yield diamictites of similar texture, internal organization, and geometry. Volcanic lava flows pro-

duce stratified deposits; stratification is generally regarded as proper to sedimentary rocks and the purview of the stratigrapher. Frozen lava flows commonly contain elongate vesicles or aligned phenocrysts that record the flow patterns existing before the lava solidified; records of flow patterns occur similarly in the sediments deposited by rivers. Volcanism on the deep ocean floor is thought to be responsible for precipitation of some bedded cherts, of interest especially to the sedimentologist. Deposits of airfall ash commonly come to be enclosed within sedimentary sequences, and their presence can be useful to the sedimentologist as a correlation tool. Large ashfall events can have severe effects on environment and ecology, as mass mortality at the base of some ancient ashfall deposits and recent human experience attest. Lava flows, ashfalls, and lahars commonly contain fossils, especially the remains of trees and other plants; sedimentologists like to think that fossils are reserved to them alone. In fact, volcanic rocks under some circumstances preserve a more complete and more accurate record of a

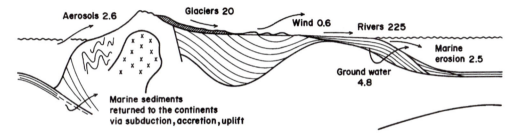

Fig. 1-2. The rock cycle (rates, 10^{14} g/year.) (After Garrels and Mackenzie, 1971, fig. 4.1.)

living community than sedimentary rocks might have—witness the erect petrified forests of Amethyst Cliff in Yellowstone Park or the wonderful but tragic preservations at Pompeii and Herculaneum. While sedimentologists might forget that volcanic rocks occasionally are fossiliferous, volcanologists and igneous petrologists might tend to ignore or disdain the fossils that they see, even though the taxonomy and taphonomy of these fossils surely bears upon such matters as age (date), violence, and suddenness of volcanic processes and modes and patterns of dispersal of volcanic materials. Volcanic processes sometimes create special sedimentary environments populated by organisms ingeniously or outlandishly designed to take advantage of them—the algae that live in the boiling waters of geysers or thermal springs, the miraculous fish that swim the hot, caustic lakes volcanically exhaled out of the East African rift; the alien food chain based on sulfur acids emitted by the "black smokers" of deep-sea spreading centers. (One imagines that the biota exists only to prove that anything is possible.) Volcanic rock bodies commonly grade laterally into sedimentary rocks; volcanologists and sedimentologists might find themselves elbow to elbow at the outcrops where these gradations occur. All of these geologists are broadening their horizons, the one expanding a vision of volcanic process and product to include interactions with air and water, the other groping toward the very beginnings of some grain collection's protracted sedimental journey.

Plutonic rocks are aggregates of crystals assembled out of swarms of loose atoms, and so are some sedimentary rocks; the textural principles by which order and rate of crystallization are diagnosed apply similarly to both kinds of rocks, and these very different rocks share many textural characteristics. In certain

peridotites, graded-bedding and cross-bedding have been observed, a result of loose, early-formed mineral grains having been transported by density currents in an encompassing magma. Here is an example of fluid transport and sedimentation of particles taking place under conditions not ordinarily attended to by the sedimentologists, though the fundamental process and the result are quite familiar to them. Sedimentary differentiation, some have argued, is responsible for much of the variety in igneous rock composition. The contrast between basalt and andesite, for example, may be due largely to sedimentary differentiation.

Metamorphism is a process inimical to a sedimentologist's investigations, adumbrating the records of sedimentary processes; but sedimentologists cope with diagenesis without undue distress, and just because these rocks have crossed the boundary between diagenesis and metamorphism (a gray line, anyway) is no reason for them to abandon their pursuit. Diagenesis is really but a prelude to metamorphism and might be viewed as a mild case of it. Mineralogical and textural changes brought about by metamorphism must be considered in relation to original sedimentary composition and texture. Cooperation between the metamorphic and sedimentary petrologist must surely advance understanding of both kinds of rock.

Those metamorphic rocks that are the products of alteration of sedimentary rocks commonly retain sedimentary features, such as compositional contrasts between beds, sedimentary rock-body geometries, various sedimentary structures, even fossils. The mineral content may remain largely unaltered; even if it does change, elemental composition commonly is unmolested. Certain grains in sedimentary rocks, such as zircons, can survive even high-grade metamorphism, retaining the shapes that sedimentary processes had imposed on them. The banding or interlayering of rock types so often observed in metamorphic terrains is to a large extent a remnant and reflection of sedimentary structure and process. The ways that fossils or cross beds are deformed are an index of the magnitude and orientation of stress fields. The degree of chemical alteration of certain fossils, such as spores and conodonts, as revealed simply by their colors, is an index of the thermal state attained by a sedimentary rock during its diagenesis and metamorphism.

As composition and texture of sedimentary detritus are inherited to some degree from igneous and metamorphic progenitors, an understanding of the petrology of these parent rocks helps us to understand the gross composition of sedimentary detritus, the significance of occurrences of minor minerals, and the sizes and shapes of detrital grains, the trace-element content of detrital grains, and the mineralogy of inclusions.

Important relationships exist between sedimentation and tectonics. The contrasts between sedimentary rocks formed on cratons and those formed at continental margins or alongside mountain ranges have long been remarked and debated. *Flysch* and *molasse* are terms that have been around for many decades; they not only describe the general lithologic and stratigraphic features of certain large bodies of rock but also imply certain contrasting associations between these rock bodies and orogens or orogenesis. The character of tectonism and its intensity determine to a large degree the composition and texture of associated sediments, rates of sedimentation, and the internal organization and geometry of sedimentary bodies. This is very evident in the sedimentary fills of aulacogens and other continental rifts. By virtue of its effects on structure and topography, tectonism controls the courses of rivers, the flow of the wind, the distribution of rainfall, and the patterns of sediment dispersal. Many lakes owe their existence to some structural element—a downwarp, fault, volcanic caldera, or crustal rift. Ancient reefs of cratonic basins, it appears, commonly were localized on faults or flexures; many modern oceanic reefs owe their existence to oceanic volcanoes. Finally, the contrast between marine and non-marine sediments and sedimentation reflects, ultimately, differences in crustal thickness and density, and forces in the mantle that push and pull at crustal masses and alter their sizes and shapes, elevations, and locations.

The converse, that sedimentary processes influence tectonic processes and structure, also is true. Sediment loading induces crustal faulting and flexure. Differences in structural competence between limestone formations and shale formations—and the absolute and relative thicknesses of these formations—control structural style in thrust belts, determining the ways in which stress is transmitted, geometries of folds, depths at which detachments occur, and thicknesses of allochthonous plates. Some tectonic-like structure is peculiarly sedimentary in origin—clastic dikes, due to buried sands with abnormal pore pressures; salt diapirs, due to subsurface flow of substantial volumes of evaporitic halite; the vertical faults and fault-bounded rock bodies created by dissolution of deeply buried beds; the listric growth faults due to compaction, subsidence, and slope failure of large sedimenting masses. Some geologic processes are equivocally sedimentary or tectonic—the sliding of a large coherent rock mass (olistolith) down a long, gentle sedimentary slope, driven only by its weight, is like a sedimentary process on a tectonic scale.

More and more in recent years, we are finding that sedimentation responds to astronomical rhythms, including daily, monthly, and annual cycles. Sedimentary rocks furnish perhaps the only record of these cycles for the distant past, and this record is precise and almost unequivocal. The longer-term effects of the earth's axial precession and periodic changes in orbital parameters seem to have influenced earth's climate; climatic cycles with periods of tens to hundreds of thousands of years are recorded in sedimentary rocks. Cosmic disturbances and catastrophes, such as bolide impacts, have made their mark on sedimentation; certain energy bursts originating even as far away as the galactic center may have left some record in the rocks. Sedimentary rocks record not only the events themselves but also the long-lasting or permanent consequences, such as mass extinctions.

Now, as the exploration of other planets and their moons proceeds, sedimentologists and other geologists are called on to help provide answers to the same questions about other planets that we have already addressed here on earth. New answers to these old questions, while expanding our understanding of the dynamics, origin, and history of the surfaces of other planetary bodies, surely will aid the earth-oriented geologists, as they are now presented with other examples of planetary surficial systems which must ultimately be compared with our own more familiar example. These systems are so alien that commonly there is some uncertainty or ambivalence about whether the process at hand is indeed sedimentary or is more appropriately considered as an igneous or tectonic process.

We suspect that we sedimentologists, while advancing our own field, have much to offer to the fields of igneous and metamorphic petrology, structural geology, volcanology, and oceanography; to ecology, meteorology, exobiology, and cosmology; and that they have much to offer to us. "It is probably quite true generally," said Werner Heisenberg, "that in the history of human thinking the most fruitful developments frequently take place at those points where two different lines of thought meet." (quoted in Fritjof Capra, 1975, *The Tao of Physics*). One of the impressions that I hope you will gain from your study of this book is that sedimentologists benefit from the contributions made by the "hard-rock" geologists and tectonicists—and by the nuclear physicists, organic chemists, fluid dynamicists, materials scientists, systems analysts, and many others—and that it is important that the modern sedimentologist have the inclination and the ability to converse productively with them. As mathematician Benoit Mandelbrot has observed (*The Fractal Geometry of Nature*, 1982, p. 27), "... boundaries between scientific disciplines are largely a matter of conventional division of labor between scientists." No more; no less.

2

Textures of Sedimentary Rocks

Texture refers to the geometrical and spatial characteristics of an aggregate of grains or crystals and to the properties that derive from the geometry of the aggregate. Among the important textural properties of sedimentary rocks are relative and absolute sizes of grains and shapes of grains. Spatial relationships among grains, such as proximity of grains to one another and their orientations, are referred to as **fabric**. Properties that emerge from texture include porosity, permeability, specific surface, bulk density, angle of repose, compactibility, and mechanical strength. These properties of utilitarian or practical value are to a large degree determined by texture, though often in quite complex ways. Two fundamental textural types occur in sedimentary rocks, **granular** (or **clastic**), and **crystalline**. Granular aggregates come together as a result of mechanical processes acting upon collections of grains, each grain responding to the processes more or less independently. The sizes and shapes of the grains are determined largely by processes that acted before sedimentation. Grain fabrics may reflect the process of sedimentation itself. Crystalline textures result from chemical processes that create crystals in situ out of dissolved atoms or ions. The sizes and shapes of the crystals are determined largely by the processes actually creating the aggregate. There is a middle ground of sedimentary rocks composed of chemically precipitated crystals that, like falling and blowing snow, were moved about after they grew—and, of course, many sedimentary rocks that are considered to be clastic contain a substantial crystalline component between the grains, which constitutes the cement.

The most complex textures occur in carbonate rocks that are composed largely of fossils or fossil fragments (skeletals). A sediment composed of skeletal grains may flaunt great textural complexity because it is a mixture of many different types of skeletals, large

grains and small, grains elongate, platy, or compact; grains with intricate webbing and ornamentation; grains smooth and simple; all having come together in the same sediment. We generally know fairly confidently what these skeletal particles looked like at the beginning, before breakage or abrasion, and can readily discern any modifications and measure their effects. Aggregates of such grains have the potential to develop complex fabrics. On the other hand, Nature churns out copy after copy of a particular skeletal, all having much the same size and shape, the same density, mechanical strength, and hardness, so that what seems improbable—a sediment containing large numbers of complex and ornate but virtually identical grains—becomes unexceptional. Chalks composed of planktonic foraminifers or coccolithophorids are important examples.

Textures and fabrics created under conditions of deposition are subject to modification under diagenesis, yielding readily in many ways to even brief or mild diagenetic processes. We explore some of these changes—due to compaction, cementation, dissolution, and recrystallization—in Chapter 11.

The Statistical Distribution of Grain Size

The diameters of sedimentary particles vary from a few meters to less than a micrometer, and this very broad spectrum is divided into three regions called **gravel**, **sand**, and **mud**:

Gravel	>2.0 mm
Sand	0.063 to 2 mm
Mud	>0.063 mm

The relative proportions of these three components serve as a basis for the gross textural classification of detrital rock materials (Fig. 2-1). Further division of

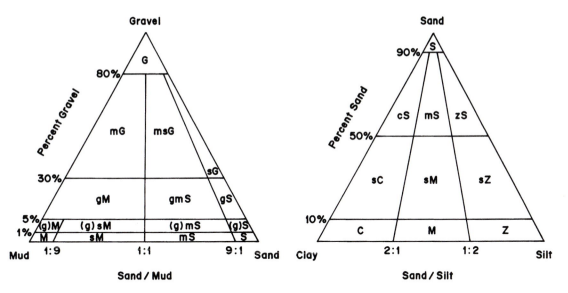

Fig. 2-1. Textural classification of granular sediments. (According to Folk, 1954, figs. 1a, 1b, with permission of The University of Chicago Press.) Scale distorted on gravel-sand-mud plot for clarity. The label (g)mS, for example, is an abbreviation for slightly gravelly muddy sand.

the grain size spectrum is based on a geometric progression and follows conventions established long ago (Table 2-1). The coarsest sedimentary particles are blocks between 1 and 10 meters across and weighing perhaps as much as 10 metric tons; the finest particles are clay mineral crystals between 0.01 and 0.1 micrometer in diameter, about the size of a typical virus. The ratio of the size of the coarsest to finest sedimentary particles is about the same as the ratio of the size of the earth to size of a beach pebble.

Because texture deals with large collections of objects, statistical methods generally are invoked as aids to description and analysis. This is especially true of grain size. Grains in sedimentary rocks are not all of the same size, so an average or mean value, and some measure of the range or spread, are useful and important as a summary of the distribution of grain sizes. Beyond mere summaries, it is useful to compare a measured grain size distribution to another or to some standard distribution model, derived theoretically, and to describe and attempt to explain differences between these distributions. Much work along these lines has been done, and departures of real distributions from standard models seem to be geologically significant and to some degree predictable.

Several statistical models have been applied to ge-

ological grain size distributions, based in part on theoretical considerations and in part on simple inspection of actual data. They are the *(Gaussian) normal*, *lognormal*, *Weibull* (or *Rosin-Rammler*), and a *log-"hyperbolic"* distribution. The gamma distribution and Pearson types I and IV distributions have received some attention; their utility in geological inquiry has not yet been adequately demonstrated, though Tanner (1958) and LeRoy (1981) have pressed the matter of Pearson's continuum of distributions as an effective descriptive system for grain size populations.

Pre-eminent among grain size distribution models is the lognormal, which is the expected distribution emerging from a proportionate effect rule:

Suppose that the size of a grain changes gradually in response to repeated breakage or continued abrasion, and that the rate of change is proportional to the size of the grain, so that abrasion reduces large grains more rapidly than small ones. As such a process operates on some ensemble of loose grains, even a collection bearing some completely random or unknowable size distribution at first, the distribution of sizes gradually (perhaps very quickly) approaches lognormality (Fig. 2-2). Natural processes that sort grains according to size apparently also create lognormal distributions, even in the absence of abrasion.

Table 2-1. Grain Size Classification

mm	μm	phi	Wentworth Class
4096		−12	
1024		−10	Boulder
256		−8	
64		−6	Cobble
16		−4	Pebble
4		−2	
3.36		−1.75	
2.83		−1.50	
2.38		−1.25	Granule
2.00		−1.00	
1.68		−0.75	
1.41		−0.50	
1.19		−0.25	Very coarse sand
1.00		0.00	
0.84		0.25	
0.71		0.50	
0.59		0.75	Coarse sand
0.50	500	1.00	
0.42	420	1.25	
0.35	350	1.50	
0.30	300	1.75	Medium sand
	250		
0.25		2.00	
0.21	210	2.25	
0.177	177	2.50	
0.149	149	2.75	Fine sand
	125		
0.125		3.00	
0.105	105	3.25	
0.088	88	3.50	
0.074	74	3.75	Very fine sand
	62		
0.063		4.00	
0.053	53	4.25	
0.044	44	4.50	
0.037	37	4.75	Coarse silt
	31		
0.031		5.00	
	15.6	6.0	Medium silt
	7.8	7.0	Fine silt
	3.9	8.0	Very fine silt
	2.0	9.0	
	0.98	10.0	
	0.49	11.0	Clay
	0.24	13.0	
	0.12	14.0	
	0.06		

A density function is a mathematical function, derivable from probability theory, that describes the expected frequency of each value of the variate (in this case, grain size). The **lognormal density function** is

$$f(x) = 0 \qquad (x \leqslant 0)$$

$$f(x) = \frac{1}{\sqrt{2\pi}\sigma x} \exp[-(\ln x - \mu)^2/2\sigma^2] \quad (2\text{-}1)$$

$$(x > 0)$$

where μ is the mean of $\ln x$ and σ^2 is variance of $\ln x$; mean and variance are called the **parameters** of the distribution. A graph of the function (Fig. 2-3) shows that it has a maximum or peak and falls off on either side of this peak, declining to zero frequency at the origin, and asymptotically approaching zero frequency at large values of the variate. Note that the function is not symmetrical but appears to lean toward the origin, favoring the smaller values of the variate.

A distribution function is the integral of a density function; in many respects it is more useful than a density function. The **lognormal distribution function** is

$$F(X) = \int_{x=0}^{X} f(x) \, dx \qquad (2\text{-}2)$$

Its graph (Fig. 2-3) is a sigmoidal curve (appropriately similar to an integral sign). (The integral is readily evaluated by a computer by reference to the so-called *error function* and its polynomial expansion.)

The lognormal distribution has some special properties that enhance its utility as a grain size model. If a distribution of grain diameters is lognormal, then so is the distribution of grain volumes and grain surface areas; it follows that, if all the grains have the same density, then the distribution of their masses also is lognormal.

It is standard practice in grain size studies to transform the grain diameter in such a way that the lognormal distribution becomes (Gaussian) normal. This operation is called the **phi transformation** (Krumbein, 1934):

$$\varphi = -\log_2 (x/x_o) \qquad (2\text{-}3)$$

where (if x is given in mm) x_o, a reference grain size, is fixed at 1 millimeter. The distribution of phi is (Gaussian) normal (Fig. 2-4). The phi transformation has three special advantages: it condenses a very broad spectrum of grain size into a much smaller range of phi values; it "opens up" or expands the important

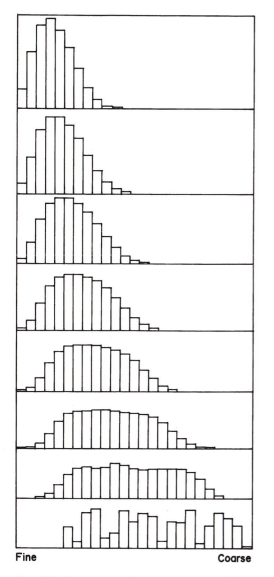

Fig. 2-2. Computer-simulated progress toward lognormality. Random histogram (bottom) becomes lognormal under successive applications of a proportional-effect rule. Mean shifts to smaller values and variance decreases.

constant. Thus, grains of 1-phi size are twice the size of 2-phi grains and half the size of 0-phi grains.

The ***mean*** is a measure of the location of the function—that is, its position with respect to the abscissa (variate)—and is the centroid or "center of gravity" of the distribution. ***Variance*** is a measure of the spread, or dispersion; the square root of variance is called ***standard deviation*** and in grain size populations is a measure of ***sorting***. The value of the variate at maximum frequency is called the ***mode***. In the normal (or phi-transformed lognormal) distribution, the mean coincides with the mode. Moreover, the ***median***, which divides the total frequency of the distribution (area under the curve) in half, also coincides with the mean and mode.

Differences between the mean and some other value of the variate are commonly given in units of standard deviation; these units are called ***t units***. The deviation in t units of value x from the mean is $t = (x-\mu)/\sigma$. The interval from one standard deviation below to one standard deviation above the mean, $\pm t$, embraces about 84 percent of the total area under the density curve; $\pm 2t$ accounts for nearly 98 percent.

Departures from Lognormality

Real grain size populations never fit a distribution model perfectly. Part of the reason for this involves the care with which the sample was taken and plain and simple errors of measurement, or false, or half-true, but unavoidable assumptions about the analytical procedure. Aside from this, there are random departures due in part to random fluctuations in the conditions under which the grain population was created, and, more significantly, systematic or non-random de-

region of very small grain sizes; and the normal distribution that emerges is symmetrical, and thus easier to work with; for one thing, mean, median, and mode of a normal distribution (or any symmetrical distribution) coincide. We note, moreover, that the *ratio* of grain sizes differing by a constant phi value also is

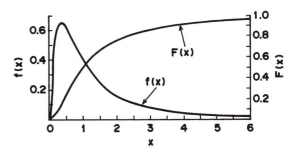

Fig. 2-3. Lognormal density function $f(x)$ and distribution function $F(x)$ with mean of $\ln x = 1$ and variance of $\ln x = 1$.

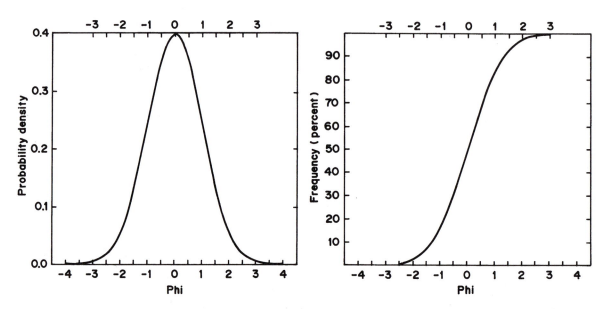

Fig. 2-4. Lognormal density and distribution of the phi-transformed variate; these are the same functions shown in Fig. 2-3. Phi mean = 0 and phi variance = 1.

partures that seem to reflect complications or pathologies in the depositional system.

Two kinds of non-random departure of real grain size distributions from lognormality, *skewness* and *kurtosis*, have been closely scrutinized. Skewness is a measure of asymmetry—a grain size frequency excess (or deficit) on one side or the other of the mode. If there is an excess on the coarse tail (that is, to the left of the mode), the frequency curve appears to lean to the right and is said to have negative (or coarse) skewness. If an excess occurs instead in the fine tail (the tail to the right), then the distribution is said to have positive (or fine) skewness. A distribution that is perfectly symmetrical has zero skewness.

Kurtosis measures the length of the tails relative to a normal distribution of the same variance, indicating an excess or deficit in the tails (that is, both tails simultaneously) as compared to the central (modal) region of the distribution. A perfect normal distribution has a kurtosis of 3.0. If there is an excess in the tails, then the peak of the distribution appears flat and the distribution is said to be *platykurtic* (kurtosis < 3). If an excess occurs instead in the modal region, then the peak is too sharp and the distribution is said to be

leptokurtic (kurtosis >3). A given frequency curve may show both non-normal skewness and non-normal kurtosis.

Multimodality is the presence of more than one peak (mode) in a distribution. When there are two only, then the term *bimodality* is ordinarily used. The distinction that sedimentologists make between grains (or framework) and matrix in many sedimentary rocks is tacit admission that bimodality exists. Multimodality nearly always results in non-normal skewness and kurtosis. Multimodality is generally regarded as a signal that the population is a composite—that is, a mixture of distinct populations—though composite populations are not necessarily multimodal.

Truncation of a population also results in skewed or kurtic grain size distributions. Truncation is the absence of all grains larger or smaller than a given size; some sedimentologists think that it occurs in many natural grain populations.

One may visually compare two (or more) distributions by plotting them on a cumulative frequency ordinate specially scaled so that a perfect normal distribution plots as a straight line. (The special ordinate scale is obtained from the inverse of the normal distribution.) Poorly sorted populations plot as lines having low slope; well-sorted populations plot as lines having high slope. Deviations from the normal distribution are plainly manifested as departures from the straight line. Fine-skewed populations make convex-upward curves (Fig. 2-5a); coarse-skewed populations make convex downward curves (Fig. 2-5b). Populations with non-normal kurtosis plot as curves with inflection points (Fig. 2-5c). Multimodal distributions display multiple inflection points (Fig. 2-5d); steeply sloping tail regions characterize truncated distributions (Fig. 2-6).

Summary Statistics

The summary statistics—mean, standard deviation, skewness, and kurtosis—of frequency distributions of phi-transformed grain size measurements can be obtained graphically or analytically; the two methods do not, in general, give quite the same results. The arithmetic **mean**, or average, of a set of values or variates $x_1, x_2, \ldots, x_k$, is

$$X = \frac{1}{k} \sum_{n=1}^{k} x_n \qquad (2\text{-}4)$$

Sedimentologists nearly always group their grain size data, dividing the total range of grain size into a set of contiguous classes or class intervals (this is usually a consequence of the analytical procedure), and prepare *histograms* and *cumulative frequency polygons* (Fig. 2-7) from which *graphic measures* can be determined. If the values are *grouped* and the groups contain $f_1, f_2, \ldots, f_k$ numbers of representatives (frequencies), then the mean is

$$X = \frac{1}{n} \sum_{i=1}^{k} f_i m_i \qquad (2\text{-}5)$$

where $n = \sum_{i=1}^{k} f_i$, that is, total number of measurements, and m_i is the phi midpoint of the ith group or class. But most of our grain size analyses are made with sieves and sedimentation tubes, which do not count grains; we are resigned to recording the f_i's as weight fractions rather than as grain counts.

The shape of a given distribution can be described quantitatively in terms of its moments. The **moments** about the mean are given by

$$\mu_r = \frac{1}{n} \sum_{i=1}^{k} f_i (m_i - X)^r \qquad (2\text{-}6)$$

r referring to the rth (central) moment. The first central moment is identically zero. The second central moment is the variance. Skewness is related to the third moment, and is defined as $\gamma_1 = \mu_3/\sigma^3$. Kurtosis, related to the fourth moment, is $\gamma_2 = \mu_4/\sigma^4$; some workers use the **coefficient of excess**, which is $\gamma_2 - 3$. If moments are to be computed from grouped data, then certain corrections might be necessary (Sheppard's corrections).

Statistical moments can be evaluated graphically as well, based on histograms or cumulative frequency polygons. To evaluate the graphical equivalents of the first four moment measures, one "samples" the cumulative frequency curve at several points, called *quantiles* or *percentiles*. The value of the variate where the cumulative polygon crosses the frequency value of 0.5 or 50 percent (Fig. 2-7) is called the 50th percentile; it is the median of the distribution. Also obtained are the 16th and 84th percentiles, the 5th and 95th, and the 25th and 75th percentiles. The graphic measures in most common use today are those defined by Folk and Ward (1957):

Graphic mean:
$$(\varphi_{16} + \varphi_{50} + \varphi_{84})/3 \qquad (2\text{-}7)$$

Inclusive graphic standard deviation:
$$(\varphi_{84} - \varphi_{16})/4 + (\varphi_{95} - \varphi_{05})/6.6 \qquad (2\text{-}8)$$

Inclusive graphic skewness:
$$(\varphi_{16} + \varphi_{84} - 2\varphi_{50})/(2\varphi_{84} - \varphi_{16}) \qquad (2\text{-}9)$$
$$+ (\varphi_{05} + \varphi_{95} - 2\varphi_{50})/(2\varphi_{95} - \varphi_{05})$$

Graphic kurtosis:
$$(\varphi_{95} - \varphi_{05})/2.44(\varphi_{75} - \varphi_{25}) \qquad (2\text{-}10)$$

Graphic kurtosis compares dispersion in the tails with dispersion in the modal region and thus differs from moment kurtosis, which measures the length of tails relative to tails of the normal distribution of same variance. Graphic kurtosis of the normal curve is 1.0, but moment kurtosis is 3.0; these two measures of kurtosis must not be directly compared.

The difference between graphic and moment measures is that the graphic measures derive their values from only a few selected points on the frequency curve, whereas the moment measures are based on all points of the curve. Commonly in geological grain size analyses, the region of largest grain sizes is not adequately represented in the sample, and the region of smallest grain sizes (silt and clay) is not accurately known; analysis of the fine tail requires special tech-

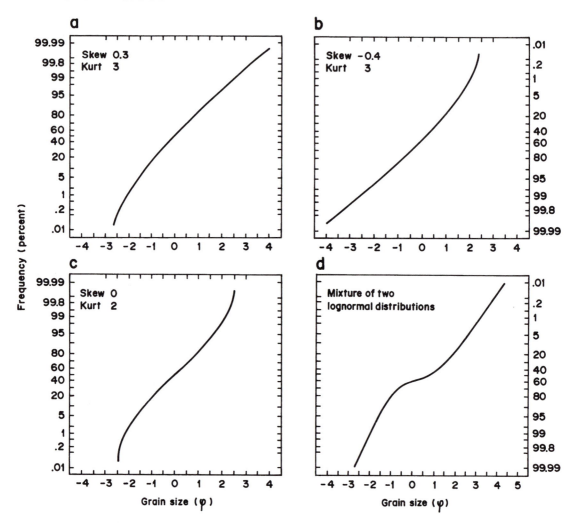

Fig. 2-5. Departures from the standard grain size model. Cumulative distributions plotted on probability ordinate: **(a)** positively skewed distribution; **(b)** negatively skewed distribution; **(c)** leptokurtic distribution; **(d)** a mixture of two lognormal distributions.

niques that perhaps were not or could not be employed. In such cases moment measures are not strictly valid.

Folk and Ward (1957) subdivided the range of each of the graphic measures, and offered a descriptive term for each division. The range of inclusive graphic standard deviation of grain size distributions is divided into seven parts:

<0.35		Very well sorted
0.35 to	0.50	Well sorted
0.50 to	0.71	Moderately well sorted
0.71 to	1.0	Moderately sorted
1.0 to	2.0	Poorly sorted
2.0 to	4.0	Very poorly sorted
	>4.0	Extremely poorly sorted

The range of inclusive graphic skewness is confined between −1 and +1; Folk and Ward divided this range into five parts:

1. to	0.3	Strongly fine skewed
0.3 to	0.1	Fine skewed
0.1 to	−0.1	Near symmetrical
−0.1 to	−0.3	Coarse skewed
−0.3 to	−1.0	Very strongly coarse skewed

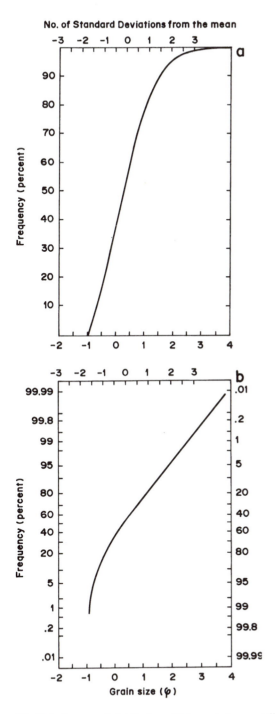

Fig. 2-6. Lognormal distribution with truncated coarse tail: (**a**) cumulative phi-transformed distribution plotted on linear ordinate; (**b**) same function plotted on probability ordinate.

Inclusive graphic kurtosis ranges from an absolute minimum of 0.41 to very large values, though natural grain size distributions rarely have kurtosis values exceeding 5:

<0.67		Very platykurtic
0.67 to	0.9	Platykurtic
0.9 to	1.11	Mesokurtic
1.11 to	1.5	Leptokurtic
1.5 to	3.0	Very leptokurtic
	>3.0	Extremely leptokurtic

Causes of Non-normality

Why do grain size distributions so often deviate from the standard model? Many geologists consider that the hydraulics of sediment transport impresses non-normal features on grain size populations; others look to the common occurrence of mixing of populations having different transportational histories; still others blame the detrital sources for supplying transportational or depositional systems with populations that are decidedly non-normal in the first place. Finally, the lognormal model might be judged inappropriate to some sedimentary processes, observation or theory better served by some other model. Some detrital grain size distributions may be Gaussian-normal rather than lognormal and thus appear to be strongly skewed when treated as lognormal. Distribution of sizes of spalled flakes might be normal, and the sizes of clay particles in a soil.

It appears that a given detrital source rock, such as a granite, can deliver to a transporting system a mixture of two or three distinct grain size populations (Spencer, 1963, p. 183)—a cobble and pebble population produced by fracturing and frost-wedging of the parent rock, a sand and coarse silt population produced by disaggregation of the rock into its constituent mineral grains, and a fine silt and clay-size population produced by cleaving and chemical alteration of the feldspars and mafic minerals to clay particles. The combination of these populations is far from normally distributed; thus non-normal features might be inherited from the detrital source.

Mixing of grain size populations causes non-normal distributions. The several tributaries of a large river bring together the detritus of widely separated and unrelated source terrains. Such mixing of distinct populations gives rise to multimodal distributions or to dis-

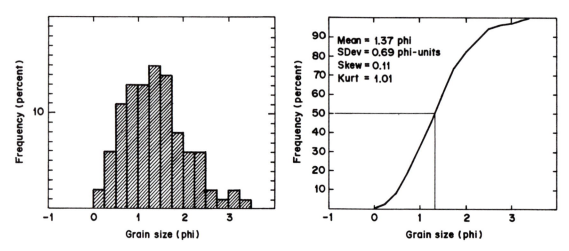

Fig. 2-7. Histogram and cumulative frequency polygon of phi-transformed grain size data. Median (50th percentile) of the distribution is about 1.3 phi; other percentiles may be similarly determined.

tributions that have non-normal moments. Folk and Ward (1957) made an elegant analysis of non-normality in the sediments of a point bar of the Brazos River in Texas; they concluded that the river was carrying essentially two sediment populations provided by two different detrital source terrains. One population was pebble gravel and the other medium to fine sand. Natural fluvial processes mixed these populations in various proportions, yielding composites whose mean size, standard deviation, skewness, and kurtosis were determined by the relative proportions of the two components in the mix.

Folk and Robles (1964) conducted a similar textural study on Isla Perez, Yucatan, of sediments that are composed not of the weathered detritus of older rocks but the skeletal debris of organisms. Organisms of different kinds (taxa) make skeletons of different sizes, shapes, and structures; under the stress of waves and tidal currents, these skeletons break into particles whose sizes are related to the form and structure of the skeletons. Sandy deposits, for example, were found to consist of varying proportions of 0-phi *Halimeda* segments and 2-phi coral grit. Sand of 0-phi mean size, as one would expect, consists mostly of *Halimeda* plates; sand of 2-phi mean size consists mostly of coral grit; these most nearly pure populations show the best sorting. Sands of intermediate mean size are less well sorted because they are mixtures of the "fundamental" populations, 1-phi sand being the most poorly sorted of all. Skewness and kurtosis also vary with

mean grain size, again because of mixing of end-point populations; sands coarser than 1 phi have positive skewness, those finer than 1 phi have negative skewness, and sands near to 1-phi mean size are most platykurtic.

Another form of mixing of grain size populations is a kind of stratigraphic leakage, fine sediment filtering into the intergranular spaces of an existing coarser deposit. The deposit thus acquires a grain size component unrelated to its own formative conditions.

A transporting system, often a complex of different mechanisms acting in concert or in succession, can resolve a raw material into several components. Wind and water currents winnow finer materials out of poorly sorted grain ensembles, leaving behind a truncated population of coarser grains as lag and depositing the finer grains elsewhere. A river transports part of its load on the bed and part in suspension. Each population may come to be deposited in a separate place. A mixed population might arise from the hydraulic equivalence (see Chap. 3) of grossly different grain sizes. Large grains that roll or slide on the bed are hydraulically equivalent in some sense to smaller grains that, within the same flow, move by saltation or entrainment (see Chap. 3). Also, it is possible that composite or "abnormal" distributions arise from a hydraulic regime that fluctuates between two levels, as a river periodically in flood stage or a seacoast alternately in storm-wave and fair-weather-wave regime.

New grain size components may originate within the transporting system itself. Consider large grains (cobbles and pebbles) mutually impacting in a stream. Their impacts cause spalling of a new population of small chips, while the sizes of the large grains change almost imperceptibly. Or, consider a population of grains that have a tendency to break in half. As breakage proceeds, the original population diminishes as a new one grows, and the grain population is a mixture of broken and unbroken components.

It may also happen that small grains come together and behave dynamically as a single grain. Certain organisms bind clay particles or carbonate mud into coherent fecal pellets that are the size of sand grains. Clay particles may adhere to the surfaces of hard pebbles; pebbles coated with sticky mud may gather other pebbles (armored mud balls; see Chap. 5) and behave as single, large, heavy objects for a time and then disintegrate. Thus the presence of aggregates surely is disruptive of grain size distributions in several different ways.

Diagenesis can have substantial effects on grain size distribution; grains can grow larger or smaller; soft aggregates that were transported as sand grains or pebbles can disintegrate into fine-grained matrix under burial and compaction. Diagenesis has effects on smaller grains that differ from the effects on the larger grains.

Finally, it cannot be denied that sampling bears on the problem of non-normal grain size distributions. Whether a sample embraces an entire cross-bed set or a single lamina no doubt affects the distribution. A sample that contains sand from one bed and mud from an overlying bed is likely to be bimodal, reflecting two successive but distinct hydraulic regimes. Presumably, inadvertent compositing is more subtle than this, but the careless or unsuspecting analyst might falsely attribute such bimodality to some fundamental interaction of geological processes rather than to processes separated in time and place but brought together by quirks of sampling.

The Rosin-Rammler and Hyperbolic Distributions

The Rosin-Rammler distribution (Rosin and Rammler, 1933), also called the law of crushing, is the distribution expected in a grain population derived from single-stage crushing or breakage of large blocks. Though the distribution was established empirically, through measurements of coal coming from a crusher, it was subsequently confirmed on theoretical grounds. It happens to be identical to the better-known Weibull distribution. In some cases it has probably been misinterpreted as lognormal; in most geological data sets, it likely is indistinguishable from a fine-skewed lognormal distribution.

The Rosin-Rammler distribution might apply to grain size of talus, grus, cataclastic debris, collapse breccias, impact breccias, and crackle breccias—that is, sedimentary material that was produced by fragmentation and that has undergone little or no transport. Presumably, if multistage breakage occurs so that a solid material is broken again and again, the law of proportionate effect comes into play; then size distribution departs from Rosin-Rammler and approaches lognormality.

Ibbeken (1983, p. 1214) found Rosin-Rammler distribution in jointed granite and granite gneiss in southern Italy; the distribution persists in the gravels at the mouths of streams 20 kilometers distant. Sands, however, though derived from the same detrital source, are lognormal. In this example, distribution of the gravel mode, according to Ibbeken, is source-rock-controlled; distribution of the sand component is transport-controlled.

The grain sizes of airborne dust, cosmic dust, and very fine grained sediments have been referred to a so-called hyperbolic distribution. Like the Rosin-Rammler, the hyperbolic distribution is an outgrowth of analysis of things broken (Gaudin and Meloy, 1962).

Bagnold (1937 and 1954) discovered that, when grain size histograms (phi-transformed) of windblown desert sands are plotted on logarithmic ordinate, the result commonly is a curve that closely resembles a hyperbola (Fig. 2-8). Bagnold and Barndorff-Nielsen (1980, pp. 200, 203) later suggested that samples of fluvial and coastal sand are similarly distributed. They reasoned that, in a transporting medium, the larger grains are segregated as a direct function of their size, but the small grains are segregated inversely with their size. Thus, the hyperbolically distributed sand of an aeolian dune is the product of larger particles having been left behind and smaller particles having been carried downwind; the size distribution would have a maximum flanked by functions largely independent of one another and decreasing monotonically away from

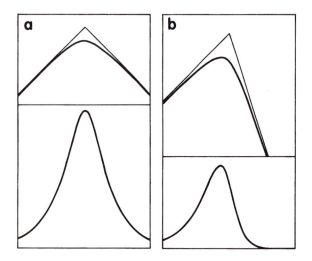

Fig. 2-8. Hyperbolic density functions: **(a)** symmetrical and **(b)** skewed, on linear (below) and on logarithmic (above) ordinate.

the maximum. Wyrwoll and Smyth (1985, p. 477), having examined the distribution as a descriptor of aeolian sediments, could find ''no clear gain'' in using this rather than the lognormal.

The Measurement of Grain Size

Methods of measurement of grain size are a function of grain size itself and whether the grains are loose or tightly bound together. Very large grains, such as boulders and cobbles, generally yield to direct measurement with scale or caliper, and removing such grains in large numbers to a laboratory is generally out of the question. Pebbles, granules, and sand can be sampled and analyzed in a laboratory. If they are not consolidated, such samples are most easily and effectively analyzed by passing them through a stack of sieves built especially for this purpose and commercially available in a wide range of mesh sizes. The sieve stack divides the sample into size groups that yield readily to construction of histograms and to statistical analysis. For materials of silt and clay size, sedimentation methods are ordinarily used, which determine grain size indirectly by measuring the speed at which grains fall through a fluid (large grains fall faster than small ones; see Chap. 3). There are many instrumental variations on this theme, some of which

are applicable to grains larger than silt and clay. In sieving and sedimentation methods, grains are not actually counted, ordinarily, but size fractions are weighed. It is an expedient that is questionable from a statistical point of view but can hardly be avoided. (One might, of course, estimate grain counts from weight fractions: 1 gram comprises about $720 \times 8^\varphi$ quartz grains of size φ.) As Sahu (1964, p. 768; 1965, p. 973) pointed out, a mean value calculated from weight ''frequency'' is not the mean size of the particles, and size distribution curves of weight frequency do not resemble distribution curves of number frequency. Other devices, such as Coulter counters, measure and count grains electronically.

For grain aggregates tightly bound together, direct measurement of individual grain cross sections on a smooth surface ground into the aggregate is resorted to; measurements obtained in this way (as from a thin section) are not the true sizes of grains and must be corrected or adjusted by some statistical method (see Wicksell, 1925, or Underwood, 1970, p. 109 ff; Kellerhalls et al., 1975, wrongly rejected Wicksell's method out of hand). Thin-section measurements underestimate mean size (give phi values that are too high), and underestimate sorting as well (that is, give standard deviation that is too high). Thus, grains are larger and better sorted than they appear to be in thin section (Fig. 2-9). Longiaru (1987) has presented computer-generated visual comparators, derived theoretically, for making estimates of sorting from thin section.

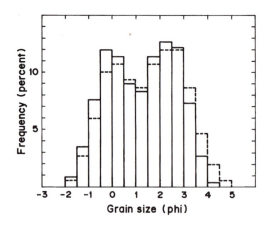

Fig. 2-9. Histogram of a thin-section-measured grain size distribution (dashed lines) and the corrected distribution (solid lines).

Because detrital grains are of irregular shapes, the size of a grain is difficult to specify clearly and consistently, and the definition of "size" generally depends on the method used to measure it. Of course, one could express the size of a grain as its volume or mass, but it is uneconomical or else difficult or impossible to measure these quantities for each of a large number of grains. Sieves measure grain size in a certain way, and if sieves are used, then one implicitly defines grain size as that quantity, whatever it is, that the sieve measures. Similarly, if one is using settling tubes or elutriators to measure grain size, then one tacitly defines grain size as that rather vague and elusive quantity—which includes aspects of size, density, shape, and cohesiveness—that settling tubes or elutriators measure.

As described above, many naturally occurring grain size populations appear to be mixtures of two or more components; composites commonly display two or more modes, or show non-normal skewness or kurtosis. If a composite grain size population is to be correctly interpreted, it must be unmixed into its components (demonstrating that the population is, in fact, a composite); the number of components and their relative proportions must be determined, as must the statistical properties (mean and variance) of each.

Harding's (1949) graphical method, which he developed in order to define and distinguish male and female components in certain marine organism populations, involves plotting the cumulative frequency data on a normal probability ordinate (so that a normal distribution would yield a straight line). Inflection points on the curve indicate multimodality or compositeness; positions of the inflection points establish proportions of the components. Positions and slopes of "linear" segments of the curve give means and standard deviations of components.

Another graphical method of finding the components of a mixture of (log)normally distributed populations was described by Hald (1952, p. 156 ff). Its application is less of an "art form" than Harding's procedure, and it yields readily to computer implementation. Hald's method is based on the fact that the logarithm of the normal density function is a parabola. If histogram data are plotted on a logarithmic ordinate, then parabolas can be fitted to various segments of the curve by least-squares regression and successively extracted (Fig. 2-10). There are simple algebraic relationships between the analytical properties of these parabolas and the normal distribution parameters.

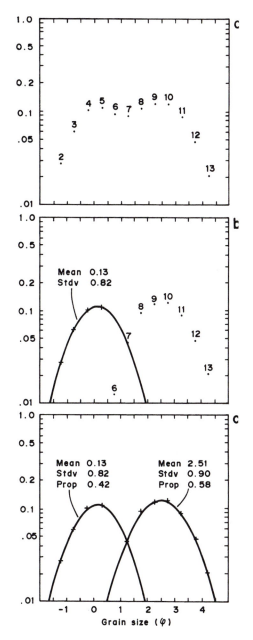

Fig. 2-10. Hald's method of decomposition of a mixture of approximately normal distributions. (**a**) A mixed population is plotted on logarithmic ordinate. (**b**) Parabola is fitted to coarse tail region by least-squares regression, then extracted from original curve. (**c**) A second parabola is fitted to the residuals. Shapes and positions of the parabolas are related to means, standard deviations, and proportions of component normal distributions.

The Shapes of Sedimentary Particles

The shape of a grain, like its size, also is difficult to quantify, and most routine descriptions of the shapes of grains are quantifications of some aspect of shape that can be measured easily, or are comparisons of certain shape aspects of grains to some set of standard images. As with grain size, *average* shapes of grain ensembles may be defined and calculated. Two shape qualities are of particular importance—roundness and sphericity.

Roundness

Some grains are characterized by sharp corners or edges; others have smooth, rounded surfaces. **Roundness** is an expression of the regularity of the margin of a grain. It is defined (Fig. 2-11), according to Wadell (1932), as

$$\rho = \left(\frac{1}{n} \sum_{i=1}^{n} r_i \right) \bigg/ R \qquad (2\text{-}11)$$

where the r_i's are radii of curvature of all (convex) "corners" of the grain margin, and R is the radius of the largest inscribed circle. Corners are convexities whose radii are less than R. (Note that the analysis is two-dimensional and applies strictly to a grain cross section or a silhouette; note also that the result is dimensionless and independent of absolute clast size.) Clearly, it would be prohibitively labor-intensive to try to measure the roundness of large numbers of grains. Thus, roundness is nearly always estimated by comparing the shape of the grain at hand to a set of standards. The most popular standard is that provided by Powers (1953), which is a slight modification of a set of images provided originally by Russell and Taylor in 1937; Folk (1955) rescaled the Wadell roundness values of Powers's images, resolving the gamut of grain roundness into six logarithmic classes (Fig. 2-12).

Sphericity

Some sedimentary particles are elongate or platy or have embayments or deep reentrants; other grains have more equant shapes. **Sphericity** is a measure of how well the gross geometry of a grain approaches a sphere. As originally defined by Wadell (1932), sphericity is the ratio of the diameter of a sphere of the same volume as the particle to the diameter of the

smallest circumscribed sphere. In contrast to the definition of roundness, which is two-dimensional, this definition of sphericity is three-dimensional. From an operational point of view it is not very useful, and some alternative is hoped for, based on simpler measurements, as follows.

Form

Generally, the gross geometry of a grain can be referred to the lengths of three mutually orthogonal axes within the grain (Fig. 2-13). The long axis must be located first, defined as the longest straight line spanning the particle; one then erects a conceptual flattest possible box whose length is parallel to this axis and into which the particle snugly fits. Short and intermediate axes are the remaining dimensions of the box. Alternatively, one might describe the short axis as a line normal to the maximum-area projection, as Sneed and Folk (1958) did (see below), but this concept is difficult to explain to a computer.

Two different operational definitions of sphericity, based on axis lengths, are in popular use (Fig. 2-14):

$$\psi = \sqrt[3]{IS/L^2} \qquad \text{(Zingg, 1935)} \qquad (2\text{-}12)$$

and

$$\psi = \sqrt[3]{S^2/LI} \qquad \text{(Sneed and Folk, 1958)} \qquad (2\text{-}13)$$

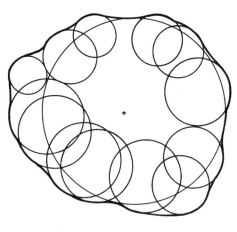

Powers roundness 2-3
Computed Wadell roundness 0.29

Fig. 2-11. Wadell's roundness of Powers's standard subangular (high-sphericity) grain image. Large circle is largest inscribed circle; the 13 smaller circles are fitted to the convex corners of the image.

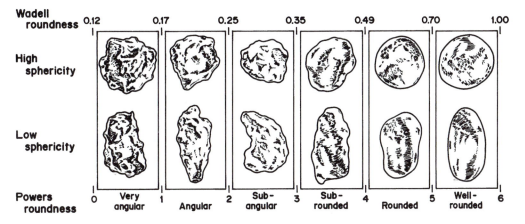

Fig. 2-12. The Wadell-Powers-Folk standard roundness images. (Redrawn from Pettijohn et al., 1987, fig. A-2.)

Under either definition, perfect sphericity is indicated by a value of unity. Sometimes, as with grains in a thin section or embedded in an outcrop, not all three axes are accessible for measurement. In such cases, a two-dimensional analog of sphericity, called *circularity*, is resorted to. Such circularity data must not be directly compared with sphericity data, because of the vagueness about which grain axes were actually measured. A more valid application of circularity is in the description of the silhouettes of loose grains resting on a horizontal surface; such grains present to view their maximum-area projection (at least very nearly, and most of the time), so that long and intermediate axis lengths are consistently and conveniently available for measurement. Because roundness and circularity are two-dimensional measures, in contrast to sphericity, an object of low sphericity may have high roundness or circularity; an example is a penny coin, which has perfect Wadell roundness and perfect circularity but (Sneed and Folk) sphericity of only 0.16.

Oblateness-prolateness index (Dobkins and Folk, 1970) is another form measure of a particle's similarity to a sphere. It is defined as

$$\overline{OP} = 10[(L-I)/(L-S) - 0.5]/(S/I)$$
$$(L-S > 0) \quad (2\text{-}14)$$
$$\overline{OP} = 0 \quad (L-S = 0)$$

This index can take on any value greater than or equal to zero.

Simple ratios among the three axis lengths, taken together, constitute *form*; the form classification of

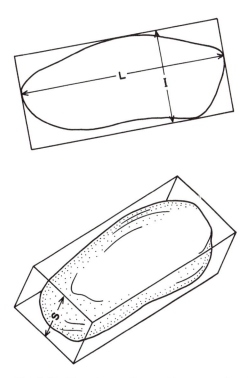

Fig. 2-13. Short, intermediate, and long axes of a sedimentary particle.

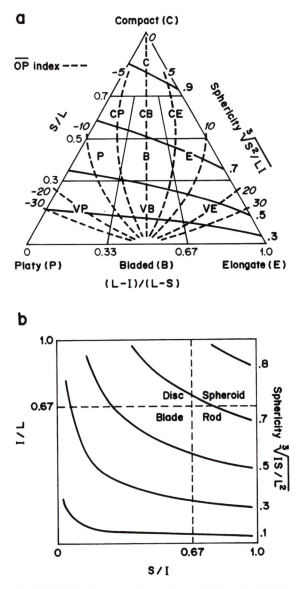

a

Compact (C)

$\overline{OP}$ index ---

S/L

Sphericity $\sqrt[3]{S^2/LI}$

Platy (P)   Bladed (B)   Elongate (E)

(L-I)/(L-S)

b

I/L

Disc / Spheroid

Blade / Rod

Sphericity $\sqrt[3]{IS/L^2}$

S/I

Fig. 2-14. Classifications of grain shape: **(a)** Sneed and Folk's (1958, fig. 2) sphericity and Dobkins and Folk's (1970, fig. 19, with permission of SEPM) oblateness-prolateness; solid curves are lines of equal sphericity, dashed curves are lines of equal oblateness-prolateness index; **(b)** Zingg's (1935, fig. 7) sphericity; curves are lines of equal sphericity.

grains was provided by Zingg (1935), and a more complicated one by Sneed and Folk (1958).

Form, sphericity, and oblateness-prolateness are interrelated quantities (Fig. 2-14), all being different functions of the three axis lengths. Numerous other shape indexes have been proposed (Table 2-2), usually to serve some special analysis, quantifying one or another of the gross shape properties of a grain or relating to the dynamic behavior of grains of various shapes in a fluid. Part of the rationale in their development certainly has been to keep the labor involved in their use to a minimum. Presumably, each of these measures of grain shape bears some special relationship to geological processes of transport and abrasion.

A more precise and more inclusive or complete quantification of grain shape (and more demanding on the analyst) involves harmonic analysis (see Ehrlich and Weinberg, 1970; Schwarcz and Shane, 1970; Boon et al., 1982) This is an application of the trigonometric series (Fourier polynomial) and is again two-dimensional. Harmonic analysis treats a grain outline or silhouette as the sum of a large number of sinusoidal waves (actually sine or cosine functions in polar coordinates, which are called lemniscates; see Fig. 2-

Table 2-2. Various Dimensionless Descriptors of Grain Shape

Form	$F = l/w$	
Elongation	$E = w/l$	
Circularity	$C_2 = \sqrt{w/l}$	$C_4 = \sqrt{A/A_c}$
	$C_2 = 4A/p^2$	$C_5 = \sqrt{D_i/D_c}$
	$C_3 = 4A/lp$	
Grain shape index	$GSI = p/l$	
Shape factor	$S_1 = p_c/p$	$S_4 = (A_c-A_i)/A$
	$S_2 = 100p/p_c$	$S_5 = 100A/A_c$
	$S_3 = A_i/A$	
Compactness	$k_1 = 2\sqrt{\pi A}/P$	$k_2 = p^2/4\pi A$
Thinness ratio	$TR = 4\pi A/p^2$	
Form ratio	$FR = A/l^2$	
Ellipticity index	$EI = \pi l^2/2A$	

where
A is area of image
A_c is area of smallest circumscribed circle
A_i is area of largest inscribed circle
D_c is diameter of smallest circumscribed circle
D_i is diameter of largest inscribed circle
l is length of image
w is width of image, perpendicular to l
p is perimeter of image
p_c is perimeter of circle having same area as image

Source: After Davis, 1986, table 1.10.

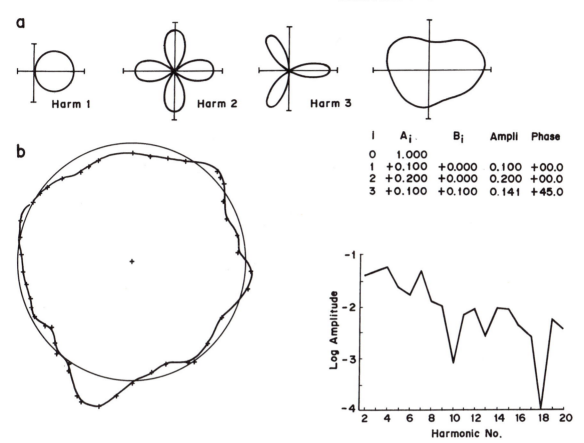

i	A_i	B_i	Ampli	Phase
0	1.000			
1	+0.100	+0.000	0.100	+00.0
2	+0.200	+0.000	0.200	+00.0
3	+0.100	+0.100	0.141	+45.0

Fig. 2-15. Harmonic analysis of grain images: **(a)** The first, second, and third Fourier harmonics, and a linear combination of these harmonics plus the 0th harmonic, which is a circle centered on the origin. **(b)** Digitized Powers standard angular grain image and a Fourier approximation. Small crosses indicate the digitized points that define the observed grain cross section; heavy closed curve is the Fourier-approximation, carried to the 20th harmonic; also shown are centroid (+), and mean radius ○; graph shows amplitudes of harmonics 2 through 20, normalized to unit mean radius; 0th harmonic has value of unity, corresponding to mean radius; 1st harmonic is identically zero when centroid is adjusted to origin.

15a); one obtains as accurate and complete a representation of a grain margin as one wants, as a set of numbers, and these numbers, called the Fourier coefficients, reveal the relative magnitudes and locations of "bumps" of all sizes (Fig. 2-15b). (Three-dimensional harmonic analysis is also possible, in terms of the Laplace spherical harmonics, but its application to grain-shape studies surely is impractical at the present time.)

Textural analysis is being overhauled and automated with the advent of new computer technologies that permit computers to "see." The computer's "eye" in these applications is the charge-coupled diode (CCD) device, which is a semiconductor chip having a rectangular array of a large number of small, closely spaced, light-sensitive electronic switches, each wired to memory-array elements in the computer. An image to be processed or analyzed is projected onto the CCD chip by a lens. As an example of how grain shape can be analyzed, consider a CCD image of a single grain, such that any picture element (pixel) outside the image is dark (or "off") and any pixel inside is lighted (or "on"). The computer can be programmed to count the pixels of a given state and to quantify the relationships between neighboring pixels. The total number of lighted pixels is a measure of the area of the image. Any lighted pixel in the array surrounded by fewer than four lighted pixels (if we do not count diagonal neighbors) is a point on the boundary or edge of the image. The number of such points is a measure of the perimeter. Other simple tricks give length and width. The average number of lighted neighbors of boundary points is a measure of roughness of the boundary.

Geological Controls on Detrital Grain Shape

Shapes of grains are controlled in part by origins of the grains or manner in which they were produced, in part by properties of the material of which the grains are made, and in part by the manner of transport and the intensity and duration of abrasional processes that accompany transport of the grains.

Shapes of detrital grains are determined initially by the textures of the parent rocks and by the ways in which these rocks break. Jones (1953) described tetrahedroid pebbles in certain gravel deposits, being corner fragments broken from joint blocks. Some workers have observed that quartz grains in sandstones tend to be elongate parallel to the crystallographic c axis (see Bokman, 1952, p. 17). Apparently this is due to the original shape of the quartz grains in the detrital source rocks. Bokman (1952, p. 21) found that quartz grains in the Stanley Formation (Mississippian?) in Oklahoma, are on average less elongate than grains in the overlying Jackfork Formation. The reason, he suggested, is that the Stanley sands were derived from plutonic source rocks and the Jackfork sands from a metamorphic source.

Blatt (1967, p. 421), having described the considerable influence of sedimentary processes on the sizes and shapes of quartz grains, commented that "It is significant that the appearance of the aggregate of quartz grains at the time of release into the sedimentary cycle bears little resemblance to the appearance of quartz grains in many sandstones in the geologic record." Sedimentary processes, according to Blatt (1970, p. 261), reduce mean size of quartz grains from approximately 0.6 phi, a typical size of single quartz crystals in newly disintegrated primary detrital source rocks (granites, gneisses, schists) to 4.0 phi, the average for all existing sediments; thus, unit quartz grains are reduced in size (diameter) by about 90 per-

cent (a volume reduction of about 1000 to 1). Moss (1966, pp. 97, 111) noted that, though granitic rocks contain much quartz in the range of -1 to -2 phi, there is a marked shortage of detritus of this size in mature detrital sands. Granitic quartz, he has pointed out, is criss-crossed with incipient cracks, and much granitic quartz, once released into a regolith or introduced into a transporting medium, is ready to fall apart into particles ranging in size down to and including silt. Moss et al. (1973, p. 510) suggested tentatively that crack-bounded volume elements range in size between 2 and 20 μm. The larger (sand-size) grains show, under high magnification, scars that are consistent with piecemeal removal of fragments 20 μm or less in size.

Some quartz grains are remarkably elongate in thin section, with length/width ratios of 5 or more (Fig. 2-16); commonly they have sharp edges as well. It seems that the origins of such grains have not, in general, been determined. Perhaps some of them are flakes made when quartz pebbles strike one another in a stream bed. Sharp flakes seem to be common in volcanic sandstones that contain abundant volcanic quartz grains. These flakes might have formed in volcanic rocks containing quartz phenocrysts; bulk fracturing of the rock, such as occurs during explosive eruptions, would produce some flakes (Fig. 2-17), which would eventually be released from the volcanic porphyry as it weathers.

Crook (1968) and Cleary and Conolly (1971, p. 2760) observed that chemical processes lead to rounding and embayment of quartz grains in soils. Potter

Fig. 2-16. Highly elongate (or platy) quartz grain in a sandstone; this polycrystalline quartz grain might have been spalled from a cobble of granitic quartz or formed during close fracturing of a granitic pluton.

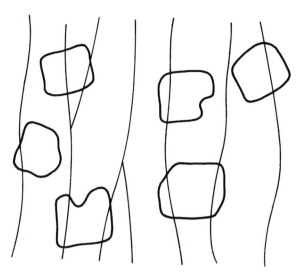

Fig. 2-17. Possible origin of some elongate quartz grains. Volcanic quartz phenocrysts are transected by numerous cracks formed during explosive shattering of the host rock. Shattering of a tightly quartz-cemented sandstone could also produce elongate grains.

(1978, p. 427) found embayed and corroded quartz grains in rivers draining terrains of humid tropical climates, and suggested that embayments are a consequence of chemical weathering in soils and not likely created by abrasion (in fact, abrasion tends to destroy embayed grains). He observed further (1978, p. 432) that the sands of tropical rivers are, in general, more rounded than the sands of temperate or boreal rivers. Thus, climate at the detrital source, insofar as it controls weathering and soil processes, influences texture.

The sphericity or form of a grain is determined in part by anisotropies in the material composing the grain, so that a grain wears at different rates in different directions. In grains larger than coarse sand, which are almost always composites of many mineral crystals, foliations or fissilities or other fabric elements control the shape. Thus, pebbles, cobbles, and boulders characteristically are elongate or flat rather than equant. Only those with isotropic textures are likely to be equant or very spherical. Some that have become rounded may subsequently fracture along some fabric-controlled weakness, yielding an object that is in part well rounded, in part quite angular.

The shapes of clay particles—and in some sediments the shapes of detrital feldspar grains, micas, amphiboles, zircons, etc.—are determined mainly by crystallography of these minerals. In small, single-mineral grains crystal habit, cleavage and fracture, hardness, and anisotropies of the mineral control the shape of the grain. Bloss (1957, pp. 218–219) found evidence of cleavage in quartz, parallel to the unit rhombohedra (both positive and negative), and possibly also basal and prismatic cleavages.

Transportation exerts a major effect on grain shape. The significance of roundness is that detrital grains commonly come into being with sharp corners and edges, and these gradually wear away as grains pass through the sedimentary cycle. Thus, rounded grains have moved farther or fared longer in the mill than have the more angular grains. Some grains that have become rounded may break in the transporting environment, so that new sharp corners appear.

Dobkins and Folk's (1970) analysis of grain size and shape at Tahiti-Nui, based on 15,000 measurements of almost 1000 (mechanically isotropic) basalt pebbles, revealed certain relationships among shape properties and sensitivity of shape to hydraulic environment. They compared the maximum-projection sphericity, roundness, and oblateness-prolateness index of pebbles in rivers, low-wave-energy beaches, and high-wave-energy beaches. Pebbles in rivers are least rounded, most spherical, and mildly prolate; pebbles on high-energy beaches are most rounded and least spherical. On average, beach pebbles are oblate, those of high-energy beaches being more oblate. Moreover, on low-energy beaches, the smallest pebbles are the flattest (most oblate); on high-energy beaches, the largest pebbles are flattest. Thus, mean oblateness and the size of the most oblate beach pebbles are indexes of surf vigor. Abrasion, clearly, is the control on shape of these mechanically isotropic pebbles—as roundness increases, sphericity decreases and OP index becomes more negative.

Waag and Ogren (1984) conducted a similar analysis on a pocket beach being nourished by a looming sea cliff of jointed granite, from which angular talus blocks had fallen directly onto the beach. The blocks tend to be bladed or elongate initially; abrasion under wave action produces boulders more compact, that is, more equidimensional (see Table 2-2), and more platy, the trend toward compactness being the stronger.

Wave action may sort pebbles on beaches into different zones or layers on the basis of their shapes. On modern beaches of South Wales, Bluck (1967, p. 135) found discoid cobbles and pebbles on the berm and upper shoreface and spherical pebbles and cobbles far-ther down the shoreface. Apparently this shape sorting reflects different mobility of grains of different shapes. Hydraulic processes may sort skeletal grains according to their shapes. Only the right valves of the oyster *Crassostrea virginica* are deposited on Florida oyster reefs described by Grinell (1974); left valves are de-posited in surrounding depressions. Cadee (1992, p. 200) observed wind-transported *Mya* shell assem-blages that contain more left valves than right ones; in closely associated littoral deposits, right valves are present in the greater numbers.

Diagenesis—specifically the precipitation of authi-genic quartz onto the surfaces of detrital quartz grains, the dissolution of material at points of mutual contact, or chemical etching—changes the sizes and shapes of detrital grains. Quartz cement ordinarily is optically continuous with the detrital host, and the boundary between grain and overgrowth commonly is difficult to discern. In many sandstones, cement grows pref-erentially on crystallographic c axes. Some detrital quartz grains contain abraded overgrowths, testament to ''reincarnation''; they are grains that resided in a previous sandstone, wherein they gained their cement overgrowths, and were subsequently recycled via me-chanical weathering and erosional processes into a newer sandstone. Thus, the grains changed shape be-tween the two episodes of transport, diagenesis in some respects undoing the work of the first cycle.

Interaction among Textural Properties

The textural properties of aggregates of particles are interrelated, or correlated in ways that we do not fully understand. Certain relationships between mean size and sorting have often been noted. Sediments having mean grain size in the fine sand range tend to be the best sorted; sediments with mean size in the very coarse sand class tend to be poorly sorted. The com-posite nature of many grain ensembles is partly re-sponsible for dependencies between size and sorting; a sediment whose mean grain size lies between two ''fundamental'' grain sizes, as described earlier, is likely a composite of fundamental populations and thus less well sorted than either of the fundamental populations alone.

Grain size and grain shape are not independent quantities either—the larger grains of a poorly sorted collection typically are more highly rounded than the smaller grains, and the sphericities of the larger grains might be different from those of the smaller ones. Ac-cording to Blatt (1970, p. 261), quartz grains smaller than 3 phi are not rounded much. If shape controls on large grains are different from controls on small grains, then grain shapes do not ''scale''—sand grains are not just miniature pebbles or boulders, and silt grains are not just miniature sand grains. Dependen-cies between grain size and shape are probably due to the fact that processes that control grain size also in-fluence shape, and that processes that determine shape are variously effective on different size ranges. Thus, attrition makes a large grain smaller but also makes it smoother and rounder. Small grains transported in sus-pension are subject to less abrasion than larger grains grinding and bounding against one another on the bed. Small grains might be created by the chipping or spall-ing of large grains; they are more angular than their parents, in part because of the way they were made and in part because they are younger. The shapes of these larger grains differ from those of the small chips derived from them (Moss et al., 1973). Quartz grains of silt and clay size are, according to Krinsley and Smalley (1973), almost invariably shaped like flat plates; the smaller the particle, the flatter it is. This seems to reflect cleavage in quartz; at the larger scales, the distribution of crystal defects controls *fracture*, but at the very small scales, comparable to or less than the distances between defects, *cleavage* controls breakage. Ehrlich et al. (1980), describing quartz grain shapes in terms of harmonic amplitude and phase, found that the shapes of grains larger than 125 μm are different from the shapes of the smaller grains. Detrital source ap-parently controls the low harmonics; abrasion mainly affects the higher harmonics. Other interactions be-tween grain size and grain shape were documented earlier by Sneed and Folk (1958), who found that, in populations of river-transported pebbles, the larger pebbles are less spherical than the smaller ones. Peb-bles of chert tend to chip and fracture, so that sharp corners and edges persist; but sand-size chert grains in the same sediment may be quite well rounded, becom-ing so more quickly than ordinary quartz grains be-cause they are softer.

Textural Maturity

The concept of **textural maturity** (Folk, 1951) refers to the fact that, under prolonged transport, the texture of detritus gradually progresses toward some end point. A disordered mixture of mud and unsorted angular fragments just released from a source rock is texturally immature as it begins its long trek through the sedimentary cycle. At first, clay content (that is, matrix) diminishes, owing to gradual unmixing of the coarser and finer materials; when it has fallen below 5 percent, the material is said to have attained **submaturity**. Eventually, the standard deviation of the distribution of sizes falls below about 0.5 phi unit (that is, sorting improves to the level of ''well sorted''). At this stage the material is said to have attained textural **maturity**. A separate process, operating more slowly, is the gradual transformation of individual sand grains, including size reduction and rounding by attrition. Rounding of quartz grains takes a long time, and if the grains, in general, have achieved a well-rounded state (Powers roundness of 3 or higher), then the material is, finally, supermature. Note that the sorting of detritus, including the removal of mud matrix, is an organizing process, in contrast to rounding, which is a grain-modifying process. Progress from **immaturity** to **supermaturity** (Fig. 2-18) is progress toward simplification: grains become more nearly similar to one another and more nearly similar to the simplest of all shapes, the sphere; grains also become smaller. Textural supermaturity is apparently a condition of maximum stability under weathering and erosion. Large grains, angular grains, and non-equant grains are more susceptible to abrasion than smaller, rounder, more equant grains. This is why grain collections tend toward fineness, roundness, and equancy and why these are the properties of maturity.

Processes at the depositional site may contribute substantially to the textural maturity of the detrital sediment that it receives. The deposit of a turbidity current is essentially a single bed, in which grains are size-sorted, both laterally and vertically. Any part of such a graded bed (see Chap. 5) is better sorted than the bed as a whole. Thus a turbidity current promotes, however briefly and incompletely, textural maturity. Many depositional environments segregate particles according to their sizes into juxtaposed or superimposed and intergradational belts or distinct geomorphically defined sedimentary bodies. This gives rise to the familiar fining upward of fluvial channel sand bodies, or the coarsening upward of shoreface sands, or the sharp distinction between channel sands and overbank muds, or between barrier-island sands and lagoonal muds. Depositional environments may be quite abrasive also, and improve the roundness of detritus that comes to them. Wave- or tide-generated oscilla-

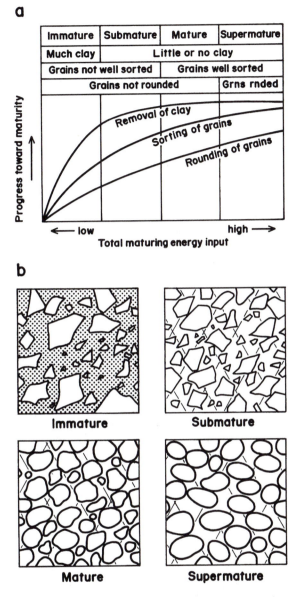

Fig. 2-18. Textural maturity in sandstones: (**a**) Degrees of textural maturity, according to Folk (1951, fig. 1, with permission of SEPM). Removal of clay matrix is the more rapid process during transport of detritus; sorting of grains is slower; rounding of grains takes a long time. (**b**) Conceptual illustrations of stages of maturity; immature rock contains poorly sorted angular grains embedded in clay matrix; in the submature rock, clay matrix has been removed and some sorting and rounding has been accomplished; in the mature rock, grains are well sorted; highly rounded grains characterize supermature rocks.

tory currents in a coastal environment, for example, can quickly change angular pebbles into well-rounded ones, even though little net displacement results. But it seems that there are special conditions and select processes that promote maturity, and that there are many transportational-depositional systems that cannot create texturally mature deposits (see Chap. 6). If sedimentation is rapid, the processes that differentiate the detrital raw materials according to grain size or that transform angular grains into round ones might not have time to act.

A *textural inversion* is an anomaly or wrong association, an apparent violation of the rules of textural maturity, such as a sediment composed of well-sorted or well-rounded grains with a clay matrix, or well-rounded but poorly sorted grains (Fig. 2-19). Some textural inversions are due to the fact that there are two distinct influences on textural maturity, namely, the detrital source and the transportational-depositional environment.

A textural inversion might be a result of inheritance, as when texturally mature sandstones become detrital source rocks, delivering to the sedimentary mill collections of grains that are well rounded from the start. Two supermature populations, each with well-rounded, well-sorted grains but with different mean grain sizes, might mix together to form a well-rounded but rather poorly sorted sand (a bimodal supermature sand; see Folk, 1980, p. 103). Some textural anomalies result from mixing of ''mature'' grains from sedimentary sources (recycled grains) with ''juvenile'' grains from primary sources, so that a sandstone containing a mixture of very well rounded quartz grains and angular quartz grains is probably a product of multiple or complex provenance. Weathering causes some

rounding of grains, even quartz grains; thus, detrital grains having no history of transport whatsoever are sometimes round, even if they are not well sorted.

Just as depositional environments promote or create textural maturity, so they can be responsible for inversions. Detritus that became well sorted and well rounded on a beach face may, during a storm, be washed or blown into a lagoon, mixing with the muddy sediments that are the characteristic deposits in those places.

A bed of well-rounded sand might acquire a clay matrix through infiltration or leakage of mud from an overlying deposit into the pores and thus become texturally inverted. In some sandstones, matrix is created by the mashing of soft framework grains under burial pressure; thus, a matrix-free sandstone containing a few soft rock fragments develops a matrix and becomes texturally inverted as a result of diagenesis.

Texture and Depositional Environment

The relationship between texture and depositional environment has received much attention, but sedimentologists debate whether the results obtained so far are useful or valid. Certainly there are obvious associations about which there is general agreement—aeolian sands are finer-grained and better sorted than most fluvial sands; glacial deposits typically are coarse and very poorly sorted; beach pebbles characteristically are discoidal, while pebbles of desert pavements are faceted. In a fluvial system we distinguish between channel deposits and overbank deposits largely (and as a good rule of thumb) on the basis of grain size, the former being in virtually all cases much coarser than the latter. Presence or absence of a mud matrix, whether in terrigenous or carbonate rocks, is almost universally taken as evidence of strength of currents (intensity of mechanical energy) at the depositional site. Such observations clearly justify our strong conviction that texture is linked to depositional environment. But many sands—even though produced in such diverse environments as aeolian, fluvial, and littoral—may appear, superficially at least, to have quite similar textures. Some sedimentologists hope that careful textural analysis of each of these sands will uncover subtle distinctions that reflect different depositional environments.

Mason and Folk (1958) distinguished between beach and aeolian sands on the basis of relationships

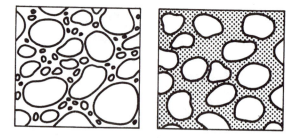

Fig. 2-19. Conceptual illustrations of textural inversions. A rock consisting of well-rounded but poorly sorted grains; a rock consisting of well-rounded and well-sorted grains with a clay matrix.

between skewness and kurtosis. Friedman (1961) was able to resolve modern river sands and modern beach sands on the basis of their standard deviation and skewness. Landim and Frakes (1968) made similar comparisons to distinguish between glacial tills and other diamictites. Moiola and Weiser (1968) discriminated among beach, aeolian dune, and river sands on the basis of skewness versus mean and standard deviation versus mean. Greenwood (1969) appears successfully to have resolved aeolian and littoral sands by applying a multivariate statistical method called discriminant analysis to large sets of statistical-moment data. Passega (1957 and 1964) plotted the 1 percentile against median size of many samples and achieved some resolution among fluvial, littoral, and turbidite sands. Visher (1965a and 1969) proposed that sands of different environments be discriminated on the basis of supposed internal truncations in grain size distributions, giving rise to cumulative distributions bearing one or more "doglegs," as revealed in grain size plots on probability ordinate.

There are as many investigations giving poor or negative results as those claiming success. van de Graaff (1970, p. 566) found that neither Friedman's, nor Moiola and Weiser's, nor Visher's methods were of any value in distinguishing among fluvial, tide-influenced delta distributary, and beach ridge sandstones of the Cretaceous Castlegate Formation of Utah. Glaister and Nelson (1974) concluded that bivariate plots of distribution parameters are, in general, unreliable. Taira and Scholle (1979) and Tucker and Vacher (1980), using discriminant analysis, concluded that grain size distribution parameters are not very useful in distinguishing river, beach, and dune sands. The problem with these kinds of analysis, aside from all the effort involved in their implementation (and aside from the fact that cemented sands do not yield readily to these procedures), is that whatever textural signatures or fingerprints do exist in the statistical "fine structure" are very likely smudged and muddled by the effects of inheritance (from the detrital source), textural overprinting as sediment moves from one environment to another, mixing of sediments having different sources or different transportational histories, and diagenesis. Rules that seem to emerge from study of a particular deposit seem not to apply to other similar deposits (McLaren, 1981, p. 611).

Probably success or failure of these methods depends also on the textural maturity of the material being tested. One suspects that materials of high maturity

yield up the least information. How closely can a sediment approach "perfect" textural maturity and still retain some vestige of non-normality bearing on matters of depositional environment?

Finally, there are the inevitable problems related to sampling of the deposit and analytical precision and reproducibility of results. Do all parts of a deposit share in the detailed textural character claimed for a given sample? Much of the burden of discrimination between distributions based on their higher statistical moments is borne by the tails, where analytical errors are greatest.

Surface Texture of Sedimentary Particles

Surface texture of grains refers to features of grain surfaces that are of such small scale as not to affect grain shape or form significantly. The surface textures of quartz sand grains from various modern depositional environments have been examined with the scanning electron microscope (SEM), and certain features seem to be useful in discriminating mode of transport. For example, medium sand grains transported by the wind commonly show dimples or cuvettes; sand grains from beaches show grooves and V-shaped pits or cuneiforms. Sand grains transported by glacial ice bear flat surfaces (facets) or conchoidal hollows (see Krinsley and Doornkamp, 1973). Larger grains (boulders, cobbles, and pebbles) occasionally show distinctive surface markings, such as crescentic marks (chatter marks) and striae, due to heavy impact or grinding. But Wolfe (1967) and Baker (1976) concluded that correlation between surface textures and depositional environments is weak, at least in sand grains.

Some clasts have markings imposed by diagenetic processes. Tanner (1963) described surface damage at points of contact of large clasts in the Beaverhead Formation (Paleocene, southwest Montana) induced by post-depositional stress. Intrastratal dissolution produces etch pits on mineral grains, whose shapes depend on the chemical environment and on crystal structure of the grain. Wolfe (1967, p. 247) found surface texture to be useful in determining the chemical environment of diagenesis of his particular sand grains.

Surface texture seems to depend on weathering, abrasion, and diagenesis. Any of these processes can overprint or obliterate the results of earlier processes, and this diminishes the utility of grain surface textures in diagnosis of the depositional history of a grain (see Krinsley and Trusty, 1986, p. 201). Moreover, grains of different sizes respond differently to the processes that mold surface texture, and, of course, different minerals behave differently by virtue of their various hardnesses, cleavages, solubilities, and so on.

Fabric of Particulate Aggregates

Fabric, the spatial relationships among grains, includes aspects of grain packing and grain orientation: proximity of grains to one another, whether framework grains are touching one another or floating in matrix or cement, geometry of grain-to-grain contacts, alignments (preferred orientations) of long axes, and imbrications of elongate or platy grains. Fabric is commonly intimately related to sedimentary structure.

Grain packing is some measure of the efficiency with which an aggregate of grains fills space. It is common experience that collections of discrete objects of various sizes and shapes, such as books or canned goods, can be packed into a cardboard box with varying degrees of space-filling efficiency. Large collections of equal-size spheres may be stably packed to some density between more or less well-defined limits, and random disturbances can readily alter the packing efficiency, sometimes increasing it, sometimes reducing it. At the higher packing densities, grains tend statistically to be closer together—that is, the spaces between grains are smaller and mean distance between the centers of grains is less; individual grains also have more nearest neighbors, and there are more grain-to-grain contacts in the aggregate. Packings of spheres of unequal sizes are fundamentally different, and packings of objects that are not spherical can be quite complex probably beyond any reasonably comprehensive theoretical analysis. If grains are deformable, then increased packing densities can result from conversion of point contacts to contacts having significant area.

It is difficult to count or measure fabric-related quantities in three dimensions, so various operational measures based on flat cross sections have been adopted, these measures presumably reflecting in some way the actual three-dimensional packing of an aggregate. Kahn (1956) examined some of the packing properties of sand-size sediments and proposed two descriptors, packing density and packing proximity,

which can be obtained petrographically (Fig. 2-20); these measures are based on linear traverses. *Packing density* (Kahn, p. 390) is the proportion of total grain intercept length in a linear traverse across a thin section. *Packing proximity* (Kahn, p. 392) is the ratio of the total number of grain-to-grain contacts to total number of grains on a traverse. *Contact index* is perhaps a more popular measure of grain packing; it is defined as the average number of grain-to-grain contacts per grain.

Some grains along a traverse may appear to make no contact whatsoever with adjacent grains—these are called *floating grains*. There are processes that can lead to truly floating grains, but in most cases these grains make unseen contact with grains outside the plane of the section. Other grains appear to make *point contact* with one or more adjacent grains. Or contacts may be extended surfaces, appearing as *long contacts*, or in some cases representing penetration of one grain into another, as *curved contacts*. Grains that have been severely deformed may show stylolitic interpenetrations, or *sutured contacts*.

Experiments by Graton and Fraser (1935) yielded 0.63 contacts per grain and 47 percent floating grains in cross sections through "random" packings of spheres. Gaither (1953) examined 14 samples of disaggregated and experimentally reassembled St. Peter sandstone (a renowned medium-grained supermature sand). He determined that an uncompacted medium-grained, well-sorted, well-rounded sand has the following fabric properties, as measured in thin section:

37 percent porosity
0.85 contacts per grain
46 percent floating grains
31 percent of grains with one contact
16 percent of grains with two contacts
6 percent of grains with three contacts
1 percent of grains with four contacts
77 percent tangential (point) contacts
17 percent long contacts
6 percent concavo-convex (curved) contacts

Keep in mind that these figures apply only to well-sorted, well-rounded sands. In a poorly sorted sand, porosity is much lower. The larger grains are in contact with large numbers of the smaller grains, so that the average number of contacts per grain is much higher and the percentage of grains with small numbers of contacts lower than for well-sorted sands. Angular sands probably have a higher proportion of point

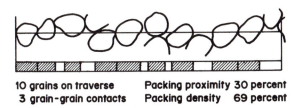

10 grains on traverse   Packing proximity 30 percent
3 grain-grain contacts   Packing density   69 percent

Fig. 2-20. Kahn's (1956) packing density and packing proximity. Bar graph shows segments of the traverse that transect grains.

contacts. In low-sphericity sands, packing characteristics are influenced by form and orientation.

Compaction, of course, changes the fabric of an aggregate of sand. Under compaction, grains move past one another into tighter arrangements, reorient, and change shape. Greater average number of contacts per grain, smaller percentage of floating grains, presence of grains with five contacts, and grains with sutured contacts all are indications that compaction has occurred (see Taylor, 1950, p. 707). Wilson and McBride (1988) studied lightly cemented sandstones in the Ventura basin in California and found that the average number of grain-to-grain contacts per grain increases by 0.69 per 1000 meters depth of burial; the average number of long plus curved plus sutured contacts per grain increases 0.89 per 1000 meters.

Orientation

If grains are not spherical or isotropic, then orientation fabrics are possible. **Preferred orientation** in a granular aggregate is a statistical alignment of the long axes of elongate grains or of the normals (poles) to the flat faces of platy grains. In some cases, long axes are aligned parallel to the direction of a current that transported and deposited the grains, these grains behaving like weather vanes or streamers. There may also be an alignment across the flow direction, though this is rarer and ordinarily weaker; grains oriented in this way probably rolled along the bed. Platy grains deposited in a current commonly are stacked like shingles or toppled ranks of dominoes, inclined upcurrent; this is called an **imbrication**. It is a common occurrence on gravelly beaches; such a beach is referred to archaically or poetically as the "shingle." Imbrication is common also in stream-bed gravels, glacial tills, and gravels transported by certain mud flows.

Walker (1975) stated that there are well-developed fabrics of two kinds in clast-supported conglomerates: (1) grains imbricated and with long axes parallel to the flow and (2) grains imbricated but with long axes transverse to the flow. The former is referred to as an **a-axis imbrication**, the latter as a **b-axis imbrication** (Fig. 2-21). Grains that have rolled on the bed characteristically have long axes transverse to the flow; clasts that were dragged along the bed by the flow or that were transported above the bed, supported by the viscosity of the flow, have flow-parallel orientations. Alignment of clasts, whether parallel or transverse to flow, seems to depend on clast concentration, according to Rees (1983, p. 446). Transverse orientations indicate a more fluid, more turbulent flow, according to Nemec and Steel (1984, p. 13).

The long axes of cobbles and boulders in glacial till are aligned parallel to ice movement (Lineback, 1971, p. 333), and there is commonly an up-glacier imbrication. Potter and Pettijohn (1977, p. 40) referred to various findings of glacial pebbles with striations (scratches on faceted surfaces) that are preferentially parallel to the long axes of the pebbles (note that this property is preserved even if the pebbles become disoriented after the glacier has melted away); evidently

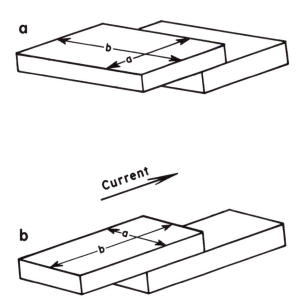

Fig. 2-21. Two patterns of imbrication: **(a)** a-axis imbrication (b axis is transverse to the flow); **(b)** b-axis imbrication (b axis is parallel to the flow).

glacial pebbles tend toward a flow-parallel orientation as they are dragged over the glacial bed. Hein (1984, p.39, 43) found flow-transverse orientations and b-axis imbrications in the low channel bars of active (modern) braided streams. In the channels between bars, the larger clasts show similar transverse orientations and b-axis imbrications, but the smaller pebbles are flow-parallel and a-axis imbricated. In clast-supported gravels deposited from deep-sea grain flows, clasts are, on average, a-axis imbricated, according to Hein (1984, p. 42).

Many conglomerates do not have a preferred orientation of clasts; apparently the flow field was chaotic and unsteady, or the clasts were carried in a highly viscous debris flow and were not free to respond to the flow field, or some subsequent process deranged an existing fabric. A so-called *disorganized clast fabric* may simply reflect short transport distance, according to Nemec and Steel (1984, p. 13), because, presumably, clast alignments take some time to develop.

Large grains on a bed respond to stronger currents than small grains; the stronger currents might be more variable, or less variable, than the weaker ones; the stronger currents might have different mean direction than the weaker ones. A gentle current might orient the smaller pebbles and leave the larger ones in their original random dispositions. Thus, mean direction and variance of imbrication or alignment of the larger grains (cobbles and pebbles) might differ in their statistical properties from those of smaller grains.

Preferred orientations and imbrications are more difficult to observe and measure in the finer-grained sediments, but some work has been done on them nonetheless. Potter and Mast (1963, p. 454) and Potter and Pettijohn (1977, p. 51) stated that the long axes of sand grains are parallel to unidirectional flows, and that the grains are imbricated up-current. Curray (1956) examined natural coastal sands deposited in the wave swash zone. He determined that, at least on a fine-sand beach, the backwash, rather than the swash, orients the grains, long axes parallel to backwash direction, and imbricated up-current with respect to a backwash. Allen (1964, p. 96) measured strong sand-grain orientation parallel to parting lineation (see Chap. 5) in Devonian fluvial sandstone of England and Wales. Turbidite sandstones also have been shown to bear a-axis imbrication, grains aligned approximately parallel to flutes and grooves (see Chap. 5).

Skeletal grains (that is, the hard parts of organisms)

respond to currents in ways that may be more complex than the response of sedimentary particles derived from disintegration of rocks. The sizes and shapes of skeletals are determined by life processes, not by source-rock textures or fluid-mechanical processes; skeletals may be more elongate or more tabular, more symmetrical or less so, or of greater surface roughness, or of lesser effective density than any grain produced by weathering of rocks could ever be. The complex shapes of these sedimentary particles make their interaction with fluid media and their response to currents and to abrasive environments either more predictable than ordinary detrital grains or less so, giving rise to fabrics and unusual preferred orientations not possible (or only vaguely expressed) in ordinary detrital rocks.

There are many examples of elongate shells or other structures becoming oriented passively after death of the organism. Pettijohn and Potter (1964, Pl. 45B, 46A) illustrated oriented *Donax* shells and parafusulinids. Sprinkle and Gutschick (1967) illustrated a blastoid come to rest in a small tidal creek, calyx like an anchor, severed column (stem) like a streamer in the current (Fig. 2-22a). Gutschick et al. (1962) illustrated tiny starfishes in shale, their little arms straining against some bottom-hugging current (Fig. 2-22b). Shales ordinarily contain fossils, however small, and many of them respond well to weak currents—small plant fragments (typically hardly more than "splinters"), graptolites, tentaculitids, ostracods, bryozoan fragments, etc. Even though these objects are small, they are nonetheless usually discernible in the field (Jones and Dennison, 1970).

Folk and Robles (1964), cited earlier, observed that coral sticks (cylindrical fragments of the staghorn coral) are aligned perpendicular to the shore, oriented by the wave backwash, and arranged "like cigars in a box" (1964, p. 270). A coral stick dropped into the surf zone in shore-parallel orientation rolls to and fro at first, but as soon as it assumes the stable shore-perpendicular orientation, it stops moving. Coral sticks on storm-wave-built subaerial beach ridges are, in contrast, randomly oriented like a pile of matchsticks.

Nagle (1967) experimented on beaches and in flumes with shells of various shapes and found that the conical shells of high-spired snails and the elongate shells of the bivalve *Mytilus* become oriented perpendicular to wave approach. In a steeply sloping swash zone, platy bivalve and brachiopod shells are oriented with umbos pointing seaward. In unidirectional flows,

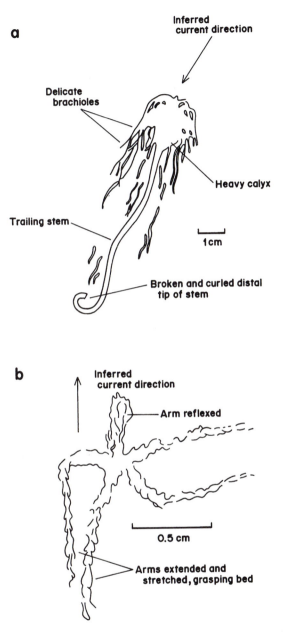

a

Inferred
current direction

Delicate
brachioles

Heavy calyx

Trailing stem

1 cm

Broken and curled distal
tip of stem

b

Inferred
current direction

Arm reflexed

0.5 cm

Arms extended and
stretched, grasping bed

Fig. 2-22 Current direction revealed by fossils: (a) Blastoid found in a tidal creek. (Redrawn from Sprinkle and Gutschick, 1967, fig. 4.) (b) Small starfish in a shale. (Redrawn from photograph by Gutschick et al., 1962, figs. 1, 2.)

the apices of high-spired snails point into the current. Seilacher (1960) found that the apices of the snail *Loxonema* and the umbos of *Mytilus* point into the flow, but elongate spiriferid brachiopods are disposed across the flow. Seilacher (1959) pointed out, and symmetry considerations require, that bipolar distributions of orientations indicate flow-parallel orientation if one mode clearly dominates and flow-transverse orientation if the two modes are of similar strength (Fig. 2-23). It is clear that different shells or other skeletals respond in ways that are not necessarily predictable and that the behavior of a shell in a current should be tested before its orientation is used to make inferences about currents.

Cup-shaped shells, such as the valves of brachiopods and bivalves, fall through standing water convex side down, like a sinking canoe, or are turned to that disposition on the bottom by scavengers, but bottom currents turn them convex side up (Emery, 1968). Orientations of skeletals of sessile benthonic organisms, such as corals and bryozoans, are controlled to some degree by magnitude and direction of currents in their environment. Such skeletons may be anchored or cemented to the substrate and be essentially immovable. They may, however, tend to grow more rapidly in one direction or another, in response to a current that buffets them throughout their lifetimes (see Chap. 5). Others may be "uprooted" and "blown over" in a certain direction. The life orientation of a sessile organism, at least with respect to gravity, is generally well known, and it is easy to know whether a skeletal is lying on its side or upside down.

It appears that several factors are involved in development of clast orientation, including viscosity and intensity of shear in the flowing medium; time and

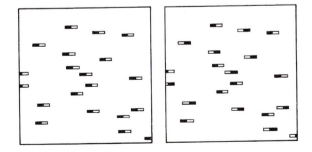

Fig. 2-23. Which fabric is flow-transverse and which is flow-parallel?

distance of transport; size, shape, and concentrations of the clasts; and even position of clasts in the flow. Compaction also affects grain orientation. It is not clear at this time whether continued study will bring some order and simplicity or will just increase the complexity of the problem.

Derived Properties

Certain bulk physical properties of particulate aggregates—such as porosity, permeability to fluids, and angle of repose (of loose aggregates)—are controlled or influenced by textural and fabric properties, including mean grain size, sorting, grain shape, grain surface texture, and grain orientation.

Porosity and Permeability

Porosity and permeability of sediments and sedimentary rocks are important properties derived from, and whose values are dependent upon, texture. **Porosity** is defined as ratio of pore volume to bulk volume, usually expressed as a percent. Bulk volume can be described as the volume enclosed by a tight-fitting wrapper around the sample; pore volume is the volume of ''empty'' space enclosed within the sample (that is, within the wrapper). There are several different ways to measure porosity, and some methods are rather indirect. The method in most common use by the petrographer is to measure total pore area within a given area of thin section (porosity is the ratio of the two areas; the result is the same as if volumes had been measured, provided that the porosity is isotropic). This is accomplished ordinarily by the *point count* or by a Monte Carlo method, both being statistical sampling techniques that yield estimates of areas of irregular or disjoint figures. Pore area in thin sections can be measured also by certain optical methods, perhaps aided by special computer interfaces. Porosity is routinely measured within oil-industry circles by determining the volume of fluid that can be caused to enter a sample of given bulk volume. This is really a measure of *interconnected* pore space, or *effective porosity*; isolated pores will not imbibe the fluid. The Boyle's law apparatus squeezes helium gas into the sample under a given pressure, then measures the gas-volume increase (into a separate chamber) when the pressure is reduced by a given amount; gas-volume increase is related to total pore volume through Boyle's law (helium behaves much like an ideal gas). Other techniques are

routinely and automatically implemented by certain down-hole well-logging devices. One technique involves measuring the electrical resistivity of rocks. Rock materials in the subsurface ordinarily are saturated with salt water, an electrically conductive fluid. High-porosity rock is more highly conductive than low-porosity rock, though certain geometrical aspects of the pore space also are influential here. Again, the measured result is effective porosity. Another method is to determine the *matrix density* ρ_m of the rock (that is, the density of the solid material of which the rock is composed) and compare this to the *bulk density* ρ_b of the sample (that is, the density of the rock as a whole, including its pore space). Porosity is related to these densities as

$$\varphi = (\rho_b - \rho_m)/(\rho_f - \rho_m) \qquad (2\text{-}15)$$

where ρ_f is density of the pore fluid.

Porosity of well-sorted sands is independent of mean grain size (Beard and Weyl, 1973, p. 369); purely geometrical arguments predict this. But finer sands commonly are more porous than coarser ones, apparently owing to a tendency of smaller grains toward lower roundness. Porosity is strongly correlated with sorting; as sorting improves, so does porosity. Poorly sorted materials are less porous because the smaller particles tend to fill the spaces between the larger grains, and the larger grains completely occupy volumes that would otherwise be occupied by smaller grains and their associated pores. Scherer (1987, p. 486 and fig. 3) derived a relationship between porosity and sorting (based on Beard and Weyl's data) as porosity $= 20.91 + 22.9/S$, where sorting, S, is given in terms of phi quartiles as 2^z, where $z = (\varphi_{75} - \varphi_{25})/2$. Angular sands appear to be, in general, more porous than more rounded sands. Grains that are very irregular in shape, such as certain skeletals, yield aggregates that are very porous.

Porosity of a grain aggregate declines gradually under burial compaction, tracking the various fabric changes described earlier, and also as a result of cementation, grain deformation, and pressure dissolution (see Chap. 11).

Permeability is a measure of the rate (volume discharge) at which a fluid can pass through a porous material. Under conditions of one-dimensional viscous flow (that is, laminar flow in a specified direction), volume discharge per unit area of the permeable medium is given by Darcy's law:

$$Q = kAg\rho/\mu \ dp/dx \qquad (2\text{-}16)$$

where ρ is density and μ dynamic viscosity of the fluid, dp/dx is the gradient of applied pressure, and k is the permeability of the medium. Permeability k has dimensions of $[L^2]$, that is, area. Permeabilities of sedimentary rocks range from about 10^{-8} cm^2 (very high permeability) to less than 10^{-12} cm^2 (virtually impervious). Permeability is more usually given in special units called the *darcy* (or millidarcy). Imagine that an incompressible fluid is being forced through a cylindrical rock sample by a piston having the same cross section as the sample; the rate at which the piston moves is called the **Darcy velocity**. A material has a permeability of 1 darcy if a pressure gradient of 1 atm/cm causes a Darcy velocity of 1 cm/sec, for water (dynamic viscosity of 1 centipoise). The darcy is very nearly equal to 10^{-8} cm^2, so the permeabilities of sedimentary materials vary roughly from a few darcies to less than a tenth of a millidarcy.

Permeability is an important property of sedimentary rocks inasmuch as it often determines success or failure of an oil or water well, disposal well, leach field, water reservoir, or solid-waste landfill site. Several methods of measuring or estimating permeability have been devised; some are performed in laboratories, other techniques are used at the project site. Laboratory measurements of permeability are carried out on cylindrical plugs of the rock; the volume-rate at which a fluid passes through the sample under a given pressure gradient is measured. A related laboratory procedure is the *capillary pressure test*, which provides an approximate analysis of the size distribution of pores and pore throats in a potential reservoir rock. A certain force or pressure is required to move a nonwetting fluid (such as mercury) through pore throats of a given size. Results of capillary pressure tests permit some estimate of the **recovery efficiency** of the rock, that is, the portion of total pore fluid (oil) that can be pumped out of the reservoir. (This is rarely more than 30 or 35 percent.) On-site measurement of permeability in an oil well is provided by the *drillstem test*, which measures the rate at which pore fluids can be sucked out of the reservoir rock and the rate at which the tested zone then recovers its natural "formation pressure." The *drawdown test* is a similar procedure in a shallow water well, providing an estimate of the rate at which the well can produce water. Permeability of soil commonly is measured simply by digging a pit in the soil, filling it with water, and then observing the rate at which the water level drops.

Mathematical models and simulations of permeability and of the flow of pore fluids through permeable materials are being developed, being an outgrowth of a branch of mathematics called *percolation theory* (see Stauffer and Aharony, 1992).

In sedimentary rocks, porosity and permeability are related to one another and to the various textural properties of the rock. In contrast to porosity, permeability is strongly controlled by mean grain size. The Kozeny equation (Scheidegger, 1957, p. 103; Bear, 1972, p. 166) indicates that permeability is proportional to the cube of porosity and inversely proportional to the square of **specific surface**, which is total grain- (or pore-) surface area per unit bulk volume (dimensions of $[L^{-1}]$); specific surface is inversely proportional to grain size. Krumbein and Monk (1942; see Bear, 1972, p. 133) found a relationship between permeability, mean diameter, and sorting. Permeability is controlled not only by total pore volume and sizes of the pores but also by shapes, surface roughness, and interconnectedness of the pores, all combined into a rather vaguely understood concept of **tortuosity**. Values of tortuosity ordinarily are obtained from electrical resistivity measurements, but Rink (1976) related it to ratio of pore area to pore perimeter in cross sections.

In many or most sedimentary rocks permeability is anisotropic, that is, it varies with direction, reflecting fabric and sedimentary structure (Mast and Potter, 1963). Vertical permeability is generally less than permeability in any horizontal direction. This is due in large part to sedimentary structure and anisotropic fabric in the rocks. Shale partings and stylolites can severely reduce vertical permeability while having lesser or negligible effect on horizontal permeability.

Observation of the shapes of pores and pore networks can be aided by means of the pore cast. A thermoplastic or molten metal is forced into (connected) pore space, and when it hardens, the rock material is dissolved away. The result commonly looks like a foam or a boxwork. Pittman and Duschatko (1970) described methods of obtaining pore casts, and they present some nice photomicrographs (see also Wardlaw, 1976).

Angle of Repose

A handful of sand poured gently onto a tabletop forms a conical pile with a definite slope angle called the **static angle of repose**. It is a property of loose, cohesionless grain aggregates that is governed by internal friction, that is, collective friction between grains in contact. The magnitude of this friction depends on the sizes or weights of the grains, their shapes, and their surface textures. Grains that are quite spherical, well rounded, and smooth or polished have low angles of repose, typically about 33 to 35°; low-sphericity, angular grains have higher angles of repose, about 40°. von Burkalow (1945) found that, for natural dry sand, the angle of repose varies only slightly with grain size and grain roundness, but that the angle of repose of poorly sorted materials is greater than that of well-sorted materials. Traction by a current alters the static angle of repose to a *dynamic angle of repose*. The inclination of a cross-bed foreset (see Chap. 5) is less than the static angle of repose, and the difference is a measure of the magnitude of a current that has swept the surface.

The Textures of Crystalline Sedimentary Rocks

The textures of crystalline rocks are a result of chemical processes and determined by mineralogy—specifically, the crystal habits assumed by the minerals involved in the precipitation—and also by the extent to which various minerals have been co-precipitated and by rates of crystallization. There are also certain controls imposed by the substrate on the crystallographic orientations of growing crystals. Crystals grow by precipitation of ions, atoms, or molecules out of gaseous, molten, or dissolved state onto a substrate. Many evaporite rocks are accumulations of crystals that formed in suspension or even floated on the surface of a concentrated brine, then sank to the bottom; currents might even have transported them before they finally came to rest. Crystals grow also by solid state diffusion or transformation processes, rather similar to metamorphism, that sedimentary petrologists refer to as *neomorphism*. Besides the crystalline sedimentary rocks per se, many particulate rocks also contain crystalline components (cements) introduced diagenetically in the spaces between the grains.

Usually it is difficult or impossible to "take apart" a crystalline aggregate so that grains or crystals can be examined individually and in three dimensions (though this has been done with some metals). This precludes many of the kinds of detailed textural studies that have been performed on particulate materials. But we have at our disposal the knowledge gained by the mineralogists and igneous and metamorphic petrologists, and by the metallurgists and ceramicists concerning the textures of crystalline materials made under conditions that are monitored and carefully

controlled. Of course concepts such as roundness and sphericity are either inapplicable to crystalline aggregates or trivial. Instead of shape terms such as *sphericity*, we use terms that refer to crystal habit, such as *acicular*, *prismatic*, or *tabular*, or to the degree of perfection of crystal form, such as *euhedral*, *subhedral*, or *anhedral*. Certain physical, chemical, and topological principles can be applied to crystalline aggregates that are meaningless or of doubtful utility in studies of particulate aggregates. Aboav and Langdon (1969) and Aboav (1972) demonstrated the important topological similarities between crystalline aggregates and foams, that is, three-dimensional bubble clusters (see also C. S. Smith, 1981).

When examining crystal fabrics in thin section, one must always keep in mind the stereological effects. Some minerals form fan-shaped aggregates of columnar or bladed crystals. If fans are more or less randomly oriented, then some of them in a thin section will appear not as fans but as aggregates of equant polygonal crystals. One might conclude (erroneously) that two different crystal fabrics exist.

As with texture of particulate rocks, texture of crystalline rocks includes aspects of grain (crystal) size, such as mean size and range (though the latter is hardly ever measured), aspects of crystal shape and orientation, and the spatial relationships among adjacent crystals. The shapes of crystal grains are controlled by atomic or chemical forces, and if the crystals of a given mineral species are allowed to grow without impediment, their shapes will all be very much alike—and mineralogy reference books offer descriptions and drawings of how they *should* look. Any anomaly in crystal form or habit, or any alteration in the shape or internal structure of the crystal subsequent to its growth (owing to mechanical stress or to partial dissolution, for example), is readily noticed. As crystals grow, they come to interfere with other crystals nearby, typically up to the point where all interstitial space is filled. The crystals thus contact their neighbors completely, or even interlock; such crystals do not, of course, retain the distinctive habits that they had when they were younger and smaller. Sometimes we find a mineral that has come to occupy the space vacated by some other mineral species, as when a cubic pyrite crystal alters to hematite or limonite, or when a rhombohedral dolomite crystal is replaced by calcite, without concomitant change in crystal form. Such objects are called *pseudomorphs*.

In some crystalline sedimentary rocks, nearly all

crystals are euhedral, and the texture is said to be *idiotopic* (Friedman, 1965); if crystals are, in general, anhedral, the texture is *xenotopic*; an intermediate state is the *hypidiotopic* texture. Size of crystals in sedimentary rocks ranges from submicroscopic to a few millimeters; only rarely do crystals exceed a centimeter (though in some evaporite rocks gypsum or halite crystals are a meter in length). Friedman's crystal-size nomenclature is as follows:

<0.004 mm	Aphanocrystalline
0.004 to 0.016 mm	Very fine
0.016 to 0.063 mm	Fine
0.063 to 0.25 mm	Medium
0.25 to 1.00 mm	Coarse
1.00 to 4.00 mm	Very coarse
>4.00 mm	Extremely coarse

An aggregate of crystals all very nearly the same size is said to be *equigranular*, in contradistinction to *inequigranular* textures. There may be two clearly defined size modes, so that certain crystals are much larger than the majority; such a texture, commonly a result of diagenetic recrystallization or partial replacement, is said to be *porphyrotopic*. The larger crystals, which perhaps are not the same mineral as the smaller crystals, bear numerous inclusions of the smaller crystals and are said to be *poikilotopic*. A single cement crystal may enclose a host of detrital grains in a sandstone or skeletal grains (fossil fragments) in a limestone; it is a crystal that grew beyond the confines of a single pore (Fig. 2-24). Presumably the pores that this type of poikilotopic cement fills were interconnected. Inclusions in crystals that formed by neomorphism or replacement may be exsolved material or unreplaced remnants of the precursor.

Bedded crystalline rocks, such as evaporite beds, may form by crystallization directly on the bed or by settling of crystals formed in suspension. Crystals that have settled from suspension might continue to grow once they have come to rest. This has been observed also in certain igneous intrusive bodies, such as the Skaergaard igneous body of Greenland; the resulting texture is said to be *accumulate* (Wager and Brown, 1967, p. 64).

Other textural or fabric elements of crystalline aggregates have important sedimentological implications. Crystallographic orientation of a crystal relative to that of adjacent crystals, crystallographic (optical) orientation relative to shape orientation (whether length-slow or length-fast, for example), and optical orientation of inclusions relative to host all have bear-

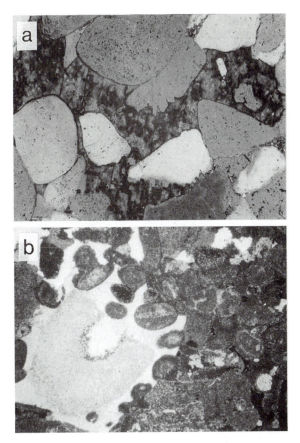

Fig. 2-24. Poikilotopic cement crystals: (**a**) anhydrite-cemented sandstone (with peculiar quartz-cement overgrowths as well); (**b**) calcite crystal, syntaxial on an echinoderm plate, encloses several adjacent pellet-like grains.

ing on whether the crystal grew from a solution or by replacement or neomorphism of some existing crystalline material. Various studies by sedimentary petrologists have shown that, up to a point, crystal size and habit, impurities and inclusions, and fabric are environmentally controlled (see Chap. 8); that temperature and salinity, acidity and oxygen level, and the chemical environments maintained by organisms can influence the textures and minor-element compositions of the crystalline components of sediments and sedimentary rocks. Textural or compositional zoning in crystals indicates a temporal change in solute concentration or composition as precipitation progressed. Some progressive changes in a pore-filling cement occur simply because the pore is getting smaller and solutions are therefore passing through more slowly.

An important textural "principle" of the igneous petrologist, one with which we all are familiar, is that slow crystallization results in coarse textures and rapid crystallization results in finer textures. It appears that when there is a rapid approach to conditions of crystallization or when substantial supersaturation occurs, many nucleations take place simultaneously and finer textures result. In a system maintained close to equilibrium, nucleations come more slowly, and they are more widely spaced. Viscosity of the fluid, which determines rates of ionic diffusion, also is important. In a crystalline rock containing two (or more) mineral species, one mineral may tend to grow faster and to form larger crystals than the other mineral does. Rate of crystal growth depends not only on the physical and chemical environment of growth but also on the elemental composition and structure of the crystal itself. Crystals of simple structure ordinarily grow faster than crystals of complex structure; disordered structures are more rapidly assembled than ordered ones. Thus, calcite, $CaCO_3$, grows faster under given physical and chemical conditions than does dolomite, $CaMg(CO_3)_2$, in part because dolomite is structurally more complex than calcite (see Chap. 8). Moreover, there are ordered dolomites, in which the calcium and magnesium ions form alternating layers, and disordered dolomites, in which calcium and magnesium ions are more or less randomly distributed in these layers; the latter kind of dolomite grows more rapidly. Slow growth permits crystals to exclude impurities; fast-growing crystals tend to incorporate many inclusions of the fluid from which the crystal grew.

Another principle of the igneous petrologist is that the euhedral or idiomorphic crystals of an aggregate are the first formed, and that the very last crystals to form take on the shapes of the irregular spaces remaining between the earlier-formed crystals and are anhedral or xenomorphic. Moreover, the later crystals may come to enclose earlier ones. These principles are certainly useful to the sedimentary petrologist also. Many crystalline aggregates begin as rather loose idiotopic aggregates, then progress toward hypidiotopic or xenotopic textures of low or negligible porosity as crystallization proceeds. Idiotopic aggregates are not efficient space fillers and have the potential of high porosity. In certain dolomites, quite spectacular idiotopic textures may appear and persist (Fig. 2-25),

Fig. 2-25. Idiotopic crystalline aggregate (dolomite).

though mineral cements may eventually fill the space between the rhombs (perhaps poikilotopically).

As adjacent crystals grow, the spaces (pores) around them become smaller, and the crystals eventually begin to compete with one another for space. Where competition is occurring, crystal boundaries can no longer be natural crystal faces governed by lattice constants, and *compromise boundaries* arise. The orientations of these interfaces are determined by the directions and rates of growth of the competing crystals. As crystallization progresses, the shapes of intercrystalline pores change from complex polyhedra to simple tetrahedra, and final closing of these pores is the result of centripetal growth of four crystals. In any cross section (as in a thin section), three of these crystals are seen (Fig. 2-26a), the late-stage pore is triangular, and closure of the pore results in a roughly equiangular triple junction of compromise boundaries.

Where three crystals come together, they may, alternatively, form a special kind of vertex, or intersection of three crystal boundaries as seen in a thin section, where one of the three angles is 180° (Fig. 2-26b). This is Bathurst's (1975, p. 423) *enfacial junction*. Such a triple junction implies that one of the crystals stopped growing, then the other two grew against it. Bathurst suggested that enfacial junctions are quite common in crystalline aggregates formed by precipitation and less common in aggregates formed by neomorphism. This seems to suggest that staggered crystal nucleation takes place in cement crystal aggregates and simultaneous nucleation in neomorphisms.

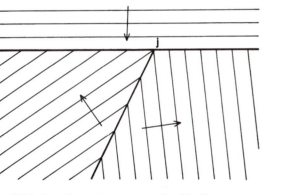

It appears that when crystals that are growing toward one another into a pore come into close proximity, further growth near the region of contact is inhibited, as illustrated by Wardlaw (1976, fig. 10E, p. 251; redrawn here as Fig. 2-27a). Each crystal that faces a pore is an ''organizing principle'' for the ions in solution in the pore. Several crystals facing the same very small pore, and each competing for the same ions, might all lose the battle. In idiotopic dolomites, it appears that dolomite rhombohedra grew until they made contact with adjacent rhombs, then stopped growing. Wardlaw (1976, p. 253) referred to this process as ***contact inhibition*** and pointed out that it is easily observed in experiments with sodium chloride solutions. Apparently, once contact (or near contact) has been made, an energy threshold must be overcome before crystal growth can continue and compromise boundaries form. Perhaps this explains why chalks (certain very fine grained limestones) commonly are so porous and poorly cemented. Bathurst (1975, fig. 302, p. 419) may fortuitously have illustrated an example of contact inhibition at enfacial junctions in a limestone (redrawn here as Fig. 2-27b). Of the three crystals **a**, **b**, and **c**, crystal **a** appears to be the oldest. Where **b** meets **a** (arrow), there is a tiny, high-index face on **b**, the younger of the two crystals. At the enfacial junction of **c** with **b**, crystal **c** (again the younger of the two

Fig. 2-26. Boundaries between crystals; thin lines represent traces of successive crystal growth surfaces; heavy lines are boundaries between separate crystals; arrows indicate directions of growth. **(a)** The compromise-boundary triple junction (at **j**) in a crystalline aggregate. A triangular intercrystalline pore gradually gets smaller, then vanishes; note that even a very inequiangular pore results in a fairly equiangular triple junction. **(b)** The enfacial junction in a crystalline aggregate; a compromise boundary meets a completed crystal face (at **j**).

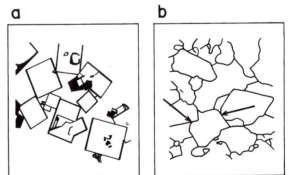

Fig. 2-27. **(a)** Contact inhibition in halite in vitro. (Redrawn from photograph in Wardlaw, 1979, fig. 2, with permission of AAPG.) When growing crystals meet, they stop growing momentarily; this can result in idiotopic aggregates. **(b)** Example of contact inhibition? (Redrawn from photograph in Bathurst, 1975, fig. 302, with permission of Elsevier Scientific Publishing Company, Inc.) Small high-index face appears at the triple junction, suggesting inhibited crystal growth at the boundary.

crystals) bears a small, high-index face. These special faces, which might be viewed as incomplete parts of ordinary low-index faces, probably are manifestations of contact inhibition.

In some rocks a diffusion process operates between two crystals in contact or close proximity that permits the larger or better-formed of the two crystals to "suck away" the smaller one (nicely illustrated by Cotterill, 1985, p. 79, involving two perovskite crystals growing in vitro, redrawn here as Fig. 2-28). Such a process probably operates during neomorphism and explains why neomorphism results in increase in crystal size and why older crystalline limestones are, as a rule, coarser than younger ones. Adelseck et al. (1973, p. 2760) observed that, in aggregates of coccoliths (very tiny single-crystal calcareous skeletal elements of certain planktonic algae) subjected experimentally to elevated temperatures and pressures, the larger coccoliths tend to grow at the expense of the smaller ones, which gradually dissolve and disappear.

The substrate provides crystal surfaces that act as templates for mineralogically similar or identical cement crystals; typically, many cement crystals, randomly oriented, are nucleated simultaneously. In this way the substrate controls cement fabrics, at least up until the time that the substrate is completely buried by the cement. Some crystals in an aggregate prove to be unfavorably oriented, in the sense that slowest growth occurs in a direction away from the substrate, and they are buried at an early stage by neighboring crystals that *are* growing most rapidly in the substrate-normal direction. This situation results in the often observed substrate-normal arrangement of columnar crystals and coarsening of a crystalline aggregate away

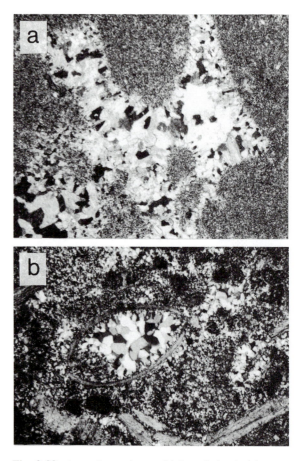

Fig. 2-29. Away-from-substrate fabric variation in (**a**) quartz cement, (**b**) calcite cement.

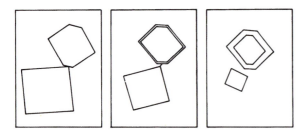

Fig. 2-28. One crystal cannibalizing another. (Redrawn from photograph in Cotterill, 1985, p. 79, with permission of Cambridge University Press.) Material in the larger crystal moves across the narrow bridge and nourishes the smaller crystal.

from a substrate (Fig. 2-29). It is also observed in pigs and other metal ingots or castings. In thin sections through more or less spherical pores, columnar fabrics commonly appear to give way to more equant crystal fabrics at pore center, but this could be a stereological effect in many cases (see Fig. 2-30).

Dickson (1993, p. 3) divided crystal fabric development and "maturation" into three stages, *isolated*, *competitive*, and *parallel*. In the first stage of isolated crystal growth, substrate is the major control. During the competitive stage, some crystals are selectively suppressed; crystals whose greatest growth vectors are most nearly normal to the substrate crowd out other crystals. This leads to a decrease in number of crystals and increase of crystal size away from the substrate

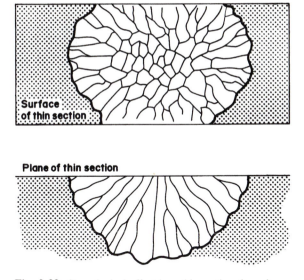

Fig. 2-30. Stereological effect in a thin section through vug cemented with columnar crystals. Crystal size and shape appear to change inward from the margin of the vug.

and regiments the crystallographic axes. The fabric and optical orientations of crystals of the mature parallel stage are consequences of the critical competitive stage.

Columnar fabrics will develop, assuming that the pore is "roomy" enough and that many crystals nucleate more or less simultaneously, no matter what the orientation of greatest growth vector. Lateral growth is soon impeded by laterally adjacent crystals, but away-from-substrate growth can continue until the pore is completely filled. Dickson (1978) pointed out that crystal fabric is independent of optic axis orientation, so columnar or fibrous crystals of quartz or calcite, for example, may be either length-slow or length-fast. Predominance of length-slow or of length-fast crystals in a given pore filling is determined during the competitive phase of cementation by orientation of greatest growth vector with respect to optic axis. In calcite crystallizing as acute (steep) rhombohedra, greatest growth vector is parallel to c axis and pore-filling columnar crystals are length-fast; thin sections parallel to the substrate show equant fabrics. In calcite crystallizing as obtuse rhombohedra, greatest growth vector is at a high angle (about 75°) to the c axis, and crystals are approximately length-slow. Thin sections parallel to the substrate show that length-slow calcite

crystals are bladed rather than columnar; this bladed fabric is revealed in thin sections perpendicular to the substrate as columns of more variable width than for length-fast aggregates.

Occasionally, a crystal changes habit at some stage in its growth. In calcite, for example, an acute rhombohedron may give way to an obtuse rhombohedron. Substrate-normal orientation of the greatest growth vector changes to substrate-parallel orientation, and a T-shaped crystal results. Growth of the obtuse form gradually buries the steep form and gives rise to peculiar zoning patterns. Partial burial of adjacent crystals may result also in enfacial junctions (Dickson, 1993, p. 11).

If pore fluids happen to be precipitating a mineral phase that is structurally different from substrate components, as anhydrite cement in a limestone or calcite cement in a quartz sandstone, then substrate control is lacking. The pore wall does not provide the appropriate templates. In such cases, the cement crystals are likely to be very large—a single crystal might fill the entire pore. Poikilotopic fabrics, wherein a single crystal occupies many contiguous pores and embraces many grains, must also indicate a dearth of templates.

In evaporitic halite deposits it is often observed that those cubic crystals at the base of the bed that happen to be oriented with corner pointing upward control the fabric of the bed as a whole. Cubes with face-up presentation are crowded out. The competitive growth advantage of the former orientation is a consequence of geometry (rather than of crystal physics)—intersections of successive faces are farther apart than the faces themselves. Thus corners advance faster than edges and edges faster than faces. The result (Fig. 2-31) is a crystalline fabric (corner-oriented crystal columns) that hardly seems proper to an aggregate of isometric crystals.

It is generally conceded that a *force of crystallization* exists (Correns, 1949), though it seems not to be well understood. We see it in certain clay minerals of soils, where formation of structured water layers in the clay-mineral crystals expands the clay particles and the soil, to the detriment of pavements and building foundations. We see it in the growth of ice crystals in narrow cracks, and in rock and concrete where chemical weathering has caused the crystallization of new minerals and consequent fracturing. Crystallization pressures and the pressures induced by hydration of existing crystalline substances can have magnitudes of hundreds or thousands of atmospheres (Winkler and

Fig. 2-31. Corner-oriented columns of evaporitic halite. (After Shearman, 1970, fig. 5, with permission of The Institution of Mining and Metallurgy.)

Singer, 1972). Maliva and Siever (1988) considered that the force of crystallization is important in authigenic replacement processes—that is, diagenetic growth of a new mineral crystal at the expense of an existing host mineral grain.

The force of crystallization evidently brings about an active exclusion of particles of foreign matter from a growing crystal, even if those foreign particles must be physically displaced or shouldered aside. We see its results in evaporitic gypsum, where nodules or single crystals growing from the pore waters within a clay mud or carbonate mud expand the mud layer and lift the surface; the gypsum crystals are largely free of inclusions.

3

Mechanics of Flow and Sediment Transport

Most natural sediment transport takes place in fluid media; thus we must deal with systems that contain mixtures of solids and liquids or of solids and gases, the solid materials being divided into particles, and the fluid phase being the medium of transport of the particles. If the solid particles are small enough, they can behave as though they were part of the fluid, a fluid that can be considered as somewhat more viscous and more (or perhaps less) dense than the pure fluid. Particles may, on the other hand, be of such large size that they do not participate in the motion but behave instead as obstacles to the flow. Finally, there are particles of intermediate size that interact substantially with the flow and that are kept in motion or intermittent motion by the flow.

The mechanics of sediment transport is a complex subject; interactions between fluids and solid particles depend on the density, sizes, and shapes of the particles, viscosity and density of the fluid, depth and velocity of the flow, and how the flow varies from place to place and from time to time. The velocity of a stream flow depends on the size of the channel and its gradient and on the roughness of the bed, but all of these quantities change from point to point on the bed, and from moment to moment; they also interact in ways not fully understood. The flow molds its bed into certain shape elements or patterns that, in turn, disrupt or modify the flow. Certain criteria are used to predict whether a given flow will initiate movement of particles on the bed, but as soon as movement begins, the criteria cease to be valid. Theory carries the dynamicist only so far; beyond this, experimental results and observations of natural systems must bring the analysis to some useful end. This is why wind tunnels, flumes, and wave tanks were invented and why natural rivers are fitted with stage recorders and flowmeters.

Solids, Liquids, and Gases

We usually discriminate solids, liquids, and gases on the basis of their mechanical properties, though they may also be defined in terms of their structure, that is, the arrangement and mobility of the constituent atoms and molecules. A *solid* might be defined as a material that obeys Hooke's law; that is, in a solid, strain is proportional to stress (Fig. 3-1). When stress is relieved, a solid recovers its original shape. But if stress exceeds a certain value (the elastic limit), the material fractures or deforms permanently. Solids typically have high shear strength and high compressive strength. A *fluid* is a material that obeys Newton's law of fluids; that is, in a fluid, strain *rate* is proportional to stress. A fluid has low shear strength. (Shear strength in a fluid is termed *viscosity* and is given by the slope of the stress-versus strain-rate line). A *liquid* is a fluid that has a high bulk modulus, that is, high incompressibility. A *gas* is a fluid that obeys Boyle's law, which states that volume is proportional to compressive stress. Gases have low bulk moduli; that is, they are highly compressible, and also have low viscosity.

There are materials having intermediate properties, such as liquids with such high viscosities as to behave much like solids. Malleable metals, such as gold or copper, behave like highly viscous liquids. Moreover, no material follows the ideal stress-strain laws exactly, and some materials depart substantially from these laws. In some liquids, viscosity varies with applied stress, so that strain rate is not linear with stress; these are said to be *non-Newtonian liquids*. A commercial product of the organic chemicals industry, sold as a toy called Silly-Putty, is a curious example of a material with variable viscosity; it behaves like a viscous liquid under low stress but like a brittle solid under

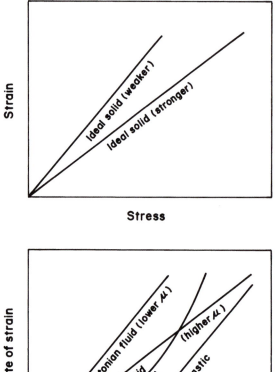

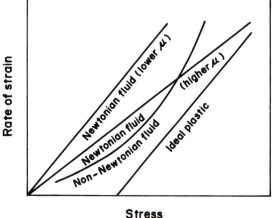

Fig. 3-1. Mechanical behavior of various kinds of materials under stress.

higher stress. It spreads and flows slowly under its own weight; hit it with a hammer and it shatters. Many materials behave like solids under low shear stress and like viscous liquids when shear stress exceeds a certain threshold; such materials, called **Bingham fluids**, are said to exhibit *plastic* behavior. A familiar example is modeling clay; moist natural clays and glacial ice under pressure are sedimentologically important examples. Certain muds, being intimate mixtures of liquids and finely divided solids, maintain a certain rigidity until some disturbance causes them to collapse into liquids; the process is reversible. Such materials, called **Casson fluids**, exhibit *thixotropic* behavior (cat-

sup is a familiar example; thump the bottle and the reluctant stuff spews forth). Many muddy sediments are thixotropic, and so are certain petroleums.

Water, or water mixed with relatively small amounts of solids, approximates Newtonian behavior very closely. It has two important physical properties, density, and viscosity, that distinguish it from other liquids and relate to the manner in which it flows and transports solid particles. **Density** is mass per unit volume, a property that varies modestly with temperature and pressure.* **Viscosity** relates to its rate of deformation (du/dt) under stress τ, and is the coefficient μ in

$$\tau = \mu \; du/dt \qquad (3\text{-}1)$$

It is called absolute, or **dynamic viscosity**; units are dyne-sec per cm^2 (called *poise*; pronounced "pwaz"). Its reciprocal is called **fluidity**. **Kinematic viscosity**, ν, is the ratio of dynamic viscosity to density; units are cm^2 per sec (called *stokes*; "kinematic" denotes independence from forces or mass). The viscosity of water varies with temperature; water is about half as viscous at 25°C as at 0°C (Fig. 3-2). In contrast to the

*Symbols (other than coefficients) and their dimensions used in this chapter are shown in Table 3-2, at the end of the chapter.

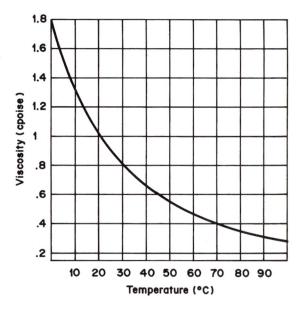

Fig. 3-2. Relationship between viscosity and temperature of water.

behavior of liquids, the viscosity of gases generally *increases* with temperature.

Another physical property that is important in some phenomena of sedimentological interest is **surface tension**. This refers to the mechanical strength of the free surface of a fluid (usually in contact with air). The surface tension of water varies with temperature only 2 or 3 percent between 0°C and 25°C but is substantially lower near the boiling point.

Fluid dynamicists combine certain variables into *dimensionless ratios*, which simplify the analysis of fluid flow. Being dimensionless, such numbers do not depend on the system of units being used, and they are invariant under change in scale. When fluid dynamicists make scale models, they can assume dynamic similarity only if the dimensionless ratios of model and prototype are similar. One such dimensionless number is the **Reynolds number**, which is the ratio of inertial forces in a fluid to viscous forces. The inertial forces cause the fluid to accelerate; the viscous forces tend to damp out these accelerations. The Reynolds number is defined as

$$Re = \rho v l / \mu = v l / \nu \qquad (3\text{-}2)$$

where ρ is fluid density, μ is dynamic viscosity, ν is kinematic viscosity, v is some characteristic velocity, and l is a characteristic length. Reynolds number (Re) may be applied to many different kinds of problems depending upon how one chooses to define velocity and length.

Particles Falling through a Fluid

One of the fundamental problems of fluid dynamics, and of sedimentology, is the behavior of a particle falling under its own weight through a fluid. This behavior depends on inertial and viscous forces, and it seems appropriate to begin our examination by constructing a Reynolds number especially tailored to the problem:

$$Re = 2r\rho v_s / \mu = 2r v_s / \nu \qquad (3\text{-}3)$$

where v_s is the velocity of the particle in the motionless fluid and the characteristic length is chosen to be the diameter $2r$ of the particle (which we will consider to be a sphere). This is called a **particle Reynolds number**. A particle falling through a fluid experiences two kinds of force, a gravitational force that tends to accelerate the particle, and a drag force that resists this acceleration. As the particle begins its fall, it acceler-

ates briefly, until the magnitude of the drag force (which depends on the speed of the particle) equals the gravitational force; at this point, acceleration ceases, and the particle experiences uniform motion thereafter. The magnitude of this motion is called the **terminal velocity** or fall velocity. The **drag**, or resistance to motion, of a sphere moving through a viscous fluid is

$$F_D = C_D A \rho v_s^2 / 2 \qquad (3\text{-}4)$$

where A is the cross-sectional area of the sphere $A = \pi r^2$, v_s is its velocity, ρ is density of the fluid, and C_D is a drag coefficient. This equation follows directly from classic mechanics, and its form was given by Newton. It is commonly known as the *impact law*, because it expresses the force, or impact, of the moving particle against the fluid (or the force of a moving fluid against a stationary particle). The origin of the drag force is in fact a far more difficult matter than the simplicity of the impact law suggests. The gravitational force acting on the sphere—that is, its submerged weight—is

$$F_W = V \Delta \rho g \qquad (3\text{-}5)$$

where g is acceleration due to gravity, $\Delta \rho$ is the difference between density of the sphere and that of the fluid, and V is the volume of the sphere $V = 4\pi r^3 / 3$. The fluid exerts an upward **buoyant force** on the particle proportional to the density contrast between particle and fluid, so its immersed weight is less than its weight in air or in vacuo. For uniform motion—that is, motion that is not accelerating—the drag force and the immersed weight are equal in magnitude (and opposite in direction); hence

$$C_D \pi r^2 \rho v_s^2 / 2 = 4\pi r^3 \Delta \rho g / 3$$

or

$$C_D v^2 = 8r(\Delta \rho / \rho) g / 3 \qquad (3\text{-}6)$$

The drag coefficient must be determined experimentally. It happens that it is not independent of the other variables in the equation; in fact, it is convenient to give its experimental value in terms of the particle Reynolds number (Fig. 3-3). The drag coefficient may also include a shape factor to account for non-sphericity of real particles. We note that for particles of given density, the net gravitational force—that is, the immersed weight—increases as the cube of particle size or radius. Variation of the drag force is more complicated; clearly, it depends on the surface area of the

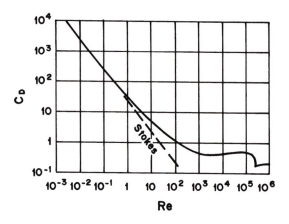

Fig. 3-3. Drag coefficient as a function of Reynolds number for smooth spheres. Note that at Reynolds numbers between 10^3 and 10^5, C_D is nearly independent of Re.

particle, which increases only as the square of the radius.

Stokes's Results

In 1851, G. Stokes investigated the case of falling spherical particles at very low Reynolds numbers, Re <1. Particles in the Stokes range behave more simply than larger or heavier grains, which cause flow separations and turbulent wakes as they fall. Stokes showed the drag force to be

$$F_D = 6\pi\mu r v_s$$

obtained theoretically. Equating this with Newton's impact law, Eq. (3-4), gives

$$C_D \pi r^2 \rho v_s^2 / 2 = 6\pi\mu r v_s$$

or

$$C_D = 24/Re$$

Now Eq. (3-6) becomes

$$24v^2/Re = 8r(\Delta\rho/\rho)g/3$$

or

$$v = (2r)^2 \Delta\rho g / 18\mu$$

This may be rearranged to give particle size ($2r$) in terms of its terminal velocity. For water at 18°C (density 0.999 g/cm³; viscosity 1.06 centipoise), particles of density 2.65 g/cm³ are in the Stokes range if their

diameters are less than 0.1 mm. For air (density 0.0012 and viscosity 0.0183 centipoise), Stokes's law applies to particles not larger than 5 μm.

Particles of Complex Shape

Grains of irregular shape exhibit falling behaviors that are far more complex than the simple motions of small spheres. Some grains rotate as they fall, others adjust their orientations so as to maximize their drag. Komar and Reimers (1978) obtained empirical relationships between drag coefficient and Reynolds number for grains of various shapes, and they measured the fall velocities of natural quartz grains (Fig. 3-4). Baba and Komar (1981) concluded that the settling velocities of natural sand grains larger than about 0.5 millimeters are appreciably lower than the measured settling ve-

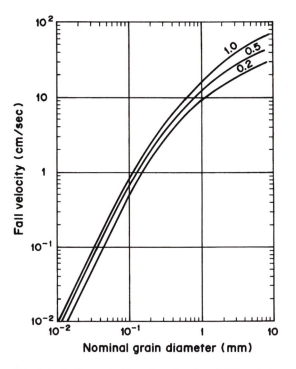

Fig. 3-4. Settling velocities of sand grains of different shapes (given in terms of Corey shape factor). Low-sphericity grains settle more slowly than high-sphericity grains (Corey factor near unity) of the same nominal diameter. (After Komar and Reimers, 1978, fig. 6, with permission of The University of Chicago Press.)

locities of sieve-size-equivalent spheres, and that these differences are due to the non-sphericity of natural sand.

Braithwaite (1973) studied the settling behavior of skeletal grains (biological shells and such) of various "odd" shapes (see also Maiklem, 1968), and found four modes of fall, or "*fall regimes*": (1) straight fall, with no oscillation or rotation but high drag, the grain presenting its maximum projection area to the direction of motion, with major concavities facing backward, that is, upward; (2) spinning fall, the grain oriented to the direction of motion similarly to that of straight fall but rotating around a vertical axis contained within itself; (3) spiral fall, the grain revolving or looping around an external axis; and (4) unstable fall, beginning in a spiral mode, then rocking and wobbling, fluttering and tumbling about different axes, this mode being mostly reserved to the larger, heavier grains.

Natural Sedimenting Systems

Sedimenting systems of geological interest mostly involve many grains, all having about the same density but different sizes and shapes, and interacting. In general, the larger grains fall faster than the smaller ones and settle out of natural suspensions sooner than the smaller grains. Sedimentary particles of different sizes tend to be deposited in different places, either because they take longer to settle out of a fluid that is transporting them or because they can settle only where fluid turbulence has fallen below a certain level. Graded beds (see Chap. 5) are a consequence of the different fall velocities of particles of different sizes. Sedimentation in air differs quantitatively from sedimentation in water, because the two fluids have contrasting densities and viscosities. Most clay particles, with their small sizes, are in the Stokes range (in both water and air), but they do not have the spherical shapes that Stokes assumed. Moreover, clay particles commonly clump together, forming flocs that are much larger than clay particles, of course, but that have lesser bulk densities. In concentrated suspensions, grains do not behave independently, and their settling rates follow the settling laws only qualitatively. The settling laws are important nonetheless, because they reveal what the variables are that control grain settling and establish at least upper and lower limits on rates of fall.

Fluids in Motion; Turbulence

Fluids in motion have mass, momentum, and energy; in many problems, one may assume that these quantities are conserved, although momentum (fluid velocity) may change into a force and energy of motion into turbulence and heat. If the direction and speed of individual fluid volume elements do not change from place to place, the motion is said to be ***uniform***; uniform flow in a channel requires that discharge, mean velocity, and depth do not change from one channel section to another. If direction and speed at a given point do not change from moment to moment, the flow is said to be ***steady***; steady flow in a channel requires that discharge, mean velocity, and depth at a given channel section do not change with time. In many cases we can examine a sufficiently small part of a fluid dynamical system that uniform flow can be assumed; we shall do just that in our analysis of flow in channels and the ways in which flows move solid particles. Small-scale non-uniformity can safely be ignored.

We also strive toward steady flow, if possible, because it is easier, though problems of unsteady flow, such as the motion of water waves, are certainly interesting and important to us sedimentologists (see Chap. 4). Sometimes we can transform a picture of unsteady motion into a picture of steady motion simply by a change in frame of reference (a so-called Galilean transformation). A turbidity current, for example, may be studied in a frame of reference attached to the standing water around it; in this fixed frame there is an abrupt increase in water velocity as the head of the muddy cloud approaches and passes, followed by a gradual decline; this is a decidedly unsteady motion. In a frame of reference attached to the turbidity current itself (an observer riding on the front end of it, let us say), a current of clear water seems to be flowing smoothly by and the big muddy cloud itself seems not to be changing very much; in this frame the motion is steady (though not exactly steady, because this is the real world).

Discharge of a flow is the volume of fluid passing through a given cross section in unit time. It is usually given as

$$q = Av \qquad (3\text{-}7)$$

where A is area of the cross section and v is average velocity through the cross section.

In his famous book *Hydrodynamics*, Horace Lamb (1932 edition, p. 663) lamented that turbulent motion is "the chief outstanding difficulty of our subject"; not much progress has even yet been made. **Turbulence** may be defined as randomly fluctuating momentum (or pressure) of the flow at a given point, or as some degree of non-correlation between fluid momentum (or pressure) at closely spaced points in the flow. In gentle turbulence, characteristic flow disturbances (eddies) are of large size, but as turbulence increases, eddies of smaller and smaller scale progressively appear, superimposed on the larger-scale patterns. Viscosity ultimately determines the size of the smallest disturbances. Thus, there is a continual cascade or transfer of momentum from large-scale eddies to smaller-scale eddies, thence to viscous deformation, operating essentially at the molecular scale, the momentum of turbulence ultimately changing to heat. Note that turbulent flow can be steady and uniform only in the mean, that is, if *time-averaged* direction and speed of fluid volume elements do not change from place to place and moment to moment.

The Reynolds number for a flow confined to a (reasonably smooth) pipe or channel is given by

$$Re = \rho v R / \mu = v R / \nu \qquad (3\text{-}8)$$

In this **flow Reynolds number**, the characteristic length is given as **hydraulic radius** of the flow,

$$R = A/p \qquad (3\text{-}9)$$

where p is **wetted perimeter** of the flow, being the total length of the margin of the cross section of the flow in contact with solid surface (that is, excluding any free surface). The hydraulic radius of *wide* channel flows is very nearly the average depth.

When the flow Reynolds number is less than about 1000, the flow can be **laminar**, meaning that individual volume elements in the flow move downstream along smooth trajectories. For higher Reynolds numbers and certainly for values above 2000, flow can be **turbulent**, meaning that individual streamlines are irregular and interlaced and vary rapidly with time. Note that a low Reynolds number requires either low average velocity, low density, or high viscosity. Local flow Reynolds numbers may be obtained to show how the magnitude of turbulence varies from place to place in a channel. Generally, turbulence near a boundary is different from turbulence in the ambient flow; a flow that is turbulent in the main may be laminar near the boundary. Moreover, turbulence near a boundary is

anisotropic, because there can be no momentum across a boundary (there can be *some* if the boundary is permeable or deformable, as in most natural channels). Turbulence developed near a rough boundary spreads upward into the main flow; thus, a rough bed may create a flow whose highly turbulent condition is not reflected in the value of the Reynolds number as obtained from Eq. (3-8). Turbulent motions in natural flows are largely responsible for the entrainment and transport of sedimentary particles. But some sedimentologically important flows are laminar—flow of ice in a glacier, and the motions of the solid particles in many mudflows and grainflows (see Chap. 4). These flows transport materials of large size and in large quantities because they are so viscous. Flow of groundwater through the systems of small pores in rocks is generally laminar, in part because the flow is very slow.

The critical value of Reynolds's number does not actually indicate the condition of the flow at which transition between laminar and turbulent flow occurs. Rather, it indicates a flow condition such that some initial disturbance of the flow is either damped out (Reynolds number less than critical) or is amplified (Reynolds number greater than critical). It is possible, as Osborne Reynolds (1842–1912) himself showed, for a flow to maintain the laminar state, under carefully controlled experimental conditions, even at Reynolds numbers of several thousands.

Water Flowing in a Channel

Open-channel flow refers to a flow (we will assume that it is water) that has a free surface—that is, a surface in contact with air—and whose motion is constrained by a channel. Steady flow—that is, a flow in which the individual fluid volume elements are not accelerating—requires equality of forces pulling the water down the channel and forces resisting this motion, just as in the case of particles falling through a fluid at terminal velocity.

The Main Flow

Consider a volume of water occupying length L of a straight channel (Fig. 3-5) of cross section A; volume is, of course, AL, and the weight of this volume, that is, the gravitational force acting on this water, is ρgAL. If the slope of the channel is S, then the down-channel

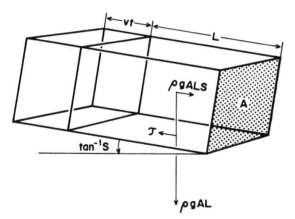

Fig. 3-5. Forces exerted on a prism of water flowing in a channel.

component of the weight of the water is $\rho gALS$. (S is gradient, that is, the tangent of the angle of slope; strictly speaking, the sine should be used, but for small angles, sine is very nearly equal to tangent.) The retarding force results from friction between the flow and its bed and is given by the product of a shear stress (friction) per unit area, τ_o, and the total area of the bed, which is pL, where p is wetted perimeter. Thus,

$$\tau_o pL = \rho gALS$$

or

$$\tau_o = \rho gRS \qquad (3\text{-}10)$$

Evidently, magnitude of shear stress (force per unit area) between the flow and its bed is a function of weight of the fluid lying on the bed. Magnitude of shear stress *within* the flow is generally assumed to increase linearly from a value of zero at the surface of the flow to the value of bed shear stress at the base of the flow.

Boundary shear can be expressed alternatively in terms of Newton's impact law, Eq. (3-4), used earlier in formulating a law of particle settling velocity, but in open channel flow, A and v have meanings that reflect attributes of the channel and the flow rather than of particles in suspension. Force per unit area of the bed, that is, F/A, is proportional to $\rho v^2/2$, according to the impact law, and we write

$$\tau_o = f(Re,k/R)\ \rho v^2/2 \qquad (3\text{-}11)$$

where we have replaced C_D, the particle drag coeffi-

cient, with the channel resistance function f; k/R (or k/D for wide channels), called **relative roughness**, is the ratio of size of roughness elements on the bed to the size of the channel or depth of the flow. Equating Eqs. (3-10) and (3-11) gives

$$\rho gRS = f(Re,k/R)\ \rho v^2/2 \qquad (3\text{-}12)$$

For rough channels, including virtually all natural channels, the resistance coefficient, f, is nearly independent of Re and very much dependent on relative roughness, so, rearranging Eq. (3-12) and combining constants, we may write

$$v = \sqrt{2g/f'}\ \sqrt{RS} = C\ \sqrt{RS} \qquad (3\text{-}13)$$

where coefficient C includes size factors (such as relative roughness) and the (approximately constant) contribution of Reynolds number and other constants. This expression is called the *Chezy equation*. Once again, as with the falling-particle problem, empirical results must come into play. In 1890, R. Manning found experimentally that the Chezy coefficient could be given by

$$C = R^{1/6}/n \qquad (3\text{-}14)$$

where n, called Manning's roughness coefficient, has been tabulated for many different channel bed and bank conditions (Table 3-1). Combining Eqs. (3-13) and (3-14), we obtain the widely used Manning's equation:

$$v = 1/n\ R^{2/3}\ S^{1/2} \qquad (3\text{-}15)$$

where R is measured in meters. Manning's formula is not dimensionally homogeneous, so a scale factor is needed, whose value depends on the system of units employed. Thus, you might see Manning's equation with a multiplier of 1.49; this is for problems wherein R is measured in feet. n is not, by the way, independent of R.

The coefficient f' in Eq. (3-13) is related to the *Darcy-Weisbach coefficient*, which has been obtained experimentally for flow in pipes. The hydraulic radius of flow in a pipe of diameter D is $D/4$. If the Darcy-Weisbach coefficient, f, is used in problems of open-channel flow, where hydraulic radius is given as mean depth D, then it must be divided by 4. Thus, Eq. (3-13) becomes $v = \sqrt{8g/f}\ \sqrt{DS}$.

The roughness of a natural channel is in part due to the fact that the bed consists of particles of non-negligible size; the bed is rough in the same way that sandpaper is rough, and this kind of roughness is

Table 3-1. Roughness Coefficients for Open Channels

Condition of Channel	n	$1.49/n$
Minor streams (width at flood stage *30 m)*		
Clean, straight; full stage	0.030	49.7
Clean, winding, with some pools and shoals; mountain stream with gravel and cobble bed	0.040	37.3
Clean, winding, with stony bed; mountain stream with boulders	0.050	29.8
Very weedy reaches	0.100	14.9
Major streams (width at flood stage 30 m)		
Regular section with no boulders or brush	0.025–0.060	
Irregular and rough section	0.035–0.100	
Flood plains		
Short grass	0.030	49.7
Scattered brush, heavy weeds; light brush and trees in winter	0.050	29.8
Light brush and trees in summer	0.060	24.8
Medium to dense brush in winter	0.070	21.3
Medium to dense brush in summer; heavy timber	0.100	14.9

Source: After Graf, 1971, tble 11.3.

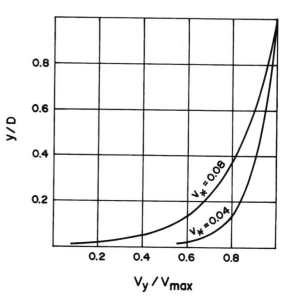

Fig. 3-6. The logarithmic velocity profile and mean flow velocity in a wide open channel for two different values of the shear velocity.

called **grain roughness**. The *relative* roughness concept recognizes that pebbles on the bed of a small channel exert a greater influence on the flow than do pebbles on the bed of a large flow, or that pebbles in a given channel exert greater influence than do sand grains. Natural channels are almost never straight, but sinuous; moreover, they bear certain other forms of roughness, called *bedforms*, that are of a scale intermediate between the sizes of grains and of channel bends. *Channel sinuosity* and *bedform roughness* increase the total drag on the flow and must be considered in predicting flow velocities in natural channels. (Note that bedforms and channel bends create nonuniform flow conditions; in the foregoing analysis, uniform flow was assumed.)

The Manning equation for average velocity in a channel cross section does not address the fact that flow velocity varies from place to place on this cross section. In fact, flow at the bed and banks is close to zero and increases to a maximum value near the surface of the flow. As a rule of thumb, the mean flow velocity in a wide channel is about 0.8 of the velocity at the surface (Fig. 3-6), or the velocity at 0.6 of the distance from water surface to bed. von Karman and others examined the problem of velocity distribution theoretically and experimentally and concluded that the local velocity v_y in a wide, rough channel is a logarithmic function of distance y from the bed:

$$v_y/v_* = c + \frac{1}{\kappa}\ln(y/k) \qquad (3\text{-}16)$$

where k is a measure of size of roughness elements on the bed, and κ, called the von Karman constant, has a value of 0.4 if the flow is clear; Nikuradse in 1933 experimentally obtained $c = 8.48$ for fully turbulent flows. The property v_*, called a **friction velocity** or **shear velocity**, is

$$v_* = \sqrt{\tau_o/\rho} \qquad (3\text{-}17)$$

It is responsible for the shear stress that the flow exerts on the bed. The shear velocity concept will be important to us later, as we pursue the matter of bedload sediment transport by fluids in motion.

A local velocity v_y relative to the maximum velocity v_{max} (at the surface) of a flow of depth D is readily obtained from Eq. (3-16) and given by

$$(v_{max} - v_y)/v_* = \frac{1}{\kappa} \ln(D/y) \qquad (3\text{-}18)$$

Kuelegan in 1938 obtained from Eq. (3-16) a dimensionless relation for *mean* flow, $\overline{v}$, given by

$$\overline{v}/v_* = 6.25 + \frac{1}{\kappa} \ln(D/k) \qquad (3\text{-}19)$$

D/k being reciprocal of relative roughness.

Secondary Flow

Secondary flow refers to components of the flow in directions perpendicular to the *primary*, down-channel *flow*; it amounts to helical or vortex motion, where the axis of the circulation is parallel to the primary flow. It influences the larger-scale aspects of bedload transport in rivers and seems to be responsible for meanders. One kind of secondary flow develops in curved channels (Fig. 3-7). Inertia of the fluid parcels at the water surface drives them toward the outer (so-called concave) bank; water parcels near the bed circulate toward the convex bank, replacing surface water that has flowed away from this region. Thus a helicoidal flow develops, which changes sense (screw direction) whenever the direction of channel curvature changes. If the bed is erodible, the secondary flow gradually amplifies the curves, and the curves, in turn, amplify the secondary flow; thus, the development and growth of river meanders. It is significant that meanders form also in channels dissolved out of limestone or melted out of ice, and in the Gulf Stream, which is a river without solid banks.

Secondary flow is present also in straight channels (Fig. 3-7), such as artificial canals, but its causes and true nature seem not to be fully understood. Some have suggested that it is due to the lateral component of turbulence in the flow. Secondary flow is from zones of high shear to zones of low shear; it has the effect of making the shear stress over a boundary more uniform and surely results from differences of primary flow velocity over a cross section.

It is possible that helicoidal flow patterns are universally responsible for longitudinal bedforms, that is, erosional and depositional features on a movable bed that are elongate parallel to the primary flow. Allen (1966, p. 168; 1968c, p. 32), referring to a rather small-scale regular pattern of helicoidal flow at the bed known as *Taylor-Gortler vortices* (Fig. 3-8), consid-

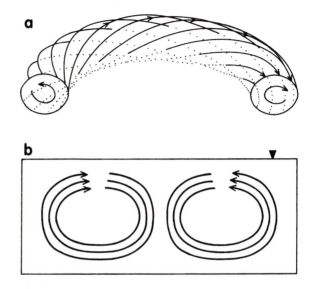

Fig. 3-7. Secondary flow in channels: (**a**) helicoidal flow in a curved channel; (**b**) typical pattern of secondary flow in a straight river channel.

ered this flow pattern to be the cause of parting lineation (see Chap. 5). Dzulynski (1965) showed similar secondary flow patterns to be responsible for the formation of certain bedforms at the base of turbidity currents. Fisher (1977, p. 1295) described, at the fronts of volcanic base surge, a cleft-and-lobe configuration that manifests the existence of vortex cells; the vortices were deemed responsible for the cutting of furrows or channels parallel to the flow. Houbolt (1968, p. 263)

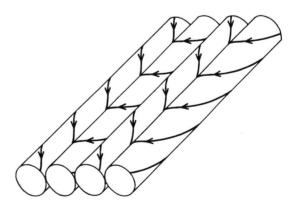

Fig. 3-8. Horizontal vortices near a smooth bed; vortex axes are parallel to the primary flow.

postulated that helicoidal flow contributes to the development of "tidal ridges," a kind of large-scale bedform elongate parallel to the primary flow direction of
tidal currents. Glennie (1970, fig. 73) and Wilson
(1972) illustrated helicoidal flow associated with aeolian longitudinal dunes. Helicoidal flow patterns similar to those near the bed of a flow are induced in lake
waters or stratified ocean surface waters by the wind;
this is called the *Langmuir circulation*.

Energy of a Flow

Every body of water has, by virtue of its depth (or
elevation) and velocity, a certain energy. Imagine a
tank standing on the ground that is filled with water.
The water has a certain potential energy that can be
expressed in terms of its height above the ground. If
the tank should burst, the water would flow out onto
the ground, and its potential energy would be transformed into kinetic energy; the water would lose elevation but gain momentum. Energy per unit weight of
the fluid has dimensions of length, and is called *head*.
In an unconfined flow such as this (or flow in an open
channel), the total head is divided between a velocity
head and a flow depth. It may happen that the velocity
head equals the flow depth, a condition known as *critical flow* (Fig. 3-9a). It is a condition of maximum
discharge (Fig. 3-9b); that is, for a given total head,
the flow cannot attain a greater discharge than the discharge at critical flow. If velocity head is less than flow
depth, then the flow is said to be *subcritical* or *tranquil*; if velocity head exceeds flow depth, then the flow
is *supercritical*, or *shooting*. The corresponding subcritical and supercritical flow depths and flow velocities associated with a given total head are called the
conjugate depths and *conjugate velocities*. The set of
conjugate depths (or conjugate velocities) is associated
with a single value of discharge. A given total head
gives rise *only* to one or the other of the two conjugate
flows; no other combination of depth and velocity is
possible.

The *Froude number* (rhymes with "food") is an
important dimensionless number that relates to energy
of the flow. It is a ratio of inertial to gravitational
forces in a fluid, or ratio of velocity head to depth of
flow:

$$Fr = v^2/gD \qquad (3\text{-}20)$$

Its value is unity at critical flow; a subcritical flow
is one for which $Fr<1$; in supercritical flow $Fr>1$. Of

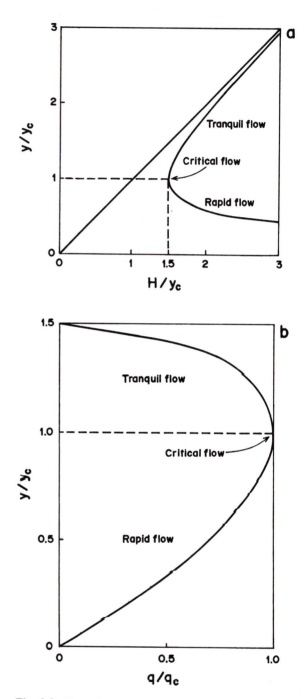

Fig. 3-9. Dimensionless plots of **(a)** specific energy and **(b)**
discharge, showing critical depth and fields of tranquil and rapid
flow in a rectangular channel.

course, the Froude number may be expressed also in terms of discharge, thus

$$Fr = q^2/gD^3 \qquad (3\text{-}21)$$

Sometimes Fr is given as its square root, that is, $v/\sqrt{gD}$. In this form it is explicitly a ratio of flow velocity to celerity of a wave on the surface of the flow (see Chap. 4).

A flow can change from one depth (and velocity) to its conjugate depth (and velocity), — that is, from subcritical to supercritical, or from supercritical to subcritical—triggered by various conditions in the channel, such as downstream change in channel roughness, width, or (especially) slope. Change from tranquil to shooting flow occurs smoothly, and the surface profile of such a *gradually varied flow* can be computed quite simply and accurately. The opposite change, from shooting to tranquil flow, though resulting in smooth surface-profile segments over parts of the reach, generates also a zone of sudden and commonly quite turbulent increase in flow depth, called the **hydraulic jump**. It is a spontaneous change from supercritical flow depth and velocity to subcritical values, accompanied in general by substantial energy loss to turbulence.

Rivers may be in upper regime over short reaches where gradients are particularly high (as in a rapids), or sometimes during flood stage, or downstream from a spillway. It is an erosive condition. Rapid (upper-regime) flow can usually be recognized easily. Of course, if a hydraulic jump is present, then the flow on the upstream side is necessarily rapid and the flow downstream is tranquil. Where the flow must climb a ''step'' on the bed, the water surface of a tranquil flow falls a bit, but the surface of a rapid flow rises over the step. Thus, water sometimes runs uphill. Generally there are waves on the surface of a rapid flow, but they are stationary, not moving either upstream or downstream; such waves, a meter or several meters crest to crest, give the water surface the appearance of a washboard. Some such waves have no crest length but are trains of mounds or ''haystacks'' of water, commonly referred to as ''rooster tails.'' The crests of some waves may curl over into the flow, like a breaking wave. Very shallow ''sheet'' flows, such as wave swash on a beach face, are typically in upper regime (though in situations where flow is very shallow, surface tension becomes an important factor in governing flow characteristics). If you touch the surface of such a flow with a stick or place an obstacle in the flow, a V-shaped shock wave forms. Irregularities on the bank of a supercritical flow set up an oblique or rhombic pattern of standing waves that we all have seen at street curbs after the rain.

Flows having different Froude numbers create different bedforms in the sediments that they are transporting. Later (Chap. 5), we shall examine relationships between Froude number and various styles of bedload transport. These relationships will permit us to estimate the hydraulic conditions of ancient flows.

Movable Beds; Scour and Sediment Transport

When water flows over a surface, it imparts some of its energy to the surface; if the surface is a bed of movable particles, then the flow may set some of these particles in motion. Whether or not this happens depends fundamentally on depth and speed of the flow and on size and weight of the particles on the bed.

Scour Criteria

A *scour criterion* describes a physical condition of a flow that permits the flow to initiate movement of some particle or collection of particles on its bed. A particle resting on the bed of a flow experiences hydrodynamic lift and drag forces that tend to move the particle and gravitational and frictional forces that cause the particle to resist this motion. The particle's immersed weight and its shape and position with respect to other particles on the bed determine the magnitude of the friction. If the bed consists of very small (silt- or clay-size) particles, then surface tension between particles and water in the small pores between particles becomes important. This analysis assumes that there are no cohesive forces, which would further increase a particle's resistance to motion.

The *hydrodynamic lift force* results from two different characteristics of the flow: (1) turbulence, which creates momentary but frequent upward flow components that can lift grains from the bed, and (2) distortions of the flow pattern near the bed, which create Bernoulli forces (the same kind of force that lifts an airfoil). A particle on the bed, of sufficient size to deform the flow, locally increases flow velocity over the particle and reduces the pressure on its upper surface; the pressure under the particle is not affected, and if the pressure difference between upper and lower surfaces produces a force that exceeds the submerged weight of the particle, then the particle is lifted into the flow. As soon as this happens, the asymmetry vanishes and the particle falls back to the bed. Even if the grain is not lifted, its effective weight is reduced, and so, too, is the magnitude of the frictional force between the grain and the bed. Thus, the presence of a lift force, even if it causes no movement, makes the drag force more effective. Southard's (1971) experiments with the transport of grains by laminar flows suggest that the Bernoulli lift is negligible for grains of sand size or finer. Values of C_D and C_L vary non-linearly with Reynolds number (refer again to Fig. 3-3).

The *hydrodynamic drag force* is due in part to friction between the flow and the particle on the bed and in part to impact of the flow on the particle. The incipient motion of a particle requires that lift and drag forces just overcome the forces due to its weight. The submerged weight of a particle [see Eq. (3-5)] is

$$F_W = k_1 \Delta \rho g d^3 \qquad (3\text{-}22)$$

The lift force on a particle is given by

$$F_L = C_L k_2 d^2 \rho v_*^2 / 2 \qquad (3\text{-}23)$$

and the drag force [see Eq. (3-4)] by

$$F_D = C_D k_3 d^2 \rho v_*^2 / 2 \qquad (3\text{-}24)$$

where k_1, k_2, and k_3 are particle shape factors; d is the size of the particle; ρ is its density; v_* is a shear velocity (as before), a velocity of the flow at (or very near) the bed; and C_L and C_D are lift and drag coefficients, respectively. It is clear from the form of Eqs. (3-23) and (3-24) that lift and drag forces are fundamentally similar functions of flow velocity and of size and shape of the particle.

Combine forces F_L and F_D that tend to move the particle; for incipient motion, this force must equal the frictional force that holds the particle in place, which is proportional to the gravitational force (immersed weight of the particle); thus

$$F_D + F_L \propto F_W$$

or

$$d^2 \rho v_b^2 / \Delta \rho g d^3 = C \qquad (3\text{-}25)$$

where C is a constant of proportionality that combines all friction coefficients, including C_D and C_L and shape factors. The value of C must depend on a Reynolds number near to the bed, or $v_b d \rho / \mu$. Shear velocity v_*

is given by Eq. (3-17) as $\sqrt{\tau_o / \rho}$; thus we may rewrite Eq. (3-25) as

$$\tau_c / \Delta \rho g d = f(v_c d \rho / \mu) \qquad (3\text{-}26)$$

which specifies critical shear stress and critical shear velocity, that is, the conditions for incipient motion of a grain on the bed (we have replaced shear velocity v_* with critical shear velocity v_c, and shear stress τ_o with critical shear stress τ_c). The quantity on the left is called **Shields's beta** (note that β is dimensionless). The quantity on the right is a Reynolds number (also dimensionless, of course); a Reynolds number containing the shear velocity is called a **shear Reynolds number**.

The functional relationship between beta and shear Reynolds number, Re_*, for incipient motion, has been obtained experimentally (Fig. 3-10). If, for a given value of Re_*, beta lies below the curve, then the grain does not move; if beta lies above the curve, grain motion does occur. Note that at values of Re_* above 400, beta has a value of about 0.06 independent of Re_*, and that at Re_* of about 10, beta has a minimum value of about 0.03. The size of a quartz grain corresponding to minimum beta is about 1 millimeter. Grains smaller than this (corresponding to smaller Re_*) do not project very far into the flow; their movement is due mainly to viscosity of the flow and is largely independent of turbulence. Larger grains (higher Re_*) are less easily moved because they are heavier. But because they project into the flow, they cause much turbulence near the bed, and this turbulence is mainly responsible for the movement of these larger particles. Leliavski (1955, p. 49) pointed out that the sloping linear segment of the left part of the Shields curve represents a condition of grains completely enclosed in a laminar sublayer— that is, a thin layer of fluid next to the bed in which flow is approximately laminar. Forces on grains in this condition are viscous forces entirely. The horizontal segment of the right part of the Shields curve, in contrast, characterizes a condition of complete disruption of the laminar sublayer and forces on grains that are due almost entirely to turbulence.

This analysis does not explicitly yield a value for critical bed shear stress, nor is the relation between incipient motion and particle size clearly expressed. To overcome this problem, we observe that $Re_*^2 = \rho \tau_o d^2 / \mu^2$, and that

$$Re_*^2 / \beta = \rho \Delta \rho g d^3 / \mu^2 \qquad (3\text{-}27)$$

We have also used Eq. (3-17), the definition of v_*.

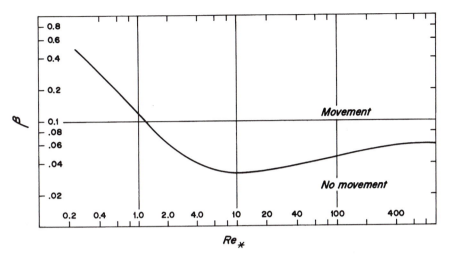

Fig. 3-10. Shields's function for critical bed shear stress. Note that β is independent of Re_* for Re_* greater than 400.

Now for a given value of beta obtain the value of Re_* from the Shields curve, or inversely, for a given Re_* obtain beta; now the value of d, the particle diameter, is readily obtained, and the value of τ_c, the critical shear stress, emerges from the definition of beta, contained in Eq. (3-26). One might compare critical bed shear stress obtained in this way with actual bed shear stress obtained from Eq. (3-10) for a given flow, to determine whether or not bed transport of a particular grain size is possible.

One also can obtain a plot of critical shear stress versus size of grains in incipient transport. For a water flow with density 1.0 g/cm³ and viscosity 0.01 poise, and quartz grains (density 2.65 g/cm³) on the bed (Fig. 3-11), critical bed shear stress for bedload of 1-mm grains is about 5 dyne/cm², and for 10-mm bedload, critical shear stress is 100 dyne/cm². Corresponding shear velocities, from Eq. (3-17), are about 2.24 cm/sec and 10 cm/sec. Leliavski (1955, fig. 17) made a similar plot, based on various experimental data. The approximately linear relationship for grains up to about 3 mm is $\tau_c = 1.66d$, where d is grain diameter in mm, and the shear stress is given in dyne/cm².

As a further example of application of Shields's criterion, consider a problem posed by Davies and Walker (1974, p. 1213), to compare rolling or sliding transport with suspension transport in the same flow. Assume a shear Reynolds number sufficiently high that critical value of Shields's beta is 0.06 (see Fig. 3-10). Thus $\beta = \tau_c/\Delta\rho d = 0.06$. Then, for a 1-cm pebble

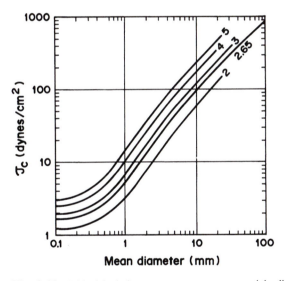

Fig. 3-11. Critical bed shear stress versus mean particle diameter for grains on the bed of a water flow; calculated from Shields's dimensionless curve for quartz grains (density 2.65) and for grains of other densities.

of quartz in water ($\Delta\rho = 1.65$), $\tau_c = 97$ dyne/cm². This gives, from Eq. (3-17), a critical shear velocity v_c of about 10 cm/sec. Now consider suspension transport of smaller grains. If a grain is to be in suspension, then its settling velocity must be less than the shear velocity of the flow, thus $v_*/v_s > 1$. The size of a spherical grain having a settling or terminal velocity of 10 cm/sec is 0.6 mm, from Figure 3-4, so that a certain flow can transport both 1-cm pebbles and 0.6-mm grains (and smaller grains) at once. Walker (in Harms et al., 1975, p. 140) made a graph of largest suspended grains versus largest rolling grains.

Shields's curve is valid for flow over a flat bed of non-cohesive granular material of uniform size; shear stress is the time-average stress rather than some maximum achieved during turbulent bursts. The curve has undergone minor modifications over the years, based on newer experimental data, or with slightly different assumptions concerning size distribution of bed material and statistical properties of time-averaged bed shear.

Hjulstrom, in 1935, obtained experimentally a curve showing the mean flow velocity required to initiate movement on a bed of quartz grains of uniform size (Fig. 3-12). It has been quite popular among geologists (though occasionally misused). Grains having a diameter of about 0.25 mm, it seems, are transported at the lowest velocities; grains larger or smaller require higher flow velocities, the larger grains because they are heavier, the smaller grains because of their tendency to adhere to one another. Hjulstrom assumed a simple relationship between mean flow velocity and velocity at the bed that applies (approximately) only to flows of depth greater than 1 m; thus Hjulstrom's

curve should not be applied to shallower flows. Hjulstrom also showed that smaller flow velocities are required to sustain movement than to initiate it, this so-called *Hjulstrom effect* being most pronounced at the smallest grain sizes. Sundborg (1956) showed experimental threshold velocity versus particle size for well-rounded, well-sorted quartz in water flows at 20°C; velocity is measured 1 m above the bed. Miller et al. (1977, p. 518) obtained, from Sundborg's data, the regression lines

$$U_{100} = 122.6d^{0.29} \qquad (3\text{-}28)$$

for mud- and sand-size beds, and

$$U_{100} = 160.0d^{0.45} \qquad (3\text{-}29)$$

for gravel beds, where d is particle size in millimeters and U_{100} is velocity (cm/sec) measured 100 cm above the bed. For muds, which are cohesive, and thus behave very differently from the materials that Shields or Sundborg investigated, Migniot (1968, p. 613) obtained critical velocity of

$$v_c = 0.5\tau^{0.5} \qquad (3\text{-}30)$$

for incipient erosion of stiff muds (shear strength $\tau > 20$ dyne/cm²), and

$$v_c = \tau^{0.25} \qquad (3\text{-}31)$$

for soft muds ($\tau < 10$ dyne per cm²).

Long-term action of a streamflow may produce a *deflation armor*, that is, a layer of material that is too coarse to be transported by the flow and lags behind on the bed. Moreover, large grains on a bed tend to hide smaller grains from the flow grains that otherwise would move readily. Deflation armors form not only on stream beds but also on shorefaces and on wind-swept flats (Fig. 3-13). Deflation in one area and deposition in another implies a lateral gradient in bed shear stress, due presumably to changing depth or velocity of the flow.

The Downstream Variation of Grain Size in Rivers

In rivers, downstream variation of critical shear stress is responsible for organizing the bedload. Long ago (1875), Sternburg measured sizes of pebbles on a long reach of the Rhine River and found that mean pebble size and size of largest pebble decline downstream. He attributed this to gradual attrition of the pebbles in transport. If the rate of attrition is proportional to size; then the relationship between size and distance trav-

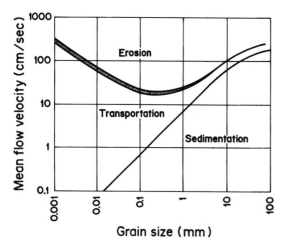

Fig. 3-12. Hjulstrom's results, mean flow velocity required to initiate movement on a flat, uniform bed, for flow depth of 1 meter. Flow velocity required to *sustain* movement is less (lower curve).

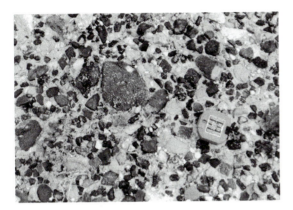

Fig. 3-13. Deflation lag on a wind-swept surface in the desert.

eled is an exponential function similar to a first-order kinetics law:

$$d/d_o = \exp[-a(x-x_o)] \qquad (3\text{-}32)$$

where d_o is size of the pebble at distance x_o, d is size at distance x, and a is a **coefficient of abrasion**, dependent on the hardness of the material in transport and on the intensity or vigor of the abrasion process. Certainly in most rivers there is a regular downstream decline in clast size, but it does not have much to do with attrition. Though one would expect roundness to improve with attrition, significant downstream trends toward greater roundness are rarely observed, and experiments suggest that particle attrition in rivers is very slow. Rivers move smaller clasts more rapidly than larger ones, so that the smaller grains soon "lead the pack" while the larger clasts bring up the rear. But this leads to a downstream grain-size decline only if all clasts entered the river at about the same time and place.

It is not differential transport rate that organizes the stream load, nor is it progressive attrition that causes a grain-size trend. It seems to be an energy gradient or bed-shear-stress gradient that is the cause of textural trends in most rivers. Grains of different sizes come to equilibrium at different points downstream. New materials, once they enter the channel, probably reach some equilibrium downstream position very quickly, then tend to remain there, held by an "energy threshold" created by downstream decline in traction competence of the stream. Thus, the "Sternberg effect" is a sorting phenomenon related to streamflow competence, which in most rivers decreases substantially in

the downstream direction; one may show this by calculating τ_0 from Eq. (3-10) for any pair of representative upstream and downstream channel sections.

Bedload Discharge

Analysis of **bedload discharge** is a hard problem but one that is perhaps of more significance to the sedimentologist than the problem of critical bed shear. As with all or most analyses of the dynamics of interactions between fluids and solids, theory must be combined with experimental results. The most difficult problem is accounting for bedload discharge of sediments of non-uniform size. Flume experiments to investigate rate of bedload transport commonly yield variations of 100 percent or more.

Most bedload equations are based on the difference between actual bed shear stress and critical bed shear stress (or the difference between actual and critical shear velocity or associated discharge). It is this *excess shear stress* that is available for bedload transport, though some of this stress may be expended against large-scale irregularities or "immovable objects" in the bed. It should be noted that bedload equations purport to yield values of *capacity*—that is, ability to transport bedload—even if that quantity of bedload is not available for transport. Many natural stream flows, in fact, operate far below capacity, actually carrying far less bedload than they are capable of moving.

Numerous bedload equations have been proposed; we will briefly examine (without development) two of them. Both are dimensionally homogeneous (some bedload equations are not), and both give weight (in air) discharge per unit width of flow (rather than volume discharge); similarly, fluid discharge also is given as weight per unit time. Finally, these equations are not directly applicable to aeolian bedloads.

Shields's dimensionless bedload equation gives bedload discharge q_s in terms of excess shear stress, fluid discharge, channel slope, and grain size and density:

$$q_s\rho_s/qS\rho = 10(\tau_o-\tau_c)/g\Delta\rho d \qquad (3\text{-}33)$$

This equation has been checked for sediments ranging in size from about 1.5 to 2.5 mm and ρ_s/ρ density ratios ranging from about 1 to 4.

Kalinske's dimensionless bedload equation gives bedload discharge in terms of bed shear stress and particle diameter:

$$q_s/(u_*d) = f(\tau_c/\tau_o) \qquad (3\text{-}34)$$

where critical shear stress is a special value given by

$$\tau_c = 0.04g\Delta\rho d \qquad (3\text{-}35)$$

which accounts for the fact that, in turbulent flows, shear stress fluctuates randomly to values exceeding the time-averaged value. These turbulent bursts are particularly effective in transporting the bedload. The functional relationship alluded to in Eq. (3-34) is based on experiments (Fig. 3-14).

If bedload moves downstream as a train of ripples or sand waves (see Chap. 5), bedload discharge per unit width of flow is a simple function of bedform cross section and bedform velocity (both quantities easily measured). If results are to be given in terms of bedload weight, then the porosity of the material must be taken into account.

Suspension Transport

Transport of particles in suspension has perhaps a broader theoretical base than bedload transport, but again there are numerous difficulties. Jopling (1966, p. 10) suggested that suspension transport begins at about 1.6 times the critical velocity for general bedload transport (as obtained from Shields's or Hjulstrom's relations) and that suspension transport of sand is well established at velocity of 2 to 2.5 times the critical velocity, or 4 to 6 times critical shear stress. Lane and Kalinske (1939, p. 640) indicated that particles whose settling velocity exceeds bed shear veloc-

ity of the flow do not occur in suspension in any substantial quantity.

If concentration of suspended sediment does not vary in time, then the speed at which particles fall through a fluid must be exactly balanced by upward motions caused by turbulence. Consider particles in suspension having fall velocity of v_s and concentration C; then the rate of downward transport (that is, discharge through a horizontal section of unit area) is $v_s C$. Upward transport by turbulence may be considered as a diffusion process, where rate of transport (that is, volume discharge, Q) through unit area is given by

$$Q = -\epsilon \, dC/dy \qquad (3\text{-}36)$$

In this application of the well-known diffusion equation, the coefficient is called the **kinematic eddy viscosity** or **coefficient of turbulence**. It is a measure of the transporting capacity of the mixing process and is applicable to the study of transport of heat, dissolved ions, and (we hope) momentum transport of suspended solids. (The minus sign indicates that we are concerned with upward transport.) Equating upward and downward transport, we have

$$v_s C = -\epsilon \, dC/dy \qquad (3\text{-}37)$$

Integration over some depth interval between y and $y=a$ yields

$$v_s(y-a) = \int \frac{\epsilon}{C} \, dC \qquad (3\text{-}38)$$

The eddy viscosity can be functionally related to shear stress in the flow, which varies linearly with distance from the bed, and to vertical gradient of flow velocity, which can be obtained from von Karman's logarithmic velocity law given by Eq. (3-16); the relationship is

$$\epsilon = \kappa v_* y(D-y)/D \qquad (3\text{-}39)$$

where D is depth of channel, and κ is, as usual, von Karman's constant (about 0.4, or somewhat less for muddy flows such as we are discussing). Substituting Eq. (3-39) into Eq. (3-38) and integrating gives concentration at distance y above the bed relative to concentration at distance a above the bed:

$$C_y/C_a = \left[\frac{1 - y/D}{y/D} \frac{a/D}{1 - a/D} \right]^z \qquad (3\text{-}40)$$

where $z = v_s/\kappa v_*$

Note that this gives concentration (relative to concentration at some point a) for sediment of a single

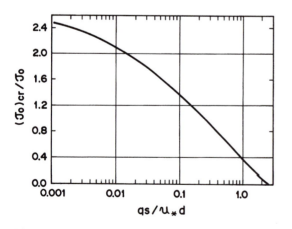

Fig. 3-14. Kalinske's (1942) dimensionless bedload discharge function.

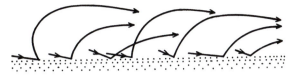

Fig. 3-16. Sand grains saltating in the wind.

size only, that is, sediment having a given terminal speed of v_s. Concentration at point a must be obtained independently, as by measurement. A dimensionless plot (Fig. 3-15) shows that concentration of suspended sediment is more uniform for smaller values of z, that is, for smaller particles or higher shear velocities. If concentration rises to such a point that suspended particles interact substantially, then the analysis breaks down.

Discharge, q_s, of suspended load having terminal speed v_s could be obtained from

$$q_s = \int C_y v_y \, dy \qquad (3\text{-}41)$$

where v_y is, again, given by the von Karman law. Vanoni (1946) showed experimentally that velocity of a suspended-sediment-laden flow is greater, and flow depth is less, than that of a clear-water flow in the same channel.

Saltation Transport

Some particles move not by sliding or rolling on the bed or by total entrainment in the flow but by bounding along, intermittently rising into the flow, then falling back to the bed; this motion is called **saltation** (Fig. 3-16). A grain resting on the bed experiences a lift force by virtue of intermittent upward components of turbulence; this lift force can pull the grain abruptly upward into the flow. But as soon as the grain loses contact with the bed, the lift force declines but the impact of the flow (drag force) on this grain increases. The grain moves forward (downcurrent) and begins immediately to fall back toward the bed; forward transport distance is typically about 10 times the maximum height of the grain trajectory (which, incidentally, is quite a complicated curve). There seems to be an additional motion: as the grain breaks contact with the bed it begins to rotate rapidly. Chepil (1965), who studied saltation in wind flows, found these rotations to range from 200 to 1000 revolutions per second! Rotation imparts an additional lift to the grain, a phenomenon known as the **Magnus effect**.

When the grain strikes the bed once again, it may rebound back into the flow or propel other grains into the flow and thus maintain the saltating motion (this is not, however, a "bouncing ball" effect). Moreover, saltating grains impacting the bed cause slow *surface creep* of grains on the bed.

Bagnold (1941) determined that aeolian sediment discharge is related to particle size and bed shear stress:

$$q_s = cd^{1/2}\rho/g(\tau_o/\rho)^{3/2} \qquad (3\text{-}42)$$

where q_s is weight discharge per unit width of bed surface, d is some measure of grain size, and c is an empirical constant. About 75 percent of this transport is due to saltation and 25 percent to surface creep.

Bagnold (1956 and 1966) described a **dispersive pressure** (see Chap. 4) in a turbulent cloud of saltating grains due to frequent, rapid collisions between the grains (this is similar to the collisions between gas molecules that cause gas pressure). Dispersive pressure influences the thickness of the saltating layer and the characteristic path length of saltating grains.

Saltation occurs in both water flows and aeolian flows but differs greatly in degree. In water, a saltating grain rises only a few grain diameters, but saltating

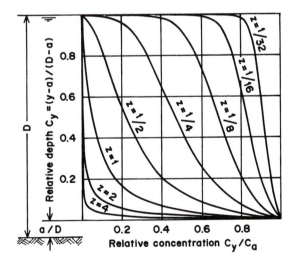

Fig. 3-15. Dimensionless suspended load concentration versus depth in a turbulent flow of depth D, relative to concentration a short distance, a/D, above the bed. Low values of z correspond to suspensions of smaller particles, or higher shear velocities.

Table 3-2. Symbols (Other Than Coefficients) and Their Dimensions[a]

Geometrical		
Length	r, l, L	L
Nominal grain size	d	L
Wetted perimeter	p	L
Hydraulic radius	R	L
Depth	D	L
Cross-section area	A	L^2
Volume	V	L^3
Slope	S	
Kinematic		
Time	t	T
Speed or velocity	v, U	LT^{-1}
Angular speed	ω	T^{-1}
Accleration	g	LT^{-2}
Kinematic viscosity	ν	L^2T^{-1}
Discharge	q, Q	L^3T^{-1}
Dynamic		
Mass		M
Force	F	MLT^{-2}
Density	$\rho, \Delta\rho$	ML^{-3}
Dynamic viscosity	μ	$ML^{-1}T^{-1}$
Shear stress	τ	$ML^{-1}T^{-2}$
Energy or work	E	ML^2T^{-2}
Power	P	ML^2T^{-3}

[a]Subscript s refers to sediment properties; no subscript refers to fluid properties.

grains in air may rise 1000 grain diameters, and saltation is the dominant mode of aeolian transport. Kalinske (1942, p. 642) showed that, for a given grain size and bed shear stress condition, the ratio of characteristic path lengths in water and in air is about 1:800. Allen (1970, p. 101) noted that, among the saltating grains in aeolian transport, many of the coarser ones rise higher than many of the smaller ones.

When strong, steady winds are blowing over a movable bed well stocked with fine sand, the myriad saltating sand grains constitute a ground-hugging cloud, typically about a meter thick. This cloud has much erosive potential, and when it encounters high-standing rock, it bites away at the lower levels of this obstacle. No doubt some hoodoos or pedestal rocks owe their fantastic forms to this process.

4

Waves and Wave-Generated Currents; Gravity-Driven Sediment Flows

Most of the waves that come to the ocean coastlines were shaped by the winds of distant storms; they carry the energy of storms at sea to the faraway shores. Some have traveled eight or ten thousand kilometers, taking two weeks or longer to make the trip. These waves beckon us to the coasts with their hypnotic rhythms, then threaten us with their spasms of violence and destruction. They hurl themselves against the reefs and beaches; they bring to bear relentless force against the jetties and sea walls and ultimately bring them down.

There are other water waves generated by earthquakes or volcanic explosions at sea (tsunami and seiche), or driven by the gravitational attraction of the sun and moon. Tsunami typically travel over the open sea at speeds of 600 or 800 kilometers per hour. The tides are up to twice as fast, circumnavigating the globe in about 24 hours; they are predictable as clockwork. In some coastal locales, their energy is funneled into stream channels, as at the Petitcodiac River in New Brunswick and the Bay of Mont Saint Michel in Normandy; a tidal bore called the Pororoca moves up the Amazon twice each day, making a great noise that can be heard 100 kilometers away.

In coastal areas, the energy of waves is transformed into turbulence and heat, sediment motion, and persistent water currents, some moving parallel to shore, some moving perpendicular to shore. The waves and wave-generated currents, in conjunction with the tides, shape the coastlines, organize and redistribute the rock materials there, and sometimes influence sedimentation far out to sea.

This chapter is largely a story of unsteady flow, in contrast to the previous chapter, which dealt as nearly as possible with steady flow. In addition to the me-

chanics of wave motion and wave-generated currents, we also examine here certain brief and intermittent (wave-like) flows, which, though ephemeral, are responsible for the transport of large masses of sediment. Among these are the turbidity currents and debris flows. They differ also from the flows treated in Chapter 3 in their typically non-Newtonian behavior.

Properties of Water Waves

Water waves are periodic variations in the height of a water surface; they are an example of non-uniform, unsteady flow, but a flow that nonetheless is organized and predictable and that yields to analysis. Most waves move across the surface of the water and are called *progressive waves*; some waves appear to remain fixed in space and are called *standing waves*. Description of the forms of water waves is similar to that of other kinds of waves, such as light or sound waves; the following terms apply generally to waves of every kind (Fig. 4-1):

Wave height, _H_	The difference in elevation between crest and trough (magnitude of the oscillation)
Amplitude, _a_	One-half the wave height; the vertical distance from still-water level to crest or trough
Wave length, _L_	Horizontal distance between wave crests, measured in the direction of wave travel
Wave period, _T_	Time taken by one wave length to pass a fixed point
Frequency, _f_	The number of waves passing a fixed point in unit time $f = 1/T$; *angular frequency* $\omega \doteq 2\pi/T$

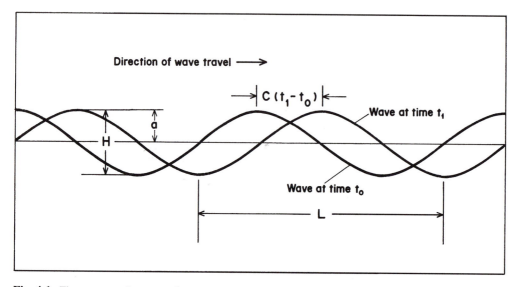

Fig. 4-1. The measure of a progressive wave.

Wave number, k   The number of waves in unit distance $k = 2\pi/L$

Celerity, c   The speed at which a waveform moves across the surface (also called ***phase speed***) $c = L/T = \omega/k$

Wave motion may be examined according to the *method of Euler* (pronounced "oiler"), by which one observes the flow at fixed points, or according to the *method of Lagrange*, by which one traces out the paths of individual water particles. Under Euler's method, one describes the instantaneous velocity (speed and direction) of a flow as a function of location; under Lagrange's method, one describes the positions of particles as a function of time. It is the difference between a weathervane/anemometer, which measures the direction and speed of the wind at a fixed point in space, and a weather balloon, which traces the direction and speed of a given "bundle" or parcel of air particles flowing from one geographical point to another. One might define a large flow field in terms of the readings of a fixed array of anemometers or in terms of the changing locations of a covey of drifting balloons.

Water surface waves are very complicated; their equations of motion are non-linear, and we do not have exact solutions for even the simplest undulations on the surface of a lake. *Airy wave theory* is an approximate analysis of water waves which assumes which amplitude is very small relative to wave length. (The theory is *exact* for waves of zero amplitude.) The form of an Airy wave (Fig. 4-2) is given by

$$\eta = a \cos(kx - \omega t) \qquad (4\text{-}1)$$

where η is the vertical distance of the water surface above (or below) the undisturbed water level, x is the horizontal distance, and t is time. The waveform is a simple sinusoid. The speed at which the waveform moves across the water is given by

$$c^2 = \frac{g}{k} \tanh(kD) = \frac{gL}{2\pi} \tanh(2\pi D/L) \qquad (4\text{-}2)$$

where g is acceleration due to gravity and D is depth of water.

Passage of a wave implies, of course, motion of water particles. It is an unsteady, non-uniform motion—the motion of a given particle varies in time, and at any instant different particles are moving at different speeds and in different directions. The water particle velocity component in the horizontal direction of wave propagation and in the vertical direction are, respectively,

$$u = a\omega \frac{\cosh k(y+D)}{\sinh kD} \cos(kx-\omega t)$$
$$\qquad (4\text{-}3)$$
$$w = a\omega \frac{\sinh k(y+D)}{\sinh kD} \sin(kx-\omega t)$$

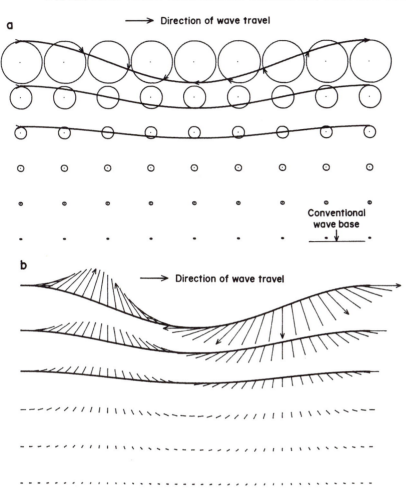

Fig. 4-2. Form of the Airy wave in deep water: **(a)** Paths (a Lagrangian frame) are circles. **(b)** Streamlines (Eulerian frame) are defined here by a field of vectors.

where (x,y) is *mean* location of a given water particle (y is negative downward).

Adding vectorially the horizontal and vertical components of the velocity, given by Eq. (4-3), we obtain

$$(u^2 + w^2) = v^2 = a^2\omega^2 A^2 \cos^2(kx - \omega t)$$
$$+ a^2\omega^2 B^2 \sin^2(kx - \omega t)$$
$$= a^2\omega^2(A^2 + B^2)$$
$$v = a\omega(A^2 + B^2)^{1/2} \qquad (4\text{-}4)$$

where we have replaced $\cosh k(y+D)/\sinh kD$ in Eq. (4-3) with A and $\sinh k(y+D)/\sinh kD$ with B. Note that x does not appear in the equation; evidently, all particles at a given depth y, no matter what their x locations, are moving at the same speed (though not, of course, all in the same direction).

Waves in Deep Water

It is clear from Eq. (4-2) that speed of a water wave depends on depth of the water over which it progresses. But if water depth is large (or small) with respect to length of the wave, then certain simplifications are possible. The hyperbolic sine and cosine of Eqs. (4-3) can be alternatively expressed (see Table 4-1) so that, for deep water, $e^{-kD} \doteq 0$, $\cosh k(y+D)/\sinh kD \doteq e^{ky}$, $\sinh k(y+D)/\sinh kD \doteq e^{ky}$. If water depth

Table 4-1. The Hyperbolic Functions

$\sinh z = (e^z - e^{-z})/2$
$\cosh z = (e^z + e^{-z})/2$
$\tanh z = (e^z - e^{-z})/(e^z + e^{-z})$
If z large, then $\sinh z \doteq e^z/2$, $\cosh z \doteq e^z/2$, $\tanh z \doteq 1$.
If z is close to zero, then $\sinh z \doteq 0$, $\cosh z \doteq 1$, $\tanh z \doteq 0$.
(Value of e is about 2.7183.)

is greater than $L/2$, then $\tanh(2\pi D/L) \doteq 1$, and Eq. (4-2) reduces to

$$c^2 = gL/2\pi \text{ or } c = gT/2\pi \qquad (4-5)$$

and we obtain also

$$L = gT^2/2\pi \qquad (4-6)$$

These are properties of a so-called **deep-water wave**. The speed (celerity) of a deep-water wave does not depend on water depth, but longer waves move faster than shorter waves.

For deep-water waves, the velocity components of Eq. (4-3) reduce to

$$u = a\omega e^{ky} \cos(kx - \omega t) \qquad y \le 0 \quad (4-7)$$
$$w = a\omega e^{ky} \sin(kx - \omega t)$$

Consider the velocity of a single particle as a function of time (the Lagrangian description). Let's look at a particle whose mean location happens to be $(0,y)$; then Eq. (4-7) becomes

$$u = a\omega e^{ky} \cos\omega t \qquad y \le 0 \quad (4-8)$$
$$w = -a\omega e^{ky} \sin\omega t$$

We note that the water particle describes a circular path (Fig. 4-2a), having a radius $r = ae^{ky}$; its orbital speed (that is, angle subtended per unit time) $v = a\omega e^{ky}$. A water particle makes one complete orbit per wave period, so its orbital speed can be written also as $2\pi r/T$.

The mean location of the particle does not change. A water particle at the surface (or a floating object) remains at the surface; as a waveform passes, this particle is carried up and forward as the crest advances toward it, then down and backward after the crest has passed. A water particle below the surface experiences a similar kind of motion, but the magnitude (amplitude and speed) of this motion is less, declining exponentially with depth.

Now consider instantaneous velocity (speed and direction) as a function of location (the Eulerian description). At a particular instant of time (call it t) we have, from Eq. (4-7),

$$u = a\omega e^{ky} \cos kx \qquad y \le 0 \quad (4-9)$$
$$w = a\omega e^{ky} \sin kx$$

A water particle at the surface of a wave crest ($x = 0, L, 2/L, \ldots$) is moving forward at speed $a\omega$; a water particle at the surface of a wave trough ($x = L/2, 3L/2, \ldots$) is moving backward at the same speed (Fig. 4-2b). Water particles at points where the water surface crosses the stillwater level are moving either straight up, on the leading slope of the crest ($x = L/4$), or straight down, on the trailing slope ($x = 3L/4$). The instantaneous direction of motion of water particles beneath the surface is the same as for particles directly above, but the magnitude of the motion falls off exponentially with depth. All of these separate vectors form a pattern—it's the pattern you would see if you took a photograph of the water at a shutter speed just slow enough that each water particle (made visible somehow) is smeared out just a bit. This whole pattern moves with the waveform. Orbital motion is revealed if a long time exposure (at least one wave period) is made.

Wave dispersion is a phenomenon that depends on the differing speeds (celerities) of waves of different period or wavelength, as given by Eq. (4-5). It is a sorting out of waves that are moving at different speeds. Suppose that waves of several different frequencies are generated simultaneously at some point. As these waves move out from that point, those of lower frequency (greater wavelength or longer period) outrun those of higher frequency. As time passes and waves advance, the long-period waves move farther and farther ahead of the short-period waves. Clearly, the distance between two waves of different period is a measure of the total distance traveled, or time elapsed, assuming that these waves began at the same time and place.

Sea is a term describing the ocean surface condition during a storm. Sea is characterized by chaotic motion of sharp- and short-crested waves of many different lengths and heights, including very large waves. As waves move out of the storm, they change gradually. The waves develop round crests; amplitudes decline slowly, attenuating by half in each 1500 kilometers. Dispersion sorts the waves according to their wavelength. Such an altered ocean surface, simpler and more organized than sea, is called **swell**. Waves of the swell can move completely across the ocean. The direction from which the waves come indicates, of course, the location of the storm that made them. The geographic pattern of swells varies with the seasons, but ocean-wide patterns nonetheless recur from year to year. On the Pacific coast of the United States, most of the waves in winter come from the north; but in the summer, many wave trains come from winter storms in the Southern Hemisphere. Native Pacific islanders have used patterns of swells in the open sea very successfully as navigation aids (beacons, in a sense), probably for many centuries, and there is mounting evidence that marine animals also depend on these patterns in their migrations.

Imagine that you are standing on the shore of a calm sea, just as waves from a distant storm begin to arrive. The first indication of the storm is the abrupt arrival of large swells of long period, crests breaking regularly against the shore at perhaps 20- or 25-second intervals. Hours later, you notice that wave arrivals are considerably more frequent and have become quite irregular, reflecting a mix of waves of many different periods. These are overlapping long trains of waves, trains so long that the fast (low-frequency) trains have not yet fully passed the slower (higher-frequency) trains. The largest waves have periods of 15 or 20 seconds, perhaps. The next day the largest waves have periods of only about 10 seconds, and there is considerably more surf—a greater total energy flux onto the coast. Another day passes, and the dominant waves have periods of 5 to 10 seconds. There is by now a noticeable decline in total "force" of the waves. Gradually, the sea returns to a calm state—no long-period waves at all, only a gentle, regular, monotonous 2- or 3-second lapping on the strand—the trailing edge of a long and complex wave train that arrived perhaps a week after the storm had waned and that took a week to pass.

Dispersion is responsible for another peculiarity of wave motion—a single, isolated wave crest is not possible. Even if some disturbance of the water surface should produce a single wave crest, this crest quickly evolves into a train of several wave crests called a group. The *wave group* (Fig. 4-3) is a property of dispersive waves. Velocity of a wave group is

$$U = \frac{c}{2} [1 + 2kD/\sinh 2kD] \qquad (4\text{-}10)$$

In deep water, the factor in brackets is close to 1,

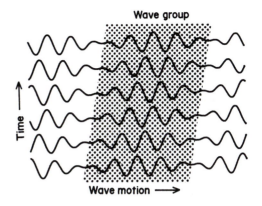

Fig. 4-3. Form of a wave group (2 1/2 groups shown). Stipple shows progress of a single wave group. Note that wave crests advance more rapidly than the group.

and the group speed is $c/2$ (see Fig. 4-4); that is, the group travels at half the *phase* speed (celerity), and waveforms progress through the group, disappearing at the leading edge of the group as new waves appear at the trailing edge. There is no wave motion at the leading or trailing edge of a group, so no energy passes into or out of the group; energy associated with a wave disturbance must therefore be propagated at group

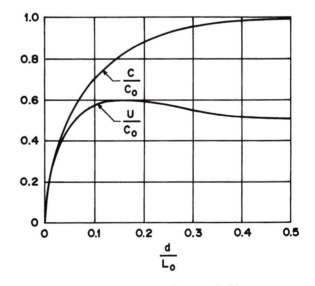

Fig. 4-4. Dimensionless phase and group velocities versus water depth, relative to these velocities in deep water. (After Le Mehaute, 1976, fig. 16-11.)

speed. In shallower water, group speed is a larger proportion of phase speed (Fig. 4-4); group speed and phase speed converge and are coincident at zero water depth.

Waves in Shallow Water

The velocity (celerity) of Airy waves was given by

$$c^2 = \frac{gL}{2\pi} \tanh(2\pi D/L) \qquad (4\text{-}2)$$

If water depth is less than about 1/20th of the wavelength ($D/L < 1/20$), then $\tanh(2\pi D/L) \doteq 2\pi D/L$, and Eq. (4-2) reduces simply to

$$c^2 = gD \qquad (4\text{-}11)$$

and since $c=L/T$, we may write also that

$$L = T\sqrt{gD} \qquad (4\text{-}12)$$

These are the characteristics of the so-called **shallow-water wave**. In shallow water, the speed and form and other dynamic properties of wind-generated waves is influenced by water depth. Wave speed is independent of wave length or period; that is to say, shallow-water waves are **non-dispersive**. Tsunami have such great wavelengths that they behave like shallow-water waves even over the abyssal depths; their speeds provided early estimates of the depth of the ocean. A tsunami typically comprises a short group of about a half-dozen crests; group speed is only a little less than phase speed, so each crest advances only very slowly through the group.

The motion of water particles in shallow-water waves also differs from that of deep-water waves. For shallow water, $\sinh kD \doteq kD$, $\sinh k(y+D) \doteq k(y+D)$, and $\cosh k(y+D) \doteq 1$; thus Eq. (4-3) becomes

$$u = \frac{a\omega}{kD} \cos(kx - \omega t)$$
$$\qquad\qquad\qquad (y \le 0) \quad (4\text{-}13)$$
$$w = \frac{a\omega}{kD} k(y + D) \sin(kx - \omega t)$$

Particle trajectories (Fig. 4-5) are ellipses. The semimajor axis is horizontal, and its length is a/kD; the length of the vertical semiminor axis is $a(y + D)/D$. At the water surface ($y = 0$), the semiminor axis of the ellipse is equal to wave amplitude; but at the bottom ($y = -D$), the ellipse degenerates into a horizontal line—that is, water particle motion is to and fro, with no vertical component. Streamlines (the Eulerian perspective) are flat, and waveforms are trochoids rather than sinusoids.

Waves propagating over water of intermediate depths have properties intermediate between the properties of deep-water and shallow-water waves. Ocean waves moving from deep water into the shallower water of coastal regions are called transforming waves or **shoaling waves**.

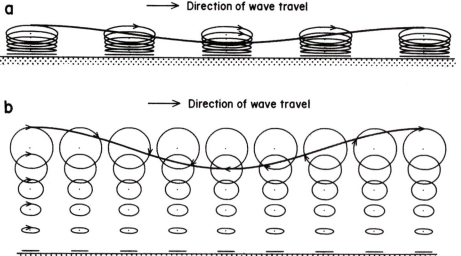

Fig. 4-5. Paths of Airy waves **(a)** in shallow water, **(b)** in water of intermediate depth.

As waves move from deep to shallow water, they slow down and their wavelength gradually decreases because frequency (or period) must remain constant. Thus the waves bunch up near the shore. The waves also change shape (owing to factors not accounted for by Airy theory); their rounded crests attain asymmetrical sharp profiles, and troughs become broader relative to crests. As wavelength and wave height change in shallow water, the *steepness*, *H/L*, changes and the angle subtended by the wave crest becomes gradually smaller. Theory predicts that the crestal angle cannot be sharper than 120°, nor can wave steepness exceed a value of 0.142. Beyond this point, the forward motion of water particles at the crest exceeds the speed at which the wave-form advances, and the wave becomes unstable and breaks; that is, swell becomes *surf* (Fig. 4-6). We conclude that surf is due ultimately to the slowing of waves as they advance into shallow water, the energy of motion changing into turbulence.

Wavelength of a shoaling wave in water of depth D is

$$L = \sqrt{2\pi D L_o} \qquad (4\text{-}14)$$

where L_o is wavelength in deep water. A *shoaling coefficient* for a shoaling wave is defined as

$$K_s = \sqrt{U_o/U} \qquad (4\text{-}15)$$

where U_o/U is ratio of group velocity of the wave in

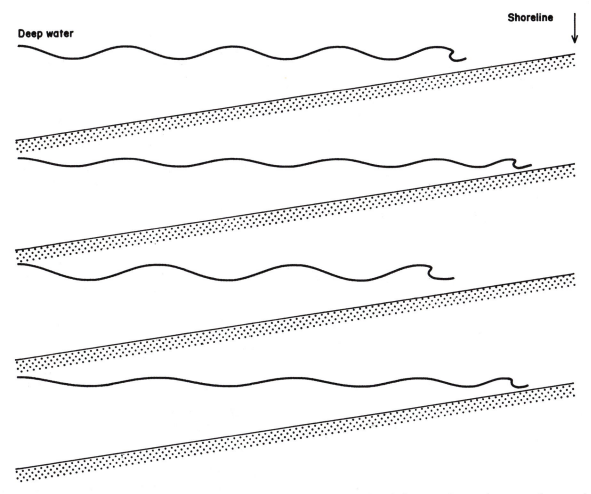

Fig. 4-6. Shoaling wave trains of different wavelengths and amplitudes. Wavelength decreases shoreward, steepness increases (approximate analysis). Note that waves of greater height break in deeper water.

deep water to its group velocity in shallow water (see Fig. 4-4).

Wave refraction is a phenomenon that results from the variation in speed of shoaling waves with depth of water. Because waves slow down in shallow water, they bend, or refract when they enter a coastal region, altering their courses toward regions of lesser depth. The phenomenon is most pronounced and simplest on long, straight coasts with gently inclined shorefaces, where the crests of waves arriving obliquely from deep water bend into gentle, convex shoreward curves as they advance over the shoaling water (Fig. 4-7); they are nearly parallel to the shore when they finally collapse onto the beach. Because the longest waves "feel bottom" first, they begin to refract earlier and in deeper water than shorter waves and are the first to begin dissipating energy to bottom friction. Refraction causes wave energy to converge upon headlands and to diverge in coastal bays (Fig. 4-7); thus the rate of energy expenditure against a ragged coast is greatest at the points, least in the pockets. Over the long term, wave energy breaks down the points and moves debris and rubble from the points into or across the pockets; thus waves tend to straighten a coastline.

A 20-second storm swell in deep water has a wavelength of about 600 meters and a wave base, therefore, of about 300 meters. The continental shelf break typically lies at a depth of about 200 meters. Thus, the whole of the continental shelf is affected by long storm swells. These long waves have already begun to refract by the time they have reached the margins of the shelves.

The energy of waves moves along wave orthogonals (Fig. 4-7); it does not move along wave crests (or at least, not very much). Thus, where wave orthogonals converge, wave energy gets compressed and wave heights increase. Where wave orthogonals diverge, wave crest lines become longer and lower, and wave energy is stretched out along these longer crests. The *refraction coefficient* indicates magnitude of this effect:

$$K_r = \sqrt{b_o/b} \qquad (4\text{-}16)$$

where b_o/b is the ratio of distance between wave orthogonals in deep water and in shallow water (the refraction coefficient, like the shoaling coefficient, varies along wave orthogonals, generally increasing toward the coast).

There is also a *friction coefficient*, K_f, describing the friction between a wave and the bottom. The three

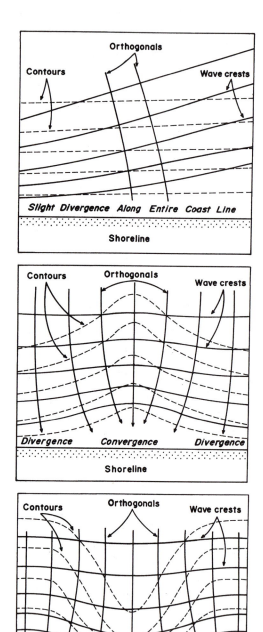

Fig. 4-7. Refraction of swell onto a coast. (After Munk and Traylor, 1947, figs. 1, 3, 4, with permission of The University of Chicago Press.)

coefficients all influence the wave height, and we may write

$$H/H_o = K_s K_r K_f \qquad (4\text{-}17)$$

where H_o is wave height in deep water.

The Energy of a Wave

The energy of an Airy wave in deep water is a function of wave height:

$$E = \frac{1}{2}\rho g a^2 = \frac{1}{8}\rho g H^2 \qquad (4\text{-}18)$$

per unit surface area. Wave energy is transported at group speed U (taken to be half the phase speed, $gT/2\pi$, so **wave power**, averaged over a complete wave form, is

$$P = EU = \frac{1}{8}\rho g H^2 g T/4\pi$$
$$\qquad\qquad (4\text{-}19)$$
$$= \frac{1}{32\pi}\rho g^2 T H^2$$

Consider a wave 1 m high, with a period of 10 sec; the power delivered by 1 m of crest length is about 10^{11} ergs/sec or 10 kw.

Capillary Waves

The ordinary wind-generated waves that we have examined are called **gravity waves**, because gravity is the major force controlling their motion. For very small waves, however, surface tension as a restoring force cannot be neglected; speed of such a wave is given by

$$c^2 = gL/2\pi + 2\pi S/\rho L \qquad (4\text{-}20)$$

where S is surface tension. The celerity has a minimum value at $L = 2\pi\sqrt{S/g\rho}$, determined by setting $dc/dL = 0$ (Fig. 4-8). Surface tension S is 74 dyne/cm for air-sea interface at 20°C; then $L = 1.72$ cm, corresponding to a wave speed of 23 cm/sec and period of 0.07 sec. Waves of length less than 1.72 cm are called *capillary waves*; those of greater wavelength but still within the range where surface tension cannot be neglected are called *ultra-gravity waves*. In capillary waves, celerity increases with decreasing wavelength, just the opposite of gravity-wave behavior; group

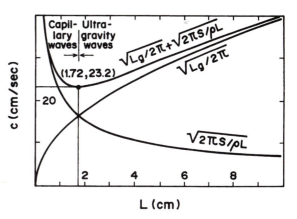

Fig. 4-8. Wave celerity versus wavelength where surface tension is important; capillary waves are shorter than 1.72 centimeters, by definition.

speed exceeds phase speed, which accounts for the peculiar bobbing motion of capillary waves.

Solitary Waves

A **bore** is a wave in such shallow water that its waveform becomes independent of (decoupled from) the waveform just ahead or just behind; it is therefore referred to as a **solitary wave**. Speed (celerity) is given by

$$c = \sqrt{g(D+H)} \qquad (4\text{-}21)$$

where H is wave height and D is stillwater depth. It is non-dispersive and does not form groups. Water particle velocity at the crest is Hc/D, and at the bottom, directly under the crest, it is just half this value, $Hc/2D$. The bore is like a moving hydraulic jump and commonly has a turbulent leading edge. In unobstructed channels, such a wave can persist for surprisingly long distances and times. A broken swell may move across a broad, shallow bottom as a bore; tidal crests also develop into bores on shallow shelves and in estuaries.

Some Non-linear Aspects of Waves

The solution of the differential equation of wave motion has many terms. Airy wave theory, which applies strictly only to waves of diminishingly small amplitude, considers only the first term and is a first-order approximation of *Stokes wave theory*. If higher-order terms (also sinusoidal, just as the first term is) are not neglected, then certain behaviors of real waves emerge that Airy wave theory does not address. More like real waves, Stokes wave crests are more peaked and troughs flatter than those of Airy waves. Particle trajectories (paths) are not exactly ellipses or circles and are not the closed figures that the Airy approximation gives; in the Stokes wave, there is a net transport of water in the direction of wave travel. If you throw a stick out onto a lake, gentle wave motion will eventually bring it back to shore.

Though Airy wave theory is only a (linear) first approximation of waves of small amplitude, it has nonetheless been very successful, even as a description of waves of considerable amplitude and waves propagating through shallow water. The linearity of the Airy approximation permits Airy waves to be added together, mathematically, so that very complex wavy surfaces can be described. *Seiche* (pronounced ''saysh''), a long-period *standing wave* in closed or semi-enclosed basins, is quite adequately described by linear theory as the sum of two waves of the same period moving in opposite directions. The seiche is thus equivalent to a wave train interfering with its reflection.

Wave-Generated Nearshore Currents

The Stokes term(s) in the equation of wave motion predicts that there is a net flux of water in the direction of wave propagation; this flux becomes particularly noticeable as waves move into shallow water. Because heavy swells (and surf) in shallow water transport substantial quantities of water shoreward, there is a zone between the strand and the breakers where mean water level is a bit higher than stillwater level (Fig. 4-9). This super-elevation of the surface, of magnitude similar to wave amplitude (that is, half the wave height), is called *wave set-up*. There is another, broader, zone beyond the breakers, where mean water level is a bit below

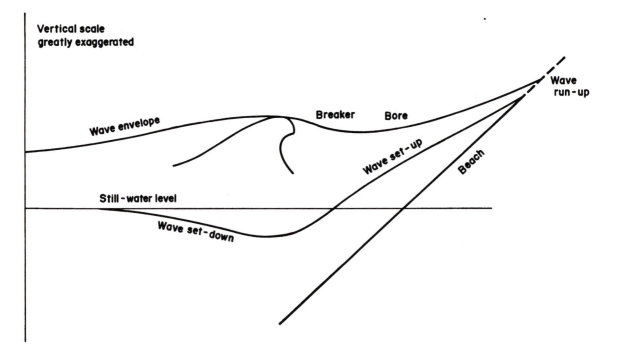

Fig. 4-9. Profile of wave set-up and set-down. (After Le Mehaute, 1976, fig. 16-15.) Under surf action, mean water level is below stillwater level in the breaker zone and above stillwater level shoreward of the breakers.

stillwater level. This zone of **wave set-down** is the source of the excess water nearer to shore. Clearly, if waves continually deliver water to the strand, there must be a return flow. This return flow commonly is concentrated in seaward jets called **rip currents**.

A **longshore current** is a shore-parallel component of wave momentum flux or *wave thrust*. Commonly, this current is confined to a **runnel**, a shore-parallel trough or channel near the strand. The current becomes gradually stronger in its downstream direction as it receives more and more water from the incoming surf and at some point must turn seaward as a rip current. Longuet-Higgins (1970) and Komar and Inman (1970) determined theoretically and experimentally that the velocity of the longshore current is related to the angle of approach and orbital velocity of the breaking waves. Longshore distance between rip currents is about four times the distance between the breaker zone and strand, according to Komar (1976, p. 244). The longshore- and rip-current system moves slowly along the shore in the direction of the longshore component of wave propagation.

Longshore currents and rip currents develop even if the incident wave crests arrive parallel to the strand, but there are certain differences. If there is no longshore component of wave thrust, then a more or less stable pattern of circulating cells develops, each cell flowing oppositely to its neighbors; converging longshore segments contribute to a rip current. Actual locations of rip currents are determined in part by longshore variations in the magnitude of wave set-up, which depends on local wave heights.

Edge waves also play a role; they are standing waves near the shore, having the same period as the incoming swells (or some multiple) and oriented more or less *perpendicular to the strand*; their amplitude diminishes seaward to zero in the zone of wave set-down. The existence of edge waves was predicted on theoretical grounds by Eckert in 1951 and demonstrated experimentally and measured on natural coasts by Bowen and Inman (1969 and 1971) and Huntley and Bowen (1973). The crests of edge waves are perpendicular to the crests of the incoming swells, and edge waves interact with swells in a way that causes alternate beach segments to experience higher or lower waves (Fig. 4-10). If edge waves have the same period as the incident swells, there are points along the beach where edge wave and swell are always in phase and such that the edge wave forms maximum crest at the

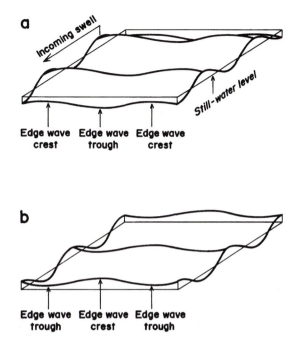

Fig. 4-10. Configuration of edge waves: **(a)** Trough of swell arrives at forward edge of block. At the center arrow, the edge wave is in phase with the swell; at arrows on left and right, edge wave and swell are one-half period out of phase. **(b)** Crest of swell arrives at forward edge of block; phase relationships are identical to those in **a**.

same instant that a swell crest arrives; at these points the edge wave and swell reinforce to make higher breaker crests. Halfway between these points on the beach are points where edge-wave phase is opposite to the phase of the swells; in this situation edge wave and swell interfere to produce lower crests. Thus, there is a steady-state longshore alternation in breaker height, and this leads to the formation of near-shore circulation cells, with regularly spaced rip currents (Fig. 4-11). At each point of highest waves there is a divergence of longshore currents, which flow toward points of low wave height, then veer seaward as rip currents. **Crescentic bars** and **cuspate shorelines** (Fig. 4-12) are sediment-distribution responses to these regular patterns of nearshore, wave-generated currents (Bowen and Inman, 1971). The bathymetric patterns that develop as a result of current patterns may tend to perpetuate and stabilize the current patterns even as wave climate changes.

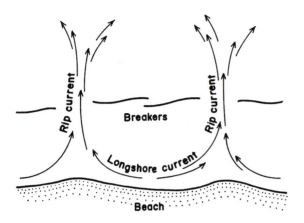

Fig. 4-11. Longshore- and rip-current system induced by shoaling waves.

Sediment Transport by Waves

Shoaling waves impart considerable shear stress to the bottom and put bottom sediment in motion. Ocean surf can create much turbulence, which lifts and suspends the finer-grained components of coastal sediments and

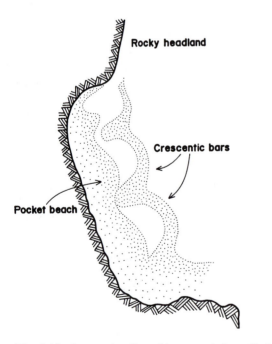

Fig. 4-12. Cuspate shoreline with crescentic bars. (Redrawn from photograph in Bowen and Inman, 1971, fig. 1.)

carries them out to sea. But even if bottom stress causes no sediment transport, it is a kind of background stress that makes other bottom currents, such as longshore and tidal currents, more effective. A concept familiar to all geologists is *wave base*, a water depth below which the waves are not stirring the bottom. Depth of wave base conventionally is taken to be half the wavelength; below this depth there is very little water motion (see Fig. 4-2). Actually, one might make various refinements of the wave-base concept, defining a "specific wave base" as water depth at which waves of a specified size (wavelength) are just capable of moving bottom sediment of a specified size.

We may define a **wave Reynolds number**,

$$Re = \rho u_b A_b / \mu \qquad (4\text{-}22)$$
$$= \rho u_b / \Delta \rho g T$$

where u_b is wave orbital velocity at the bottom, $\Delta \rho$ is difference in density between sediment particle and fluid, and the characteristic length, A_b, is maximum bottom orbital displacement, which can be obtained from Eq. (4-13). This is a shear Reynolds number for waves.

Shear stress at the bottom varies cyclically with wave period; direction of the stress vector is shoreward under a crest and seaward under a trough. Magnitude of maximum bottom shear stress associated with a wave of frequency ω is

$$\tau_b = \sqrt{\rho \mu \omega} \, u_b \qquad (4\text{-}23)$$

There is also a *relative roughness* concept for waves, analogous to that for flow in channels; it is defined as A_b/d, where d is some measure of grain size of the bed. Just as in the case of steady flows (Chap. 3), bottom shear stress and the threshold of grain movement for wave orbital currents on the bottom depends on a Reynolds number when this number is of low value but on relative roughness otherwise. Madsen and Grant (1975) found Shields's relation (see Chap. 3) to be a reliable criterion for threshold of sediment movement under wave action, provided that shear stress given by Eq. (4-23) is used. Komar and Miller (1973, p. 1105) used a Shields-like dimensionless bed-shear criterion for sediment transport under wave action, and later (1975) presented a plot showing critical shear stress versus grain diameter for quartz (Fig. 4-13).

When a real wave advances over shallow water, there is a burst of strong forward motion on the bottom

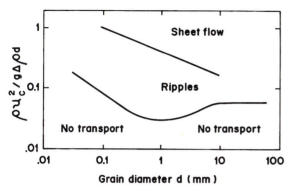

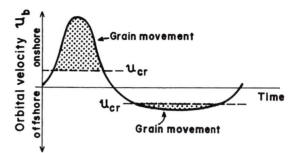

Fig. 4-14. Time variation of bed shear stress under a wave with threshold velocity for a particular grain size. (After Komar, 1976, fig. 8.)

Fig. 4-13. Dimensionless critical bed-shear stress under wave-induced oscillating current versus grain size, for quartz grains. (After Komar and Miller, 1975, fig. 3, with permission of SEPM.) If critical stress is exceeded, then sand begins to move, and ripples form on the bed; higher shear stress washes out the ripples.

as a wave crest passes, followed by a gentler backflow of longer duration as the trough passes (Fig. 4-14). Larger particles behave differently from smaller particles under asymmetrical current oscillations. For a given grain size, the forward flow may exceed the critical value needed for transport, but the backflow might not (even though the critical velocity would be a bit less for downslope seaward particle motion). Thus, there is a null-point depth for a particle of given size (and waves of a given size) at which the particle experiences no net movement, moves onshore at lesser depth, and moves offshore at greater depth. The coarser particles experience net onshore movement, the finer particles experience net offshore movement. The result is a coarsening of the bed shoreward and a concentration of pebbles on the beach. Water depth and particle transport interact in such a way as to produce an equilibrium shoreface profile (Greenwood and Mittler, 1984, p. 93). Presumably, if wave conditions remain constant for some time, then all points on the coastal profile attain equilibrium and there is no net particle motion at all thereafter.

A wave ultimately collapses onto the beach face as a thin sheet of water that flows up the slope of this surface, then reverses direction and flows back into the sea. Bascom (1951) found a strong correlation between grain size and angle of slope of the beach face. The forward wave swash is driven by momentum; the return flow, driven by gravity, is weaker. Moreover, the return flow is diminished because some of the

swash water infiltrates the beach face. Coarse material has higher permeability than fine-grained material, so there is very little return flow on a coarse-gravelly beach face. Thus, on a gravelly beach face, bed shear stress is directed almost entirely onshore and the beach face is steeper; on a fine-sandy beach, a strong seaward-directed backwash alternates with the onshore wave swash and the beach face slopes much more gently.

The longshore- and rip-current system is responsible for much coastal sediment transport, returning toward the sea sediments that wave stress had moved landward. These currents scour channels into the shoreface; sediments delivered to shore-parallel channels by the wave stress drift along the shore for some distance and then are returned to deeper water, perhaps even below wave base, by the rip currents.

Besides the longshore drift of sediment resulting from the longshore currents, there is another littoral drift caused by wave swash on the beach face. If waves are entering the coast obliquely, then the wave swash pushes sand grains obliquely up the beach face; the return flow pulls the grains back down the beach face (Fig. 4-15). Thus, each wave cycle causes a net particle motion parallel to the strand.

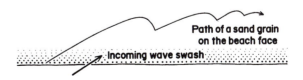

Fig. 4-15. Conceptual wave swash and beach-face particle trajectory after four wave arrivals.

The waves and currents of the coastal waters give rise to distinctive deposits. The sand and pebbles on the beach, carried to and fro by the wave swash for long periods, gradually take on the textural characteristics of particles transported great distances, even though net displacement might be minuscule. Pebbles tend toward discoidal shapes. Shoreward variation in wave stress results in a general coarsening-upward of shoreface sediments. Because larger waves can cause sediment transport at greater depths than can smaller waves, the deeper parts of coastal sediment prisms show mainly the effects of the larger waves; the higher levels of the deposit typically record the effects of smaller waves. Also, there is a succession of different bedforms along the coastal profile (see Chap. 5).

The Astronomical Tides; Tidal Currents in Shallow Waters

The astronomical tides are bulges raised in the hydrosphere (and atmosphere and lithosphere as well) by the gravitational forces of moon and sun. The solar tide-raising force is about 46 percent as strong as the lunar force. A tide is a *forced wave* whose speed is governed not by its wavelength nor by water depth but by angular speed of the moon (or sun) relative to the ocean surface, which equates to a linear speed of about 1750 kilometers per hour on the equator, greater by far than the speed of any other (unforced) water wave. Daily fluctuation of water level is a result of earth's rotation under moon and sun. Local patterns of tidal cycles are determined by the periodic variations in earth-moon and earth-sun distance and the relative position of earth, moon, and sun. The magnitude of the tides varies with latitude and with time of month and time of year, reflecting the complexity of orbital motions. The tidal wave everywhere "feels" the ocean floor and so exerts a shear stress there. (This shear couple has caused a long-term decline in earth's rotational frequency and an increase in moon's orbital speed.) Lonsdale et al. (1972) determined that tidal shear creates ripples and larger bedforms even at oceanic depths of 2000 meters.

The tides are modified in complex ways by coastal physiography and bathymetry, and the tides, in turn, modify wave-generated coastal currents and influence coastal sedimentation and erosion. When the tidal wave crest impinges upon a coast, it creates currents that typically are of greater magnitude than any of the other currents in the coastal waters. Many of the differences between coasts are attributable largely to differences in the tides and tidal currents (see Chap. 8). Where the range between high tide and low is great, coastal physiography may be dominated by a broad, muddy tidal flat, and sands of intertidal and shallow subtidal zones are organized into longitudinal bars that are perpendicular to the coast. In contrast, where tidal range is small, bars are characteristically parallel to the coast, and barrier islands may be the distinctive physiographic feature.

Gravity-Driven Sediment Flows

Sediment flows, though ephemeral, are responsible for the transport and deposition of great quantities of sediment, both on dry land and on the floors of lakes and oceans. Ordinarily they are a result of loss of mechanical strength of some sediment mass already deposited; the mass then moves downslope and, in some more stable configuration, is *resedimented*. We have heard of the devastation they wreak—the headlong plunge of a snow avalanche or the stubborn advance of a volcanic mudflow—they block river drainage, splinter forests, ruin roads and dwellings, and bury or sweep away local populations.

Four types of gravity-driven sediment flows have been recognized (Middleton and Hampton, 1976), distinguished on the basis of the mechanism that supports the grains in the flow (Fig. 4-16)—*turbidity current*, wherein grains are supported by turbulence of a dominating fluid phase; *fluidized sediment flow*, in which grains are supported by upward motion of escaping pore fluid, which is the minority phase; *grain flow*, in which a dispersive pressure is created by rapid mutual impacts of the grains in motion and in which a fluid phase need not be present; and *debris flow*, which is supported by its high viscosity. These four types of flow might be viewed as benchmarks on a continuum, as there are transitional types and natural flows that progress or evolve from one type to another.

Turbidity Current

A turbidity current is a fast flow driven downslope by its high density relative to the ambient fluid, this high density owing to abundant particulate matter in suspension, that is, turbidity. Motion imparts turbulence to the flow; this turbulence maintains the turbidity,

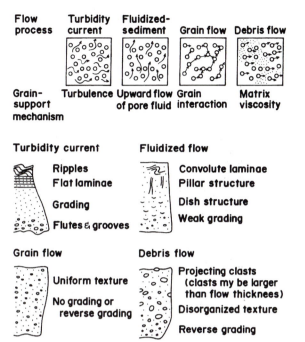

Flow process	Turbidity current	Fluidized-sediment	Grain flow	Debris flow

Grain-support mechanism	Turbulence	Upward flow of pore fluid	Grain interaction	Matrix viscosity

Turbidity current

Ripples
Flat laminae
Grading
Flutes & grooves

Fluidized flow

Convolute laminae
Pillar structure
Dish structure
Weak grading

Grain flow

Uniform texture
No grading or reverse grading

Debris flow

Projecting clasts (clasts my be larger than flow thicknees)
Disorganized texture
Reverse grading

Fig. 4-16. Classification of subaqueous gravity flows and general nature of the deposit made by each. (After Middleton and Hampton, 1976, figs. 1, 9.)

which, in turn, provides the driving force that keeps the flow in motion. Turbidity currents are abrupt and short-lived phenomena; many of them are triggered by earthquakes or storms at sea. They can travel several thousands of kilometers down very gentle submarine slopes. Probably most or all are confined at first to a channel, such as a submarine canyon. Beyond the mouth of the channel, flow spreads and weakens, and the turbidity current eventually collapses and deposits its suspended load, contributing one more layer to a deep-sea fan (see Chap. 8) or abyssal basin. This collapse probably also is rather abrupt. Turbidity currents seem to be the great unseen process of deep marine sediment transport and deposition.

An incipient turbidity current, given sufficient time and distance, evolves into a fast-moving, turbulent mass having a head, body, and tail, the head being the thickest part of the flow (at least in the faster flows), the body of rather uniform thickness, and the tapering tail a region of diminishing thickness and sediment concentration (Fig. 4-17). The head advances less rapidly than the forward flow in the body, so suspended

sediment from the body moves forward to the head; then it sweeps upward and back again into the body. This upward motion at the front is familiar to anyone who has witnessed a snow avalanche. Velocity of the head is

$$v_h = 0.7\sqrt{(\Delta\rho/\rho)gD_h} \qquad (4\text{-}25)$$

where D_h is thickness of the head, ρ is density of the ambient fluid, and here $\Delta\rho$ is density contrast between turbidity current and ambient fluid. Froude number $Fr = v/\sqrt{(\Delta\rho/\rho)gD}$ (the so-called **densimetric Froude number**), so Eq. (4-25) states simply that Froude number at the head has a constant value of 0.7. Flow in the body is faster, and if it may be considered as a steady, uniform flow, then its velocity (averaged over the thickness of the flow) is given by a Chezy-type equation (see Chap. 3):

$$v = \sqrt{8g/(f_b+f_t)}\,\sqrt{(\Delta\rho/\rho)DS} \qquad (4\text{-}26)$$

where S is slope, f_b is (Darcy-Weisbach) coefficient of friction at the base of the current, and f_t is friction at the top of the current. For smooth beds f_b depends on Reynolds number; for rough beds, f_b depends on relative roughness (by now a familiar concept, surely). An increase in slope results in increase in velocity in the body and a corresponding decrease in thickness; but the head *increases* in thickness, and velocity of the head correspondingly changes (increases) to maintain the Froude number at 0.7 (Migniot, 1968, p. 608; the value varies somewhat with the viscosity of the current). On slopes greater than $1.24°$ ($S=0.022$), the head is thicker than the body; on gentler slopes, the body tapers gently toward the head. The upper surface of the body of the turbidity current is a boundary between two fluids of contrasting (but similar) density, in relative motion. Waves develop at such boundaries, so the body of a turbidity current is not really a steady, uniform flow at all. van Andel and Komar (1969, p. 1183) suggested that turbidity currents are supercritical on continental slopes but experience hydraulic jump to subcritical flow as they descend onto the lesser gradients of the continental rise or abyssal plain (see Hand, 1974 and 1975; Komar, 1975). Their mathematical models indicate that a jumping turbidity current experiences a sevenfold increase in thickness, a velocity reduction to a quarter of its prejump value, and a 40 or 50 percent dissipation of energy of flow into turbulence at the jump. Turbulence at the hydraulic jump puts much previously deposited material (de-

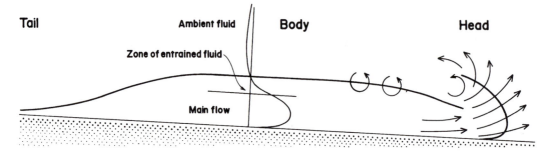

Fig. 4-17. Flow structure of turbidity current; graph shows approximate envelope of velocity vectors. (Scale distorted.)

posited by earlier turbidity currents, perhaps) back into suspension.

It appears that the coarser sediment is concentrated in the head. The head seems to be an eroding region of the current, producing flutes and grooves on the bed (see Chap. 5), which are almost immediately filled and buried by outfall of suspended finer-grained material from the body and tail of the flow, and this results in deposition of a graded bed.

Useful distinction has been made between high-density and low-density turbidity currents. The former, with 50 to 250 g/l suspended material ($\Delta\rho/\rho$ about 0.03 to 0.15), are very fast (up to 70 km/hr); they are probably initiated by transformation of slumps or debris flows. The low-density turbidity currents, <2.5 g/l, are suspensions of clay and silt, commonly hundreds of meters thick, moving 10 to 50 cm/sec. They are probably identical to, or the cause of, the so-called *nepheloid layer* of suspended fines commonly observed in oceanic bottom waters. The nepheloid layer seems to be a more or less permanent feature in many places and may be just the last gasp of numerous faraway high-density turbidity currents.

Base surge is a fast-moving grain suspension in air, created by an explosion, part of the energy of the explosion being rapidly transported in the atmosphere, along the surface, and away from "ground-zero." It is similar to a turbidity current. Base surge is associated with nuclear explosions (Moore, 1967), volcanic explosions (Mattson and Alvarez, 1973), and, undoubtedly, large meteorite impacts. It is probably driven at first by expansion of highly compressed gas (air), then by momentum, typically moving away from the blast at speeds of 10 to 40 meters per second (35 to 150 kilometers per hour). It appears that base surge can form also in airless space, as on the lunar surface,

in which case it is supported, presumably, by dispersive grain pressure. (On the other hand, vaporization of projectile and target material during meteor-moon impact might provide plenty of rapidly expanding gas.)

Base surge has great erosive power at first but later collapses into a blanket of sediment. Base surge created by volcanic explosion erodes channels or furrows that radiate from the center of the blast; they are U-shaped or parabolic in section and are as much as 30 meters wide and 15 meters deep (Mattson and Alvarez, 1973, p. 564; Fisher, 1977). The bright rays and crescentic hills arrayed around lunar impact craters are probably base-surge-deposited or -modified crater ejecta.

Grain Flow

All of us are familiar with slopes of loose sand or talus that, when disturbed even gently, develop into a thin sheet of grains moving down the slope, seeming to take on a life of its own. The so-called grain flow is maintained by a dispersive pressure (Bagnold, 1954), which is due to momentum exchange of grains interacting. A fast-moving cloud of small particles, supported by chaotic mutual impacts of its myriad grains, can behave like a fluid, even in the absence of a fluid phase; grain flows presumably exist on the lunar surface. The quantitative aspects of dispersive pressure are not well known.

A loose granular layer under incipient shear experiences a tangential shear stress T and a normal stress N, being the tangential and normal components of weight stress. The two stresses may be combined into an ***angle of internal friction***, α, such that

$$\tan \alpha = T/N \qquad (4\text{-}27)$$

Angle α is also referred to as *angle of repose*. A grainflow results from some momentary increase in slope angle beyond the angle of repose, an oversteepening due to deposition at the top of the slope or undercutting at the base. A loose granular aggregate under shear develops a dispersive pressure that causes dilation of the aggregate of interacting grains, the dilation effect first described in detail by O. Reynolds in 1885. The magnitude of the dispersive pressure (see Lowe, 1976a, p. 189) depends on particle mass (inertia), viscosity of the intergranular fluid, grain concentration, and strain rate. In a steady flow, the shear stress and dispersive pressure are exactly balanced by tangential and normal components, respectively, of the weight of overlying material; this condition is achieved only if the angle of slope of the moving layer is equal to the angle of internal friction. On more gentle slopes, there is no grain motion. Dispersive pressure cannot increase without limit, of course, so there is some depth below which dispersive pressure cannot support the weight of overlying material, and this determines thickness of the grain flow. In upper levels of the flow, shear stress may fall short of shear strength of the aggregate, and a rigid (non-shearing) plug develops at the top of a grain flow, in which grains are not in relative motion. Such a rigid plug (Fig. 4-18) rides passively on the shearing layer beneath.

Lowe (1976a, p. 193) showed graphically the relationship between grain flow velocity and thickness of the flow (for grain flows in air) for different grain sizes. Coarser grain flows are faster and thicker than finer grain flows. Typical sandy grain flows are less than 2 centimeters thick and have velocities less than 1 meter per second. Subaqueous grain flows also are generally less than a few centimeters or decimeters in thickness.

Grain flows are not turbulent or only minimally turbulent; that is, there is not much mixing between deeper layers in the flow and shallow layers. Large grains *do* appear to migrate to the top of a flowing bed, possibly a *kinetic sieve* effect, so there may be a reverse grading (see Chap. 5). A subaqueous grain flow may change gradually into a debris flow (see below) if the intergranular fluid becomes muddy, and into a turbidity current if the muddy fluid becomes turbulent. So-called **modified grain flows**, which are intermediate between true grain flows and debris flows, do not behave quite like true grain flows and may be thicker and faster than true grain flows.

A single grain flow results in a thin bed with steep depositional slope. Grain flows occur intermittently on the slip faces of aeolian dunes and on subaerial talus slopes. Every lamina in a cross-bed set is a grain flow deposit. Grain flows are important also in glacial moraines, and in small, sandy, or gravelly Gilbert-type deltas and fan deltas (see Chap. 8). Shepard and Dill (1966) observed them in sands of submarine canyons (through the porthole of a submersible vessel). Modified grain flows are common in some fore-reef slope deposits and in the upper (proximal) parts of deep-sea fans.

Fluidized Flow

Fluidized sediment flows result from the upward escape of intergranular fluid, which momentarily supports the grains against gravity and results in a grain aggregate of very low strength. Thus the sediment behaves like a fluid. As pore fluid escapes, the sediment subsides and freezes into a grain-supported mass, more tightly packed than before. Lowe (1976b, p. 304) noted that in natural sediments and in carefully controlled laboratory or industrial solid-liquid systems, upward movement of escaping water tends to be localized as vertical pipes or sheets. This is manifested in natural sediments as dish or pillar structure (see Chap. 5). He also suggested that fluidized flows are not a very significant mode of sediment transport, though gas-fluidization may be important in ignimbrite transport. Lowe (1976b; 1979, p. 76) urged that

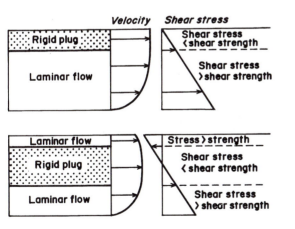

Fig. 4-18. Shear stress and velocity profiles, and the rigid plug, in a grainflow or debris flow.

distinction be made between the fluidized flow and *liquefied flow*, in which sediment is settling through its pore fluid, which, as a result, is displaced upward. In a liquefied flow, sediment is only partially supported by upward-moving pore fluid. Many workers consider that cyclic (vibratory) stresses, such as those associated with earthquakes (seismic waves), are necessary for complete liquefaction to occur. Rapid charging of the sediment with water from below (making a "quicksand") or mechanical failure of a weak or loosely packed mass accumulated on a slope also may result in liquefaction (Lowe, 1982, p. 295). Liquefied flow, it appears, can evolve from a grain flow or a fluidized bed.

Debris Flow

Debris flow or mud flow is a downslope movement of materials lubricated by intergranular water. Water mixed with small particles behaves like a viscous fluid; larger particles are held in the flow by their buoyancy and by the high viscosity, yield strength, or cohesiveness of the fluid phase rather than by turbulence or dispersive pressure or hydrodynamic lift. Lowe (1979, p. 80) noted that, within many debris flows, clasts are not fully supported but are rolled or pushed by the flow as bedload. In some subaerial mudflows, clasts as large as 1/2 meter are transported in this way; clast diameter may even exceed flow thickness. Lowe (1982) considered that turbidity currents begin and end as debris flows, the major distinction being that the former is turbulent and the latter laminar.

Hampton (1972) considered debris flows to be the precursors of turbidity currents. He developed a theory of subaqueous debris flows; there seems to be no reason why the theory does not also apply to subaerial flows, though there are some differences. Consider a debris flow moving down a slope S. As usual, assume steady uniform motion. Then, for unit width and unit (downslope) length of the flow, the downslope component of weight of overlying debris at depth y below the surface of the flow is

$$w_y = \rho g y S \qquad (4\text{-}28)$$

where ρg is the submerged weight of the stratum having thickness y and unit area. This is the downslope driving force (at depth y) per unit area of the stratum (thus w_y is a stress—that is, a force per unit area). There are two stresses opposing this driving force, the internal friction τ_y, which increases linearly with depth, and the shear stress τ_t at the interface between the flow and the overlying fluid (that is, where $y = 0$). For steady, uniform motion,

$$\tau_y = w_y - \tau_t \qquad (4\text{-}29)$$

The shear stress at depth y in the debris flow is either positive or negative (Fig. 4-18), depending on the relative magnitudes of the gravity-induced stress w_y and the shear stress τ_t at the top of the flow.

Now let's compare the internal shear stress with the mechanical properties of the debris-flow material. Hampton considered that a debris-flow has a certain non-zero strength (let's call it τ_s) at low shear stress (like a Bingham or Casson fluid), so that flow (shear) occurs only when shear stress, τ_y, exceeds this strength, that is

$$|\tau_y| > \tau_s \qquad (4\text{-}30)$$

Substituting Eq. (4-29) into Eq. (4-30), and rearranging, we have

$$y_c = (\tau_t \pm \tau_s)/\rho g S \qquad (4\text{-}31)$$

That is, in general there are two depths y_c between which shear stress falls short of strength of the debris "fluid," and a rigid plug exists, just as in the grain flow.

Reverse grading (see Chap. 5) is characteristic of the basal layer of debris flow deposits. Naylor (1980, p. 1113) suggested that this is due to "strain softening" of the clay mud in such flows, being a loss of strength that clay mud sustains upon deformation (thixotropy). Shear stress being greatest at the base of the flow, the mud is weakest there, and large clasts in the basal layer fall out and are left behind upslope. Large clasts higher in the flow, that is, in the rigid plug where shear is not taking place, remain suspended. Thus, there is a deficit of larger clasts in the basal part of a debris flow deposit.

Subaqueous debris flows differ from subaerial ones. Nemec and Steel (1984, p. 16) found in debris-flow deposits of subaqueous fan-delta settings a tendency toward better internal organization, with well-developed grading—either normal, reverse, or reverse-to-normal—and a marked increase in matrix content toward the tops of beds. A subaerial debris flow tends to become thinner as it enters standing water and to break up into several distinct lobes. Other debris flows, of course, *originate* under water. Subaqueous debris flows tend to incorporate water as they flow downslope and to change gradually into turbidity currents.

Nemec and Steel showed that correlations between maximum clast size and thickness of mass flow deposits may serve to distinguish between (cohesive) debris flows and (cohesionless) grain flows. They estimated maximum particle size (MPS) as the arithmetic mean of the ten largest clasts observed. Many beds in a succession of mass flow beds are plotted (Fig. 4-19), and a best-fit straight line is obtained by regression. In some analyses the regression line passes near to the origin; in others the regression line has a y intercept value of 6 or 8 centimeters, compared to greatest bed thicknesses 3 to 10 times this value. Small intercept values are interpreted as representing cohesionless flows and larger values as cohesive flows, the y intercept value related to cohesive strength of the flow. The matrix content of the gravel bed, in contrast, does not seem to be a reliable criterion; that is, there are cohesive flows with only a very little mud and muddy flows that are essentially cohesionless (moreover, mud can be washed into or out of a gravel deposit). Scatter in the plot results from misidentification of bed boundaries and partial erosion of beds before the overlying bed was emplaced; the method also assumes that the succession does not contain a *mixture* of cohesive flows and cohesionless flows and that the large clasts were available to all flows.

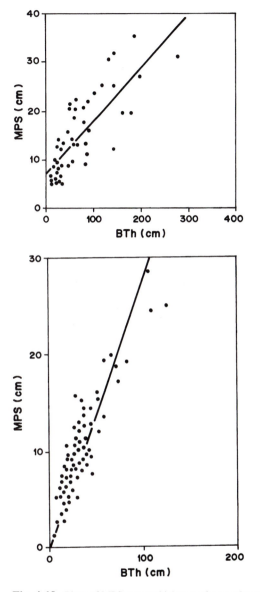

Fig. 4-19. Plots of MPS versus thickness of a set of mass-flow deposits. (After Nemec and Steel, 1984, figs. 20, 21.)

5

Sedimentary Structure

Sedimentary structures are records of the dynamics of the depositional environment. They are of such scale as to be examined easily in the field, and they have received much attention. Some sediments seem to be very homogeneous and without structure but reveal structure when samples of them are examined with medical x-ray machines or fluoroscopes. Sedimentary structures are the unequivocal mark of sedimentary process. We see them occasionally in igneous and metamorphic rocks; we know without doubt that their presence in a plutonic or volcanic rock indicates that some kind of fluid transport and deposition took place, though at temperatures and pressures not ordinarily associated with sedimentary processes. When we see them in metamorphic rocks, they remind us of the rock that must have existed before the metamorphism. In terrains where structural deformation has been severe, sedimentary structures often may be the only convenient criterion of "way-up."

Perhaps the simplest of sedimentary structures is the ubiquitous bedding plane, a surface that forms the boundary *between* beds and indicates some rather abrupt change in sedimentation. Other sedimentary structures are features contained within beds or imprinted upon bedding planes. Some structures are formed by currents in the overlying fluid, and they appear during sedimentation; others result from mechanical processes acting upon sediments shortly after their deposition; and still other sedimentary structures, formed during or shortly after deposition, result from the activities of plants and animals busily pursuing their livelihoods in this sometimes noisy theater of the depositional site. Sedimentary structures (in conjunction with body fossils, if they exist) constitute the major evidence in our diagnosis of depositional environment.

Beds and Bed Boundaries

The simple bedding plane is of great value to stratigraphers because it establishes a natural frame of reference that indicates the direction of time and divides large rock volumes into single pages or small packages, little bites, easily chewed. Bedding planes occur in detrital rocks and in chemically precipitated rocks alike. But the origin of bedding planes is not always clear. Obviously they reflect variable conditions at the depositional site and episodic accumulation of sedimentary materials. Most bedding planes (also called *diastems*) seem to represent intervals of zero sedimentation rate, interposed between intervals of rapid sedimentation. It appears that, in many depositional systems, sedimentation occurs as a sequence of brief pulses, much as rainfall or snowfall does. The cause? Asymmetrical response to fluctuating conditions of sedimentation or thresholds or critical values in the conditions that govern sedimentation. Below some threshold, nothing happens; once the threshold is exceeded, rapid response is triggered. In many or most depositional environments, bedding planes take more time to make than the beds themselves. We are quite certain that some single beds, even some beds a meter or more thick, were deposited almost instantaneously, within hours or even seconds. Such a bed may represent an event that occurs only once in a hundred or a thousand years. The turbidite, certainly, is one of the best examples—sedimentation events are brief compared with the long intervals between sedimentation events. Bedding is the most obvious and most striking characteristic of turbidite successions (Fig. 5-1). It should be kept in mind, when we attempt to diagnose an ancient depositional environment, that a body of rock composed of such beds is *not* a record of the

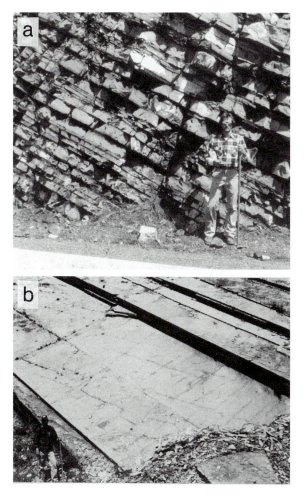

Fig. 5-1. (a) Bedding in a turbidite; (b) remarkably flat bedding surfaces (and joints) in a black shale.

ordinary conditions at the depositional site but a record of the rare or unusual.

Some bedding is a manifestation of variations in composition or texture of the sediment rather than actual breaks in sedimentation. This might be a result of cyclic variations in the conditions under which sedimentation occurs—seasonal change, for example, or the tidal ebb and flow. *Varves*, which are beds that can be demonstrated to have resulted from annual rhythms, are akin to tree rings. Many of them are lacustrine and show a "summer layer" and a "winter layer" that have contrasting color, composition, and texture. Glacial ice is varved, showing quite distinct

winter and summer layers that are commonly a meter or more in thickness. Some evaporite deposits are varved (Anderson et al., 1972; Kushnir, 1981); the thicknesses of these layers may reflect some seasonal or annual mean temperature or rainfall, and thickness patterns commonly can be exactly correlated across localities, just as tree ring patterns can be matched from tree to tree. Cyclic layers occur on some tidal flats, river flood plains, and some bathyal or abyssal oceanic deposits. Such layers may be annual in some cases, or they are a response to longer or shorter cycles (see Chap. 12).

Many beds grow by accumulation of sediment falling or precipitating out of an overlying fluid; the thickness of the bed reflects rate and duration of sediment fallout. But other beds grow sideways, as a current transports sediment along the surface of the bed to its edge; an example is the tabular cross-bed set. The thickness of a bed that grows by lateral accretion is determined by current strength, and possibly by depth of water (or topography of the surface just ahead of the advancing edge) and quantity of sediment in motion (sediment discharge). Typically, many such beds are growing at once. There are, of course, examples of bed growth of an intermediate character, where a bed grows upward and outward simultaneously, so that as a bed becomes thicker it also covers more area; such beds are wedge-shaped.

Bed Thickness

Pettijohn (1975, p. 103) thought that bed thicknesses of terrigenous rocks are lognormally distributed; this seems to have been demonstrated for graded beds and for cross-bed sets. Bokman (1953, p. 155) observed that turbidite sandstone beds and interposed shale beds of the Stanley Formation in Arkansas and Oklahoma are each approximately lognormally distributed; Dott (1963, p. 121) and Ricci-Lucchi and Valmori (1980, fig. 3) found similar patterns in other turbidites. Many workers, such as Schwarzacher (1953), Potter and Siever (1955), Pelletier (1958), Pryor (1960), and Shackleton (1962), have presented histograms of sandstone cross-bed thickness that appear to be lognormal. Lumsden (1971, p. 595) indicated that limestone beds too are approximately lognormal. But in spite of these and other analyses, lognormality of bed thickness is by no means an established fact, and indeed theory suggests other distributions for some kinds of bed successions, distributions that may only resemble the lognormal. Nishiwaki (1979), for example, proposed that bed thickness is similar to waiting times in queues and thus follows a gamma distribution. Bak et al. (1988) described complex systems that tend toward a state of *self-organized criticality.* A depositional system might be nearly always on the verge of making a new bed; as soon as the bed "happens," the system becomes temporarily subcritical but gradually builds once again toward criticality. The thin grain flows that create cross-bed laminae are good examples. Systems having this characteristic cause events whose magnitudes follow a *rank-size law,* that is, $r = (1/m)^b$, where m is magnitude (such as bed thickness), b is some constant $(1 < b < 2)$, and r is rank, or position of an event in a list of all events ordered according to magnitude (see discussion of the *Pareto distribution* by Simon, 1978). Another approach that might be useful in understanding bed-thickness distributions is Hurst et al.'s (1965) *rescaled range analysis* (see Chap. 7).

As with grain size, it is convenient to *assume* that the lognormal model applies, and transform bed thickness measurements so that the Gaussian model can be used. A logarithmic transformation was suggested by Bokman (1957):

$$\theta = \log_2(t/t_o)$$

where t is bed thickness in inches and t_o is the reference thickness of 1 inch.

Ingram (1954) advanced a geometric scale of bed thickness, based on the ratio of $\sqrt{10}$; reference measure is 1 centimeter; beds thinner than 1 centimeter are referred to as laminae:

	Very thick bed
100 cm	------------------------------
	Thick bed
31.6 cm	------------------------------
	Medium bed
10.0 cm	------------------------------
	Thin bed
3.2 cm	------------------------------
	Very thin bed
1.0 cm	------------------------------
	Thick lamina
0.3 cm	------------------------------
	Thin lamina

Nodular Bedding

Bedding planes in limestones commonly are knobby, the surfaces highly irregular, with relief of 2 or 3 centimeters; similar features occur in some sandstones. If individual beds are quite thin, then the irregular bedding surfaces intercept or merge with one another, giving rise to "nodular" bedding. Pettijohn and Potter (1964, Pl. 13 through 16) illustrated several kinds of nodular bedding. Some bedding surfaces may be coated with films or partings of clay (shale), enhancing the nodular effect. The causes of knobby or nodular bedding are not yet fully known. McCrossan (1958) described a kind of nodular bedding in thinly interbedded limy and shaly rocks, which he interpreted as due to early compaction and labeled as sedimentary "boudinage." Knobby bedding surfaces in some fine-grained detrital rocks are a result of loading. Some nodular bedding, especially that which is accompanied by mottling of beds, is due to bioturbation—that is, the disruptive, sediment-churning activities of animals living on or within the sediment. Wobber (1967) described nodular bedding in limestones of the Lias in Wales that he considered to be a result of bioturbation accentuated by loading. Goldhammer and Elmore (1984, p. 1131) considered certain red nodular limestone beds to be paleosols, the nodules apparently created by weathering processes.

Chalky carbonate rocks of pelagic (deep marine) environments commonly are nodular. Fischer and Garrison (1967) described numerous occurrences of calcareous crusts on ocean-floor sediments, associated with manganese oxide nodules and coatings, and due apparently to long-term dissolution and reprecipitation in places where sedimentation is essentially at a standstill. Garrison and Fischer (1969) examined nodular chalks (Knollenkalk) in the Alpine Jurassic, the nodules described (p. 26) as "discrete clast-like remnants of former beds." The nodules appear to be relics of beds partly lithified shortly after they were deposited, then partly dissolved or eroded. Some nodules appear to have been swept up by turbidity currents and re-sedimented in regions distant from their origin. Nodules in the Lower Magnesian Limestone (Permian) in Yorkshire, England are, according to Kaldi (1980), a result of early cementation up to a half meter below the sediment-water interface. They are early diagenetic nodules, but some clearly were disturbed and resedimented, probably by subaqueous debris flow (Kaldi, p. 56). Mullins et al. (1980) described nodular bedding forming today in water depths of 375 to 500 m on Bahaman slopes, where submarine cementation, disrupted by burrowing, is occurring about a half meter below the sediment surface. Development of nodular limestones might be compared to (and contrasted with) the formation of "chickenwire" structure in gypsum beds.

Kennedy and Garrison (1975) examined the most famous chalk formation of all, the Upper Cretaceous pelagic chalks of the South of England (the rocks for which the Cretaceous Period was named), and determined that the nodular bedding of these chalks is early diagenetic and closely tied to development of **hardgrounds**:

When sedimentation ceases for a time, other processes may nonetheless continue and leave their mark at the bedding plane. But now these processes, such as bed-grazing and burrowing by organisms, and chemical reactions between bed and overlying fluid, concentrate their attack on a "dormant" surface rather than a surface continually refreshing itself by ongoing sedimentation. A hardground begins with an **omission surface** (Heim, 1924; see Bromley, 1975, p. 400), a discontinuity owing to minor interruption of sedimentation. The surface develops a hard, smooth, or knobby character; it comes to be cemented with aragonite or radial-fibrous high-magnesium calcite (see Chap. 9); commonly it is stained black by iron and manganese salts and suffused with glauconite or calcium phosphate (collophane). It may be abraded, surface-micritized, bored, or encrusted with serpulid worm tubes or certain molluscs; the organic burrow *Thalassinoides* is commonly associated with hardgrounds.

Just beneath an omission surface, a slow, erratic, early marine cementation takes place. There are scattered centers of cementation below the sediment-water interface, and nodules of cemented chalk grow out of these centers. As they grow, they coalesce into continuous networks or dense packs and may eventually give rise to a hardground. If erosion occurs, the nodules amalgamate into "condensed beds" as the softer sediment is swept away.

Graded Bedding

The **graded bed** is a bed in which there is an upward decline in grain size. It seems to be a result of sedimentation from a grain suspension. Volcanic ashfall beds are graded—grains of many different sizes are released suddenly into the atmosphere as a result of volcanic explosion, and the larger grains, having the greater terminal velocities, fall out first; the finer grains, falling more slowly, eventually come to rest on top of the larger grains already deposited. A similar sorting out of grains occurs when a storm strikes a coast, raising large waves that put much bottom sediment into suspension. When the storm passes, the suspended material returns to the bottom, the larger, heavier grains first. The turbidity current results when a rapidly created suspension flows down a slope; driven by its higher density relative to the ambient fluid, it can move at speeds exceeding 30 meters per second. As long as it continues to move at substantial speeds, it can keep its constituent material in suspension and even erode more material from its bed. But as it finally comes to rest, it drops its load as a graded bed.

Dott (1963, p. 121) found a positive but weak correlation (correlation coefficient 0.29 in a log-log plot) between the diameter of the largest clast and bed thickness in turbidites. Potter and Scheidegger (1966) documented a statistical correlation between grain size and thickness of graded beds, showing that thicker beds are coarser-grained. These results seem to establish some dependence between competence and capacity—flows that carry the most material can also carry the largest or heaviest material. This is not true of all types of flows, and certainly many flows *could* transport heavier particles that simply do not happen to be available.

There is a special kind of grading, called **coarse-tail grading**, wherein the coarser grains show the strongest vertical segregation, the finer grains showing the least; that is, the finer grains are dispersed throughout the bed. Coarse-tail grading apparently forms from grain suspensions of high concentration or of substantial thickness. Fine grains that happen to be near the base of a thick flow presumably reach the bed sooner than larger grains higher in the flow, which must fall farther. Carey (1991, p. 42) found coarse-tail grading in many pyroclastic flows.

Reverse grading occurs in the basal part of many mudflows, the upper part being graded normally (or not graded at all). This is a distinctive characteristic of debris-flow and grain-flow deposits. Sparks (1976, p. 148) observed that certain volcanic flow deposits called *ignimbrites* likewise show a reverse-graded basal layer; moreover, in the main part of the deposit, large embedded pumice clasts are reverse-graded, but other large lithic clasts (having higher density than pumice) are normal-graded. The cause of reverse grading is not known. Naylor (1980) suggested that it is due to large clasts dropping out of the highly sheared and thus weak basal layer of a debris flow, these clasts being left behind. Clasts higher in the flow remain suspended. Some have suggested that it is a **kinetic sieve**

effect, smaller grains falling through the spaces be-tween the larger grains and accumulating beneath the larger grains; this might be important in a grain flow. Laminae of a wave swash zone are reverse graded (Clifton, 1969, p. 554). Hunter (1977a, p. 381) ob-served reverse grading in the laminae of aeolian dunes, and Hunter (1985) in subaqueous slip-face laminae as well. Sallenger (1979) considered that reverse grading in grain flow deposits (such as slip faces) is a conse-quence of dispersive pressure.

Current-Generated Sedimentary Structures

Current-generated sedimentary structures, not quite exclusively reserved to particulate beds, include rip-ples, flutes and grooves, dissolution flutes and scal-lops, current shadows or crescents, flat lamination, rill marks, parting lineation, and cross-bedding. These sedimentary structures, and their particular sizes and shapes, are responses to specific hydraulic conditions at the depositional site—such as depth of flow, veloc-ity, viscosity, and turbulence. But grain size, grain density, and other characteristics of the particles in transport also are influential. Certain flume experi-ments (described later) have shown that sediment transport in flowing water can take place in either of two distinct and contrasting styles, and the associated contrasting flow conditions are called *lower flow re-gime* and *upper flow regime*. There are differences in steadiness and rate of transport as well as nature of interactions between bed and flow. Different *bed-forms*, that is, shapes molded into the surface of the bed by the flow, develop under different flow regime levels. Cross-bedding, the most important of the cur-rent-generated sedimentary structures, results from the

migration of bedforms, whose shapes may or may not be preserved with the cross-beds.

Local flow magnitudes and directions are revealed by the sizes, shapes, and orientations of bedforms, and local currents are commonly parts of geographically extensive patterns of currents that may attain even con-tinental proportions. Thus, the study of bedforms and their internal structure can contribute toward a knowl-edge and understanding of ancient transportational and depositional systems and the ways in which detritus has been dispersed over continents and ocean basins in the past.

Ripples and other Small-Scale Structures

Ripples are a bedform that appears almost spontane-ously on the surfaces of beds of sand and silt as a result of unidirectional flow of air or water in low regime or bidirectional flow associated with passage of waves in shallow water. They are a series of corrugations (crests and troughs) or mounds and hollows typically spaced uniformly a few centimeters apart. In plan view, ripple pattern shows considerable variety (Fig. 5-2). Some ripples are long and gently sinuous, with parallel crests and troughs; others are rather more complicated asym-metrical mounds having short, discontinuous, curved crests. *Ripple length* or *chord* is the crest-to-crest dis-tance, measured perpendicular to direction of the crests (Fig. 5-3). *Ripple height* is the relief of the sur-face, or difference in elevation between crest and trough. *Ripple index* (or *vertical form index*) is ratio of ripple length to ripple height. (Compaction of rip-pled sediments presumably increases ripple index.)

It is clear that ripples are a result of tractive forces exerted by a flow on its bed, that the ripple is an ob-stacle to the flow, and that the ripple perpetuates itself by interacting with the flow. Probably the most im-

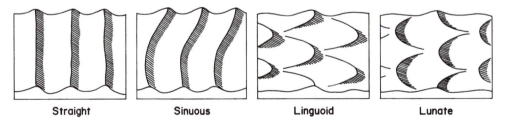

Fig. 5-2. Ripple crest patterns. Straight (or two-dimensional) ripples form in deeper water or at lower current speed or in the wind; three-dimensional ripples are associated with shallow water and higher speed.

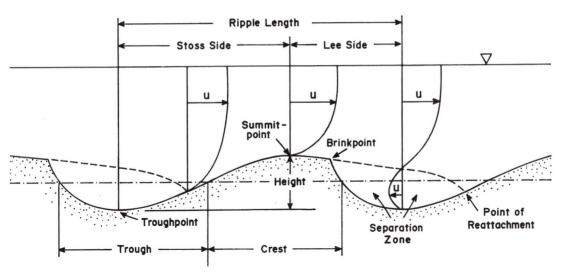

Fig. 5-3. The ripple profile. (After Allen, 1968c, fig. 4.1.)

portant interaction is the creation of local separations between the flow and the bed. Within a zone of separation, small vortices appear, with horizontal axes disposed across the flow direction, these vortices locally scouring the bed, locally piling up bed material, creating bedforms that, in turn, perpetuate the vortices. There may be a resonance linking flow and movable bed that causes bedform periodicity and controls chord length. Ripples apparently never form in sands coarser than about 0.6 millimeter (0.75 phi) or finer than about 0.01 millimeter (6.5 phi). Ripples can form at any flow depth; if the flow is deeper than a few times ripple height, the crests tend to be straight and continuous (two-dimensional ripples). In very shallow flows, ripples interact with waves on the surface of the flow and ripple crests tend to be arcuate and discontinuous (three-dimensional ripples). A fast, shallow flow can shear off the crests of ripples formed under more congenial flow conditions.

The magnitude and directional aspects of the tractive force determine the geometry of the bedforms and serve to organize or unify them into extended spatial patterns. Ripples align themselves across the direction of flow. Those generated by unidirectional flows (Fig. 5-4a) move slowly downcurrent, by erosion of grains from the upcurrent (stoss) surface of the ripple, and deposition of these grains on the downcurrent (lee) side; these ripples are steeper on the lee side than on the stoss. According to Harms et al. (1982, p. 2-17),

ripples formed by currents of the lower velocities have sharp crests and triangular cross sections; higher velocities tend to produce ripples having round crests and sigmoidal foresets. Velocity of the flow also seems to bear on ripple crest patterns, the lower velocities forming ripples with the longer and straighter crests (Harms, 1969, p. 365). Ripples are stable only if shear Reynolds number (Re_*) is less than about 10.

Harms found that average height and spacing of water-formed ripples in flumes does not vary appreciably with current speed, and that ripple index, ranging between 12 and 22, is not consistently related to current velocity. Ripple index seems to be a function of Shields's beta, that is, the dimensionless shear stress on the bed (recall that β is not a function of velocity). Yalin (1964, p. 113) determined that the wavelength of subaqueous ripples is about a thousand times mean grain size of the constituent material.

The ripple index of aeolian ripples commonly exceeds 15, reflecting low amplitudes, whereas the index of subaqueous ripples is typically less than 10. In aeolian ripples, coarse grains generally are concentrated on the crests, opposite to the usual arrangement in subaqueous ripples (Bagnold, 1941, p. 152). Aeolian ripples are nearly always long-crested (Fig. 5-4b).

Flemming (1988) and Lancaster (1988) examined the height-versus-spacing relation for subaqueous and aeolian bedforms, respectively (not just for ripples but for larger forms as well, with spacings even exceeding

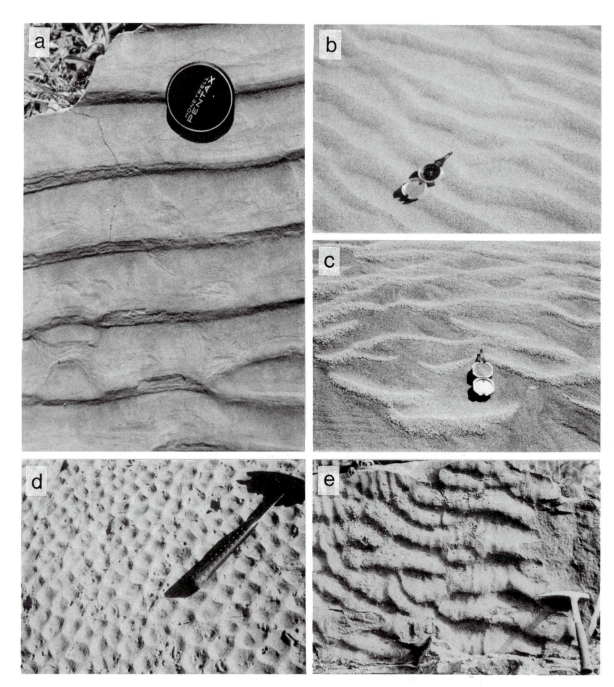

Fig. 5-4. Various current-generated bedforms: (**a**) current ripples; (**b**) aeolian ripples; (**c**) starved ripples; (**d**) interference ripples; (**e**) ladderback ripples.

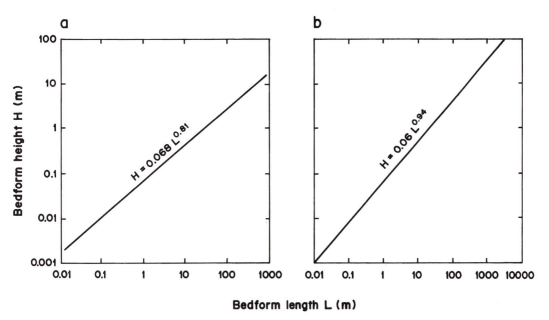

Fig. 5-5. Height versus spacing relationships for various bedforms: **(a)** Flemming's (1988) results for subaqueous bedforms; **(b)** Lancaster's (1988) results for aeolian bedforms. (After Ashley, 1990, figs. 9A, 9B, with permission of SEPM.)

one thousand meters). On the basis of nearly 1500 examples, Flemming determined that height H is related to spacing L according to

$$H = 0.068 \, L^{0.81} \qquad (5\text{-}1)$$

for water-made features (Fig. 5-5a), where H and L are measured in meters. Lancaster (see Ashley, 1990, fig. 9b) found that aeolian ripples and dunes are, in the aggregate, slightly different, following the relation

$$H = 0.06 \, L^{0.94} \qquad (5\text{-}2)$$

that is, giving a somewhat steeper line on a log-log plot (Fig. 5-5b).

Ripples show internal structure that records their growth and progress over the bed. The details of this structure depend on crest pattern, direction of rippleform migration, and rate of sedimentation. Ripples can form under conditions of zero net sedimentation, in which case all sand delivered to the lee side of a rippleform is sand that was eroded from the stoss side. *Starved ripples* (Fig. 5-4c) consist of isolated crests without the intervening trough parts of the rippleform; they look like an ordinary ripple train truncated horizontally, the lower slice discarded. They are produced when there is not sufficient material in transport to

make complete ripples. But ripples commonly form under aggradation, or net sediment accumulation. As the depositional surface rises, it buries and preserves the old rippleforms, resulting in a pattern of stacked ripples, or *climbing ripples*. Rate of climb depends on rate of aggradation relative to rate of downcurrent migration of the bedform. Under the most rapid sedimentation, deposition may occur on both sides of a ripple crest, though somewhat more rapidly on the lee side. This can result in rippleforms that climb upward through a deposit more rapidly than they migrate downcurrent. There is a critical aggradation rate below which only the lee-side laminae are preserved and above which the complete rippleform is preserved. Ashley et al. (1982) demonstrated the process in flumes, and Rubin and Hunter (1982) obtained theoretical relationships between bedform geometry and rate of sediment transport and bedform migration. Hunter (1977b) classified the various types of rippleform lamination in terms of rate of climb (Fig. 5-6). Kocurek and Dott (1981) proposed that a rather peculiar type of subcritical form, with reverse size-graded laminasets, is diagnostic of aeolian transport.

Shallow-water waves, or waves "feeling" the bottom, we have learned (Chap. 4), cause oscillating or

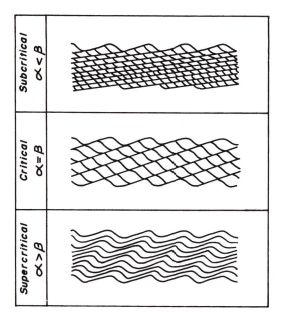

Fig. 5-6. Spectrum of climbing ripples. (After Hunter, 1977b, fig. 4, with permission of SEPM.)

alternating current flow on the bed, and such bidirectional flows also form ripples. Wave-generated ripples tend to be long-crested forms with round crest profiles, and they remain quite stationary or move slowly in the direction of wave propagation. They have the smallest ripple indexes (R.I.) (Tanner, 1967, p. 97); Harms (1969, p. 372) obtained R.I. = 6 for wave ripples in flumes and noted that natural wave ripples have values ranging from 6 to 10, as compared to the substantially higher values for the ripples of unidirectional flows. Komar (1974) found that for low, short-period water waves, ripple spacing (chord) is 0.8 of bottom orbital diameter, and also that ripple spacing increases with increasing grain size of the bed. Miller and Komar (1980, p. 178) found that ripple spacing is 0.65 of bottom orbital diameter but that the relationship is more complicated for large ripple spacings. Harms et al. (1982, fig. 2-10, p. 2-30) suggested that ripple spacing varies with wave period and maximum orbital speed. Moreover, lower orbital speeds seem to give rise to long, straight-crested ripples; higher orbital speeds produce short-crested ripples.

Typically, the current strength is a bit greater in the direction of water-wave propagation than in the retrograde direction. This asymmetry is important in rip-

ple formation and migration; one might view a wave-generated ripple as a ripple that forms under net forward flow, like unidirectional-flow ripples, but a forward flow that is periodically interrupted by a weaker backflow. There seems to be a continuum between ripples formed under strictly symmetrical (bidirectional) flow and those formed under purely asymmetrical (unidirectional) flow. Harms (1969, p. 366) found the geometries of such ripples to be transitional between unidirectional and purely oscillatory ripples.

Ripples of different wavelengths, amplitudes, phases, or directions can intermingle. This is sometimes due to complex fluid motions, sometimes to simple superposition as the hydraulic environment changes or drifts. Ripples with lower-amplitude secondary crests in the troughs of larger crests are occasionally preserved. Intersecting wave-generated ripples (Fig. 5-4d), sometimes called *interference ripples* (perhaps inappropriately), might form when waves approach a coast from different directions or when waves, obliquely incident upon some steep obstacle, reflect back across the water surface in a different direction; the ripples on the bed mimic the pattern of intersecting waves on the water surface. Another type of intersecting ripples commonly occurs on tidal flats. Called *ladderback ripples* (Fig. 5-4e), they consist of a larger ripple set formed during the tidal flow and a smaller, perpendicular set restricted to the troughs of the larger set; the smaller ripples are formed during the ebb tide, by a return flow much weaker than the flood, which was channeled by and trained to the direction of the larger crests. Note that flood-tide ripples form under conditions of deepening water; ebb-tide ripples form under conditions of shoaling water.

In some environments, as on tidal flats or on river floodplains, rippled sandy layers are deposited intermittently with muddy layers; the latter may be very thin relative to the sandy layers, and discontinuous, just coating the troughs of the ripples. These mud films form by sedimentation of suspended matter during quiescent periods, as between storms on a shoreface or between high tides on an intertidal flat, or between floods on the floodplain of a stream. Ripple bedding with these clay seams is called *flaser bedding* (Reineck and Wunderlich, 1968, p. 100). In *lenticular bedding* (Reineck and Wunderlich, 1968, p. 102), clay is much more abundant and the ripple-bedded sand forms are isolated lenses in the clay (Fig. 5-7).

There are some even smaller-scale features, superficially resembling ripples and commonly associated

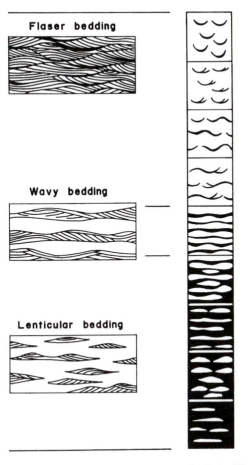

Fig. 5-7. Classification of flaser and lenticular bedding. (After Reineck and Singh, 1980, figs. 183, 184.)

ated by the turbulent wake behind a local obstacle to the flow. Commonly the obstacle is preserved in place, along with the bedform. In carbonate environments, there may be lenses of mud and sand deposited in the lee of coral heads or large shells. In river deposits, veneers, lenses, or wedges of finer sands among the coarser beds form at low stage in the sheltered hollows on the downstream sides of the larger bedforms that formed at high stage; some such wedges are no doubt made by the wind.

Cobbles and smaller clasts in stream beds may arrange themselves into *cluster bedforms* (Brayshaw et al., 1983; Brayshaw, 1984). A complete cluster contains three parts (Fig. 5-9)—an obstacle clast, which ordinarily is the largest clast in the cluster; several smaller clasts on the upstream side of the obstacle, typically imbricated and more or less progressively smaller upstream; and a collection of small grains on the downstream side, in the wake of the obstacle, these grains being progressively smaller in the downstream direction. The whole cluster varies in length from 10 centimeters to over 1 meter. The length of the obstacle clast, oriented flow-transverse, defines width of the cluster bedform.

Rill marks and *swash marks* are ephemeral patterns on the gently sloping surface of sand in the zone of wave runup or swash, the thin rush of water (in upper regime) which is the last gasp of a breaking wave. The final, lobate upper edge of this sheet makes the swash mark. Sometimes the return flow engages the smooth surface of the sand and erodes short, shallow rills perpendicular to the strand. Rill marks can form wherever

with them. They are the *wrinkle marks* of Hantzschel and Reineck (1968; see Reineck and Singh, 1980, p. 65), little corrugations, typically curved and roughly parallel, much like ripples, but with wavelengths of only a few millimeters. They result from nonerosive shear on a cohesive surface, as when the wind blows over wet mud. *Adhesion ripples* and *adhesion warts* are small bedforms that result when dry sand is wind-blown over a damp sandy surface. Some of the sand grains stick to the damp surface and build up little ''buttons'' or ''microridges'' (Hunter, 1969, p. 1574). Reineck and Singh (1980, p. 66) illustrate modern ones in plan and in cross section; Kocurek and Fielder (1982, figs. 7–9) show some ancient examples.

Current crescents (Fig. 5-8) are local bedforms cre-

Fig. 5-8. Current shadow in a windswept backshore sand.

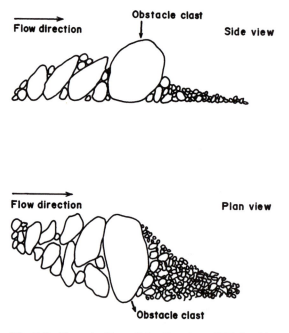

Fig. 5-9. Cluster bedform. (After Brayshaw, 1984, fig. 2.)

a thin layer of water, typically only 3 or 4 millimeters thick, drains over a smooth, gently sloping sandy surface. Reineck and Singh (1980, pp. 68–69) illustrate several different forms, some like combs, others with branches and meanders. Many of them look like miniatures of erosional patterns developed at scales thousands of times greater (as represented on topographic maps and aerial photographs). Many rills have thin *accumulation tongues* or tiny fans (Reineck and Singh, 1980, p. 60) at their mouths. These structures are not often reported in ancient rocks.

Larger-Scale Bedforms

Sand waves (Fig. 5-10a) are large-scale bedforms, rather similar to ripples but grossly asymmetrical and having crest-to-crest distances greatly in excess of their amplitudes. They do not form in sediment of less than about 0.1 millimeter mean grain size. Probably the individual waves are independent of one another; it might be inappropriate to refer to a wavelength in this situation. Sand waves can be viewed as the surface of a succession of ordinary beds deposited by lateral accretion on their downcurrent edges. They seem to

occur most commonly in shallow water flows, such as the beds of rivers or tidal channels or flood-tidal deltas. The lee face (or slipface) is inclined nearly at the static angle of repose of the water-saturated granular material of which they are made. The very gently inclined stoss face commonly is ornamented with ripples. The next wave upstream advances over this surface with little or no attendant destruction of the surface overridden, and it may happen that an upstream sand wave outruns the one just below. Migration of a sand wave generates a tabular cross-bed set. A *reactivation surface* is a sloping, minor disconformity in a tabular cross-bed set which indicates that sand-wave advance was interrupted for a time, during which some degradation of the slipface occurred; then slipface advance resumed. Such surfaces may result from change of stage in a stream, as in an ephemeral or flashy stream; they occur also on the low bars built by tidal currents in shallow water, a situation in which hydraulic regime changes on a regular basis. deMowbray

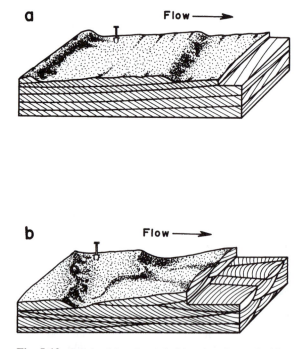

Fig. 5-10. Tabular (**a**) and trough (**b**) styles of cross bedding, associated with sand-wave migration and dune migration, respectively. (After Harms et al., 1975, figs. 3-1, 3-2, with permission of SEPM.)

and Visser (1984, p. 816) described reactivation surfaces in tidal deposits resulting from change in current direction.

Dunes (Fig. 5-10b) are large-scale short-crested bedforms that develop near the upper limit of lower flow regime. Like sand waves, they do not form in sediment of less than 0.1 millimeter mean grain size. Harms and Fahnestock (1965, p. 91) described dunes in the Rio Grande (border between the United States and Mexico) as elongate mounds and troughs, as much as 10 meters long, aligned parallel to the flow. The mounds move by tractive erosion of their stoss faces and sedimentation on their lee faces, typically prograding into trough-shaped depressions just ahead. Seldom is ripple-mark developed on them. The lee face is typically less steeply inclined than the lee face of a sand wave developed in the same sediment, reflecting the lesser dynamic angle of repose associated with the stronger current. Moreover, the lee face is typically concave. Migration of dunes results in trough cross-bedding. Clifton et al. (1971, p. 658) observed similar bedforms in the zone of wave transformation on gently sloping sandy shorefaces.

Flat-bed (Fig. 5-11a) is a sedimentary structure that forms under flows just strong enough to eradicate ripples or under rapid (upper-regime) flow; typically it is quite remarkably flat. It is of common occurrence in ancient rocks, in some places so abundant that it is commercially exploited as flagstone. Like ripples, flat-bed can form at any flow depth. Lower-regime flat-bed in sand less than 0.6 mm is unstable—any small irregularity on the bed grows into a ripple; in coarse sand, ripples cannot form, and lower-regime flat-bed with grain movement does occur and is stable. Upper-regime flat-bed is restricted to sand that is less than 0.3 or 0.4 millimeters mean size; it is stable in the sense that small disturbances wash out rather than grow into some other bedform. In upper-regime flat-beds, elongate grains tend to align parallel to the flow, and this, coupled with long, straight, barely perceptible ridges or furrows on the flat surface, gives rise to flow-parallel ***parting lineation*** (Pettijohn, 1962, p. 1474). This faint feature in rocks is best observed with a concentrated light source (such as the sun) illuminating the bed surface at a grazing angle. Fabric associated with parting lineation apparently influences cementation patterns, thereby accentuating the lineation. Parting lineation occurs in the surf zone and breaker zone of sandy coasts, forming under the rapid

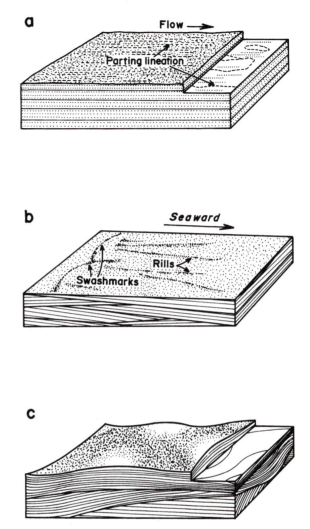

Fig. 5-11. (a) Upper-regime flat bed with parting lineation; (b) swash cross stratification; (c) hummocky cross stratification. (After Harms et al., 1975, figs. 3-3, 5-4, 5-5, with permission of SEPM.)

sheet flow created by breaking waves. Flat-bed and associated parting lineation is also common on sandy tidal flats, and in some river channels. Potter and Mast (1963) examined the relation between parting lineation and sand grain orientation, observing that preferred orientation of elongate grains is consistently to the left of the current lineation. It is a peculiar phenomenon; Bouma (1962, p. 84) observed that in turbidites, grain orientations are "always turned more or less to the right" of current directions indicated by flutes and grooves. Probably there are two modes in most cases, symmetrically disposed with respect to the lineation and reflecting the helicoidal current pattern of Taylor-Gortler circulation (see Chap. 3).

Bridge (1978, p. 7) suggested that some lamination is produced by turbulent bursts in the ambient flow, which have frequencies of a few seconds to a few tens of seconds in rivers. (Flat-lamination is also made in surf zones, where turbulent bursts are regular and frequent.) Jackson (1976, p. 551) noticed that on certain beds there are flow-parallel "lanes" of clear fluid and slower, sediment-laden fluid; he offered this as a possible cause of parting lineation.

Antidunes are large-scale wave-like bedforms that develop in upper flow regime, in sediments ranging from fine sand to coarse gravels. In profile they have round crests and round troughs; they are roughly symmetrical and have wavelengths of a few tens of centimeters. Antidunes are like waves on the bed, but they are stationary or sometimes regressing slowly *upstream* by erosion of material from the *lee* face and deposition of that material on the *stoss* face of the waveform just downstream. In this respect, they behave just the opposite of dunes. There is strong interaction between the bedform and the flow, and it seems necessary that the flow have a well-defined upper boundary or free surface that is not far from the bed (relative to the size of the bedform); the free surface maintains waveforms that are of the same wavelength and phase as the bedform. Antidunes may grow to such height as to cause the waves on the surface of the flow to break; this leads to obliteration of the antidunes themselves, which reform after a minute or several minutes. Antidunes probably have lower preservation potential than other bedforms, but Skipper (1971) illustrated ancient antidunes preserved in a turbidite in Quebec (here, the free surface presumably was the top of the turbidity current). They have been described in modern stream gravels (Shaw and Kellerhals, 1977), and in ancient ones as well (Hand et al., 1969). Blair (1987, p. 10) described gravel antidunes on a recently deposited alluvial fan made by a catastrophic flood in Rocky Mountain National Park, Colorado; they are short-crested features with wavelengths of 5 to 25 meters.

Langford and Bracken (1987, p. 867) described antidune bedforms in Medano Creek, which flows across aeolian sands in Great Sand Dunes National Monument in Colorado. These bedforms are preserved as lenses of gentle backset or foreset laminae embraced by flat-lamination above and below. Barwis and Hayes (1985, p. 914) described similarly subtle structures in an ancient sandstone, likewise attributing them to antidunes.

Mattson and Alvarez (1973) described antidunes formed by "ultra-high" flow regime of volcanic base surge, "a fluid moving too fast for its bed" (p. 555). These antidunes are asymmetrical, having the steeper slope *facing the flow*. Flow-transverse antidunes described by Fisher (1977, p. 1289) have both foreset and backset laminae; these bedforms occur in U-shaped furrows that apparently were cut by volcanogenic base surge. Similar features are associated with maar eruptions in Death Valley, California (Crowe and Fisher, 1973). Dune-like and antidune-like bedforms made by base surge of a phreatic eruption of Taal Volcano in the Philippines (Fisher and Waters, 1969) have wavelengths of nearly 20 meters near the crater; wavelength decreases progressively with distance from the crater.

Erosional Bedforms in Cohesive Sediment

Flutes are small, spoon-shaped scours on the surface of a bed, typically a few centimeters or a few tens of centimeters in length and a centimeter or two in depth. The deep, blunt end of the flute (the part of the spoon near the handle) is upcurrent. Flutes are largely restricted to muddy beds but are best observed as convex casts on the underside (sole) of an overlying sandstone bed. Sandy fill material commonly shows inclined lamination. Flutes seem to be a result of local erosion by small, ephemeral vortices, their axes parallel to the main flow. Allen (1968b, p. 602) considered that they are a product of a flow of substantial power and turbulence interacting with a cohesive bed. The deposits of turbidity currents characteristically display these bedforms (Fig. 5-23e).

Grooves or *tool marks*, and *prod marks* are long, remarkably straight, flow-parallel furrows or trains of small depressions or dimples, "machined" into a soft bed surface (typically in mud) by sticks, shells, or other objects partly suspended or saltating, rolling, or sliding in the flow. Again, they are best observed as sole marks. They are commonly associated with flutes, as in turbidites. Sanders (1965, p. 207; see also Ricci-Lucchi, 1969, p. 224) suggested that flutes are made by hydraulic conditions such as might obtain in the head of a turbidity current, and grooves by conditions in the tail.

Glide surfaces and the grooves that gliding blocks leave on them are occasionally noted. A block of reef rock, for example, might slide down a muddy slope into deeper water, leaving in its wake a long straight groove that may get preserved. If the block happens to rotate as it slides, the grooves that it makes are more complex, of course. The smooth surfaces made by glacial ice in either hard or unconsolidated substrates contain grooves created by stone tools embedded in the ice. Recent grooves in Great Slave Lake, Northwest Territories, Canada (Weber, 1958) are very large (up to 33 meters wide and 4 or 5 kilometers long), and apparently inscribed in the soft lake-bottom sediments by wind-driven icebergs or floes. Iceberg scours in the former Lake Agassiz in southeast Manitoba, Canada (Woodworth-Lynas and Guigne, 1990, p. 220), are 50 meters wide and 5 or 6 kilometers long; they are revealed on aerial photographs as prominent criss-crossing lines, typically quite straight. Dionne (1969) described long furrows plowed by ice blocks drifting across a tidal flat.

Cross-Stratification

Cross-bedding or *cross-lamination* is structure internal to a bed that reflects bedform migration. It is associated with ripples, sand waves, and dunes. Different kinds (or styles) of cross-bedding are distinguished on the basis of size of sets, number of sets in a coset (see below), geometry of the lower bounding surface, angular relation of foreset laminae to lower bounding surface (whether concordant or discordant), and homogeneity of the sediment that composes the sets.

Commonly in outcrops only vertical sections are available for direct observation, but the plan view is important also. The scale and other geometrical aspects of cross-lamination are controlled, of course, by the sizes and shapes of the bedforms that gave rise to it. Aside from the small-scale cross-lamination associated with ripples, four kinds of cross bed are of particular importance; these are tabular, trough, swash, and hummocky.

Tabular cross-bedding (Fig. 5-10a) is associated with sand waves (Harms and Fahnestock, 1965, p. 105). A bed that bears this kind of internal structure is laterally extensive and of more or less uniform thickness. Individual inclined laminae, called *foresets*, that comprise the **cross-bed set** are flat (not curved),

and meet the upper and lower boundaries of the set at roughly the same high angle, typically about 30°. *Trough cross-bedding* is associated with dune bedforms (Harms and Fahnestock, 1965, p. 103). A trough cross-bed set (Fig. 5-10b) bears a lower boundary that is shaped something like a canoe or a scoop and cuts across (truncates) bed sets beneath. The foreset laminae are curved, meeting the boundary more or less tangentially and tending toward the shape of the lower bounding surface. The shapes of trough cross-bed sets only vaguely and incompletely reflect the shapes of the dune bedforms. McKee and Wier (1953) recognized a form of cross-bedding in which bed sets are wedge-shaped (Fig. 5-12), each wedge-shaped laminaset truncating sets beneath. It seems that most such cross-bedding is aeolian.

Swash cross-stratification consists of slightly divergent sets of very gently inclined planar laminae (Fig. 5-11b). It is created by wave swash on the gentle slope of the beach face. The laminae are hydrodynamically similar to upper-regime flat lamination and may show parting lineation; they may be ornamented also with swash marks or rill marks. There is, presumably, a preferred dip direction toward the sea, but inclinations sometimes are so low that this can be difficult to discern.

Hummocky cross-stratification (HCS) typically forms lenticular bed sets a few centimeters thick (Fig. 5-11c). Bundles of more or less parallel but gently undulating laminae truncate similar bundles beneath; laminae are not preferentially inclined in any particular direction (Harms et al., 1975, p. 87). Hummocky sets characteristically occur in coastal environments as isolated layers of sand or silt embedded in shale. They are thought to be a result of the chaotic currents created on the bottom by storm waves and form, apparently, just above storm-wave base. They are not likely to be

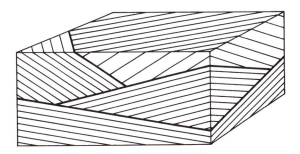

Fig. 5-12. Wedge-shaped cross-stratified sets.

preserved at depths above *fairweather* wave base, however, where they would tend to be reworked by ordinary waves. Commonly, HCS sandstones are embedded in shale and have sharp basal contacts with shale. This suggested to Hamblin and Walker (1979, pp. 1680–1681) that sand is entrained in very shallow water by storms of hurricane proportions and moves to deeper water as a density current, where it is then molded by waves of the same storm into HCS.

Some cross beds do not dip down-current but are inclined perpendicular or obliquely to the current; such beds or laminae are a result of lateral accretion *across* (more or less perpendicular to) the direction of flow. It is important that they be distinguished from the "well-behaved" cross beds. They are present in fluvial point bars, in certain aeolian dunes, and probably in flow-parallel tidal ridges. It is not unusual for inclined laminae of this type to be covered with ripples that indicate strike-parallel flow (Fig. 5-13). Single sets of **lateral-accretion cross-stratification** may exceed a meter in thickness. Other thick foresets result from progradation of Gilbert-type deltas, or of shorefaces, or the building of coastal spits. Inclined laminae formed by upper-regime flows commonly dip up-current. Again, it is important to distinguish these from ordinary cross-laminae.

Experimental Hydraulics of Sediment Transport

Most of our understanding of current-generated sedimentary structures comes from experiments performed

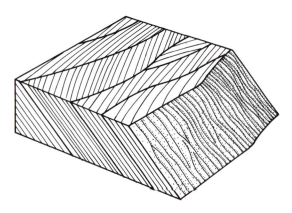

Fig. 5-13. Strike-parallel flow on a slip face indicated by ripple crests parallel to dip. (After Harms et al., 1975, fig. 3-5, with permission of SEPM.)

in laboratory *flumes* and *wave tanks*, counterparts of the wind tunnels of the aerodynamicist. Much empirical work has been done that is directly applicable to interpretation of fluid-dynamical conditions in natural environments, modern or ancient. In most problems of geological interest, the fluid phase is water, and the solid phase is particles of density close to 2.65 g/cm^3 (the density of quartz). Under these constraints, three variables control bedform development, namely depth of flow, velocity of flow, and sediment grain size (Southard, 1971; Southard and Baguchwal, 1990). In flume experiments, depth and velocity of flow can be varied at will, so that the effects of these variations on a bed of material of given grain size can be observed. These experiments can then be repeated for beds in which grain size is different. Several generalizations emerge: (1) with increasing flow velocity, there is a sequence of equilibrium bedforms (Fig. 5-14) from ripples to lower-regime flat bed to sand waves to dunes to upper-regime flat bed to antidunes; (2) except at very shallow flow and except for antidunes, depth of flow does not greatly influence bedform development—that is, it does not determine whether a bedform develops, though depth probably does influence a bedform's size and shape; and (3) sediment grain size controls the range of flow velocities in which certain bedforms may appear (Fig. 5-15) and influences the size of bedforms. Note particularly that in silt-size sediment, ripples develop at low flow velocity, then give way directly to upper-regime flat-bed at higher velocities. For sand-size sediment, in contrast, sand waves and dunes intervene between ripples and upper-regime flat bed. Ripples do not grow into sand waves or dunes as flow regime increases; rather, the ripples vanish and the new bedforms emerge in their place. Thus, sand waves and dunes are *not* just ripples of larger-scale; they are distinct bedforms.

One of the first thorough experimental analyses of bedforms was conducted by the United States Geological Survey in large flumes at Colorado State University in Colorado Springs (Simons et al., 1965; Guy et al., 1966; Harms's work on ripples also was conducted there). The results are well known among geologists and are considered to be useful in the hydraulic interpretation of bedforms, ancient and modern. In this analysis, bedforms were related to **stream power**, the product of mean velocity and bed shear stress (which depends on velocity and depth of the flow). At stream power below a certain critical value (lower flow regime), individual grains move in discrete steps—that

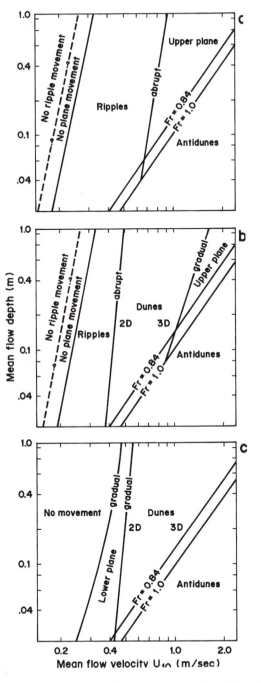

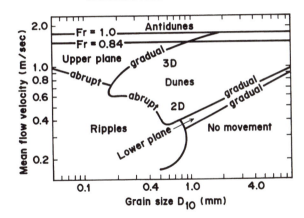

Fig. 5-15. Bedform stability fields in relation to grain-size and depth of flow. (After Southard and Boguchwal, 1990, fig. 8, with permission of SEPM.)

Fig. 5-14. Depth-velocity diagrams for beds of different grain sizes. (According to Southard and Boguchwal, 1990, fig. 6, with permission of SEPM.) (**a**) Fine sand (3 φ); (**b**) medium to coarse sand (1 φ); (**c**) very coarse sand (−0.5 φ).

is, they are sometimes in motion and sometimes at rest. As stream power increases from very low values, certain bedforms appear; they then vanish as they are replaced by other bedforms. Ripples are first in the succession (Fig. 5-16), dunes come next. (Though the flumes were large, they apparently were not large enough to permit the appearance of sand waves.) The water surface develops waves that are about 180° out of phase with the dunes; thus water wave crests overlie dune troughs. At intermediate values of stream power (transitional flow), dunes wash out and a plane bed (or "upper" flat bed) appears, adorned with parting lineation; grains are in continuous motion. Froude numbers at transitional flow seem to vary over a wide range, from values as low as 0.3 to values exceeding 1.0 (the critical Froude number), depending largely on grain size of the bed material (Simons et al., 1965, p. 40). At even higher stream powers (upper flow regime) antidunes appear; in contrast to dunes, these bedforms are in phase with water surface waves. Kennedy (1963, p. 538) determined (from theory) that the transition from flat bed to long-crested antidunes occurs at Froude numbers between 0.84 and 1.0; the short-crested forms can develop at Froude numbers exceeding unity. At the highest stream powers, antidunes too wash out as the flow develops into a chaotic system of "chutes and pools," resulting from oscillation between tranquil and shooting flow.

A major distinction between lower and upper flow regime is that in lower regime, particles on the bed are in intermittent motion, but they are in almost contin-

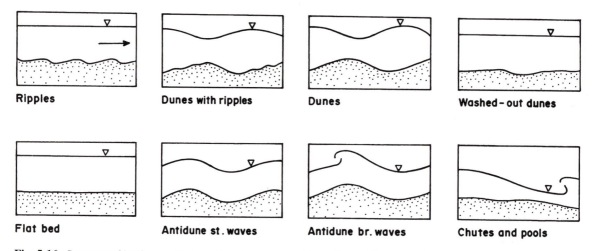

Ripples **Dunes with ripples** **Dunes** **Washed-out dunes**

Flat bed **Antidune st. waves** **Antidune br. waves** **Chutes and pools**

Fig. 5-16. Sequence of bedforms with increasing stream power. (According to Simons et al., 1965, fig. 3.)

uous motion in upper regime. Bedforms developed in upper regime are in phase with water surface waves, in contrast to lower-regime bedforms, which are out of phase with surface waves. Lower-regime flow does not ordinarily transport pebbles; these and larger particles do not participate much in the formation of lower-regime bedforms such as dunes, but some antidunes are composed largely of pebbles.

Angle of inclination of foreset laminae is related to flow velocity, the faster flows giving the gentler foresets. Simons et al. (1965, p. 44) observed that, with increasing flow velocity, dunes became longer and lower, making gradual transition to flat bed. One expects that sand waves, forming at lower velocities than dunes, have somewhat steeper foresets than dunes. There are some indications that this is true, but at the

same time there are some complications. Sorting and other textural characteristics of the bed material influences bedform steepness (Simons et al., 1965, p. 50). The foresets of dune bedforms generally are curved in the vertical plane, and it appears that this is a result of retrograde flow, that is, a backflow in the zone of separation (Fig. 5-17). Jopling (1965, fig. 4) noticed ripples forming in the troughs of sand waves and dunes and migrating *upstream*, a consequence of this backflow. Simons et al. observed backflows in the zone of separation in dune troughs, the velocity of these backflows fully one-half to two-thirds the value of mean downstream velocity, and increasing the angle of the slip face even to values exceeding static angle of repose. Where flow is *parallel* to an erodible bank, angle of repose is less; thus, faster flows give rise to lower

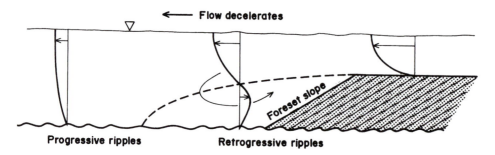

Fig. 5-17. Flow over the edge of a foreset layer in a flume, with backflow in the zone of separation of the main flow. (After Jopling, 1965, fig. 4, with permission of SEPM.)

slope angles on the lateral margins of a longitudinal bar.

Clifton et al. (1971) pointed out the similarity between the fluvial hydraulic regimes of Simons and his associates and the wave-related hydraulic regimes of sandy beaches, as reflected in sequence or distribution of bedforms. This resemblance exists in spite of the fact that wave-generated flows are periodic or oscillatory, that is, decidedly unsteady. The succession of bedforms (Fig. 5-18)—from ripples in the offshore zone, to dunes in the zone of wave transformation, to plane bed in the surf zone—suggests a gradual increase in flow regime, in the sense of Simons et al., corresponding to a gradual increase in maximum bottom orbital velocity due to wave motion. Close to the strand there is apparently a reversal from upper regime to lower regime in the surf zone, then an abrupt return to upper-regime hydraulics in the swash zone.

Larger-scale bedforms—such as the point bars and braid bars of rivers, the tidal ridges and wave-built longshore bars of coastal regions, the barchans and seifs of aeolian environments, and the drumlins created by glacial ice—are too big to be reproduced or modeled under the controlled conditions of the laboratory flume, and their dynamics and limiting conditions are less well known than those of the smaller bedforms. These basically are composites of structured elements (cross-bed sets, etc.) that formed under a unified hydrodynamic system of spacial and temporal extent greatly transcending the single elements. These bodies are built brick by brick; often, one can determine the way in which such a body was constructed by observing the manner in which the "bricks" are stacked. (We examine some of these larger sedimentary bodies at some length in Chap. 8.)

Statistics of Directional Properties

Current-generated sedimentary structures have *directional* or *vector* properties, which record the direction of the flow that generated them. The statistical properties of ensembles of these vectors are useful in our analysis of environments of deposition and of patterns of sediment dispersal. But statistics of vectors differs in some respects from statistics of scalars, such as grain size. A fundamental difference is that, with vectors, the values of the variate are constrained to a specific range, such as 0 to 360°, or 0 to 180°, and this range "wraps around," so that 0 is equivalent to 360 or 180. Vector properties that are three-dimensional are distributed over spheres or hemispheres. A special distribution model, designed for analysis of vector data in two dimensions, is the **von Mises distribution**; it is

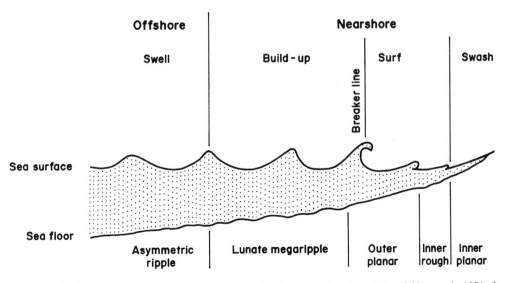

Fig. 5-18. Hydraulic zones and bedforms on non-barred high-energy shoreface. (After Clifton et al., 1971, fig. 15, with permission of SEPM.)

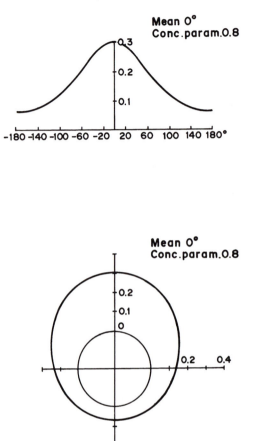

Fig. 5-19. Von Mises density function in rectangular and in polar coordinates.

the normal error law for angular data. (Some workers use the term "circular normal distribution," but this corrupts the distinction between the von Mises and the bivariate Gaussian normal distribution on the circle, which is not in any way related.) The von Mises density (Fig. 5-19) is

$$f(\theta;\ \mu,\kappa) = \frac{1}{2\pi I_0(\kappa)}\ \exp\ [(\kappa\ \cos(\theta\ -\ \mu)]$$
$$0 < \theta \leq 2\pi \qquad (5\text{-}3)$$
$$0 \leq \mu < 2\pi;\ \kappa > 0$$

where μ is the **mean**, κ is a **concentration parameter**, which corresponds to variance in the normal distribution, and $I_0(\kappa)$ is the modified Bessel function of the first kind of κ. This distribution model is appropriate

to populations of cross-bed dip directions. With minor modifications, it is applicable to 180° data, such as ripple-crest or parting-lineation trends. Magnitude of angle of dip of cross-stratification presumably is distributed according to a *folded* von Mises distribution, that is, the distribution that results when frequencies on the left side of the distribution are added to frequencies on the right side (Fig. 5-20). This is true because a dip (or plunge) angle of $+x$ is indistinguishable from a dip angle of $-x$.

The *sample statistics* of a set of vector data—that is, computations made with measured values and used as estimates of the parameters of the theoretical distribution—are the **vector mean** and the **consistency ratio**. (We could go beyond this, to measures involving third and fourth moments, but this is rarely done; see Batschelet, 1981, p. 285, 288.) The vector mean is

$$X = \arctan \sum_{i=1}^{n} \sin x_i / \sum_{i=1}^{n} \cos x_i \qquad (5\text{-}4)$$

where x_i is the ith measured value, and n is the number of measurements. The consistency ratio is

$$L = [(\sum_{i=1}^{n} \sin x_i)^2 + (\sum_{i=1}^{n} \cos x_i)^2]^{1/2}/n \qquad (5\text{-}5)$$

As with grain size data, sedimentologists commonly group their cross-bed measurements. (They often make their groups far too narrow. A 5 or 10° interval commonly yields ragged histograms that do not smooth out measurement errors or "noise" in the population; a 30 or 40° interval is much better.) For grouped data (Fig. 5-21), the vector mean is

$$X = \arctan\ [(\sum_{i=1}^{k} f_i \sin m_i)/(\sum_{i=1}^{k} f_i \cos m_i)] \qquad (5\text{-}6)$$

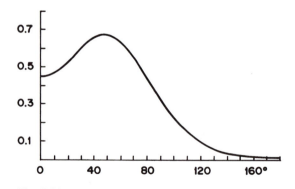

Fig. 5-20. A folded von Mises distribution.

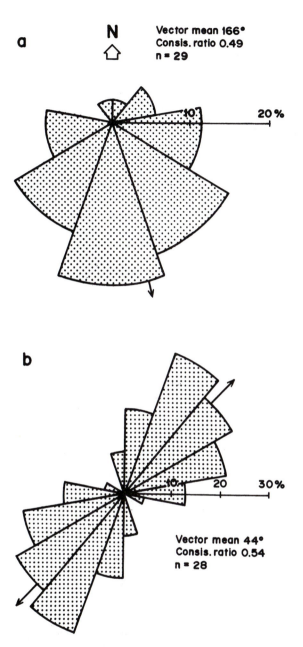

Fig. 5-21. Circular histograms: **(a)** cross-bed dip directions; **(b)** ripple-crest directions.

where m_i is the midpoint of the ith class, f_i is the number of measurements (frequency) in the ith class, and k is number of classes. Consistency ratio for grouped data is

$$L = [(\textstyle\sum_{i=1}^{k} f_i \sin m_i)^2 \\ + (\textstyle\sum_{i=1}^{k} f_i \cos m_i)]^{1/2}/k \quad (5\text{-}7)$$

Some workers record consistency as $100L$. Consistency ratio, a sample statistic, may be transformed into an estimate of κ, the concentration parameter, through

$$L = I_1(\kappa)/I_0(\kappa) \quad (5\text{-}8)$$

where I_0 is, again, the modified Bessel function of the first kind, order zero, and I_1 is the first-order function. Obtaining κ from L, that is, the inverse of (5-8), is not possible analytically, so some graphical or inverse interpolation procedure must be employed. (Incidentally, the Bessel functions have series expansions that make their computer evaluation very easy.)

In spite of the existence of a robust statistical theory for vector data, many workers nonetheless use the Gaussian normal statistics for vectors. This may be adequate for vector ensembles of low variance, but if variance is high, the results can be misleading. Potter and Pettijohn (1977, p. 376) advised that *"The best procedure always is to compute the vector mean"* (emphasis theirs).

Sedimentological vectors commonly are misoriented owing to structural deformation of the rocks that host them. It is necessary in such cases to restore the measured vectors to the orientations that they had before the deformation. Restoration can be accomplished by means of the stereographic net (see Potter and Pettijohn, 1977, p. 371) or more readily by mathematical matrix methods in this, the computer age.

Inclined laminae ordinarily dip down-current, and their variability, either areal or temporal, is probably an indication of variability of the current. Some workers have claimed that the degree of variability is a criterion of depositional environment (e.g., Miall, 1974). Bimodal current-direction distributions surely are characteristic of tidal systems; in some the modes are about 90° apart, in others 180° apart (bipolar). Fluvial, deltaic, aeolian, and turbidite systems nearly always give unimodal patterns. Marine shelf sands commonly yield the most disordered, multimodal distributions.

Certain current-generated bedforms, such as parting lineation, have low variability because they are produced by currents that are intrinsically of low variability. Glacial ice makes bedforms with high directional consistency over large areas. Pryor (1960, p. 149) suggested that large-scale areal variability of vector properties reflects magnitude of depositional slope, at least in large-scale fluvial systems; greater slope leads to lower directional variance. McGowan and Groat (1971, pp. 34–35) suggested that consistency of cross-bedding reflects relatively high depositional gradient in the alluvial sediments they studied.

There may be two different current regimes, each dominant at different times. On delta slopes, dominant current may be downslope during seasonal high discharge; at other times, contour currents (see Chap. 8) may dominate. Allen (1966, p. 184) concluded that directional variability in fluvial systems is hierarchal. For example, a certain variability, and a certain suite of sedimentary structures, is associated with the meander belt and channel sinuosity, and a different kind of variability is associated with channel bars. McGowan and Garner (1970, p. 87) pointed out that meandering stream channels are straighter at flood stage than at normal stage. Bluck (1974) indicated that the variability of vector properties in braided stream deposits is lowest at high stage and increases during times of normal (lower) stage. Pebble imbrication, for example, has consistently lower directional variability than does cross-bedding in sand. The imbrication forms during flood, when variability derives mainly from channel sinuosity, which is low in braided channels. But sand sedimentation occurs mainly during falling stage, when flow patterns are changing rapidly. On the other hand, Schwartz (1978, p. 250) found, in the Red River in Oklahoma and Texas, that falling-stage deposits have *less* dispersion and are more closely aligned with local channel trend than are flood-stage deposits.

Variability of vector properties at a given locality also can be a result of overlapping flow patterns or a flow pattern in translation (Fig. 5-22). Adjacent turbidity currents might overlap and create successions of beds in which flow vectors are widely divergent, even though the axes of mean flow of these turbidity currents are parallel (Potter and Pettijohn, 1977, p. 174). An advancing glacial lobe is an example of a translating flow pattern that otherwise does not change much. A single lobe, although moving in a constant direction, can create striae on a bedrock surface having a range of orientations (Potter and Pettijohn, 1977, p. 173).

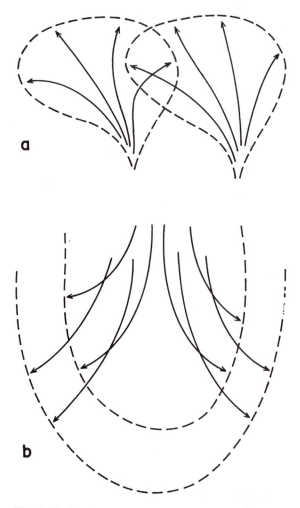

Fig. 5-22. Overlapping flow patterns: **(a)** adjacent turbidity currents or fan lobes; **(b)** divergent flow lineations created by a single glacial lobe or turbidity current. (After Potter and Pettijohn, 1977, figs. 5-12, 5-13.)

Large, elongate bodies of rock may contain cross beds whose dips are essentially parallel to the trend of the rock body, as in a stream-channel sandstone, or perpendicular to the trend of the body, as in a wave-built coastal bar sandstone. Moreover, the cross beds themselves may bear smaller-scale vector properties, such as parting lineations or preferred grain orientation, that bear some directional relationship to the larger structures. The larger the structure, the greater its significance and the smaller its variance, because it was formed by a stronger and larger current, according to Potter and Pettijohn (1977, p. 138). Large structures, such as tidal ridges or aeolian dunes, respond slowly to changing current direction, and their orientations probably represent long-term, time-averaged current directions. The smaller bedforms, such as the ripples and flutes, presumably have the higher resolving power, revealing the small-scale, local elements of flow patterns, and they respond rapidly to high-frequency changes of current direction. Grain fabrics might show even higher variances; single grains probably do not respond as accurately or as sensitively to flow direction as bedforms do; bedforms may reflect actual flow direction with the greatest fidelity.

Certain consistent trends in directional properties may occur in vertical profiles. The lower parts of a channel fill may contain lateral accretion deposits; dip directions typically deviate from local channel orientation by about 90°; cross-beds higher in the deposit have dips approximately parallel to channel orientation. In shoreface deposits, cross beds in the lower shoreface dip landward, responding mainly to the shoaling swells, but those in the upper shoreface dip seaward, landward, or alongshore, reflecting the complex wave-generated currents and the various bars and channels and troughs that these currents make. Cross-bedding in the lower parts of aeolian dunes might be less variable than in the upper parts, where slip-face reversals might be occurring.

Post-depositional, Non-transport-Related Structures

Sedimentary structures created by physical processes acting shortly after deposition include desiccation cracks, rainprints, fluid escape structures, and structures resulting from sediment loading or gravitational slumping or flow. Some workers consider these features (and bioturbation) to be diagenetic, since they are post-depositional, but as long as depositional environment is the dominant influence on their production, these features should be considered as depositional. Sedimentation is not really complete as long as such structures are forming—sediment particles are still in relative motion.

Desiccation cracks (Fig. 5-23a,b) or mud cracks form when muddy sediment dries out and shrinks. As soon as such a crack appears, it tends to propagate

lengthwise until it meets some other crack; it also propagates downward into the bed, pursuing the desiccation front, and perhaps ultimately reaching some surface of detachment, such as a bedding plane between the shrinking mud layer and an underlying layer of sand. The cracks divide the muddy layer into polygonal plates with right-angle corners; the sizes of the plates are controlled by thickness of the desiccating layer, coherency or tensile strength of the mud, its capacity for shrinkage, and rate and degree of desiccation. Generally there are hierarchies of plates, large plates further subdivided into smaller plates by narrower, shallower, younger cracks (cf. the Koch fractal constructions in Mandelbrot, 1982, pp. 57, 65). Since desiccation begins at the surface, the plates undergo greater shrinkage at the surface than farther down, and this causes the plates to curl upward, forming shallow bowls.

Shrinkage is not always a result of desiccation; some cracks develop while sediments are fully submerged; some concretions have cracks that formed during or shortly after their precipitation from aqueous solutions. Such cracks are a result of *syneresis*, or shrinkage owing to expulsion of fluids from colloidal suspensions or gels. Concentrated brines created by evaporation are hygroscopic and remove water from any mud layers with which they are in contact; this creates shrinkage cracks that are, according to Sonnenfeld (1991, p. 162), indistinguishable from those formed subaerially.

Megapolygons are larger systems of cracks created by expansion and contraction of (carbonate) muddy sediment subaerially exposed in tropical or subtropical supratidal environments. They are 2 to 10 meters across and involve as much as a 5-meter thickness of sediment. Eugster (1969, p. 9) observed large desiccation polygons 2 to 50 meters across in the siliceous sediments of Lake Magadi in Kenya, noting that plate margins commonly are overthrusted. Associated with these polygons are many intricate little folds and reticulated surfaces and other curious structures. Muir et al. (1980, p. 54) described large polygons in Recent ephemeral lakes (billabongs) around the Coorong Lagoon in South Australia, observing that they increase in diameter away from lake centers. They are an active and evolving feature—mud that fills the cracks between the drying and shrinking plates in the winter months extrudes in summer as plates absorb moisture and expand against one another. Thus small plates fuse

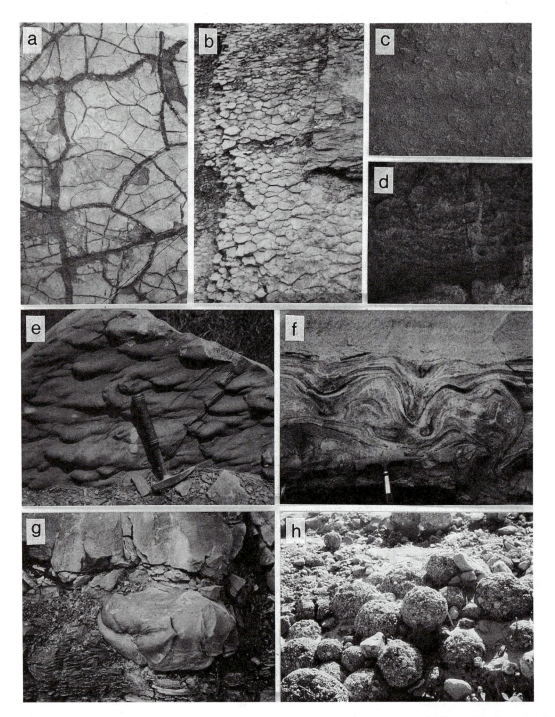

Fig. 5-23. Various sedimentary structures of diverse origins: (**a**) mud cracks in a shale (argillite); (**b**) mud cracks in a laminated dolomite; (**c**) rainprints (buttons in aeolian sandstone); (**d**) dish structure; (**e**) flutes on underside of sandstone bed; (**f**) convoluted bed; (**g**) ball-and-pillow structure; (**h**) armored mud ball.

106

together into larger plates with active margins, and margins of the larger polygons become sharp-crested anticlines, called **tepees**. One might compare them with the pressure ridges of ice floes. Tepee structure results generally from displacive growth of evaporite minerals within the upper layers of a muddy sediment, or hydration of beds or nodules of anhydrite, which causes both lateral and vertical expansion of these beds.

Even larger polygons, up to 300 meters across and bounded by fissures 1/2 meter wide and 5 meters deep, occur in playas of the Great Basin in the western United States (Neal et al., 1968). The major contraction that creates these giant polygons is thought to occur not at the surface but in the capillary fringe just above the water table, which may lie at several tens of meters depth. Desiccation cracks sometimes develop around a phreatophyte, that is, a plant whose root system seeks the water table. As the plant draws water from the aquifer, it creates a water-table drawdown (cone of depression); local dewatering causes shrinkage and development of a **ring fissure** around the plant. Such fissures also form around water wells in the desert. Their diameters are related to depth to water table.

Anhydrite beds commonly show a nodular structure aptly called **chickenwire**. The nodules, typically a few centimeters across and composed of felted or swirling aggregates of small anhydrite laths, appear to have been squeezed together and are delineated by thin, dark seams of clay (the "wires"). Nodules or isolated crystals of gypsum form in arid, intertidal or supratidal environments, just beneath the surface of a muddy (carbonate or clay) bed. As these masses of gypsum grow, they displace the host mud to their surfaces and remain, to a large degree, mud-free. Later, under burial to depths of 1000 meters or so, the gypsum dehydrates and recrystallizes to anhydrite (see Chap. 11) but may retain the nodular structure. Nodular anhydrite apparently also forms when large bottom-precipitated selenite crystals alter to anhydrite under deep burial; such nodules commonly are vertically elongate and have carbonate chevron ghosts that reflect fishtail twinning in the former selenite crystals. The **enterolithic structures** of laminated anhydrite or gypsum deposits are small-scale disharmonic folds in the evaporitic layers caused by lateral expansion related to hydration; in vertical sections, these folded layers have the appearance of entrails.

The **crystal cast** is an impression or sediment-casting of a crystal, usually of a very soluble evaporite mineral, such as halite, that grew on or within a loose sediment (displacing or impressing it) and then dissolved. Crystal casts of ice (frost) are not unknown in ancient rocks.

Another structure, seldom preserved, is the **raindrop impression** (rainprint). Rainprints are small impact craters made by drops of rain (or even hailstones; see Rubin and Hunter, 1984) on soft sediment, especially mud. They are randomly arrayed, commonly overlapping dimples, circular or slightly elliptical, of various small sizes. A raindrop falling on dry sand makes a little "button" of sand held together by surface tension (Fig. 5-23c). Such structures have been observed in ancient aeolian sandstones. Some markings that are called raindrop impressions might in fact be a result of surfacing air bubbles that rose through the sediment.

Neither mudcracks nor rainprints are climatic indicators—they merely reveal the weather on a particular afternoon or weekend (in fact, mud cracks and rainprints commonly occur together). Modern examples are familiar to all of us—we have seen them on desert playa, tropical beach, and temperate mudhole. Paradoxically, rainprints probably are most frequently preserved where rainfalls are least frequent. Mudcracks and rainprints do, of course, indicate emergent or subaerial conditions.

Some sediments initially contain gas. This gas may be entrapped atmospheric gas (air) or gas evolved internally due to life processes, including organic decay. Infaunal organisms can carry air into their burrows or respire gas into the sediment that surrounds them. If the host sediment is not too coherent, the entrapped gas collects into buoyant bubbles that rise toward the surface. As a bubble rises through an aggregate of loose grains, those grains forming the roof of the bubble fall through to the floor. The grains that have fallen through are less tightly packed than they were in the undisturbed sediment, so that the floor of a bubble rises a bit faster than the roof; as a result the bubble diminishes in size to the point where it is no longer sufficiently buoyant to rise further, or it vanishes into the interstices in the loose pack. Commonly a rising bubble encounters some impermeable or coherent stratum that prevents its continued upward motion. The gas then collects under this stiff layer, inflating lenticular openings there. Cloud (1960) described and illustrated structures formed in this way, which he called **gas accretion sites**.

Wave surf can pump air into the sand of a beach face. Hoyt and Henry (1964) described spongy fabric in beach sand, owing to injection of air by surf action and tidal inundation; the formation and later collapse of air bubbles destroys the lamination that characterizes beach-face sands. Fryberger et al. (1988, p. 36) found vesicular sands beneath algal mats on interdune surfaces in Baja California, Mexico. Bull (1964b, p. A31) and Chandler (1972, p. 231) described air bubbles caught up in mud flows. They may be so abundant as to render the mud a scoriaceous fabric.

Sediments that are rapidly deposited commonly contain excess water; this water maintains an internal pressure (pore pressure) in the sediment that prevents, or delays, the constituent grains from achieving a stable packing arrangement. If, subsequently, this excess pore fluid is allowed to escape, its motion through the loose sediment creates certain fabric disruptions, such as **dish structure**. Lowe and LoPiccolo (1974) examined dish structure generally, and an associated feature called **pillar structure**. Dishes are concave-upward meniscus-like clayey seams in a sandstone, typically a few centimeters across, presumably segments of originally continuous clay partings (Fig. 5-23d). Pillar structures are vertical, cylindrical columns of homogenized sand, commonly sharply bounded by argillaceous smears; pillars may be very short tubes a millimeter in diameter, or diapirs a meter across and several meters high. Both types of structure seem to result from dewatering of a rapidly deposited bed (they are common in turbidite sands). Dishes form as escaping water is forced to flow laterally beneath a low-permeability lamina (such as a clay parting); dishes may give way upward to pillars, which form as a result of "explosive" dewatering of unconsolidated sand. Pillars, in turn, may grade into convolute laminae (see below) at the top of the bed. Lowe and LoPiccolo observed that dish and pillar structures occur in silt-, sand-, and gravel-size deposits and even in carbonate rocks. Hein and Walker (1982, figs. 6,7) offer some nice illustrations of dishes and pillars in the Cap Enrage Formation in Quebec, Canada.

Sediments with high pore pressures are weak sediments, because the constituent grains are in loose mutual contact; pore pressure reduces the effective weight of the grains so they do not "bear down" upon one another as heavily as they could, and the internal friction of the deposit is less than it would be in the absence of fluid pressure or overpressure. Such a sediment (sometimes said to be **underconsolidated**) may maintain a quasi-stable condition such that, if disturbed even gently, it can spontaneously release its excess fluid. The motion of the fluid momentarily fluidizes the sediment as a whole, which then flows laterally (downslope) or makes some vertical adjustment in response to non-uniform distribution of overburden, perhaps. An overlying layer, if it is coherent, may detach from its overpressured foundation and slip downslope, wrinkling or rolling up like a rug, and forming a **convoluted bed** (Fig. 5-23f), such as Horowitz (1982, p. 173) described, and van Loon (1992, figs. 10, 11, 16). Or it may partially collapse into the fluidized layer beneath, forming billows or even isolated pods totally engulfed in the underlayer. The resulting structures are called **loading structures** or, more picturesquely, **ball-and-pillow structures** (Fig. 5-23g). Convoluted beds form in some shoreface sands and in oversteepened damp aeolian or fluvial sands.

The flow of a buried layer can result also in *diapirism* if overlying layers are not too strong; it is a result of the instability that exists when a low-density layer underlies layers of higher density. Such diapirism gives rise to the **mud lumps** of the Mississippi delta, large (up to a kilometer across) blobs of mud that have risen to the surface from a buried layer. Perhaps better known are the **salt diapirs** (salt domes), occurring abundantly on the Gulf Coast of the United States, where Jurassic evaporitic salt rises through semiconsolidated Cretaceous and Tertiary sediments; and in the Paradox basin of Colorado and Utah, where Pennsylvanian salt rose slowly through Mesozoic sediments, even as these sediments were being deposited. In Iran, there are places where salt diapirs break the surface and salt flows out as "salt glaciers."

Strong currents in the fluid overlying a weak deposit may drag or shear the deposit en masse, causing it to deform plastically. This can create little cuspate fabric discontinuities, called **flame structure** inside the affected layer, that all curl or lean down-current, like blades of grass or tongues of fire in the wind. Many flame structures seem to be just the subjacent counterpart of small, squamose load casts that have been sheared slightly. Currents shearing an unconsolidated cross-laminated bed give rise to **overturned cross-lamination**, wherein the upper edges of the foreset laminae have been recurved into recumbent folds (see Doe and Dott, 1980, pp. 795, 808).

Some Other Structures

In some sense a sedimentary structure, the **mud clast** is a flat pebble, typically a centimeter or two in longest dimension, even as thin as a millimeter or two, and well rounded, occurring in some sandy or silty beds. Mud clasts appear when a thin layer of clay, deposited during some period of slackened currents, is ripped up during an ensuing stronger flow. Clearly, such a clast cannot survive much transport; Smith's (1972) experiments suggest that they can get no farther than a few tens or hundreds of meters from their point of origin before they disintegrate. Some mud clasts are no doubt dried-out, curled-up, small mud-crack polygons, fully detached from the substrate, perhaps rotated or transported just a bit. Mud clasts commonly occur in substantial numbers at or near the base of fluvial cross-bed sets; they also are common in shoreface and tidal-flat deposits, typically associated with flaser bedding. Similar plates, composed of laminated carbonate mud, occur in certain marginal marine limestones or dolomites, and are of similar origin (see Chap. 9). There are other carbonate rocks that contain platy clasts of lime mud, even 10 centimeters or more in diameter, and rounded and imbricated, showing that they were transported, though again probably not more than a few tens of meters. Mud clasts are common also in some turbidites and are not necessarily an indicator of shallow-water or desiccative environments.

Occasionally, in fine-grained lake sediment or deep oceanic ooze, one finds a singular large clast—a pebble or a cobble—and there is evidence that the host sediment near to this object has been disturbed. Clearly, the exotic clast fell through overlying standing water and dropped into the mud; it is called a **dropstone** (see Thomas and Connell, 1985; Gilbert, 1990). It was rafted into deep water, far from any shore, probably, in most cases, by icebergs, then released when the ice melted. Icebergs are pieces of glacial ice that have calved into standing water (a lake or ocean). Dropstones are generally taken as evidence of glaciation, though they can also be dropped from drifting plant matter or by certain animals, such as sea otters, which carry stones around with them to use as anvils for opening shellfish. Instances of logs or coconuts drifted into deep water, then waterlogged, and embedded now in deep marine sediment also have been described.

The **armored mud ball** (Bell, 1940, p. 1) is an unusual aggregate created in certain fluvial and beach environments. Typically it begins as a block of mud that has fallen out of an eroding muddy bank or been torn from a stream bed. It rolls with the current, snowballing more sticky mud and accreting small stones that get pressed into the sticky surface. A heavy coat of armor inhibits further growth. Armored mud balls may roll several kilometers and grow to as much as 1/2 meter in diameter; they may be quite spherical (Fig. 5-23h). Bull (1964, p. A30) described them from semiarid fans in California. Fritz and Harrison (1983) described a very large one that was transported on the surface of a recent volcanic mudflow of Mount St. Helens in the state of Washington. Reports of mud balls from ancient rocks are rare, but Little (1982) found a Jurassic example.

Some rock surfaces bear forms made by dissolution or sublimation to a fluid in contact. **Scallop** surfaces on the walls of ice caves and limestone caves, or on the undersides of river ice, consist of sharp, asymmetrical peaks and spoon-shaped troughs that resemble a field of short-crested ripples. The peaks lean down-current; peak-to-peak distance is typically a few centimeters, though Goodchild and Ford (1971, p. 52) indicated chord lengths of 2 millimeters to 2 meters. Forms with longer crests are called **flutes** (unfortunately). Scallops and flutes form on surfaces that are undergoing dissolution or ablation to moving water or air. Flume experiments of Curl (1966), Goodchild and Ford (1971), and Blumberg and Curl (1974) indicate that bedform length (chord) varies with velocity of the flow, the shorter bedforms created by the faster flows. Gale (1984) used scallops and flutes in the determination of hydraulic conditions in limestone caverns. Other features of limestone surfaces, developed in the subaerial weathering environment where occasional rainwaters wash over a surface, are the **rillenkarren** (and other karren), or **lapies**, the sharp crests and peaks having formed by dissolution (Esteban and Klappa, 1983, figs. 23, 30).

Some sedimentary features have been explained as resulting from ephemeral or transient cementation by ice. Ahlbrandt and Andrews (1978) described snow buried in aeolian dunes, and sedimentary structures created when the snow melts. Dillon and Conover (1965) described blocks of ice-cemented sand on a Rhode Island beach in winter. Laminae within many of the blocks showed contortions (perhaps a cryoturbation). Illich et al. (1972) suggested that blocks of cross-laminated sandstone embedded in horizontal-laminated sandstone in the Pilcher Quartzite

(Precambrian) in Montana might have been ice-cemented blocks. Presumably these blocks, in order to have maintained their shape and internal structure, would have to have been buried (in unfrozen sand) before they thawed.

Paleohydraulics

Paleohydraulics attempts to reconstruct the properties of hydraulic systems of the past—the speed, depth, viscosity and density, direction, duration, and variability of a flow. The success of such endeavors depends on the ability of sedimentary materials to respond to their hydraulic environment and on preservability of this response. It also depends on the sedimentologist's ability to know what gets recorded—the extreme flow, the average flow, or the last state of flow, and how this relates to the hydraulic environment as a whole. We know from common experience that stronger currents are required to transport the larger or heavier grains, and we have attempted to quantify the relationship between mean grain size and current strength (Chap. 3). We know that quiet water gives rise to sediment fabrics that differ from fabrics of sediments deposited in flowing water and that mud-free sediments are products of stronger currents than associated muddy sediments. We have already examined some of the relationships that have been established between flow conditions and bedform geometry, that flowing water creates bedforms that are different from those created by the wind, and that shallow flows and deep flows, gentle and strong, steady or intermittent, make their different marks on the deposit, and that sedimentary structures commonly record the direction of the flow.

A typical natural stream maintains essentially constant velocity from source to mouth (contrary to popular myth), indicating that channel size, roughness, and gradient are in balance—the river maintains a certain long-term equilibrium and is said to be "at grade" or "in regime." The regime concept implies that any segment of a stream is adjusted to its other segments. Thus, when a reach of a natural stream at grade is artificially altered, as in stream channelization projects, there usually are unwanted and perhaps unforeseen changes upstream and downstream of the project reach. Discharge changes from reach to reach, in most natural streams increasing greatly from source to mouth. Channel size (that is, width and depth), gra-

dient, and roughness (including sinuosity) are tuned to discharge. Numerous empirical relationships involving channel width, depth, slope, discharge, and sediment characteristics have been established through measurements of modern rivers (Schumm, 1968 and 1972; Leopold et al., 1964). Schumm (1960a; 1960b; 1963) suggested that the channel width, channel depth, and sinuosity of modern rivers are related to percentage of silt and clay on the bed and banks, and that mean discharge, channel slope, and meander length also are related to channel size. Regime equations, which purport to describe these relationships in a quantitative manner for streams (or channel reaches) in general, typically are regression lines having the form $x = kQ^b$, where x refers to width or depth or gradient or sinuosity, Q is discharge, and k and b are empirically determined constants. Their graphs are straight lines on log-log plots, but their simple form does not reflect the considerable degree of scatter of the data that they are intended to describe and summarize. Of course, discharge changes from time to time in all natural rivers (mostly with the seasons), and grade or regime must be considered as a time-averaged condition. Engineers successfully apply regime equations to the design of canals (which are in practically all respects simpler than natural streams), and attempts have been made to apply the regime concept to fluvial paleohydraulics. The regime equations (all purely empirical), coupled with the relationships between bedform and hydraulics and between grain size and hydraulics, as revealed by flume studies, are the basis for estimating the hydraulics and channel characteristics of ancient fluvial systems.

Southard and Boguchwal's (1990) diagrams relating bedforms to mean flow velocity, flow depth, and grain size seem to be of great value in paleohydraulics. For an observed grain size and bedform (or related sedimentary structure), a corresponding range of mean flow velocity, or ratio of velocity to depth, is obtained. With flow velocity obtained from Southard's diagrams and an estimate of bed roughness based on sediment texture and bedforms, some indication of flow depth and gradient emerges from the Manning equation or other Chezy-type formula.

In paleohydraulics analysis of ancient fluvial deposits, one of the most difficult quantities to determine is the size (width and depth) of the channel (Ethridge and Schumm, 1978, p. 705). Depth can be viewed as among the most fundamental properties of a river. We have already pointed out that water depth does not much influence the types of bedforms that appear. Allen (1967, p. 441) suggested that relationships between water depth and bedform *height* may be of some value in certain circumstances, but that the relations are noisy and the results accompanied by much uncertainty. Allen (1968a, fig. 11) determined that the *wavelength* (or some other characteristic length) of dunes, antidunes, and river meanders is related to depth of flow, but these quantities typically are difficult or impossible to measure in ancient deposits. Measurements made of modern rivers indicate quantitative relationships between meander size and channel size; thus the radius of scrolls in an ancient fluvial deposit may give an estimate of channel size. (Scroll radius is inaccessible in most ancient fluvial deposits, however.) One ought to be able to obtain depth (and width) simply by measuring the channel deposit, and this is sometimes possible. But channel deposits ordinarily are much larger than the active channel that made them, because rivers aggrade and channels move laterally. The normal-stage channel depth of a meandering river channel is probably given with reasonable accuracy by (compaction-corrected) thickness of the lateral accretion deposit of point bars (see Chap. 8). These deposits may consist of a single set of inclined laminae that has an erosional base and is size-graded (fining upward); it is the epsilon cross-bed of Allen (1963, p. 102). Bridge and Deimer (1983, p. 614) estimated ancient channel cross-section geometries on the basis of thickness of lateral-accretion deposits. Angle of inclination of the internal laminae is closely related to channel width; Allen (1966, p. 166) indicated that horizontal component of down-dip length of the inclined lamina is about two-thirds the channel width. Leeder (1973, p. 268) found that, for high-sinuosity streams, width and depth (in meters) are related by $W = 6.8D^{1.54}$, obtained from many measurements of modern streams, but there is much scatter in Leeder's plots. Meander loops may get cut off from the active channel; isolated channel segments (oxbows) fill with a muddy deposit whose dimensions are generally an accurate reflection of channel size.

Bridge (1985) searched for criteria for determining ancient channel patterns (whether meandering or braided) and considered the following to be the most useful: (1) volume of channel fills relative to lateral accretion deposits, the ratio increasing with increase in degree of braiding; (2) mean grain size of the channel fill relative to that of the lateral accretion deposit, the ratio decreasing with increasing channel sinuosity;

(3) paleocurrent variance, which may reflect sinuosity; and (4) paleohydraulics reconstructions, based on texture and sedimentary structure, the results of which can then be used with empirical relations such as the regime equations. Bridge advised that there is a continuum between meandering and braided rivers, the geometry of any particular reach controlled mainly by the flow characteristics obtaining during periods of high discharge.

Malde (1968) studied an ancient catastrophic flood in northern Utah and southern Idaho, possibly 30,000 years ago, due to rapid discharge of some 1760 cubic kilometers of Lake Bonneville water into the Snake River. On the basis of water levels in constrictions, peak discharge was calculated to be about 4×10^5 m^3/sec, about three times the mean discharge of the Amazon. The so-called "channeled scabland" of eastern Washington state, described in great detail by J. Harlan Bretz in the 1920s, and shown by him to be a result of catastrophic flooding related to deglaciation, was re-examined from the hydrodynamicist's point of view by Baker (1973). Baker documented, on the basis of high-water marks, channel sizes and slopes, large-scale bedforms, and other evidence, maximum water discharges exceeding 750 million cubic feet per second (21×10^6 m^3/sec) and sediment discharges of a ton per second per foot width of channel (about 3000 kg/sec/m). Baker described giant ripple-like bedforms composed of coarse gravels, with wavelengths up to 200 meters and heights of 10 meters. Estimated Froude numbers of the flows that made the bedforms are between 0.5 and 0.9, which is about the range in which dune and antidune bedforms develop (in much finer sediments) in flumes.

In ancient coastal deposits, tidal range and wave climate can be estimated from the distribution and dimensions of bedforms and sedimentary bodies and the sizes of sedimentary particles. Again the studies of modern systems and the results of flume and wave-tank experiments come into play. Klein (1971, p. 2589) suggested that the thickness of fining-upward successions in tidal deposits gives tidal range. Ancient wave conditions have been estimated from chord lengths of wave-generated ripples (see Komar, 1974; P.A. Allen, 1981 and 1984; Clifton and Dingler, 1984; Dupre, 1984).

Gross stratigraphic relationships are useful in estimating depth of standing water. Stratigraphic distance below the top of a set of offlapping marine strata, after compaction correction, correctly gives water depth if it can be demonstrated that the top of the set formed at or near sea level. Examples of studies in which marine water depths were determined in this way are Newell et al.'s (1953) analysis of the Capitan reef (Permian) and Delaware basin in New Mexico and Texas, Swann et al.'s (1965) of the Borden delta (Mississippian) in Illinois, and Campbell's (1971) of the Gallup (Cretaceous) beach sandstone in New Mexico.

Biogenic Structures; Ichnology

Biogenic sedimentary structures are records of the *activities* of organisms or some aspect of their behavior (ethology). They are evidences of ancient life and thus are fossils. But they differ from body fossils in certain respects that give them special utilities—most have long time ranges and narrow facies ranges; they are almost never transported (or transportable); they can be formed by organisms not disposed to make body fossils and commonly occur in rocks lacking body fossils (what are the oldest rocks in which they occur?). Whereas carcasses are scavenged or bacterially degraded, trails and burrows ordinarily are not, though burrows commonly are overprinted by other burrows. As most of them are disruptions of the texture or structure of a bed, they are inherently as preservable as the bed that contains them; diagenesis and weathering accentuate them. Insofar as behavior and activity are molded by environment, biogenic structures reflect the nature of depositional environments. Trace fossils are the products of *living* organisms; body fossils are the products of *dead* organisms (included among the body fossils are dead parts that living organisms have sloughed away).

Classification and Origins

Trace fossils are classified binomially (genus and species), following procedures similar to those applied to body fossils (though there are no families, orders, or phyla). Some genus names, such as *Corophioides* or *Arenicolites*, imply the maker of the burrow (*Corophium* sp. and *Arenicola* sp., respectively, a crustacean and a worm), but ichnologists today generally avoid classification based on the maker because, in many or most cases, it is not known. Moreover, a given organism can create several different traces, and different organisms probably can create the same trace. No doubt some organisms use the burrows made and sub-

sequently abandoned by, some other organism, perhaps even modifying them to their own purposes. Further complicating the classification of traces is the plasticity of form that many of them display, and the possibility of some burrows being incomplete. Is a long, straight, vertical tube or a J-shaped tube but an incomplete U-shaped tube, the maker having abandoned the project in mid-construction? Many or most burrows are made by a single organism; but some are the collective effort of large communities (consider an ant gallery). Modern callianassid shrimps are very active on some tidal flats in the tropics, communities of them building mounds of sedimentary material excavated from the burrows beneath; these mounds are rather like large (up to 1/2 meter high) anthills.

Much of the original work on biogenic structures was done by German researchers—R. Richter, W. Hantzchel, O. Abel, and A. Seilacher—and the term *Lebensspuren* (a German word meaning life traces) has come to be applied to them. They are also known as *trace fossils* or as *ichnofossils*, and the science of trace fossils is called *ichnology*, or *palichnology*, to make clear that ancient traces are being studied, rather than modern ones. Seilacher (1953) classified trace fossils according to the essential life activity that was involved in their creation (the last entry in this list was added by Simpson, 1975, p. 49):

Cubichnia—resting traces, impressions of organisms resting for a time on a soft (deformable) substrate.
Domichnia—dwelling structures, burrows and borings made by organisms as a shelter or place to live.
Fodinichnia—feeding traces, marks made as an animal searches for food.
Pascichnia—grazing traces, marks made as an animal exploits the nutrient content of its substrate.
Repichnia—crawling traces, tracks or trails of animals in locomotion.
Fugichnia—escape traces, structures made as an animal, suddenly buried beneath new sediment, attempts to regain the surface.

This classification implies that one can know the intentions or purposes of the organism that created the trace or structure at hand, and this is probably not always possible. It should be clear also that there are considerable overlapping regions (as Seilacher recognized; Fig. 5-24) in this ethological classification, that grazing, for example, involves locomotion, or that the place where an animal dwells is, in many cases, also where it eats. The classification does not address traces made during or for the purpose of laying eggs;

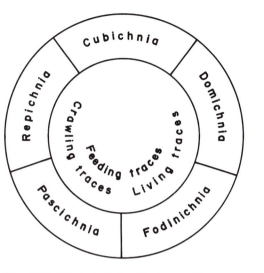

Fig. 5-24. Ethological classification of biogenic structures. (According to Seilacher, 1953, fig. 6.)

nor does it consider courting, camouflage, play or struggle—or animal "artifacts," such as middens and hoards (some animals collect and store things), excretions (quite common), regurgitations (very rare in the rock record), and gastroliths.

Raup and Seilacher (1969) presented an elegant computer simulation of certain fossil grazing traces, based on an earlier analysis of foraging behavior by Richter. Richter examined the trace fossil *Helminthoida labyrinthica* and determined that the pattern of this trace was controlled by three kinds of reactions in the worm that made it: (1) *strophotaxis*, a propensity of the animal to make 180° turns; (2) *phobotaxis*, an aversion to crossing its own track; and (3) *thigmotaxis*, a compulsion to keep close by its own previous track (see also Seilacher, 1967b). Coding these rules for computer and allowing certain variations in them, and a certain random element, Raup and Seilacher were able to simulate several different "species" of naturally occurring grazing trace (Fig. 5-25).

Some traces are formed within beds (by an infauna) and are called ***endichnia***; ***epichnia*** are formed on the upper surface of a bed and are mainly the "tracks and trails" of animals in locomotion. Consider a track or trail made by an animal such as a worm, a trilobite, or a bird on the surface of a bed of mud; such a structure was impressed into the mud, and the trail is an epichnion. If, later, the mud bed should be covered by a

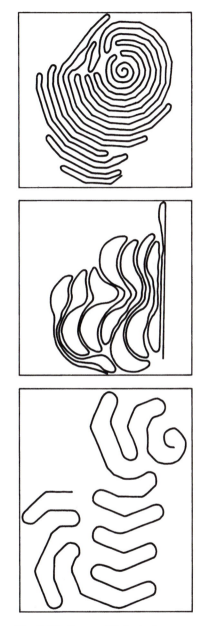

Fig. 5-25. Raup and Seilacher's computer-simulated trace fossils (1969, figs. 1, 2, 3, with permission of AAAS).

bed of sand, this bed will bear the "negative" of the trail (it is a cast). Because it is borne on the bottom of a bed, it is referred to as a *hyporelief*; its counterpart, the *epirelief*, is in the top of the underlying bed of mud. Hyporeliefs and epireliefs may be *convex* or *concave*, depending on whether they protrude from or intrude into the surface under examination. Thus, a concave epirelief has its counterpart in the overlying bed as a convex hyporelief. Some workers are confused about what is a mold and what is a cast. A mold comes first; it is a vessel (either convex or concave) that receives a cast. The mold determines the shape of the cast.

Some trace fossils consist of tubes (burrow passages) excavated into a substrate, then actively backfilled by the burrow maker itself. Some tubes are closely associated or interconnected with clay smears or webs splayed out from the tube in arcuate or spiral patterns caused by repeated lateral displacements of the tube. These features are called *spreiten* (sing., spreite). Most spreiten record the activities of infaunal organisms mining the substrate for nutrients. Some spreiten probably reflect increasing tube length to accommodate normal growth of the animal. Other spreiten patterns reflect the influence of sedimentation or erosion of the substrate of an organism during its lifetime. In *Diplocraterion*, for example, which is a U-shaped tube, arcuate spreiten above and below the curved segment of the tube (Fig. 5-26) indicate that the organism adjusted the tube upward or downward, presumably to maintain a favorite tube length against the disruptions of erosion and sedimentation. Spreiten formed by upward adjustments to sedimentation are said to be *retrusive*; those formed during erosion of the substrate are *protrusive*.

Some tubes have "haloes" that reflect a chemical environment altered locally by the burrow-maker (but some haloes can form diagenetically). *Paramoudras* are elongate masses of chert that occur in some chalk formations, and appear to have formed around large (or long) vertical burrows (see Clayton, 1986).

Besides those trace fossils made in unconsolidated sediment, there are others scraped out of hard rock, shells, bones, or wood (Fig. 5-27). These are called *borings*. (One should keep in mind that borings can be made in rocks formed eons earlier.) Predation marks are a related kind of trace fossil, made on living skeletal parts; also, in some fossil vertebrate skeletons, bone pathologies, such as osteoarthritis, are in evidence and might be considered as a trace fossil of a

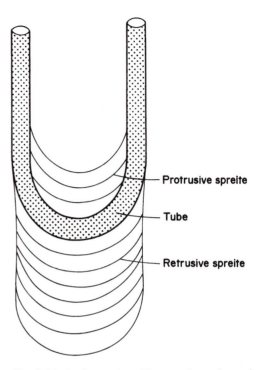

Fig. 5-26. *Diplocraterion* with protrusive and retrusive spreiten.

bacterium or virus. Certain sponges, barnacles, worms, gastropods, bivalves, and even fungi and algae are borers. These and other similarly adapted organisms are responsible for the destruction of much skeletal matter in the depositional environment and the production of much lime mud. Algal and fungal borings are very small (commonly referred to as *microborings*), and are a cause of *micritization* of skeletal and other carbonate grains (see Chap. 9). Micritization is clearly observed with the petrographic microscope, but the scanning electron microscope (SEM) is much more revealing. Golubic et al. (1975) presented numerous startling SEM photomicrographs of these microborings (see also Golubic et al., 1984).

Some trace fossils are made by terrestrial plants. Lichens, those lobate encrusting alga-fungus symbioses that dwell on rock surfaces, actually decompose and consume their rocky substrate. Especially on limestone surfaces in semiarid climates, there are tiny pits and anfractuous rills or creases where lichens have been. They have been observed preserved on ancient rock surfaces. They are generally considered, not with-

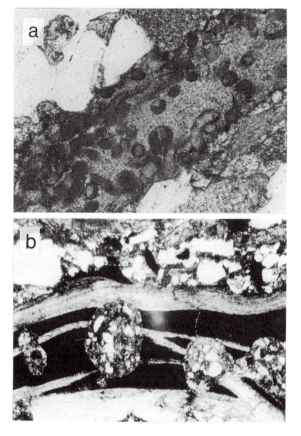

Fig. 5-27. Borings in (**a**) echinoderm plate; (**b**) mollusc shell.

out justification, to be weathering phenomena, but they are in fact trace fossils of lichens. ***Rhizoliths***, ***dikaka***, and ***alveolar structures*** are structures made by the roots of small plants, such as grasses. Klappa (1980, p. 615) classified rhizoliths into root molds, casts, and tubules, rhizoconcretions, and petrifactions.

Trace Fossils and Depositional Environment

Trace fossils can be useful in determining the depth of water and related characteristics of the environment in which a sediment was deposited. One presumes that the microborings of algae can be made only in shallow, well-lighted waters. Some petrographers have noted that carbonate skeletal grains deposited in shallow water are more heavily micritized than similar grains in deeper water. But some microborings are

made by fungi, which are not light-dependent. Weimer and Hoyt (1964) pointed out the utility of *Ophiomorpha* as an indicator of marine littoral environments. Seilacher (1967a) identified a series of five marine ichno-assemblages, each characteristic of a particular depth interval (Fig. 5-28). This sequence reflects a behavioral trend in organisms occupying different depths. In the very shallow environments (beach, intertidal), organisms build protective burrows, which typically are vertical tubes, commonly quite deeply penetrating. These burrows help to insulate their makers from the stress of fluctuating temperatures and tidal currents and hide them from predators. In the shallow waters turbulence keeps food particles in suspension, so many animals are filter feeders, and their burrow shapes reflect this mode of life. Moreover, there are rapid sedimentation and erosion events in these shallow environments, to which burrowing animals must respond. Examples are *Skolithos*, *Arenicolites*, *Diplocraterion*, and *Ophiomorpha*. Some of these burrows, constructed in loose or shifting sands, are lined with mucus or pellets to prevent them from caving in. On rocky coasts boulders, pebbles, and shells are bored. In deeper waters, food particles settle out and become part of the substrate. Here, burrows are horizontal and branching or wending, as organisms confront generalized volumes of sediment with admixed nutrients or graze the surface of a bed or some particularly nutrient-rich layer just below the surface. Animals such as trilobites or snails that crawl about on the sediment surface make trails that can be obliterated in the presence of strong currents. Burrows characteristic of this environment include *Cruziana* and *Zoophycos*. In the very deepest waters, patterned grazing tracks, such as *Helminthoida* and *Nereites*, formed at the sediment-water interface, are the rule. Here food drizzles down from the overlying waters; it is scarce and valuable; most of it is quite fine-grained. Rates of inorganic sedimentation are very low, so that organic detritus rarely gets buried; bottom-dwelling animals consume it long before that could happen. They work the sediment surface, systematically grazing their allotment and not bothering to probe the barren deeper layers.

Ekdale (1988) warned us of "pitfalls" in ichno-assemblage-based paleobathymetry. Trace-making organisms do not measure the depth of overlying water; trace fossils indicate water depth only insofar as depth controls or influences other properties of the environment, such as light quality, salinity, oxygen content,

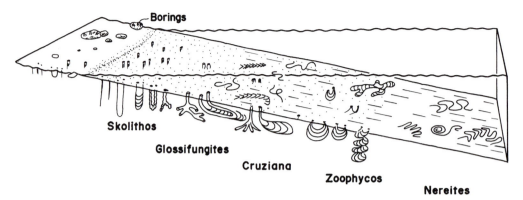

Fig. 5-28. Seilacher's depth-dependent ichno-assemblage succession. (After Seilacher, 1967a, figs. 2, 3.)

nutrient quality and quantity, substrate character, sedimentation rate, and variability of the environment. In an ideal world, these qualities might vary progressively or monotonically with water depth, but (Ekdale's point) the correspondences are not so simple or universal in the real world.

There are many examples of directionality in biogenic sedimentary structures. Infaunal bivalves of tidal flats appear to orient their incurrent and excurrent siphons in line with the flow. U-shaped burrows, it appears, commonly are oriented with the excurrent tube "downstream" of the incurrent opening. Ghost crabs on modern beaches excavate short burrows that incline away from the ocean or away from the prevailing wind. Orientations of resting and grazing traces may be influenced by currents. Animal trails, such as *Cruziana* (made by trilobites), occasionally tell the story of the trail maker's struggle against a bottom current; moreover, a current might degrade a trail in a way that indicates current strength and direction.

Bioturbation is the reworking or mixing of sediment by organisms, which partially or completely destroys the sedimentary fabric and structure that may have been present before. Extensive bioturbation in a bed indicates vigorous activity of organisms and suggests high trophic level or slow sedimentation. Total absence of bioturbation may indicate a poisonous or anoxic environment inimical to life, a sediment devoid of nutrients, or an environment in which the mechanical energy level is too high to permit habitation by organisms or that intermittently reworks a bioturbated bed, restoring (inorganic) sedimentary fabric and structure. Droser and Bottjer (1986; 1990; 1991) made charts showing various degrees of bioturbation (Fig. 5-29) to be used as standards in stratigraphic descriptions.

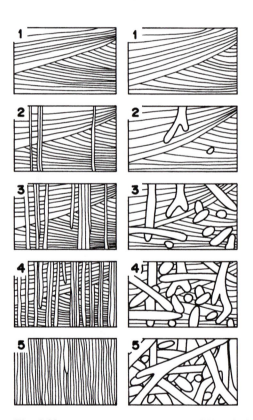

Fig. 5-29. Reference charts for degree of bioturbation. (After Droser and Bottjer, 1990, fig. 1, with permission of SEPM.)

Vertical variations in the degree or extent of bioturbation within a single bed or sequence of beds are related in part to biological factors and in part to sedimentological factors. All organisms have limits on the depths to which they are willing or able to penetrate a substrate. Some burrowing organisms are rather sedentary, excavating a single small burrow in which they remain for long periods; other organisms pursue a more active existence, churning up large volumes of sediment. Bioturbation ordinarily begins at the top of a freshly deposited bed, and a "bioturbation front" progresses slowly downward; the degree of bioturbation naturally dwindles toward such a front. Should the organisms sense that a new bed has just been deposited above the one that they are currently exploiting, they may feel the need to retreat upward, perhaps frantically, into or through this new bed, leaving the old bed not fully bioturbated. Many organisms are substrate-specific and may work downward to the bottom of a sand bed, for example, and totally avoid an underlying bed of clay. Other organisms burrow vertically through a sand bed in search of some clay layer to exploit; once they encounter such a seam, they stay with it, much as a miner might "high-grade" a gold vein or coal seam.

Bromley and Ekdale (1986) described *tiering* of burrows in sediment layers, due to different ichnotaxa living at different levels below the sediment-water interface; this they compared to the multilevel canopy of the tropical rain forest. Preservation within the several tiers varies. The shallow burrowers occupy the soupy, uncompacted upper levels of the bed and are destined to "fuzzy" preservation. The deeper burrowers reside in the firmer lower levels of the bed, where compaction is already partly "over with." These deeper burrowers cross-cut and overprint the shallow burrows, and they penetrate to levels beyond the bustle and fury of the shallow burrowers; in the deep tiers, burrows are sparser. The deep tier is, therefore, usually the best preserved, least compacted, and most prominently displayed.

Bromley and Ekdale (1984) showed that the trace fossil *Chondrites* is an index of anoxia (low oxygen level) in the sediment containing this burrow. Savrda and Bottjer (1986) expanded on this theme, suggesting that the maker of the trace *Planolites* is among the least tolerant of low oxygen levels and that the *Thalassinoides* maker is somewhat more tolerant, followed by *Zoophycos*, then *Chondrites*. They also suggested that oxygen-starved organisms are smaller and make burrow tubes that are shorter and of smaller diameter.

Sediments deposited in anaerobic environments (dissolved oxygen < 0.1 ml/l) are laminated, that is, not bioturbated at all. Pratt (1984, p. 1151) found a relationship between the composition of organic matter and the degree of bioturbation in a core of the Greenhorn Limestone (Cretaceous) in central Colorado. Organic matter in the most highly bioturbated layers showed the lowest hydrogen indexes and highest oxygen indexes (see Chap. 10), suggesting both that burrowing activities were more active in well-oxygenated bottom waters than in dysaerobic waters and that burrowing tends to aerate and irrigate newly deposited sediments and to promote oxidation of incorporated organic matter. Pratt presented evidence that stratigraphic variations in organic matter composition and in degree of bioturbation were caused by climatic and hydrologic factors affecting the rather deep marine waters under which the Greenhorn was deposited.

Burrows and bioturbation have been used to estimate relative rates of sedimentation; in a lithically homogeneous sequence, variations in intensity of bioturbation may reflect variations in rates of sedimentation. The escape burrow is almost certainly an indication that the bed containing it was deposited almost instantaneously. (Escape burrows commonly occur in tabular cross-bed sets, such as those created by a tidal current; the bed grows laterally rather than vertically and can advance rapidly over an unsuspecting creature even in the absence of sensible change in hydraulics.) Burrows truncated at a bedding plane indicate that erosion occurred before deposition of the overlying layer.

Flattened horizontal tubes or corrugated vertical tubes indicate the degree of compaction of the sediment that contains them. The depth and condition of a footprint indicates the original mechanical strength (consistency) of a substrate (and weight of the maker); burrows lined with pellets or mucus were probably excavated in soft or shifting sediment; hard substrates are bored rather than burrowed, though the differences between a boring and a burrow in ancient rocks are not always clear. Burrows probably indicate current directions, and even strengths of currents, to a degree not fully appreciated. Bioturbation may dilate a sediment, influencing its final porosity; the clay smears of burrow spreiten reduce final permeability. Bioturbation might increase the vertical permeability of a sandstone body by breaking up clay partings or flasers.

Chamberlain (1975) described some of the abundant and diverse organisms that inhabit non-marine aquatic environments and their habits, and traces, including the chambers they make for pupating, brooding, and hibernation or estivation (see also Bracken and Picard, 1984). In some lakes, the long, millimeter-thin, tangled burrows of *Tubifex* worms have invaded the soft muds (Zahner, 1968). Retallack (1984) described trace fossils made in an Oligocene paleosol by beetles and bees. Burrows and bioturbation are common in aeolian sands. The insects and other arthropods, the snakes and lizards, birds and rodents that inhabit the desert burrow into the sand to obtain moisture and food or to protect themselves from heat, cold, and predators. McKee and Bigarella (1979b, p. 193) observed that tracks of animals are common in the (aeolian) Coconino Sandstone (Permian) in Arizona, including sharply defined, well-preserved tracks of reptiles, mammals, insects, scorpions, and millipedes.

Stromatolites

An artifact of the growth of certain primitive organisms is a rock called the ***stromatolite*** (Fig. 5-30), containing laminae or sets of laminae deposited by films or thin mats of cyanobacteria (formerly referred to as blue-green algae) or other bacteria (the term is also used to designate the sedimentary structure itself). The deposit is made either by active biological secretion of carbonate or (rarely) some other material, or by passive entrapment or adhesion of small suspended detritus drifting by.

Algal stromatolites commonly begin as essentially flat layers, but some tend to develop local irregularities that are self-perpetuating or self-amplifying, so that small wrinkles gradually become prominent domes or fingers having heights of several centimeters or tens of centimeters. The geometry of stromatolites seems to reflect certain aspects of the depositional environment, such as water depth and strength of currents, tide range, and frequency of desiccation; possibly the taxonomy of the organisms involved in their construction has some influence. Associated faunas may modify or limit the growth and development of stromatolites.

Stromatolites are generally considered to be products of shallow marine environments, especially intertidal; indeed, most of today's active stromatolites are of this environment. Shark Bay in Western Australia—a shallow, hypersaline, tide-influenced lagoon—is the most famous example of a modern stromatolite-making environment. Living intertidal stromatolites have been recently described also by Reid and Browne (1991) in the

Fig. 5-30. Stromatolites.

Bahamas. These are laminated lithified domes up to about a meter high, composed of carbonate peloids and small skeletal fragments held together by acicular aragonite cement and by high-magnesium-calcite-encrusted algal filaments. Lamination is manifested by differential cementation, but lamination is largely obliterated in the deep interiors of the stromatolite domes by intensive boring by sponges, worms, and bivalves.

If algae (or cyanobacteria) are responsible for the structures, then lighted (photic zone) conditions are mandated. However, some ancient stromatolites are thought to have formed in the lower fringes of the photic zone (Playford and Cockbain, 1969; Donaldson, 1976), and the so-called manganese nodules of the deep-sea floor also might be stromatolitic, formed by certain non-photosynthesizing bacteria (see Chap. 10). Some of the largest, most diverse, and most complex stromatolites occur in Precambrian rocks. An example is those in the Proterozoic Pethei Formation, cropping out around Great Slave Lake in Canada's Northwest Territories, that grew in deep water as well as in shallow.

Domal stromatolites exposed at Great Slave Lake are directional, the domes elongated in plan, parallel to wave-generated or tidal currents (Hoffman, 1974). Individual domes grew faster into unidirectional or strongly asymmetrical currents, appearing to lean into these currents. (Stromatolites that appear to lean may also indicate that they grew on a sloping substrate, like trees on a hillside.) Hoffman (1969) showed that the very strong preferred orientation of these structures is coincident with current directions inferred from ori-

entations of associated ripples. Eriksson (1977, p. 234) observed similar elongation and oversteepening in stromatolites of the Lyttleton Formation (Precambrian) in South Africa. Some of the modern stromatolites of Shark Bay in Western Australia are oriented perpendicular to the strand. One suspects that stromatolite orientation is a self-amplifying process— once a stromatolite becomes elongate in a current, it becomes a current-regimenting "vane."

The onkoids (unattached spheroidal stromatolites) with concentric layering (Fig. 5-30) must "roll over" now and again if they are to grow. (They say that a rolling stone gathers no moss.) Wave-generated or tidal currents probably are important, but fish and other animals also may play a role, accidentally or intentionally rolling the onkoids as they probe the sediment surface for food or shelter. Computer simulations of stromatolite growth (Fig. 5-31) might be useful in helping us to understand the complex interactions that surely are involved in stromatolite growth and development.

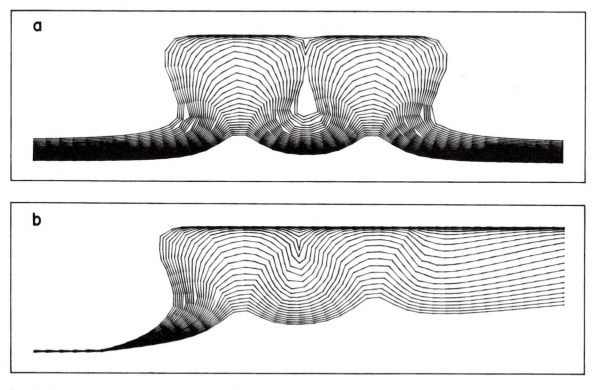

Fig. 5-31. Computer-simulated stromatolites: **(a)** flat bottom; **(b)** sloping bottom.

There are well-developed stromatolites and onkoids in some lakes, ancient and modern. Dean and Fouch (1983, p. 114ff) present numerous illustrations of these. Many lacustrine environments lack alga grazers, and this is conducive to stromatolite growth in such environments.

Stromatolites appear to have been far more abundant in the distant past than they are today. Some consider that the decline of the stromatolites is due to alga-grazing herbivores, which, once evolved, began cropping stromatolitic algae faster than these algae could build their impressive laminated structures (Garrett, 1970; Awramik, 1971). (In the Recent stromatolite-making environment of Shark Bay, high salinity keeps the grazers away.) Pratt (1982b) questioned Garrett's hypothesis, suggesting that stromatolites are more common in Phanerozoic rocks than generally believed and that these often coexisted with grazing and burrowing organisms. Nonetheless, Phanerozoic stromatolites are less abundant, less diverse, and less complex geometrically than their Precambrian cousins. This, Pratt suggested, is due to stratigraphic "dilution" of stromatolitic rocks by other carbonate rocks, competition for space and habitat by the higher organisms, and the "fouling" of substrates by large quantities of mobile skeletal and peloid debris that make for less favorable conditions for the establishment and growth of stromatolites.

Classification

A classification of stromatolites (Fig. 5-32), based on their geometries, was developed by Logan et al. (1964). They recognized that stromatolites are built up of convex layers stacked one upon another. Some are vertically stacked into separate or isolated fingers or domes, referred to as *stacked hemispheroids* (SH); in some stromatolites, the layers bridge the gaps between adjacent domes, producing *laterally linked hemispheroids* (LLH). Other stromatolites are not attached to a substrate, and consist of layers more or less spheroidally arranged; these objects are called onkoids or *spheroidal structures* (SS). Among the SH types are those in which all layers in the stack have about the same curvature—*mode C*, or constant-curvature stacks; and those in which radius of curvature of higher laminae in the stack significantly exceeds that of laminae at the bottom of the stack—*mode V*, or variable-curvature stacks. The LLH structures may be closely spaced—*mode C* lateral linkage; or widely

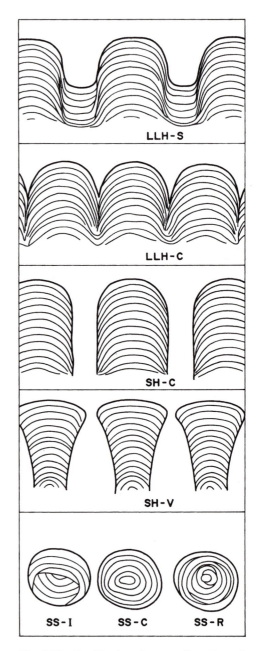

Fig. 5-32. Classification of stromatolites. (According to Logan et al., 1964.)

spaced—*mode S*. Among the onkoids or SS stromat-olites are those consisting of randomly stacked hem-ispheroids—*mode R*; those consisting of concen-trically stacked spheroids—*mode C*; and the bipolar or inverted types consisting of a set of convex-down-ward hemispheroids overlain and overlapped by a set of convex-upward hemispheroids—*mode I*. Aitken (1967, p. 1170) proposed that the (planar) cryptalga-laminate structure (see below) be designated in this system as *mode P*.

Also, there are compound forms, wherein a stro-matolite has a macrostructure of one type and a fine structure of another type, or the stromatolite displays some vertical transformation from one type to another (Fig. 5-33). For example, it is not unusual for a large SH dome to be built up of small LLH fingers (desig-nated SH/LLH), or for a community of upward-ex-panding SH-V stromatolites to merge at some level into LLH-C forms above (SH→LLH), or for LLH stromatolites to bifurcate upward into discrete SH fin-gers (LLH→SH).

Aitken (1967), pointing out the need for distinction between the skeletal remains of calcareous algae and those deposits formed by or mediated by non-calcar-eous (non-skeletal) algae, proposed the term *cryptal-gal* for the latter. A cryptalgal sedimentary rock or rock structure is one that is ''believed to originate through the sediment-binding and/or carbonate-pre-cipitating activities of non-skeletal algae'' (Aitken, 1967, p. 1163). Aitken recognized that some cryptal-gal structures are successions of essentially flat lami-nae, called *cryptalgalaminate* structures. Aitken of-fered numerous criteria for distinguishing cryptalgalaminate carbonate rocks from sediments whose laminae are of fluid dynamic or chemical ori-gin. Among these criteria are (1) small-scale discon-formities; (2) bubbles or little lenticular openings (''birds-eyes'' or fenestrae), typically cemented, and tiny burrows and desiccation cracks; (3) thin breccias composed of fragments of single laminae; (4) traces of poorly preserved algal filaments; and (5) close as-sociation with algal stromatolites. But by no means are all these criteria necessarily fulfilled in a particular cryptalgalaminate succession.

Hoffman (1976) identified six algal mat types in Shark Bay. Three of these types—*smooth*, *pustular*, and *colloform*, can grow into domal and digitate forms; three other types—*blister*, *tufted*, and *gelati-nous*—form only flat laminae (stratiform mat depos-its). Each mat type comprises a distinct community of

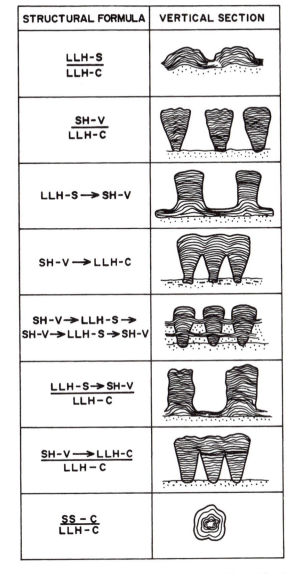

STRUCTURAL FORMULA	VERTICAL SECTION
LLH-S / LLH-C	
SH-V / LLH-C	
LLH-S → SH-V	
SH-V → LLH-C	
SH-V → LLH-S → / SH-V → LLH-S → SH-V	
LLH-S → SH-V / LLH-C	
SH-V → LLH-C / LLH-C	
SS - C / LLH-C	

Fig. 5-33. Some compound forms of stromatolites. (After Lo-gan et al., 1964, fig. 5, with permission of The University of Chicago Press.)

alga species, and the distribution of the various algae (and of mat types) is controlled by degree of desic-cation, which varies across the littoral zone. Colloform mats of subtidal places make multiconvex laminae with numerous lenticular openings (fenestrae; see Chap. 11). Strong wave-generated turbulence inhibits the formation of extensive mats but instead stimulates the growth of columns or domes that broaden upward. In lower intertidal belts, smooth mats grow into columnar stromatolites with simple convex laminae. Pustular mats of the upper part of the intertidal zone in Shark Bay make deposits that are not laminated.

Other Algal Structures

Some domal or columnar stromatolites appear not to be laminated but have a spongy structure of little formless clots; these are the *thrombolites* of Aitken (1967, p. 1171). Thrombolites appear to be gradational with the (laminated) stromatolites, and some are hemispheroidal or spheroidal (like onkoids). But others are loaf-shaped bodies with dimensions of several meters or tens of meters. Commonly the clots are mixed with skeletal grains; there may be organic burrows as well, or other openings (fenestrae) created perhaps by entrapped gaseous biochemical products. We examine the matter of little carbonate clots at greater length in Chapter 9.

Some workers distinguish *spongiostromes* from true algal-mat deposits; they are, according to Johnson (1961, p. 210), irregular crusts composed of alternately compact and loose layers of dark granules or peloids, pierced by numerous short, branching algal tubes. They occur in limestones of Devonian to Pennsylvanian age.

6

Composition and Provenance
of Terrigenous Sedimentary Rocks

Sedimentary rocks can and do contain any and all materials present at the earth's surface, including the products of degradation of igneous and metamorphic rocks and of other sedimentary rocks; the solid, liquid, and gaseous products of living systems; atmospheric and oceanic precipitates; and materials that fall to earth from outer space. Composition may be defined at various levels, such as elemental or oxide content, mineral content, or grain types. Each level of description has its uses, and there are certain correspondences, as set forth in so-called *normative mineral calculations*, for example. Rock composition determined remotely from bulk properties such as hardness, sonic velocity, transparency to radiation or nuclear particles, etc., is routine in the petroleum industry. Other technologies have been applied to the remote determination of rock composition on the Martian and other planetary surfaces.

Sedimentary materials are classified fundamentally according to their modes of origin:

Terrigenous materials—the solid particles resulting from the breakdown of older rocks; most such materials form on land. Volcanogenic particles, such as ash, and larger particles formed during explosive volcanism, are included here. One might choose also to include the detritus of terrestrial plants.

Allochemical materials—solid, typically structurally complex particles formed de novo at the depositional site by aggregation, by chemical precipitation of dissolved material, or by biological activities. Fossils or fossil fragments, ooids, and fecal pellets are examples.

Orthochemical materials—mineral precipitates, either crystallized directly on or within the substrate or appearing as suspended flocs, destined for deposition. Biological intermediaries may or may not be involved or required. Included here are evaporite minerals, lime mud, and chemical cements precipitated perhaps long

after deposition of the host aggregate of terrigenous and allochemical materials.

Probably all sediments contain at least small amounts of each of these materials. Rocks in which terrigenous materials are dominant (i.e., that contain more than 50 percent of terrigenous material) are called *terrigenous rocks* (Fig. 6-1). Sandstones and clay mudstones are examples. In an *allochemical rock*, more than 10 percent of the non-terrigenous material is allochemical; otherwise the rock is an *orthochemical rock*. Many or most limestones are allochemical rocks. Some limestones, gypsum, and rock salt (halite) generally are orthochemical. If the quantity of terrigenous material in an allochemical rock exceeds 10 percent, then it is an *impure allochemical rock*; similarly, there are *impure orthochemical rocks*.

This chapter is devoted to the composition and provenance of terrigenous rocks. The essential components of allochemical and orthochemical rocks form by very different processes, and to a large degree, terrigenous minerals and chemical minerals are mutually exclusive sets. Description of the chemical rocks is reserved to Chapters 9 and 10.

The composition of detritus initially reflects the composition of the source rock, but sedimentary processes cause materials of different sources to mix together, or materials of a single source to unmix, and materials of different sources to look more and more similar. *Sedimentary differentiation* is an accumulative segregation or separation of components resulting from the varying response of these different components to sedimentary processes acting repetitively. It is a gradual unmixing of homogenized or linked rock materials and a gradual selective or differential de-

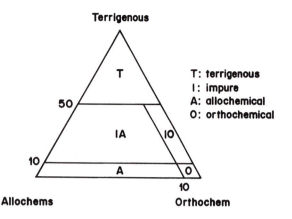

Fig. 6-1. Fundamental classification of sedimentary rocks. (According to Folk 1959, fig. 1.)

struction of components. It can result in sediments of great compositional purity and generally destroys information about the provenance of detritus.

The determination of *provenance*, or source of detritus, implies knowledge of igneous and metamorphic rocks, since virtually all detritus is derived, ultimately, from such rocks. Igneous and metamorphic rocks are the raw material of sedimentary detritus, and the composition of detritus can only be some distillate or extract or other modification of the raw material. Thus, sandstones contain quartz and feldspars in abundance because these minerals are abundant in primary source rocks. Shales or mudstones, in which clay minerals are the dominant component, are abundant because clay minerals are the major product of weathering of feldspars.

Weathering of Rocks

Most terrigenous particles are derived from weathering of rocks. Rock weathering is a complex phenomenon that involves mechanical processes and chemical reactions between rocks, natural waters, atmospheric components, and biological systems. Other processes are locally important, such as volcaniclasis and fault cataclasis, the impact of falling rock upon rock, and the grinding and plucking action of glacial ice against its bed. Weathering is ordinarily a slow process, but fast enough that it can degrade human constructions and monuments, often to the extent of destroying the esthetic or structural integrity of these works of humankind, even within a single human generation.

Many chemical weathering processes have been hastened in recent decades by our alterations of the composition of air and water. Rates of erosion and sedimentation are limited by rates of weathering.

Mechanical degradation of rock materials at or near the earth's surface is promoted mainly by expansion: the expansion of freezing water, of minerals undergoing hydration or other chemical alteration, of growing plant roots, and of hard rock experiencing relief of overburden pressure during exhumation. Thermal expansion may play a role in weathering, though various observations and a few experiments have been offered in rebuttal. Tectonic stresses transform monolithic rock masses into millions of fitted blocks, bounded by joints that give access to the freezing waters, the growing roots, and the acids and other natural reagents that further reduce the large blocks to cobbles, pebbles, sand, mud, and solutions. The mechanical, chemical, and biological processes work together, augmenting and supplementing one another in complex ways.

Chemical weathering leads to the development of residual soils or regoliths. The effects of chemical weathering are most intense at the surface and diminish with depth, though weathering effects commonly can be instrumentally detected far below the obvious effects that we see within and near to soil profiles. The products of chemical weathering are more mobile (either because they are in solution or because they are very finely divided), are of lower density, occupy more space, and are chemically more stable in sedimentary environments than the unweathered rock materials from which they are derived. Chemical weathering of the more reactive components of a granite or other polymineralic rock releases the less reactive components (such as quartz) as detrital grains. Nahon (1991, p. 53) considered that there are two chemical weathering pathways, transformation of a parent mineral into a chemically related solid product that replaces the parent and total direct dissolution of the parent.

Though the chemical reactions of weathering are very complex, their net result is typically a fairly simple collection of ionized, oxidized, or hydrolyzed materials. Under neutral, weakly acidic, and strongly acidic conditions, the hydrolysis of a typical carbonate mineral, magnesite, is, respectively,

$$MgCO_3 + H_2O \rightarrow Mg^{2+} + OH^- + HCO_3^- \quad (6\text{-}1)$$

$$MgCO_3 + H_2CO_3 \rightarrow Mg^{2+} + 2HCO_3^- \quad (6\text{-}2)$$

$$MgCO_3 + 2H^+ \rightarrow Mg^{2+} + H_2CO_3 \quad (6\text{-}3)$$

Weathering (dissolution) of carbonate minerals occurs because the carbonate ion, CO_3^{2-}, readily combines with H^+ to form the bicarbonate ion, HCO_3^-, which is quite stable in water.

The hydrolysis of silicate minerals makes silicic acid, H_4SiO_4, a very weak (very stable) acid. Under neutral, weakly acidic, and strongly acidic conditions, the weathering of a simple silicate, such as forsterite, is described by

$$Mg_2SiO_4 + 4H_2O \rightarrow \\ 2Mg^{2+} + 4OH^- + H_4SiO_4 \quad (6\text{-}4)$$

$$Mg_2SiO_4 + 4H_2CO_3 \rightarrow \\ 2Mg^{2+} + 4HCO_3^- + H_4SiO_4 \quad (6\text{-}5)$$

$$Mg_2SiO_4 + 4H^+ \rightarrow 2Mg^{2+} + H_4SiO_4 \quad (6\text{-}6)$$

Note that acid is consumed in the weathering of silicates (or stronger acids replaced by weaker acids), so that a solution in contact with silicate becomes more alkaline. When rain falls onto the rocks, it has an acidic pH of about 5.7 or less, due mainly to its takeup of atmospheric CO_2; as it passes over and through the silicate rocks exposed on continental surfaces, its pH gradually rises and the waters are basic by the time they return to the ocean via the rivers.

Weathering of aluminosilicates, such as orthoclase, usually results in formation of certain other aluminosilicates, the clay minerals:

$$4KAlSi_3O_8 + 22H_2O \rightarrow \\ 4K^+ + 4OH^- + Al_4Si_4O_{10}(OH)_8 \quad (6\text{-}7) \\ + 8H_4SiO_4$$

In this reaction, the solid product is the clay mineral kaolinite. The reaction is almost certainly more complex than this, with various intermediate steps involving, especially, hydrolysis of aluminum ions. Recombination of hydrolyzed aluminum with hydrolyzed silicon and various ions in true solution creates kaolinite clay minerals that are chemically more complex.

Weathering of any ferrous mineral exposed to oxygen results in precipitation of very insoluble ferric hydroxide or oxide, such as hematite. For example, the net reaction for the weathering of fayalite is

$$Fe_2SiO_4 + \tfrac{1}{2}(O_2) + H_2O \rightarrow Fe_2O_3 + H_4SiO_4 \quad (6\text{-}8)$$

or, in the presence of carbonic acid,

$$Fe_2SiO_4 + 4H_2CO_3 \rightarrow \\ 2Fe^+ + 4HCO_3^- + H_4SiO_4 \quad (6\text{-}9)$$

$$2Fe^{2+} + 4HCO_3^- + \tfrac{1}{2}(O_2) + 2H_2O \rightarrow \\ Fe_2O_3 + H_2CO_3$$

One way to follow the progress of chemical weathering in a soil is the **gain-loss diagram**, which graphically compares the composition (obtained by chemical analysis) of fresh rock with that of its weathered residue (regolith). Presumably the more mobile constituents are carried away in solution; the less mobile constituents become relatively more abundant in the soil (Fig. 6-2). For comparison purposes (as one parent rock with another or one climate with another), it is useful to consider one of the constituents as constant, so that all gains and losses are with respect to this one constituent; this is called a **constant-oxide diagram**. One prefers to choose the least mobile component as the constant; usually this is Al_2O_3, but TiO_2 or Fe_2O_3 are also used occasionally.

The role of plants and plant residues in chemical weathering is very complex but can hardly be overemphasized. The roots of the vascular plants are surrounded by electrical fields of charged particles that hasten the ordinary inorganic chemical reactions between soil particles and soil fluids. Micro-organisms in soils greatly accelerate the chemical breakdown of rocks and minerals. According to Weaver (1989, p. 137), bacteria, algae, fungi, lichens, and mosses all play a major role in the chemical weathering of fresh rock. Lichens extend into the rocks their little roots, called hyphae, which excrete organic acids and chelates that can dissolve any mineral. Some plants, especially desert plants, excrete to the soil special enzymes and other substances that discourage other plants from taking root nearby, plants that would compete for the same scarce moisture and nutrients; these substances may be effective in the degradation of rocks. When plants or plant parts die and decay, they release a host of organic acids to the soil. Water-soluble anions of organic acids are important complexing agents and are abundant in most soils. Much of the iron in river waters, according to Weaver, is complexed with fulvic acid. Humic acids bring silicon, aluminum, and iron to concentrations of ten times or more their concentrations in the absence of these complexing agents (Huang and Keller, 1970). In tropical forests, regoliths typically are very thick and mineralogically simple, due in part to rapid interactions between

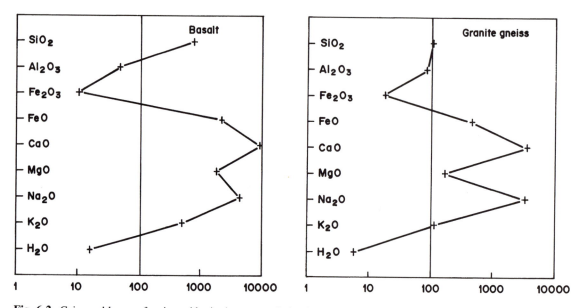

Fig. 6-2. Gains and losses of major oxides in the course of chemical weathering of a granite and a basalt. (From data compiled in Garrels and Mackenzie, 1971a, tables 6.5, 6.6.)

the rock materials and organic matter, which disappears quickly.

Various plants accumulate and concentrate certain chemical elements of the soil, such as selenium or copper; ore prospectors occasionally take advantage of this fact, sampling plants in their exploration programs instead of rocks or regoliths. Certain reeds accumulate potassium in large amounts, and plants of arid climates take up lithium from saline groundwaters. Reeds, sedges, horsetails, and bamboo contain as much as 10 percent of silica (dry weight). Lovering (1959) described the accumulation of silica in tree tissues (it is a major component of the ash that piles up in your fireplace). A tropical forest accumulates a ton of silica per hectare in a year. This silica, brought up from the deeper soil layers by the roots, is added to the top of the soil as leaf litter or deadfall decays on the forest floor.

**The Major Detrital Components
of Terrigenous Rocks**

Most of the mineral components of terrigenous rocks are silicates. This is due in large part to the fact that silicates are the major component of the fundamental source rocks. Also, silicates are among the most durable minerals in sedimentary environments.

Quartz

Silica occurs naturally in several different forms, the most common of which by far is α-quartz or low quartz, having the rhombohedral (trigonal) structure. It is the most stable of the polymorphs of quartz at earth-surface temperatures and pressures, having the greatest hardness and lowest solubility, and it occurs abundantly in most igneous and metamorphic rocks of continents. Most detrital grains of quartz and most silica cement in terrigenous rocks are of this form (Fig. 6-3).

The low-quartz polymorph does not exist at temperatures above 573°C (atmospheric pressure), where a higher-symmetry (hexagonal) form called β-quartz or high quartz is the stable polymorph. It inverts spontaneously to low quartz as the temperature falls below 573°; but if the higher-temperature form bears crystal faces, these are retained during the transition, and a *paramorph* results. Such quartz grains are very distinctive, typically limpid and strain-free in thin sections, commonly with hexagonal or squarish outline (reflecting cross sections, roughly normal, or parallel,

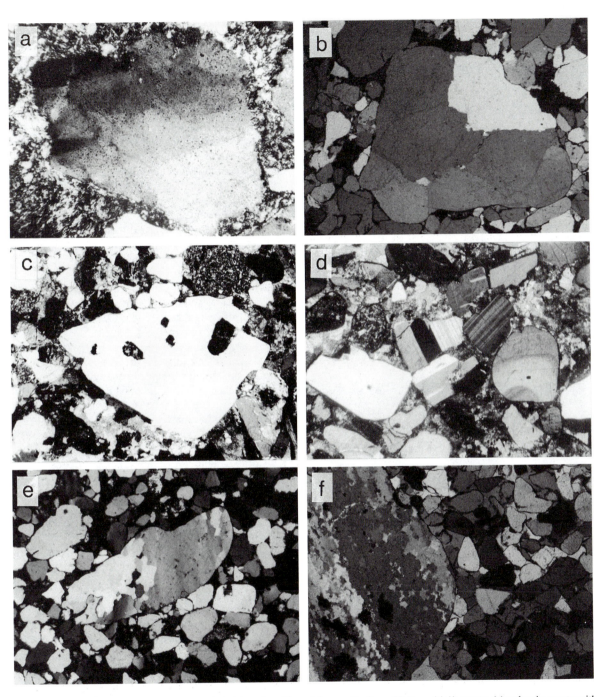

Fig. 6-3. Special types of detrital quartz grains: **(a)** grain with undulose extinction; **(b)** ''granitic'' quartz; **(c)** volcanic quartz with embayments or inclusions; **(d)** volcanic quartz associated with volcanic plagioclase; **(e)** ''gneissic'' quartz showing bimodality of grain size; **(f)** small pebble of gneissic quartz showing stretched fabric.

127

respectively, to c axis of the hexagonal dipyramid); commonly they are fractured, due to anisotropic volume change during inversion; or they are embayed, due to partial resorption into hot volcanic lava, where probably all high-quartz paramorphs are created.

Unit quartz, or detrital grains consisting of a single crystal unit, may be derived from any quartz-bearing rock. The grains may show some degree of undulose extinction, resulting from plastic deformation at some time in their history. In view of the fact that strain may be inherited from a plutonic or metamorphic provenance or induced diagenetically in the detrital grain residing in a sandstone (Conolly, 1965, p. 126), undulose extinction is not a useful guide to provenance (Blatt and Christie, 1963, pp. 569, 574; Blatt, 1967, pp. 408, 410, 413, 417). Basu et al. (1975), however, have suggested that degree of strain in ensembles of detrital grains, in combination with polycrystalline or composite character (see below) of the quartz, can be of some value in diagnosis of provenance (Fig. 6-4). Tortosa et al. (1991) provided some additional documentation.

A *composite*, or *polycrystalline*, *quartz* grain is a detrital grain comprising two or more distinct crystallites having widely differing optical orientations. In a thin section, some grains of composite quartz appear to consist of one or two large units and several much smaller units, sinuously interlocked. Others appear to comprise several units of similar size; each unit shows straight or gently undulose extinction; typically there are few inclusions, maybe a few vacuoles. These kinds of composite quartz are most likely of plutonic heritage.

Other composite quartz consists of many small, polygonal domains, equant or elongate, with straight or slightly undulose extinction; and there is composite quartz comprising elongated (sheared or "stretched") domains with crenulated or granulated boundaries and strongly undulose extinction. These are products of dynamothermal-metamorphic terrains.

Vein quartz, or the quartz of pegmatitic and hydrothermal veins, commonly consists of individual crystallites only slightly misoriented with respect to one another; these are sometimes referred to as *semicomposite quartz*. Fluid-filled vacuoles may be very abundant, giving the quartz a milky appearance or brownish cast in thin section. Other semicomposite quartz derives from gneisses.

A quartz-cemented sandstone can break down into composite quartz grains that are sometimes difficult to distinguish from the composite quartz of plutonic heritage. The pattern of bubble trails is useful; in granitic or gneissic quartz, bubble trails commonly cross grain boundaries and exhibit a directional consistency that transcends grain boundaries; this is not the case in sandstone fragments.

"Compositeness" of a detrital grain implies that the grain is sufficiently large to accommodate the several constituent domains. If sedimentary processes reduce the size of a composite grain, it eventually becomes a unit-quartz grain (or several such). There is substantial evidence (Blatt and Christie, 1963, p. 570; Blatt, 1967, p. 421) that composite and strongly undulatory quartz is less durable than "straight," unit quartz, tending to disappear gradually as it progresses through the sedimentary cycle.

Much detrital quartz contains mineral inclusions visible and perhaps identifiable under the petrographic microscope. Their presence and identities are reflected in the trace element content of the quartz grains. To some extent, inclusions and trace element content reflect provenance. Gilligan (1920, p. 260) cited Mackie's 1897 study of the quartz of granites, gneisses, and schists of Scotland as among the earliest uses of inclusions in provenance analysis. Schnitzer (1957) ex-

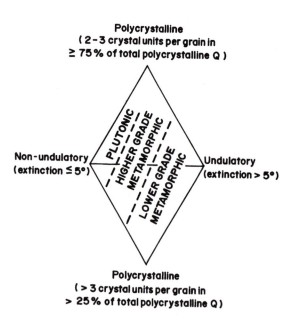

Fig. 6-4. Possible relationship between quartz types and provenance. (According to Basu et al., 1975, fig. 6, with permission of SEPM.)

amined detrital quartz from the Buntsandstein (Triassic) in East Bavaria. Various granitic plutons in the detrital source area released quartz grains having distinctly different colors, and these differences show up in the sandstones derived from the various terrains. Schnitzer reviewed the causes of the various colors of quartz—fluid inclusions, numerous tiny fractures, various mineral inclusions such as carbon particles and hematite particles, and various chemical impurities, such as ferric hydroxide. He noted that in some cases, quartz colors are more useful than heavy minerals in stratigraphic, provenance, and paleogeographic studies, but he cautioned that diagenesis affects quartz color.

Dennen (1964, p. 243) determined spectroscopically that titanium is almost universally present, at levels of about 45 parts per million, in quartz of intrusive and extrusive igneous rocks, but that it is virtually undetectable in the quartz of most pegmatites, veins, and metamorphic rocks. Suttner and Leininger (1972), suggested that trace titanium, iron, and magnesium in quartz is of some value in discriminating plutonic, volcanic, and metamorphic sources. The isotopic composition of single quartz grains (and other mineral grains) is useful in provenance analysis. Isotopic dating of accessory mineral grains is one approach (isotopic dates of detritus and source should match); another is the matching of stable isotope signatures in detrital grains to isotope signatures in potential source areas (Heller and Frost, 1988). Heller et al. (1985) presented an example of the latter approach that gave surprising results, apparently in conflict with the results of more conventional techniques.

Cathodoluminescence (CLM) also is useful in the study of provenance; many minerals emit visible light when exposed to cathode rays. Quartz luminescence is generally quite dull, either red or blue. (Other detrital materials in the rock, in contrast, luminesce quite brightly—K-feldspar is bright blue; plagioclase is bright green, red, or blue; zircon and apatite commonly are yellow.) According to Zinkernagel (1978, p. 9), the luminescence color of quartz is uniform in any given crystalline rock body. Bluish and reddish luminescence characterizes quartz that crystallized at high temperature (above 273°C) and quartz that has been subjected to high temperature in contact metamorphic aureoles, then quickly cooled; quartz of low-grade metamorphic rocks luminesces brown. Hydrothermal quartz displays zoned greenish luminescence; diagenetic quartz is non-luminescent. Thus the color

or intensity of the luminescence of detrital grains can reveal the different origins (provenances) of the grains. Moreover, as Sippel (1968) showed, composite and unit quartz are readily confused; syntaxial cement on quartz grains is easily overlooked, and the sizes and shapes of detrital grains, as distinct from their syntaxial cements, commonly are not as they appear to the petrographer using ordinary petrographic techniques. These problems tend to degrade a provenance analysis.

Chert, chalcedony, and opal are forms of detrital silica that are generally classified as rock fragments (described later).

Feldspars

The chemically and structurally diverse feldspar minerals are, like quartz, a major constituent of most igneous and metamorphic rocks and are of common occurrence in detrital sedimentary rocks as well. Certain feldspars, because of their restricted paragenesis, are quite provenance-specific. Sanidine, for example, is a high-temperature alkali feldspar that has retained its high-temperature structure because it cooled quickly; sanidine is almost exclusively derived from volcanic rocks, and its appearance in detrital sediments indicates a volcanic source. Microcline's pericline-albite ("tartan") twinning records its inversion from high-temperature monoclinic form to lower-temperature triclinic form during slow cooling, such as occurs in large plutons or after regional metamorphism but never in volcanics. Thus, microcline in detrital rocks indicates that old shields or the cores of orogens have been unroofed. The composition of plagioclase, whether calcic or sodic, reflects the overall composition of the intrusive or extrusive igneous rocks or the rank of the metamorphic rocks in which the plagioclase was formed.

Most feldspar grains or crystals bear various optical heterogeneities resulting from either chemical or structural variation—perthitic structure, twinning, compositional zoning, inclusions, and alterations. Rimsaite (1967) examined nearly 1000 diverse Canadian rocks and distinguished nine types of optical heterogeneity, each favoring certain types of crystalline rocks. Compositional zoning, for example, occurs primarily in igneous rocks of intermediate composition and their gneissic equivalents but rarely in granites or gabbros or their extrusive equivalents. Moreover, oscillatory zoning is characteristic of plagioclases that have grown from igneous liquids, whereas normal or progressive zoning characterizes metamorphic feldspars. Perthitic structure, due to ''molecular'' unmixing (exsolution) of potassic and sodic feldspars that were in solid solution at some high temperature initially, manifests a variety of patterns and scales, the coarser perthites having formed in the larger, more slowly cooling plutons. Volcanic rocks may contain perthites of such fine scale as to be revealed only by x-ray methods. Myrmekites and ''graphic'' intergrowths of feldspar with quartz, present in the detrital grains of some sedimentary rocks, were derived from pegmatites and granophyres. Intrastratal dissolution of feldspars is very important in some sandstones (Heald and Larese, 1973), sanidine being particularly susceptible and microcline least; this differential dissolution can muddle a provenance analysis based on feldspars. There is another difficulty—possible confusion of detrital feldspar with authigenic feldspar (see Chap. 11).

In crystalline source rocks, the feldspars are comparable in size to or larger than single quartz grains, but—by virtue of their cleavage, twinning, and perthite lamellae—are mechanically reduced more rapidly than unit quartz. This results in concentration of feldspars in the finer fractions of a sandstone and in the finer-grained beds of a stratigraphic succession. Odom (1975) and Odom et al. (1976) examined this matter in Cambrian sandstones of the upper Mississippi valley, noting that detrital feldspars are concentrated in the 3- to 5-phi size fractions (Fig. 6-5). In closely associated coarse and fine-grained sandstones,

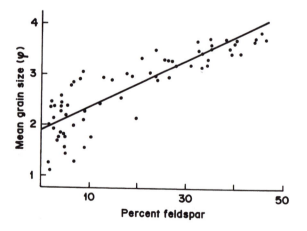

Fig. 6-5. Feldspar content versus grain size in some sands of the upper Mississippi Valley. (After Odom, 1975, fig. 8, with permission of SEPM.)

the finer-grained rocks may contain substantial quantities of detrital feldspar even if the coarser rocks do not.

Rock Fragments

Rock fragments are little samples of the source rock, preserving some aspects of texture and mineral assemblages (Fig. 6-6). Generally it is easy to distinguish among fragments of plutonic, volcanic, sedimentary, and metamorphic rocks on the bases of their mineral contents and their textures; obviously, larger rock fragments are more easily diagnosed than smaller ones (see the analysis of Boggs, 1968). Fragments of volcanic glass occasionally have sharp edges and concave sides, indicating that they have been broken from a pumiceous or scoriaceous source rock; these little objects are called **bubble-wall shards**. Many or most such fragments have been air-carried to their depositional site. Many volcanic rock fragments are porphyritic or vitrophyric and contain (at least parts of) zoned plagioclase phenocrysts or other smaller plagioclase or hornblende laths embedded in a very fine grained or glassy groundmass. Such rock fragments are unequivocally volcanic or hypabyssal, though certain fine-grained volcanic rock fragments are difficult to distinguish from sedimentary chert (artificial staining helps; CLM also is useful). Quartz-feldspar rock fragments with phaneritic textures are mainly a product of plutonic and high-grade metamorphic terrains. As they are transported, they quickly disintegrate into monomineralic quartz grains and feldspar grains. Fine-grained metamorphic rock fragments—namely slate and phyllite fragments—commonly are difficult to distinguish from shale fragments; slate fragments are a bit harder than most shales, and clay particles more strongly oriented; phyllite is coarser. Cameron and Blatt (1971, p. 574) observed that schist fragments are mechanically very unstable and disintegrate almost immediately upon entering a mountain stream, if they have not already disintegrated in soils mantling the source rock. Clayey or micaceous rock fragments are susceptible also to mashing under burial pressure; if a sandstone contains such rock fragments mixed with quartz grains, the rock fragments may, under compression, extrude to adjacent intergranular space and become matrix.

Anyone who has spent much time on the outcrop knows that some shales weather into small (3- to 5-mm) equant chips; other shales weather essentially

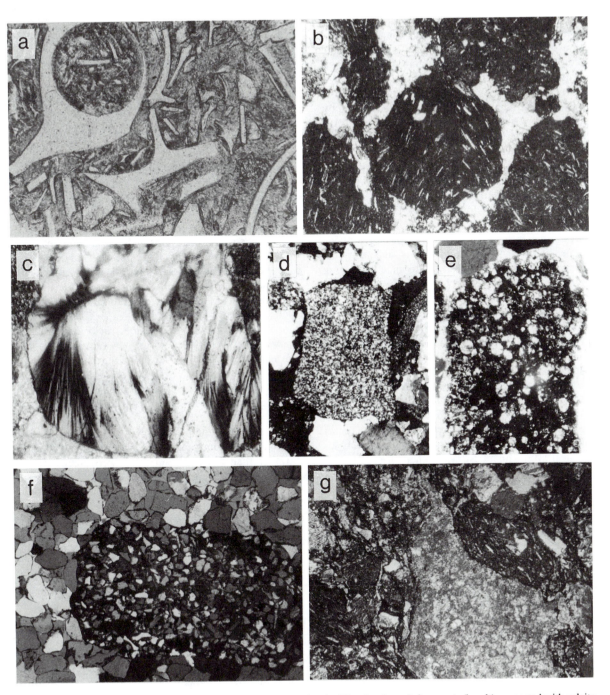

Fig. 6-6. Various types of rock fragments: (a) bubble-wall glass shards, (b) volcanic rock fragments (basalt), cemented with calcite, (c) chalcedony fragment, (d) chert fragment, (e) spicular chert fragment, (f) siltstone fragment, (g) mixture of volcanic rock fragments and limestone fragments.

into mud. Climate plays a role in this, but certainly the texture and composition of the shale or mudstone are important, too. Thus, some shales, under certain conditions, yield rock fragments; other shales or other conditions produce lots of clay and silt but no rock fragments.

Limestone fragments are present in some sandstones and are a major component of some terrigenous rocks; such rocks are called *calclithites* (see Folk, 1980, p. 143). Again, climate is important; there are many rugged carbonate terrains, especially in arid or semiarid or cold locales, that produce abundant carbonate rock fragments; one finds them accumulating as talus at the bases of cliffs, littering the stream beds, and collected in the hollows. Because they are soft and soluble, such fragments cannot survive much transport, no matter what the climate. Chert fragments are another product of carbonate terrains, and many calclithites contain them. Many carbonate rock fragments, large and small, are generated by surf against reef margins; these particles accumulate as large detrital aprons on the flanks of the reef.

Chert is a very fine crystalline form of quartz, individual crystals appearing under SEM as tightly fitted polyhedral blocks with smooth, gently curved faces, much like a foam. The structure of tiny, equant grains generally is resolved by petrographic microscope; but when the individual crystals are substantially smaller than the thickness of the thin section (30 micrometers), the chert looks quite dark between crossed polars. Chert commonly contains numerous small bubbles, giving it a brownish cast. Most chert in sedimentary rocks forms by replacement of carbonate materials or of evaporitic sulfate minerals; skeletal materials—especially echinoderm, brachiopod, and bryozoan skeletals—and wood, commonly are replaced or partially replaced by chert. Nodules of chert are quite abundant in some carbonate rock units (see Chap. 11), and these may become residual detritus, available for transport when the host rock weathers and dissolves. Pebbles of such chert may have a weathered rind of porous silica called porcellanite. In some regions, chert fragments derive from pelagic bedded cherts; such chert fragments commonly contain sponge spicules, radiolarians, or diatoms.

Chalcedony is low quartz with a fibrous habit, made quite evident by the petrographic microscope (Fig. 6-6c). It occurs rarely as detrital grains in sedimentary rocks, more abundantly as a chemical cement, as replacement of skeletal materials or of evaporite minerals (see Chap. 11), and as an alteration component of opal. The scanning electron microscope (SEM) reveals the presence of myriad small water-filled bubbles.

Chert and chalcedony are less durable than macrocrystalline (unit) quartz because they are softer, contain numerous bubbles, and are chemically less pure than ordinary quartz. Chert is quite brittle, and detrital pebbles of this material fracture readily.

Most chert fragments probably are first-cycle. Certain of them are so distinctive that their source is known unequivocally. Most chert is gray or black or brown, but some is green or red, and the unusually colored cherts may be of value in provenance studies. The dark cherts of the "Phosphoria" Formation (Permian) of Idaho and Montana made their mark far and wide as detrital grains in Jurassic and Cretaceous sandstones; one spots these grains immediately as black specks among the colorless quartz grains, like pepper mixed with salt. Sandstones older than the Phosphoria characteristically do not contain these chert grains.

Where coarse and fine-grained rocks are closely associated, the coarser rocks may contain abundant rock fragments, while the finer-grained rocks do not. Davies and Ethridge (1975) found that the quartz content of detrital rocks is related to mean grain size, quartz being most abundant in rocks whose mean grain size is about 1.5 phi. The different sandstone beds show differences in composition that seem to have resulted from size-reduction and size-sorting mechanisms rather than from differences in provenance. Boggs (1969) noticed that the relative abundance of various types of rock fragments in pebble populations varies with pebble size and suggested that it is due to different behavior of different rock types under weathering at the detrital source and under abrasive transport.

Other Detrital Grains

Certain orthochemical or allochemical sedimentary minerals occasionally participate to some degree in transport, thus briefly playing the role of a detrital mineral. Transported cement fragments occur in the aprons of carbonate debris that surround coral reefs. It appears that single dolomite rhombs or gypsum crystals and pyrite crystals or framboids occasionally are transported, though probably not very far. Amsbury (1962) noted that Ellenberger Dolomite (Ordovician) and Glen Rose Formation (Cretaceous) are weathering into loose carbonate grains that are an important constituent of the detrital load of modern rivers in central Texas. He described also detrital dolomite in certain Lower Cretaceous sandstones and limestones in central Texas, being single rhombs damaged by transport; commonly they occur as dark, cloudy, rounded grains with clear cement overgrowths. Scholle (1971, p. 238) and Freeman and Rothbard (1983, p. 545) observed abraded dolomite crystals that, once come to rest, developed cement overgrowths. Detrital gypsum is fairly common in or near certain evaporite environments, even accumulating into cross-bedded and rippled deposits, as nicely illustrated by Hardie and Eugster (1971) in the Miocene evaporite deposits of Sicily. Schreiber et al. (1976) described gypsum turbidites and chaotic debris flow deposits composed of gypsum in these same strata. Kinsman (1969, p. 836) observed that wind-dispersed abraded polycrystalline anhydrite sand is common in sabkhas near the Persian Gulf. The aeolian sands of the White Sands National Monument in New Mexico are composed of gypsum eroded out of ancient evaporite deposits. Volcanic rocks, whether vesicular flow rocks or volcaniclastics, commonly contain large quantities of cement, including opal, chalcedony, prehnite, and a host of zeolite minerals. Weathering of such rocks can release these unusual cements as detrital grains.

A host of other minerals occur in small quantities as detrital grains, many of them considerably denser than the more common detrital minerals, and are referred to as *heavy minerals* (described later).

Clay Minerals

Because they are so complex, so common in soils and in sedimentary rocks, so consequential in foundation and slope stability and other engineering matters, and so varied in their properties and applications, clay minerals have been set apart from other minerals as a separate discipline. Their identification and study generally requires special equipment, such as the x-ray diffractometer and electron microscope. More than most other minerals of sedimentary rocks, the clays are susceptible to various chemical and structural alterations under weathering and diagenesis. All of the major groups of clay minerals can occur in sedimentary rocks as detrital particles or as authigenic or diagenetic materials as well.

The clay minerals include a variety of chemically and structurally complex aluminosilicate compounds having in common that they form only very small crystals, generally only a few micrometers. All but a few uncommon clay minerals are monoclinic phyllosilicates, their structures based on layers, parallel to {001}, of tetrahedrally coordinated cations composited with layers of octahedrally coordinated cations (Fig. 6-7).

The simplest of the common clay minerals (both chemically and structurally) is kaolinite, $Al_4Si_4O_{10}(OH)_8$. The basic structure comprises a tetrahedral layer/octahedral layer couplet. The tetrahedral layer consists entirely of Si^{4+} ions coordinated by O^{2-} ions. The octahedral layer consists of Al^{3+} ions coordinated by OH^- and O^{2-} ions, but only two out of every three coordination octahedra actually contain an aluminum ion; the other one-third of them are empty. Thus kaolinite is said to be *dioctahedral*. The repeat distance (basal spacing) is about 7 angstrom units (AU; 1 angstrom unit = 10^{-10} m = 0.1 nanometer). Kaolinite and structurally similar clay minerals are commonly referred to by clay mineralogists as *kandites*.

Illite is a clay mineral that may be considered as a non-stoichiometric variant of the mica muscovite, which has the composition $K_2Al_4Si_6Al_2O_{20}(OH)_4$ (doubled, for comparison with below). The variable composition of illite may be written as $K_{2-x}Al_4Si_{6+x}Al_{2-x}O_{20}(OH)_4$, where x varies between about 1 and 1.5. The illite (and muscovite) structural unit consists of an octahedral layer sandwiched between two tetrahedral layers. In muscovite, one out of every four tetrahedra is occupied by an Al^{3+} ion instead of the usual Si^{4+}; this creates a deficit of positive charge in the tetrahedral sheets, which deficit is redressed by K^+ ions between the unit layers. In illite, a bit fewer than one out of every four tetrahedral sites is occupied by an aluminum ion, and there are correspondingly fewer potassium ions. Like kaolinite, the illites (mostly) are dioctahedral. The basal spacing of illite is about 10 AU.

In illite, we have just observed, there is interchange of Si^{4+} and Al^{3+} ions in the tetrahedral sites; in smectite, there is substitution of some octahedrally coordinated Al^{3+} by Mg^{2+}, either instead of or in addition to substitution at tetrahedral sites. This again results in a deficit of positive charge, which is balanced by Na^+ or Ca^{2+} between unit layers. It is not easy to write a chemical formula for smectite, but perhaps this one, $(Ca,Na)_x(Al,Mg)_4(Si,Al)_8O_{20}(OH)_4 \cdot nH_2O$, is suffi-

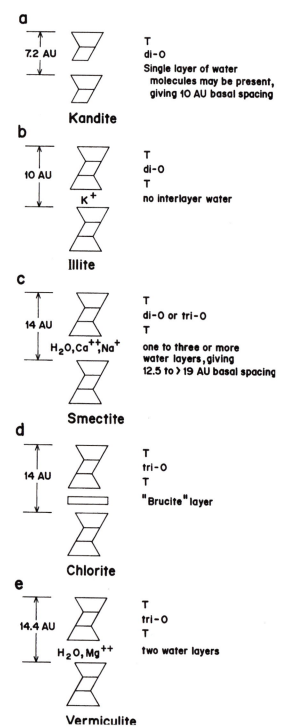

a

7.2 AU

T
di-O

Single layer of water
molecules may be present,
giving 10 AU basal spacing

Kandite

b

10 AU

K⁺

T
di-O
T

no interlayer water

Illite

c

14 AU

H_2O,Ca^{++},Na^+

T
di-O or tri-O
T

one to three or more
water layers, giving
12.5 to > 19 AU basal spacing

Smectite

d

14 AU

T
tri-O
T

"Brucite" layer

Chlorite

e

14.4 AU

H_2O, Mg^{++}

T
tri-O
T

two water layers

Vermiculite

Fig. 6-7. Structures of major clay-mineral groups.

ciently general, where x depends on the kinds of ions, and their proportions, contained in tetrahedral (Si, Al) sites and in octahedral (Al, Mg) sites. Some smectites are dioctahedral, but others are *trioctahedral*, wherein all (that is, three out of three) octahedrally coordinated sites are occupied by Al^{3+}, rather than just the two out of three of the dioctahedral clay minerals. Besides the interlayer cations, there may also be one or more structured layers of water molecules. The basal spacing of smectite varies from about 14 AU to greater than 20 AU, depending mainly on the number of water layers contained. Smectite flakes are very small and commonly not even clearly resolved by SEM.

Chlorite clay is a monoclinic or triclinic clay mineral similar to the well-crystallized metamorphic mineral chlorite. Its structure was worked out by Linus Pauling in 1930; it consists of illite-like (or talc-like) trioctahedral units alternating with brucite-like layers, that is, layers of magnesium ions octahedrally coordinated by hydroxyl ions. Replacement of some of the Si^{4+} ions by Al^{3+} in the mica-like layers creates charge imbalance that is taken up by partial replacement of Mg^{2+} in the brucite-like layers with Al^{3+}. Basal spacing is about 14 AU, similar to that of many smectites.

Vermiculite is another clay mineral containing mica-like trioctahedral units, but here they are separated by structured layers of water; ordinarily these water layers are two molecules thick. Silicon-aluminum ionic substitution brings Mg^{2+} and Ca^{2+} ions into interlayer positions. Basal spacing is about 14 AU.

Most clay minerals of sedimentary rocks are *mixed-layer clays*, that is, clays that consist of "molecular" layers of one kind of clay mineral alternating with layers of another kind of clay mineral. It is quite common, for example, for a clay particle to comprise illite layers interleaved with smectite layers, and the interlayering may be regular or random. Such clay particles, on entering a marine environment or under burial diagenesis, tend to become gradually more illitic and less smectitic.

Glauconite is an authigenic iron-rich clay mineral structurally similar to illite. Its striking green color makes for easy identification; bright green grains in a sandstone or limestone can hardly be anything else (except maybe chamosite; see below). It occurs as little pellets (probably an alteration of fecal pellets), or as fillings in foraminiferal tests or other small shells, or as coatings on phosphatic lumps or nodules. Petrographic evidence suggests that glauconite can form from almost any precursor—it is often observed to have partially replaced volcanic rock fragments, feldspars, micas (especially biotite), even quartz. Commonly, glauconite grains are rather highly concentrated in certain layers in a sedimentary succession and absent in other layers. Glauconite is an almost foolproof index of marine environment (but not too far from continental shores); glauconite on the present sea floor is most abundant at depths of 200 to 300 meters (Odin and Matter, 1981, p. 627). Transgressive environments seem to be most favorable to glauconite formation.

Glauconite seems to form most readily in confined, chemically reducing microenvironments (contained in an otherwise oxygenated marine shelf setting), either by transformation of detrital micas or clays or by authigenic growth on a substrate of detrital micas or clays (Odin and Matter, 1981, p. 631 ff; Bornhold and Giresse, 1985, p. 662). The reducing environment seems, in general, to be provided by the interiors of the tests of decaying foraminifers or within fecal pellets. Partial chemical isolation from seawater is also provided by borings in echinoderm plates, fractures in detrital quartz or feldspar grains, or cleavages in biotite. Odin and Matter observed a range in glauconite composition and structure from potassium-poor smectite-like to potassium-rich illite-like material; they noted also that iron fixation precedes incorporation of potassium. Similar observations were made earlier by Ehlman et al. (1963, p. 95). Burst (1958a, 1958b) determined that glauconites of younger rocks are illite-smectite mixed-layer clays, and that older glauconites are essentially iron-rich illite. Presumably the differences are due to slow diagenesis. Like other authigenic minerals, glauconite probably can be recycled under some conditions.

Chamosites are certain iron-rich clay minerals, some related to chlorite (with 14-AU basal spacing), some related to kaolinite (7 AU); the latter are referred to as berthierine. They occur as greenish gray or greenish brown coatings on detrital grains or as compact aggregates (peloids), most abundantly in certain iron-rich sedimentary rocks, and are apparently a result of early diagenesis under mildly reducing conditions. Porrenga (1965, p. 401; 1967, p. 498), working in the Niger and Orinoco deltas, showed that chamosite is widely distributed in zones of warm and shallow marine waters, while glauconite forms in cooler, somewhat deeper zones. According to Odin and Matter (1981, p. 627), berthierine is restricted to shallow water of tropical seas, shoreward of glauconite occurrences. Like glauconite, chamosite typically is poorly ordered in young surface sediments, but diagenesis "improves" it. The chlorite-like chamosites may be a diagenetic alteration of (kaolinite-like) berthierine, according to Velde (1989, p. 7), and, according to Odin and Matter, burial diagenesis changes berthierine into chlorite.

Anderson et al. (1958) observed that clay minerals lose magnesium and gain potassium as they pass through the digestive tracts of sediment-feeding organisms; kaolinites appeared not to be affected. Pryor (1975), who examined the pelletizing activities of modern callianassid shrimps, determined that the digestive tract of these animals "wreaks havoc" on ingested clay minerals, totally or partially destroying the brucite layers of chlorite clay and mixed-layer minerals, removing potash from illite, and gradually disordering the crystallinity of kaolinite and illite clay particles, maybe by removing aluminum from the octahedral layers. With the passage of time, diagenetic processes reconstitute these degraded clays or alter them to glauconite.

Cosmogenic Materials

Extraterrestrially derived *cosmogenic* particles in sediments generally are either not looked for, not detected, or misidentified (it is easy to confuse them with certain volcanic particles and smelter-derived particles, or with altered pyrite crystals, etc.). Aside from the very rare large *meteorite* body, there are the much more abundant but very much smaller *micrometeorites*, generally less than 100 micrometers and typically 5 micrometers or less in diameter. There are three broad categories of meteorites: the *irons*, *stony irons*, and *stones*, the latter being by far the most common. The stones are further classified into the *chondrites* and the less common *achondrites*, depending on whether or not they contain small (about 1-millimeter) spheroidal bodies called *chondrules*. The major minerals of meteorites are the iron-nickel alloys kamacite and taenite, the iron sulfide troilite, and the silicates olivine, certain ortho- and clinopyroxenes, and plagioclase. There are no high-pressure minerals (other than those that might have formed during shock of impact), and this indicates that meteoroids did not form in the interior of any large (earth-like) parent body. In fact, the composition and texture of meteorites suggests that most of them came from a half-dozen to a dozen distinct objects having diameters of only 200 to 300 kilometers. A few meteorites, it appears, are fragments of comets and asteroids and other orbiting bodies of the solar system (including the earth's moon and the planet Mars). Some cosmogenic materials appear to have come from interstellar space, but most seem to be essentially local primordial materials left over from the major planet-forming event early in the development of the solar system.

Among the cosmogenic materials in terrestrial sediments are other small grains (typically less than 1 millimeter but some a centimeter or more in size) called *spherules*, which are essentially little beads formed by rapid solidification of molten drops. A distinction is made between *magnetic spherules* and *glassy spherules*. The magnetic ones are composed of olivine or magnetite, some with a core of metallic iron-nickel-cobalt alloy. They are considered to be products of ablation of larger meteorites that have entered the earth's atmosphere. The glassy spherules, called *tektites* and *microtektites*, are globular, tear-shaped, or dumbbell-shaped objects that are quite non-uniformly distributed, occurring in *strewnfields*; all individual tektites in a given strewnfield are of similar age. They seem to be made of target material rather than projectile material. There has been protracted debate about whether tektites are splashed out of the moon by meteor-moon impact or are formed from earth materials by meteor-earth impact; in view of the composition of samples gathered from the lunar surface in the 1960s, the latter hypothesis now seems the more likely. The ages of certain strewnfields seem to correspond to times of reversal of the geomagnetic field; Muller and Morris (1986) offered an explanation.

Probably the best place to look for cosmogenic particles of all sizes is in the glacial ice of Greenland or Antarctica, where they are essentially the only non-ice component and are thus easy to spot, and one can easily concentrate the smaller ones simply by melting the ice and decanting the water. The micrometeorites are detected without much difficulty (though difficult to distinguish from dust of terrestrial origin) in fine-grained pelagic (ocean floor) oozes and evaporitic salt beds.

Classification

Sandstones show great variety in texture and composition (Fig. 6-8). Modern classification of sandstones

(and other sedimentary rocks) demands that texture and composition be considered as independent properties, so that a rock name typically consists of two parts, one part designating composition and another part that relates to some important textural quality. Terrigenous rocks are considered, for classification purposes, to be composed of grains or *framework*, and *matrix*, or terrigenous silt and clay. Dott (1964) established some definitions that refer to the texture of the rock: if the rock at hand contains less that 15 percent of matrix (defined as terrigenous material finer than 30 micrometers, it is an *arenite* (Fig. 6-9); if the quantity of matrix lies between 15 and 75 percent, it is a *wacke* (pronounced "wacky"); and if more than 75 percent of the rock is clay mud, then the rock is *mudstone*. Ordinarily the term arenite (coined by Grabau in 1904) is reserved specifically to grains of sand size; *rudite* (also from Grabau) is used instead if there is an important granule and pebble component; and *lutite* (from Grabau) if the grains are generally smaller than sand grains or if the rock is quite muddy.

Folk (see 1980, p. 127) conceived a classification of the composition of the grains; this is in wide use. It is a *ternary* classification, wherein certain framework components of the rock are ignored and the others are assigned to one or another of three categories, or poles, of the ternary classification.

The Q pole subsumes all kinds of quartz grains, including composite quartz (but not chert). All detrital feldspars and all quartz-feldspar composite grains of presumed plutonic or gneissic provenance are assigned to an F pole. All other composite grains, such as fragments of volcanic and metamorphic rocks, fragments of limestone and shale, and chert, are relegated to an R pole (referred to by some petrographers as RF, or as L, for lithic). The relative proportions of these three types of components determine part of the name of a sedimentary rock (Fig. 6-9). If more than 95 percent of the framework grains (that is, those framework grains actually considered) are of type Q, the rock is a *quartzarenite* or *quartzwacke* (depending on the quantity of matrix, of course). Rocks containing less than 75 percent of type Q grains are called *arkose arenite* or *arkose wacke* if type F grains predominate over type R grains (the terms *feldsarenite* or *feldswacke* are also used). If type R grains are predominant, then the rock is a *litharenite* or *lithic wacke*. Sandstones of composition intermediate between the true arkoses and the true lithic sandstones are called *lithic arkose arenite* or *lithic arkose wacke*, and *arkosic litharenite* or *arkosic lithic wacke*. The *subarkosic*

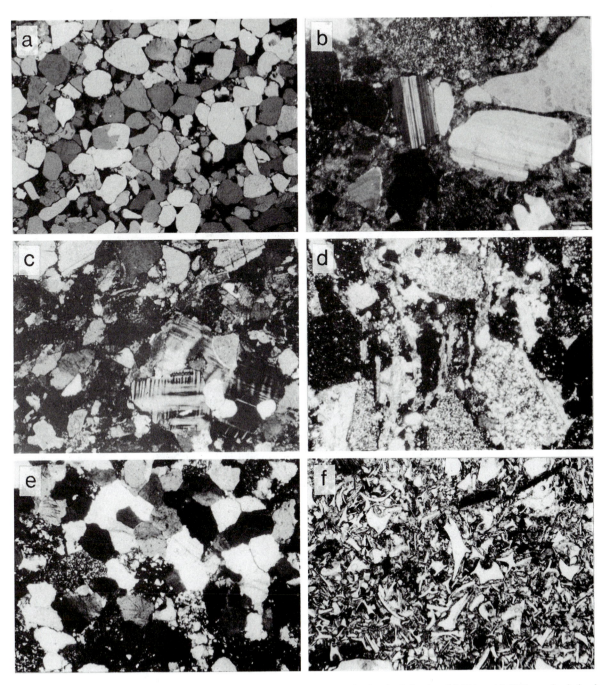

Fig. 6-8. Various sandstones: **(a)** quartzarenite, **(b)** feldspathic lithic arenite of volcanic derivation, **(c)** feldspathic lithic wacke derived from plutonic/metamorphic terrain, **(d)** lithic wacke with chert fragments, **(e)** lithic arenite, **(f)** vitric tuff with glass shards.

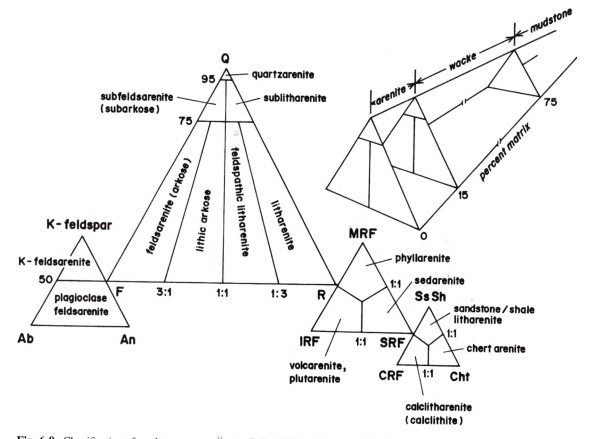

Fig. 6-9. Classification of sandstones, according to Folk (1980), with textural definitions of Dott (1964, fig. 3). Note pre-eminence of quartz, at the apex of the triangle.

and *sublithic* sandstones contain between 5 and 25 percent detrital quartz. Thus, the triangular region is divided into seven fields, each of which defines a framework composition for an arenite or wacke. One may require certain refinements, such as specifying whether the feldspathic grains are predominantly potassium feldspars or plagioclases, and resolving the R pole into specific rock fragment types (Fig. 6-9). (The term *graywacke,* recognized by some classifications, is obsolete.)

Dickinson (1970, p. 702) referred to an interstitial paste formed by deformation of weak detrital grains under burial pressure as *pseudomatrix*, and he offered several criteria by which this might be distinguished petrographically from true matrix. Matrix that varies in texture or color from place to place or that shows

contrasting fabric orientations from place to place is probably not true matrix. If pseudomatrix forms during a compaction that post-dates the earliest cements, it comes to overlie such cements.

Some sandstones have a lime mud matrix rather than the clay mud matrix of the ordinary wacke. The two kinds of mud play the same textural role; presumably they behaved similarly in the fluid-dynamics sense. Some carbonate that looks like lime mud may in reality be very fine crystalline carbonate cement, and lime mud may recrystallize diagenetically, so taking on the appearance of carbonate cement. A grain-supported sandstone containing fine-grained carbonate might be a true arenite or a wacke with lime mud matrix.

Some lithic sandstones are composed mostly or entirely of volcanically derived materials. These so-called *volcaniclastic* rocks generally are mixtures of volcanic glass, fragments of fine-grained volcanic rock, and crystals or crystal fragments embedded in a matrix of volcanic ash (perhaps altered to clay or carbonate). Pettijohn et al. (1987, fig. 6-10) have classified these rocks as *vitric tuff*, *lithic tuff*, and *crystal tuff* (Fig. 6-10) and have described some of the special kinds of observations that must be made in studies of such rocks. In an attempt to standardize the terminology, Schmidt (1981) pointed out the distinction between *pyroclasts*, which are crystals or crystal fragments, glass fragments, and rock fragments formed as a direct result of volcanic activity and not modified by subsequent transport. In contrast, *epiclasts* are volcanic rock materials liberated from existing volcanic rock by weathering, then transported and deposited by sedimentary processes.

Compositional Maturity

As weathering and abrasion are prolonged, minerals chemically less stable or mechanically less durable gradually disappear into dust or solution. Even the most subtle difference in durability between two grains finds expression in the sedimentary mill. After gentle buffeting applied millions of times, one grain finally reveals its mechanical shortcomings. Thus, the

more stable or durable minerals become concentrated. Quartz, kaolinite, and illite are the most stable of the common detrital minerals, and rocks composed almost entirely of these materials are not uncommon. Once formed, a quartz grain or clay particle may last as long as it remains in the sedimentary realm, which, for some, is a billion years or longer (diagenesis dissolves or alters some of them). Other very stable but not abundant detrital minerals are zircon, tourmaline, rutile, and magnetite. Detrital rocks composed mostly of feldspar or of olivine or other labile grains are rare.

Gradual destruction of labile grains is one of the processes leading to the differentiation of sedimentary rocks, and the extent to which ''purification'' has occurred is called *compositional maturity*. Compositional maturity is commonly quantified in terms of relative amounts of stable and less stable components, such as the ratio of quartz grains to non-quartz grains, or the ratio of total quartz to polycrystalline quartz, or the proportion of ultrastable grains in the heavy mineral assemblage. Compositional maturity can be expressed also in chemical terms, SiO_2/Al_2O_3 ratio being a useful measure. Strained quartz apparently is less stable than unstrained quartz. This is indicated by the paucity of strained quartz in the most mature quartz-arenites; it is a striking feature of these rocks that sets them apart from sandstones of ''average'' maturity.

Cleary and Conolly (1971) examined some of the important textural and mineralogical changes that take place during residual soil development. In soils of the southern Appalachian piedmont (Georgia and the Carolinas), they found an upward increase in percent of quartz as proportions of feldspars and rock fragments decreased. Polycrystalline quartz grains, abundant near the base of the soils, disintegrated with time into unit quartz grains, which are by far the dominant quartz type at the top of the soils. Johnsson (1990, p. 724) observed that tropical-climate chemical weathering of sands on alluvial plains quickly advances their maturity. Franzinelli and Potter (1983) measured a downstream increase in compositional maturity of sand in the Amazon River system (see their fig. 3), a trend that they thought might not obtain in temperate or boreal rivers. Suttner et al. (1981) and James et al. (1981), having examined Holocene detrital feldspars of humid and arid (or semiarid) climates, suggested that climate initially controls the composition of detrital sand, but that the climatic imprint fades as detritus moves downstream and into marine environments.

Many petrologists have suggested that quartzare-

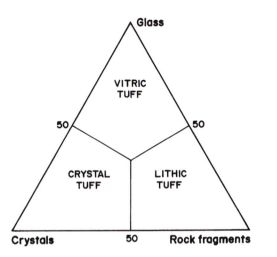

Fig. 6-10. Classification of volcanic sandstones. (After Pettijohn et al., 1987, fig. 6-10.)

nites are mostly a product of multicyclicity (Suttner et al., 1981, p. 1242) and that first-cycle quartzarenites are (p. 1244) only "remotely possible." By this they imply that sedimentary processes cannot ordinarily bring about compositional maturity in a single cycle of weathering, erosion, and deposition. But there are many quartzarenites that, based on stratigraphic relations, are undoubtedly first-cycle. Silurian sandstones in the foreland basin of the Appalachian range are an example:

Compositional maturity increases, often quite rapidly, away from a detrital source, so that a single deposit, though immature on the proximal margin, is quite mature in its more distal parts. Consider the Silurian clastic wedge that formed in the aftermath of the Taconic orogeny in the eastern United States. It comprises conglomeratic lithic sandstones of the Shawangunk Formation (pronounced "SHAWN-gun") in eastern Pennsylvania, grading westward into fine-grained, well-sorted quartzarenite of the Tuscarora Sandstone 100 kilometers to the west. There is much multicycle detritus, to be sure, but according to Epstein and Epstein (1972, pp. 16, 22, 23), source rocks for the Shawangunk included low-grade metamorphic rocks and quartz-feldspar granites and gneisses. It is unlikely that the Shawangunk and conterminous Tuscarora had different detrital sources or that they are of significantly differing ages. Why, then, are sandstones of the Tuscarora so much more mature?

Folk (1960) examined some of these Silurian sandstones in great detail. The Tuscarora, he noted, consists of a lower red part and an upper white part. The red Tuscarora is lithic wacke formed in deltaic and estuarine marine environments characterized by generally low mechanical energy. Quartz in these rocks is a mixture of angular and round grains, reflecting a source terrain supplying both first-cycle and multicycle detritus. The white Tuscarora, in stark contrast, is a supermature quartzarenite formed on a beach where much mechanical energy was being expended against the detritus. There are no lithic fragments and no angular quartz grains; apparently, the energetic depositional environment destroyed the lithics and wore the angular quartz grains into round ones, making them indistinguishable from the multicycle grains that presumably were still being supplied by the source terrain. Thus, an energetic environment was all that was necessary to transform immature detritus to supermature detritus, and it accomplished this in a single sedimentary cycle. Close association of the more mature

and the less mature sands suggests that source rocks and multicyclicity are not necessarily the major controls on maturity; within the Tuscarora, depositional environment, whether high-energy beach or quiet lagoon, controlled maturity.

Johnsson et al. (1988) described first-cycle quartzarenites forming today in the Orinoco River basin from plutonic and metamorphic parent rocks. They are remarkably pure and are appreciably better rounded than some multicycle sands nearby. First-cycle quartzarenites are produced in tropical soils by intense chemical weathering, under conditions that are "neither unique nor particularly unusual" (Johnson et al., 1988, p. 274).

Compositional and textural maturity generally go hand in hand; this indicates that processes promoting textural maturity also promote compositional maturity. Grain size and composition of detritus are not fully independent, so textural sorting can result in compositional differentiation and an increase in compositional maturity. Progress toward textural maturity, we have learned (see Chap. 2), entails gradual grain size reduction, and this necessarily destroys rock fragments, converting them into monomineralic grains. Thus much compositional maturity is achieved via comminution and mechanical sorting of detrital material in transport.

There are, however, examples of conflicts between the two maturities—compositionally mature rocks that are texturally immature and vice versa. These are rather like textural inversions (see Chap. 2), indicating some departure from "normal" sedimentary processes and begging explanation. There may be some unusually wide disparity between the intensities of mechanical weathering, which promotes textural maturity, and chemical weathering, which promotes compositional maturity. Many workers attribute compositional maturity in part to climate. Certain climates surely are more conducive to rapid chemical degradation of mineral grains in soils, and climate influences the quantity and quality of organically derived chemical weathering agents in the soil. In a humid lowland, detritus may attain compositional maturity even in the absence of the transport that would ordinarily bring about textural maturity. Thus, chemical weathering acting alone may "clean up" a collection of detrital grains but leave the compositionally mature residue of quartz grains largely unsorted and unrounded. Some sandstones contain much composite quartz but little feldspar. Feldspar may be quickly destroyed by chemical weathering in soils, but composite quartz may survive because low relief, perhaps, limited the vigor of processes that cause mechanical disintegration. Topography (relief) influences compositional maturity of detritus also because it controls residence time of mineral grains in soils. The longer a collection of grains remains in the soil, the more mature it becomes, purely as a result of chemical processes.

In desert environments mechanical weathering outperforms chemical weathering, so that desert detritus achieves textural maturity more rapidly than it achieves compositional maturity. In the Great Sand Dunes of Colorado, aeolian sand is texturally mature, but it contains over 50 percent of volcanic rock fragments and only about 28 percent of quartz (Andrews, 1981, p. 283) and thus is compositionally quite immature.

Provenance

As detritus is transported, so is information about its source; moreover, information about the paths taken by the detritus also is preserved. Some information resides in the individual grains; some is contained in the aggregate. Alas, information content fades or becomes less definitive as grains move ever farther from their places of origin, and the analyst's job becomes ever more challenging. Though most sedimentological provenance analysis is performed on mineral grains and rock fragments, provenance of transported plant matter, such as spores and pollens, has been accomplished, and provenance of pore fluids also, including petroleum.

There are various approaches to detrital provenance analysis. One can examine single grains or grain ensembles at a given locality and attempt to match the characteristics of these grains to rocks subcropping at unconformities of the same age as the deposit, these unconformities representing the possible source areas of the detritus. Such a provenance analysis typically is based on distinctive chemical or physical characteristics of certain "indicator grains." Some provenance studies today are based on bulk modal composition of a detrital deposit rather than the characteristics of individual grains. This approach attempts to match different kinds of sandstone with different kinds of large-scale tectonic elements. Alternatively, one can map the dispersal system, tracing out the regional patterns of currents, as revealed by current-generated sedimentary

structures (Fig. 6-11a); one assumes that detritus has moved with the currents. Such vector mapping works even if detrital grains are all the same and not very distinctive. Scalar properties also can be mapped; certain properties of detritus, such as grain size or textural and compositional maturity, vary progressively with distance of transport. Maps of these properties (Fig. 6-11b) help to identify detrital sources because they are, in effect, maps of proximity to source.

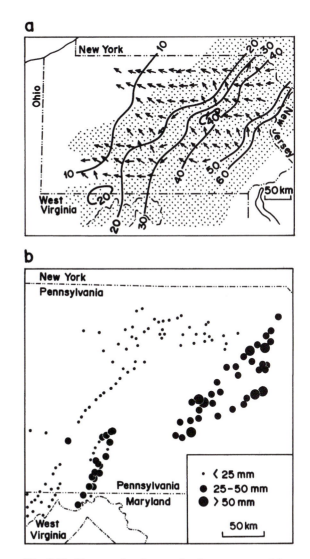

Fig. 6-11. Vector and scalar mapping for provenance: (**a**) paleocurrents and (**b**) size of largest grain in the Pocono Formation. (After Pelletier, 1958, figs. 14, 16.)

Krynine (1940) made a chart (Fig. 6-12) showing how detrital grains flowed from various source terrains to the Third Bradford sand, an oil-reservoir sandstone in the subsurface Upper Devonian Series in western Pennsylvania. Potter and Pryor (1961) presented a flowchart (Fig. 6-13) for Phanerozoic detrital rocks of the upper Mississippi Valley, again showing numerous pathways of detritus from the several source terrains to the numerous present (but not so final) resting places. These charts are based mainly on characteristics of detrital grains. The importance of sedimentary recycling is clearly revealed and indicates the need to distinguish between proximate and ultimate provenance. Complexity, rather than simplicity, seems to be the rule in such charts.

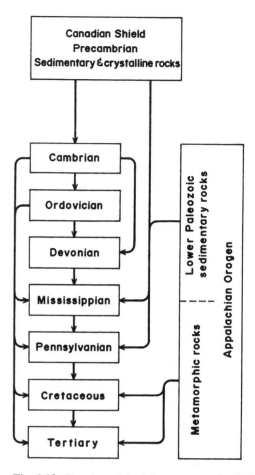

Fig. 6-13. Flowchart of detrital provenance of early Paleozoic sandstones of the upper Mississippi Valley. (After Potter and Pryor, 1961, fig. 14.)

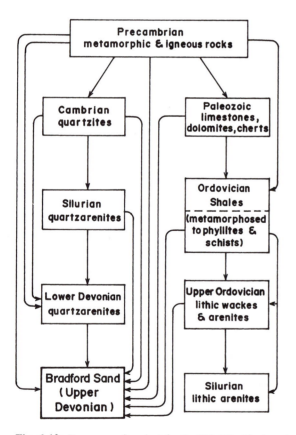

Fig. 6-12. Provenance flowchart for the Third Bradford sand. (After Krynine, 1940, fig. 3.)

Provenance Based on Detrital Grains

Matching of detrital grains to detrital source rocks may in some cases be a simple side-by-side comparison as long as the actual source terrain still exists and is accessible. Distinctive mineral grains—ones with unusual color, luminescence, inclusions, or isotope signature—can be of use in provenance analysis. Absolute dating of some detrital grains indicates the age of some of the source rocks, and some rock fragments contain fossils (especially micro- and nanno-fossils) that can be useful in identifying detrital source. Examination of the detritus tells us only something *about* the source terrain rather than actually identifying it. Thus, characteristics of detrital grains may distinguish between plutonic, volcanic, metamorphic, and sedimentary source terrains. From collections of detrital grains, we might estimate the topography, climate, and tectonics of the source terrain; we might also estimate the distance that the detrital grains have traveled. We scan the horizon for terrains that fit the description—maybe several suspects can be located, perhaps none can be found, perhaps the actual source terrain no longer exists. Some detrital sources have been subducted, and there are instances where a detrital deposit and its source, once side by side, are now an ocean apart, owing to continental fragmentation. A somehow ''definitive'' collection of detrital grains might be shown petrographically to be identical to, or at least indistinguishable from, materials contained in one of a number of plausible source terrains, other suspected or plausible terrains rejected.

Detrital rocks are limited in their provenance information content, and some kinds of grains are more useful than others. Many or most detrital grains in a typical sandstone give minimal or ambiguous information about their parent rock. Larger grains are generally more useful than smaller ones. *Presence* of a given grain type in a detrital rock is more important than its relative abundance (although provenance certainly controls relative abundances, all sedimentary processes—including weathering, abrasion and sorting under transport, and diagenesis—alter these abundances). Absence of a certain grain type may bear on provenance, but it is risky to rely on negative evidence; absence of a ''stable'' grain type in the presence of ''unstable'' grain types might be revealing.

Rock fragments preserve far more compositional and textural evidence of detrital source rocks than single-mineral grains do. A rock fragment contains more provenance information than do any or all of its component grains considered in isolation. Most of the pebbles in a detrital deposit are rock fragments, and diagnosis of the sources of the pebbles involves a simple matching of rock types. One can reasonably infer that the sand and mud that accompany the pebbles in the deposit were derived from the same terrain. Of course, pebbles and larger clasts ordinarily are not far removed from their source, so the provenance analysis of coarse detritus is doubly easy. Wilson's (1970) study of the Beaverhead Conglomerate in southwestern Montana, which was deposited during the Laramide orogeny in Late Cretaceous and Paleocene time, illustrates the specificity that is possible in provenance studies of

gravel deposits. In Beaverhead outcrops between the towns of Lima and Monida (about 20 kilometers apart), Wilson distinguished on ordinary visual lithologic characteristics more than 75 different rock types among the pebbles and cobbles and apparently was able to assign great numbers of these clasts to specific formations, ranging from Precambrian to Cretaceous.

Heavy Minerals

Though nearly always of diminutive proportion in sediments, heavy minerals have been of great value in provenance analysis. Their analysis makes use of natural ''tracers,'' distinctive mineral grains that are nearly always present in terrigenous rocks and whose presence (or perhaps absence) signifies something distinctive about provenance. Their abundance is controlled in part by their abundance in detrital source rocks and in part by the degree to which sedimentary processes destroy or concentrate them. These grains are called ''heavy'' because their densities (or specific gravities) are greater than those of the overwhelmingly more abundant ''light'' mineral grains, such as quartz, the feldspars, clay minerals, and the common carbonate minerals. This property is, incidentally, the basis for separating these grains from the light minerals, so that they may be efficiently and effectively studied (Fig. 6-14).

There are certain frequently encountered associations of heavy minerals that reflect compositions of source terrains. Thus a hornblende-epidote, kyanite-

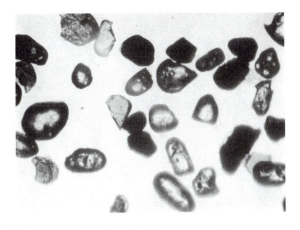

Fig. 6-14. Heavy mineral grains concentrated from a quartzarenite; mostly zircons, tourmalines, and magnetite.

Table 6-1. Heavy Mineral Associations Indicative of Provenance

Provenance	Heavy Mineral Suite
Sedimentary	Rounded zircon, tourmaline, rutile, sphene, magnetite
Low-grade metamorphic, contact metamorphic	Andalusite, staurolite, chondrodite, corundum, topaz, tourmaline, vesuvianite, zoicite, wollastonite, chlorite, muscovite
Higher-grade metamorphic, dynamothermal metamorphic	Garnet, epidote, zoicite, staurolite, kyanite, sillimanite, andalusite, magnetite, ilmenite, sphene, zircon, biotite
Acid igneous	Monazite, sphene, zircon, tourmaline, rutile, magnetite, apatite, muscovite
Basic igneous	Ilmenite, magnetite, anatase, brookite, diopside, rutile, chromite, olivine
Pegmatitic	Tourmaline, beryl, topaz, monazite, cassiterite, muscovite

Source: Modified from Feo-Codecido, 1956.

garnet, or zircon-tourmaline assemblage each points to a particular kind of source rock (Table 6-1). A very complicated heavy mineral assemblage indicates either an intricate source terrain or the mingling of detritus from several unrelated terrains. Simple assemblages either derive from a simple source or result from maturity-promoting sedimentary processes.

Many heavy minerals—such as zircon, tourmaline, rutile, apatite, and magnetite—begin their existence in igneous or metamorphic rocks as euhedral crystals, commonly about the size of sand or silt grains. When the rock weathers, these grains can "pop out" intact and enter the sedimentary cycle as minority black sheep among the fair flock of quartz and feldspar grains. Zircon grains, by virtue of their great hardness and chemical stability, wear very slowly. The extent to which they preserve their original crystal habit is an indication of the length of time that they have resided in abrasive transporting systems. Zircons survive the rigors of multiple burials and exhumations while in the sedimentary cycle. Apatite, on the other hand, is rather soft, and its presence in a sedimentary rock suggests an igneous or metamorphic provenance not far away in time or place.

Zircons can reveal their great age (that is, time elapsed since their crystallization), by the extent to which they have deteriorated as a result of their own chemical makeup. Most zircons contain small amounts of radioactive elements, such as thorium, which, as they decay, release nuclear particles that are energetic and heavy enough to disrupt the crystal structure of their host. Over the long term, zircons turn pink or purple and lose some of their birefringence as a result of this; then they become black or cloudy. Such altered zircons are called *malacons*, and their disrupted structures are said to be *metamict*. Because it takes a very long time for a zircon to become a malacon, most malacons in a sedimentary rock probably came ultimately from a source rock of Precambrian age (Tomita, 1954; Beveridge, 1960, p. 533).

The crystal habits of zircons seem to be related to the types of rocks in which they formed. Not much work has been done on this, and differences in crystal habit are subtle. Poldervaart (1956) pointed out that the temperature of zircon crystallization is such that zircon is late in a basaltic crystallization but early in a granitic crystallization sequence. Thus, zircons in the basaltic igneous rocks have irregular shapes, but those in granitic rocks are mostly euhedral. Zircons of a granitic pluton, according to Poldervaart, commonly show peculiarities of size and shape that are consistent throughout the body and in syngenetic or comagmatic plutons, but distinguishable from the zircons of other, unrelated plutons in the orogenic belt. Poldervaart pointed out also that zircons may occur also as xenocrysts in igneous rocks, that is, contaminants introduced into a melt during assimilation of country rock but that have survived the assimilation intact. These zircons generally are small and round (xenomorphic), their shapes to a large degree inherited from a sedimentary cycle. Winter (1982, p. 426) observed that different bentonite (volcanic ash) beds in the Eifel-Ardennes region (Belgium and Germany) have zircons of distinctly different habits.

Callender and Folk (1958) correlated zircon idiomorphism in the lower Tertiary sands of central Texas with "relative volcanicity," as estimated from abundance of glass shards, detrital plagioclase, volcanic quartz, biotite, apatite, and interbedded bentonites. A zircon-based *volcanicity index*, being the percentage of idiomorphic and hypidiomorphic grains among all the zircons in a sample, seemed to follow closely the other sedimentological evidences of pyroclastic volcanic activity.

Hubert (1962) documented an interdependence between gross mineral composition of sandstones (compositional maturity) and *ZTR index*, the proportion of zircon, tourmaline, and rutile in the non-opaque heavy mineral assemblage. There seems to be a progressive increase in ZTR as proportions of feldspar and igneous and metamorphic rock fragments decrease. The data seem to show, moreover, that modification of the heavy mineral fraction proceeds more slowly than modification of the light mineral fraction (reduction of feldspar and rock fragment content).

Tourmalines have chemistries that are quite variable, and their compositions and sources are reflected in their colors. Schorlites (iron tourmalines), with colors ranging from green to brown or black in thin section, occur in granites and granite pegmatites. Dravites (magnesium tourmalines)—colorless, bluish, or greenish—are characteristic of metamorphic or metasomatic aureoles surrounding granitic plutons. Elbaites (lithium tourmalines) are pegmatite minerals and are blue (indicolite) or pink (rubellite). Krynine (1946) determined that tourmalines of granites are typically dark brown, green, or pink with a greenish cast and commonly are full of bubbles; that pegmatitic tourmalines are blue and pleochroic in shades of mauve and lavender and largely free of inclusions; that tourmalines of slates, phyllites, and schists are variable in color and quite small; and that sedimentary authigenic tourmaline (as overgrowths on detrital grains) is colorless or very pale blue and is rather more common than is generally supposed. Power (1968, p. 1084) indicated that hydrothermal tourmaline is yellow or brown.

Rutile is formed under metamorphic conditions, either in kyanite or sillimanite facies of regional dynamothermal metamorphism or in the highest grade of burial metamorphism (Force, 1980, p. 485). It is virtually non-existent in igneous rocks, and detrital rutiles subjected to low-grade metamorphism do not survive beyond the chlorite facies (due to retrograde metamorphic reactions), so its presence as detrital grains in a sedimentary rock is almost incontrovertible evidence of high-grade metamorphic source terrain. Rutile is, of course, subject to sedimentary recycling.

Apatites occur in all igneous and metamorphic rocks. Their CLM colors are useful in provenance analysis; Smith and Stenstrom (1965, p. 629) noted that apatite from basaltic rocks luminesces yellow, but other apatites luminesce lavender or green. Cryptocrystalline apatite, called collophane, originates in sedimentary environments, probably always as skeletal material or biochemical precipitate.

The color of amphiboles, particularly the hornblendes, is diagnostic of provenance. In metamorphic rocks, amphibole color seems to be related to the thermal grade of metamorphism. Engel et al. (1961, p. 313) found bluish-green hornblende in the higher-temperature hornblende-granulite facies. Colorless or nearly colorless amphiboles—such as anthophyllite, grunerite, and tremolite-actinolite—are almost exclusively metamorphic, so their presence in detrital sediments is a virtually certain indication of metamorphic source. The oxyhornblende of high-temperature hypabyssal and volcanic rocks is brown or red and strongly pleochroic.

The color of micas also is diagnostic of source rock. Engel and Engel (1960, p. 34) observed that biotites change from greenish-brown to red or reddish-brown with increasing degree of metamorphism. Micas of the granites and other acidic rocks are green; micas of basalts and marbles are yellow (Fig. 6-15). Muscovite occurs in a wide range of metamorphic and plutonic rocks but rarely in volcanics. Chlorite is a guide to low-grade metamorphic source areas (some is diagenetic). Pegmatitic micas contain high rubidium and lithium content; the phlogopites of marbles and skarns have higher fluorine content than the phlogopites of basic igneous rocks (Rimsaite, 1964, p. 181).

Garnets and spinels are associated in most minds with metamorphic rocks, but they are also characteristic of mafic igneous rocks; their compositions (re-

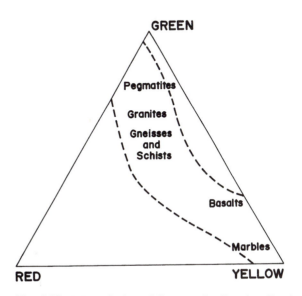

Fig. 6-15. Colors of micas of diverse rocks. (Based on Engel and Engel, 1960, fig. 7.)

flected in their colors and refractive indexes) vary widely. Presumably these variations are of value in provenance analysis. Wright (1938) and Barth (1962, fig. IV-4) verified that the garnets of different parageneses have different compositions (Fig. 6-16a), the garnets of granites and pegmatites are spessartite and almandite, the garnets of peridotites are pyrope and those of biotite schists are almandite, and grossularite

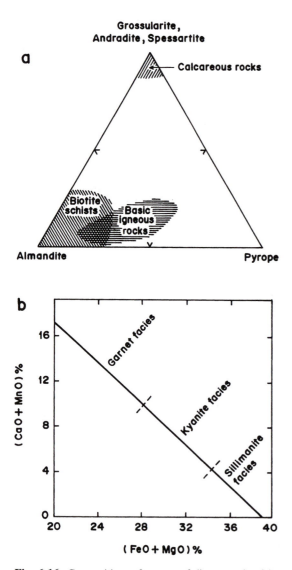

a

**Grossularite,
Andradite, Spessartite**

— Calcareous rocks

Biotite schists   Basic igneous rocks

Almandite   **Pyrope**

b

$(CaO + MnO)$ %

Garnet facies

Kyanite facies

Sillimanite facies

$(FeO + MgO)$ %

Fig. 6-16. Compositions of garnets of diverse rocks: **(a)** according to Barth (1962, fig. IV-4); **(b)** according to Nandi (1967, fig. 1, with permission of The Mineralogical Society).

and andradite are characteristic of calcareous contact metamorphic aureoles. Nandi (1967) showed that garnets of different metamorphic grades have different compositions (Fig. 6-16b).

Many other heavy minerals—epidote, andalusite, kyanite, dumortierite, etc.—occur in detrital sedimentary rocks; Krumbein and Pettijohn (1938, chap. 17) present a careful description of detrital grains of many mineral types and numerous very effective drawings of typical grains.

The average heavy mineral grain may actually be of silt size. Many petrologists who have used heavy minerals in their provenance analyses have noticed that they commonly are concentrated in the shales of a sand-shale succession, and that heavy mineral suites from the shales show greater variety than the suites in the associated sandstones. There might be several causes of this, not the least of which is that intrastratal dissolution impoverishes and alters the heavy mineral suite in sandstones but has far less effect in the less permeable shales (Pettijohn, 1941; Blatt, 1985). Blatt and Sutherland (1969, p. 600) suggested that shales rather than sandstones should be the focus of heavy mineral analyses. A practical advantage of using shales is that heavy minerals are more easily extracted from these rocks, with less risk of physical or chemical damage, than from tightly cemented sandstones.

Some heavy mineral grains, having already undergone abrasion in transporting systems, can develop authigenic overgrowths during subsequent residence in a sediment. If, still later, the host sedimentary rock is exhumed and becomes the source of detritus for a new sedimentary cycle, this overgrown grain again undergoes abrasion. A mineral grain that bears an abraded overgrowth is a grain that has been *recycled* from a sedimentary (or perhaps metasedimentary) source rock.

Spores and pollens (collectively referred to as palynomorphs) are abundant in many sediments and provide a means of provenance analysis where a mineralogically based analysis is impossible. In marine terrigenous muds, for example, heavy mineral grains and other conventional sources of provenance information may be non-diagnostic or altogether lacking. Palynomorphs might reveal the climate or general elevation of a source terrain. Palynomorphs are subject to sedimentary recycling, just as mineral grains are; the presence of recycled spores and pollens, as revealed by their biostratigraphic ages and their physical condition, might constitute proof that the associated mineral grains also were recycled. (It has also been determined that the color and intensity of fluorescence of palynomorphs varies with their age; see Manten, 1966, p. 322.) In the vitrinite reflectance studies undertaken by the petroleum industry, bimodal distributions in reflectance histograms generally are taken to be indication of recycling. Batten (1991) described examples of the use of reworked plant microfossils in provenance analysis and the difficulties involved.

Provenance from Modal Composition

A magmatic arc, thrust belt, or uplifted cratonic block each presumably yields up sandstones of distinctive composition (see Chap. 12). Some provenance studies have taken on the problem of identifying and characterizing gross tectonic elements, based on bulk modal composition of their detrital derivatives (e.g., Dickinson et al., 1983, 1985; Valloni and Mezzadri, 1984). Thus, arkoses are a product of basement uplift; various sublithic sandstones are derived from belts of deformed sedimentary rocks; and certain lithic-feldspathic sandstones indicate magmatic arc derivation. But it is clear from these studies that two different source terrains can produce sandstones that have the same modal composition, and that a given detrital source or a given type of tectonic element can produce different kinds of sandstone. Because of limitations of drainage basin size in such terrains, any single sandstone might not be representative of a large tectonic element, and different drainage basins are likely to produce different sandstones.

If the relationship between modal composition and provenance has proved to be somewhat elusive, these studies have at least increased our awareness of the influence of climate, topography, transport distance, rate of basin subsidence, etc., on sandstone composition. Sedimentary processes alter the modal composition of detritus (Fig. 6-17) and distort and degrade provenance information—soil processes and diagenetic processes selectively alter or destroy detrital grains; transport processes break rock fragments into grains that are not classified as rock fragments; depositional processes mix grains from different sources or hydraulically unmix grains from a single source.

Provenance from Dispersal Patterns

Provenance analysis by mapping of dispersal patterns commonly relies on measurement of some vector

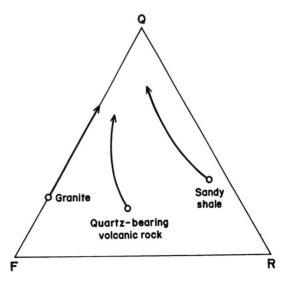

Fig. 6-17. Conceptual effects of sedimentary processes on modal composition of detrital source rocks and detritus. Composition "matures" toward the quartz pole; detritus from different sources becomes more similar.

property of sediments. It is perhaps surprising, and certainly not obvious to the geologist in the field, that a local cross-bed set that he or she is viewing, covering only a few square meters, might be part of an extensive pattern of even continental proportions. Only by making maps of cross-bed dip directions (or of other oriented sedimentary structures) can the geologist discover these regional patterns. The vector field that emerges shows the paths taken by detritus. Of course, the map does not show the whole path; in general, not all parts of the path are accessible or preserved.

Potter and Pettijohn (1977), drawing on a large literature, present numerous examples of basin-wide paleocurrent studies from many different kinds of depositional systems, including fluvial-alluvial, glacial, aeolian, and deep-sea fan. Paleocurrent directions are obtained from many local measurements of cross-bed dip direction, orientation of flutes and grooves or striations, or even rock fabric orientations. Different transport mechanisms and different kinds of depositional basins have given rise to different paleocurrent patterns, so the mapping of paleocurrent vectors informs not only detrital provenance and dispersal but also environments of deposition and sizes and shapes of depositional basins.

In many detrital basins, the areal pattern of certain scalar properties, such as clast sizes, reflects distance from detrital source. Scalar maps of intensity or concentration can be every bit as useful in analysis of provenance and dispersal as any vector map. It is much like sniffing out a gas leak or a dead fish; one is homing in on the noisome source as long as the perceived intensity of the odor keeps increasing. Some of the best examples were produced by students of F. J. Pettijohn: the Pocono Formation in the central Appalachian Mountains (Pelletier, 1958, fig. 16); Tuscarora Formation, (Yeakel, 1959, fig. 14), also in the central Appalachians; Pottsville (Meckel, 1967, fig. 6); Quaternary upland gravels in Maryland. (Schlee, 1957, fig. 12). The Pocono and Tuscarora are deposits in which fluvial processes were important. Systematic variation of clast size related to distance from detrital source are almost universal in such deposits; clast size, moreover, commonly is reflected in some way in the regional thickness patterns. Regional patterns are far less pronounced or even insignificant in aeolian and glacial dispersal systems, but concentrations of certain kinds of detritus show important and useful trends—detrital elements that do not ''travel well'' but wear out to a degree that depends on the distance that they have traveled. For the same reason, there might also be trends in roundness of some detrital element.

Mineral Provinces

A large body that is unified under some distinctive suite of detrital minerals, that is, a distinctive *mineral association*, has been referred to as a *mineralogic province* (see Suttner, 1974, p. 75). A mineralogic province is defined in terms of a detrital source that is distinguishable, in the deposit that it has made, from contemporaneous or contiguous detrital sources. Two sources nourishing some depositional basin thus manage to make deposits that are distinct from one another by virtue of their contrasting detrital mineralogies. They almost certainly are mixed together, however, at their common boundary, and the body of sediment bearing the compositional imprint of both detrital sources might be considered as a third mineral province. Such a mixed province might be a result of the two detrital sources being close together or of two widely separated detrital sources having made their deposits close together. One of the sources might have been very near to the depositional basin and another very far away. The differences between adjacent mineralogic provinces may be quite subtle, and a given stratigraphic formation (formal rock unit) might embrace any number of mineralogic provinces. Moreover, a mineralogic province may cross stratigraphic horizons or boundaries at high angles.

The geometry of a mineralogic province—that is, its geographic extent, whether it is elongate or lobate or fan shaped, its thickness and continuity, the fuzziness of its boundary and its relationship to surrounding provinces, and the way in which the province relates to rock bodies defined on the basis of other facies elements—all obviously are a reflection of sizes and distribution of source terrains, patterns of dispersal, agents of transport, and survivability of the various detrital minerals under abrasion or chemical alteration in transporting environments. Forming some part of the boundary might be the detrital source terrain itself, but a mineralogic province may be far removed from its source terrain, and there are examples of source rocks completely vanished, having sacrificed themselves totally to the generation of their detrital descendants. Some detrital plumes or haloes feed upon themselves; this autocannibalism can cause a geographic and temporal (stratigraphic) trend in maturity and an unconformity (or several of these) locally punctuating the detrital deposit. Probably there are many examples of a detrital source expanding onto its own sink. A rising crystalline basement terrain, for example, sheds immature first-cycle detritus onto its flanks. As uplift continues, early detrital deposits become part of the source terrain, so that juvenile detritus now mixes with reworked detritus.

Orogenic highlands are vigorous producers of detritus, but near to such a terrain there typically are numerous very effective sediment traps. The coarser detritus piles up at the mountain front or in successor basins and intermontane valleys, partly because it is too heavy to be moved. Foreland basins intercept much of the finer detritus shed from developing orogens, but as soon as the basin is filled, incoming detritus passes through to more distant sites. The foreland basin itself eventually is deformed and uplifted, changing, perhaps quite abruptly, from a sediment sink to a sediment source. Large depositional basins commonly have numerous unconformities in the unstable border zones, these unconformities dying out basinward (and merging into a single major unconformity on the detrital source land). This suggests that the coarse clastics characteristic of border zones are more liable to destruction by erosion than the finer-grained rocks of basin centers. (In many cases it results also in the ripping up of non-marine successions and their redeposition as marine successions.)

There have been times, especially in the Paleozoic, when continents were almost totally submerged in the sea. At such times detrital sources were tiny and weak, and not much detrital sedimentation could occur. The shallow sea likely becomes a site of carbonate sedimentation, which buries old detrital sources. Ensuing regression exposes a carbonate terrain. Carbonate terrains, while helping to conceal and isolate silicic source rocks, contribute very little detritus on their own; only after the carbonate veneer has been stripped away and silicic rocks re-exposed can detrital sedimentation resume. This happens gradually—small, simple source areas slowly emerging and growing into larger, more complex ones.

The areal patterns of mineral provinces are controlled or influenced by selective mechanical and chemical destruction of minerals. Certain detrital minerals are much less durable than others and may not have the potential of traveling any great distance from the source. Thus, there is progressive change in the composition of the detrital mineral suite away from the source terrain, and the abundance of some transport-sensitive mineral is an index of distance from source. Because different minerals tend to have different fundamental sizes, the regional differences in heavy mineral suites may reflect sedimentary sorting processes rather than different provenances.

Mineral provinces have important vertical (stratigraphic) relationships also. There may be a certain succession of heavy mineral assemblages that suggests a progressive *unroofing* at the source. For example, erosion of volcanics in a highland source area yields up its hornblende, apatite, and magnetite to a detrital apron in an adjacent lowland; dissection of the volcanics eventually exposes old sedimentary deposits, which begin to shed recycled zircon and tourmaline; finally, an ancient crystalline basement is exhumed, and this releases garnet and kyanite to the lowland deposit. Of course there may be differences in other detrital components, and textural contrasts as well. In the new detrital deposit, there is an inversion of the succession that existed in the source—grains of older, deeper layers of the source occurring at the top of the deposit. Wilson (1970, p. 1850) found that the pebbles and cobbles of the Beaverhead Conglomerate are arranged in inverse order of their stratigraphic parentage, the younger formations represented by clasts in the lower layers of the conglomerate, the older formations represented in the upper layers of the conglomerate. The clarity of the stratigraphic inversion suggests a certain structural and denudational simplicity that source terrains do not necessarily possess. The unroofing of a long-buried cratonic basement or the granitic core of an orogen may make an abrupt and unmistakable mark on the detritus. The age of the bed in which the telltale detritus first makes its appearance indicates time of unroofing, probably as clearly and accurately as possible by any means; in fact, the source area itself, a place of rock destruction, might provide no insight whatsoever into the matter.

Vertical trends in sandstone composition and texture (encompassed under the property that we call maturity) commonly have been interpreted as reflecting gradual change in provenance, such as occurs during gradual burial (onlap) of a detrital source. In the lower Paleozoic of the upper Mississippi Valley, for example, the younger sandstones are the more mature, according to Potter and Pryor (1961). This trend is a result of changing provenance. The oldest sandstones were derived from the crystalline rocks of the Canadian shield, but these sandstones gradually blanketed and isolated the shield to some degree, diminishing it as a source of detritus, so that the younger sandstones were derived less and less from the crystalline basement and more and more from the detrital blanket. Recycling reduced border-zone coarse clastics to finer-grained clastics and improved the textural and compositional maturity of the detritus. In the Pennsylvanian sandstones of southern Illinois, in contrast, maturity *decreases* upward (over a stratigraphic interval of some 600 meters, according to Potter and Glass (1958). This was attributed to gradual unroofing of the Canadian shield. Thus, what was blanketed and concealed in early Paleozoic time came back to light in late Paleozoic time.

Abrupt stratigraphic change in composition of detritus may reflect a reversal in direction of dominant dispersal. In times of tectonic quiescence, detritus is transported from the center of a craton to its edge. But suppose now that an orogen appears at craton margin; dispersal from this new source terrain is toward craton center. Detrital deposits of foreland basins commonly betray this chain of events. For example, Cambrian sandstones of the Appalachian basin of the eastern United States bear heavy mineral assemblages and cross-bedding patterns that indicate eastward dispersal of detritus from the interior. Taconic orogeny in Ordovician time elevated the cratonic margin; in Silurian time, the Taconic highland shed large quantities of sand and gravel westward, back toward the craton. No doubt some of this detritus was reworked from the Cambrian sandstones, so that it is detritus which, after a long sojourn, was returned toward its place of origin. What is now the Amazon basin of South America once drained to the Pacific. Now it is a great foreland basin of the Andes orogen, and detrital transport is toward the opposite (Atlantic) margin of the continent.

Volcanism causes abrupt changes in the composition of detritus in coeval sediments. Not only does the onset of volcanism bring about sudden and drastic changes in detrital source terrains, but eruptive processes can also deliver volcanic materials directly to sedimentary basins without any intervening stages of weathering and sedimentary transport. Detrital sediments may contain the signs of volcanic or intrusive events that are not otherwise evidenced or suspected. In an unusual provenance problem, Mudge and Sheppard (1968) suggested a connection between certain geophysical anomalies in the Sawtooth Range in western Montana and pebbles in Lower Cretaceous sandstones just to the east of the mountain front. In Cretaceous conglomerates in the foreland, Mudge and Sheppard found igneous pebbles and cobbles that they thought could not have been transported more than about 30 kilometers, but no source terrain could be identified. Aeromagnetic and gravity data, however, indicated the existence of two small plutons, now buried beneath some 3000 meters of overthrusted limestones and sandstones. These tectonically buried plutons, they suggested, were at the surface in Early Cretaceous time and shed detritus to east-flowing streams.

Detrital sedimentation at a given locality eventually ceases. This may result from depletion of the source, its burial, or its submergence; derangement of the dispersal system, so that detritus goes somewhere else; or interruption of conditions favorable to deposition at the given locality (and sedimentary bypassing to a new or displaced depocenter). Tectonics is a controlling factor, both at the source area and at the depositional site.

The Mudstones and Siltstones

Most of the total volume of sedimentary rock is mudstone. Mudstone might be viewed as a pervasive medium in which bodies of sandstone and limestone occasionally are deposited and come to be embedded, though there are places where shales are not abundant at all. Because of their small size, clay particles are highly mobile and may enter virtually any depositional environment as a diluent or contaminant. Much mud is associated with river floodplains and deltas and with lakes and tidal flats. The mud was deposited from suspension where running water, having drained into a basin or flat or debouched into standing water, stopped flowing. Some mud is deposited as a thin film where a standing suspension evaporates or leaks through a porous substrate. Mudflows are very concentrated suspensions that can leave thick muddy deposits in certain continental environments. Turbidity currents deposit large quantities of mud in deep marine environments.

Much clay moves through the sedimentary cycle not as single, isolated particles but as little aggregates, ranging in size from a few tens or hundreds of micrometers (flocs) to the size of ordinary sand grains (peloids and rock fragments), all of these aggregates being much larger and heavier than single clay particles. The clay matrix of some sandstones is not mud, but sand—a product of diagenetic loss of definition of shale rock fragments originally present in the deposit. It looks like matrix but was deposited as framework. Benthonic animals pelletize mud by ingesting it, then excreting it as fecal pellets; even certain filter-feeding organisms make peloids out of mud that they have strained out of suspension. Many of the pellets manufactured by organisms are about the size of sand grains and behave hydrodynamically much as sand grains do.

The importance of pelletization of clay was amply demonstrated by Pryor (1975), who scrutinized the activities of the filter-feeding infaunal shrimp *Callianassa* and the polychaete worm *Onuphis* on modern tidal flats and shorefaces of the Atlantic and Gulf coasts of the United States. Pryor noted that there can be as many as 500 *Callianassa*s per square meter of bottom, which collectively could pull out of suspension almost 2 kilograms of solid material per day. The pellets move along the bottom in traction transport, behaving like medium or fine sand.

Inorganic pelletization of mud also has been described. Pryor and van Wie (1971) examined the peculiar "sawdust sands" of the Wilcox Group (Eocene) of western Tennessee and Kentucky, these being sandy shales consisting partly of sand-size aggregates of silt and clay. These rocks, some of which are cross-bedded, could have formed, according to Pryor and van Wie, by flocculation of suspended clays in an agitated fresh- to brackish-water environment. Bowler (1973) described aeolian dunes composed of clay that was transported as peloids (see also Huffman and Price, 1949). The peloids form on seasonally exposed mud flats, apparently by inorganic processes, then are moved during the dry season by strong winds. The mud peloids accumulate into transverse ridges that do not migrate. (Anomalously, windward slopes are steeper than leeward slopes.)

Classification and Properties

Mud is defined (Folk, 1954, p. 346) as terrigenous detritus finer than 63 micrometers embracing all silt and clay-size material; a *mudstone* is a rock containing a substantial quantity of mud. Potter et al. (1980, table 1.2) made a classification of the *fine-grained* terrigenous rocks, in which more than 50 percent of the detrital material is mud, building upon one presented by Folk (1954, p. 348).

Recognizing the importance of bedding and of the distinction that Folk made between silty rocks and clayey rocks, Potter et al. divided the fine-grained rocks into six types (Fig. 6-18). If the silt/clay ratio is greater than 3/1, then the rock is called either *bedded siltstone* or *laminated siltstone*. Rocks with a silt/clay ratio between 3/1 and 1/3 are called *mudstone* or *mudshale*, depending on whether beds are greater than or less than 1 centimeter thick, and those with a silt/clay ratio less than 1/3 are called *claystone* or *clayshale*. Siltstone could, of course, be classified in the same way that the coarser sandstones are classified, though the distinction between framework and matrix may be obscure. Certainly there are some cemented siltstones that are largely free of clay matrix. In many or most mudstones, quartz silt is so finely divided that it cannot be detected in thin section, and other techniques, such as x-ray diffractometry, must be resorted to.

Besides the characteristics that form the basis of this classification, others that are almost certainly important in any mudstone study are the fossil content, character of bioturbation, the clay mineralogy of the rock, and the quantity and composition of admixed carbon-

Percentage clay-size constituents		0-32	33-65	66-100
Field Adjective		Gritty	Loamy	Fat or Slick
NONINDURATED — Beds — >10 mm		BEDDED SILT	BEDDED MUD	BEDDED CLAYMUD
NONINDURATED — Laminae — <10 mm		LAMINATED SILT	LAMINATED MUD	LAMINATED CLAYMUD
INDURATED — Beds — >10 mm		BEDDED SILTSTONE	MUDSTONE	CLAYSTONE
INDURATED — Laminae — <10 mm		LAMINATED SILTSTONE	MUDSHALE	CLAYSHALE
METAMORPHOSED — Degree of metamorphism LO		QUARTZ ARGILLITE	ARGILLITE	
METAMORPHOSED — Degree of metamorphism		QUARTZ SLATE	SLATE	
METAMORPHOSED — HI		PHYLLITE OR MICA SCHIST		

Fig. 6-18. Classification of mudstones. (According to Potter et al., 1980, tbl. 1.2.)

ate minerals, iron oxide and hydroxide minerals, organic matter, etc.

Some of the properties of shales and mudstones are very easy to discern; other properties require special instruments. The texture of mudstones often can be studied with the aid of the petrographic microscope, but the scanning electron microscope yields important results that ordinary microscopy does not reveal. Mineralogy of the clays and other components is routinely determined by x-ray diffraction (for a thorough treatment see Moore and Reynolds, 1989). Heavy mineral grains, plant macerals (kerogen particles), and microfossils are readily concentrated from disintegrated shale samples. In the field, color seems to be a useful property; shales seem to be more varied in this property than the other sedimentary rocks. Grays or black indicate abundant organic matter or finely disseminated metallic sulfides; reds, purples, and browns indicate ferric oxides; greens and yellows suggest greater purity, or the presence of ferrous oxides. Potter et al. (1980, p. 55) related shale color to ratio of ferrous to ferric oxide and organic carbon content. Many mudstones are calcareous—this property is easily diagnosed by the usual simple tests; many mudstones contain sand or silt that one can easily detect by grinding a bit of the rock between one's teeth; and other mudstones are abundantly micaceous, and sparkle in the sun. Gamma radioactivity is a property of mudstones or shales that is routinely measured in oil wells and can be measured easily in outcrops as well. It relates intimately to composition. The way that a mudstone breaks—as observed on weathered outcrops and in well-cuttings samples—commonly is distinctive, controlled mainly by texture or fabric. Fissility is a megascopic manifestation of clay particle alignment. Many mudstones are quite brittle and break into small, sharp, equant chips; other shales break into splinters or "pencils." Some shales, upon weathering, disintegrate directly into mud; mudstones containing abundant expandable clays may develop a distinctive "popcorn" surface (Fig. 6-19).

Mudstones commonly contain lenses or streaks of coarser detritus—sand, silt, or calcareous particles (skeletals and such) that may exhibit sedimentary structures, such as ripples or cross-bedding, that are unlikely or impossible in the associated mudstones but relate to the origin of the mud (a sort of guilt by association). Many depositional regimes have at their disposal a mixture of sand and mud but sort it out and organize it quickly into fairly distinct sandy deposits and muddy deposits, perhaps thinly interbedded. There can be little doubt that these closely associated deposits are also closely related genetically. Petrography of the thin sandstone or limestone layers in shales thus reveals as much about the origin of the shales as it does about the origin of the rocks actually examined.

Composition

Bodies of mudstone and their incorporated pore waters are complex chemical systems, and they can form in and respond mineralogically to practically any chemical environment. Thus, we find calcareous shales, siliceous shales, sodic or potassic shales, and shales with abundant authigenic feldspars, shales rich in organics, and phosphatic, iron-rich, and evaporite-bearing shales. Shaw and Weaver's (1965) result that Phanerozoic mudstones of North America contain an average of 61 percent clay minerals and Picard's (1971) finding that the average clay content of modern muds is but 40 percent suggest that there is much detrital quartz and feldspar not contained in sandstones, much carbonate not contained in limestones, and much organic matter not contained in coals. (There is also

Fig. 6-19. "Popcorn" surface on a smectite-rich deposit.

much clay not contained in mudstones.) Mudstones or shales seem to be the least differentiated of the major rock types.

The composition of mudstones is, like that of the other detrital rocks, controlled partly by provenance—the composition of the source rock and the relief, climate, and vegetation of the source terrain. The most important aspect of composition of the parent rock is the content of alkalies, especially sodium and potassium, and alkaline earths, such as calcium and magnesium (Grim, 1968, p. 504). Thus, basic igneous rocks, with their abundant calcium and magnesium, are more apt to make smectite clays, while granites and syenites, rich in the alkalies instead, favor the production of illite. Smectite clays in deep-ocean sediments in many cases seem to have been derived from nearby volcanics rather than from continental sources. Climate, relief, and vegetation play an important role, by influencing the moisture content and general direction of water movement in soils and their parent materials; clay mineralogy is also influenced by temperature, pH, and organic chemistry of the soils as well as residence time of clay particles in the soils. The suite of clay minerals carried by the Amazon River out of its tropical forest basin is different from the suite of clays issuing from the mouth of the Mississippi, draining a basin of temperate forest and grassland, or the arctic tundra-derived clays of the Yukon River. These differences are reflected in the distribution of terrestrially derived clay minerals in oceanic abyssal sediments (see Chap. 8).

But the influence of parent rock and of the weathering environment diminish as the clays make their way out of their places of origin. Important changes are imposed by depositional environment. Clay minerals "degrade" in certain weathering and transporting environments, then they are reconstituted in marine depositional environments. Degraded clays deposited in marine waters absorb potassium and magnesium from the water, and become better-crystallized illites and chlorites. Clay minerals formed in soils and in transport in rivers have been observed to change into other clay minerals within a few years of entering the marine environment, though adsorbed organic matter on three-layer clay particle surfaces apparently can suppress these mineralogical changes.

Finally, as shales age under diagenesis and enter progressively harsher chemical and physical conditions under deep burial, their mineralogies change in ways that are more or less predictable. Thus, clay minerals can indicate diagenetic grade, analogously to the various minerals that define the various grades of metamorphism. Also under diagenesis, shales may eventually expel certain reactive fluids that cause chemical alteration of adjacent sandstones and limestones. (We examine these matters further in Chap. 11.)

Texture and Fabric

The texture and fabric of mudstones and shales is controlled or influenced by factors different from those that control textures of the coarser rocks. The sizes and shapes of clay particles are controlled by crystal chemistry and structure rather than by breakage and abrasion in the transporting environment. Clay particles are of such small size that electrical charges on their surfaces are an important particle-to-particle organizing force, causing flocculation of suspended particles. This hastens their sedimentation and imparts special fabrics to the deposit.

The fabric of a mud deposit, especially the degree to which platy clay particles are aligned and the manner in which they are packed together, is probably determined by degree of flocculation at time of deposition. The nature of admixed non-clay particles also must be important. Aggregates of flakes or tabulae can show *face-to-face*, *face-to-edge*, and *edge-to-edge* relationships between adjacent particles (Fig. 6-20). Dominance of one or another of these grain-to-grain relationships results in oriented, or "house-of-cards," or "honeycomb" fabric, respectively. Bioturbation and organic pelletization disrupts these fabrics (see O'Brien, 1987), and diagenetic compaction brings grains closer together, reorients them, and blurs the boundaries between clusters or flocs (Reynolds and Gorsline, 1992, p. 51). Odom (1967) examined the fabric contrasts between fissile shales and other mudstones displaying no kind of fissility. X-ray diffraction traces revealed that clay particles are more strongly oriented in fissile shales than in non-fissile mudstones, as expected. That fissile and non-fissile mudstones commonly are interbedded suggests that burial compaction is not necessarily the controlling influence on degree of clay-particle orientation. Rather, it seemed to Odom that physical and chemical factors in the depositional environment are responsible, especially insofar as these factors control clay-particle flocculation. High organic matter content seems to enhance the development of preferred orientation or fissility in

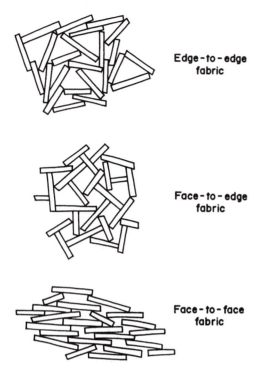

Edge-to-edge
fabric

Face-to-edge
fabric

Face-to-face
fabric

Fig. 6-20. Clay particle-to-particle relationships.

shales; the presence of other components, such as carbonate, appears to inhibit development of fissility.

Oriented fabric produces rather striking extinction contrast in shale thin sections rotated between crossed polarizers. At maximum extinction, which normally occurs when bedding is parallel (or perpendicular) to plane of polarization, only those clay particles are visible that are *not* parallel to the bedding. "Mass" dichroism in some shales seems to be caused by abundant kerogen, its large molecular chains or sheets aligned parallel to the clay fabric.

Sedimentary structure in mudstones includes ordinary bedding, lamination, varves, size grading, flutes and grooves, organic traces and burrows, mud cracks, and rainprints. Some shales contain bentonite beds (altered volcanic ash); characteristically they are laterally persistent light gray or yellowish streaks or bands, some may be only a few millimeters thick. There may also be concretions (see Chap. 11). Many sedimentary structures of shales depend on or relate to contacts with other, coarser, sediment layers. Many shales are deposited as long, gentle clinoforms, easily missed in

small or disadvantageously oriented outcrops; the bed set might be so thick that there is no convenient horizontal reference. But inclined bedding in shales commonly is plainly evident in subsurface log sections or seismic profiles.

Earlier we learned that sedimentary processes can separate sand grains mineralogically. Clay particles also can experience mineralogical differentiation. Kaolinite and smectite, it appears, flocculate in marine environments more rapidly than does illite, which is thus carried farther to sea. Edzwald and O'Melia (1975, p. 43) showed that illite is deposited farther downstream in the Pamlico River estuary in South Carolina and suggested that this is because illite suspensions are more stable than kaolinite suspensions. Pryor and Glass (1961, p. 50) found evidence of sedimentary differentiation of clay minerals in Cretaceous and Tertiary sediments of the upper regions of the Mississippi embayment, owing to differential settling, which they attributed to differences in particle size, flocculation rate, and floc size. Many factors surely are involved, as patterns of clay mineral occurrence vary. Whether or not flocculation occurs depends on the clay mineral type, concentration of the clay particles in suspension, and concentration of organic surfactants and salinity (O'Brien, 1970, p. 241; Reynolds and Gorsline, 1992, p. 51).

Provenance

Clay minerals are almost entirely the products of weathering of non-quartz silicate rock materials contained in igneous and metamorphic rocks. But most clay particles in a typical mudstone probably have been recycled from sedimentary rocks. In view of their softness, mudstones or shales are eminently recyclable, and the overwhelming majority of clay and silt particles in sedimentary rocks probably are multicyclic.

Most provenance analyses of shales or mudstones have been based on the non-clay mineral components (see Blatt, 1985). Many shales contain heavy mineral grains that can be useful in provenance analysis. Blatt and Totten (1981) and Jones and Blatt (1984) used the percentage and grain size of admixed detrital quartz to determine direction of detrital transport in certain shales. In another shale, Blatt and Caprara (1985) found a rapid decline in detrital feldspar away from the source.

Shales commonly display directional properties

useful in paleocurrent studies. Some shales contain small plant fragments or skeletal grains that have been oriented by currents. Current-induced anisotropies in a shale may be revealed in the field as oriented ellipsoidal concretions or elliptical reduction spots on bedding planes.

Breccias and Conglomerates

Rocks composed of angular clasts, typically rock fragments of granule or pebble size or much larger, are a special kind of sedimentary rock, called **breccia**, that the ordinary classification schemes do not adequately address. Many breccias can be looked upon simply as ''proto-conglomerates,'' that is, deposits of detrital grains that were not sufficiently transported to become rounded. Minor breccias of this kind are liberally interspersed among the limestones, shales, and sandstones and created in the depositional environment by glacial plucking, heavy surf against a sea cliff or a reef, slope failure, or desiccation. Many chert breccias are a residue of the weathering of carbonate terrains. Other breccias have unusual or special origins; there are breccias caused by plutonic, volcanic, and tectonic processes, and by impact by extraterrestrial objects; some form diagenetically. In many such breccias, clast motion has been negligible, only a few meters or even no more than a centimeter; clasts may have moved in concert, like a school of fish. It is commonly a single, instantaneous event.

Various textural and compositional characteristics of breccias help us to distinguish among different breccias and to determine their respective origins. One should make the following observations and distinctions:

1. Clasts of single lithic type (**monomict** or **oligomict**) or of widely variable types (**polymict**); lithic types in clasts similar to or different from rocks bounding the breccia body; durable rock fragments only, or mixture of hard and soft clasts.
2. Clasts closely fitted, loosely fitted, or spatially independent.
3. Clast shape and clast-to-clast contacts indicating either plasticity of clasts or brittle behavior at time of sedimentation.
4. Spaces between clasts filled with cement or with detrital matrix; matrix support or clast support.
5. Individual clasts with weathering rinds or coats all around, or developed only on one side, or no indication of weathering.

6. Clast size and size grading, evidence of bedding; imbrications or other spacial alignments.
7. Overall size and shape of the deposit; sharp lower contact or sharp upper contact; discordant or concordant relations with bounding rocks.

Richter and Fuchtbauer (1981) recognized 11 genetic types of breccia and pointed out that breccias are far more diverse than are conglomerates. Breccias were classified (Fig. 6-21) according to degree of relative movement of the constituent clasts (clasts rounded as opposed to clasts fitted) and degree of intermixing of clasts (monomict as opposed to polymict).

Morrow (1982) made a descriptive field classification of *carbonate* breccias that distinguishes between **packbreccia** and **floatbreccia**, depending on whether the breccia is clast-supported or matrix-supported, and between **crackle**, **mosaic**, and **rubble breccias**, depending on whether the clasts are fitted, mildly disoriented, or totally independent (Fig. 6-22). Some packbreccias have a particulate matrix and some are cemented. Morrow conceded that characteristics other

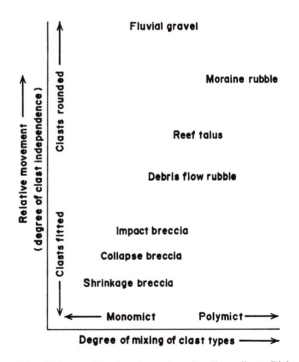

Fig. 6-21. Classification of some breccias. (According to Richter and Fuchtbauer, 1981, table 1, with permission of Deutsche Geologische Gesellschaft.)

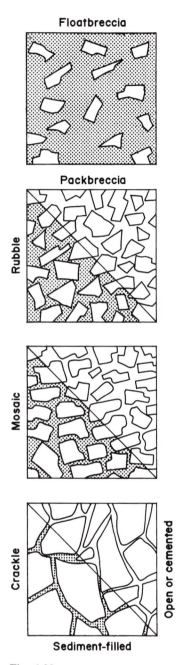

Fig. 6-22. Classification of breccias. (According to Morrow, 1982, fig. 1.)

than fabric—such as size and shape of the breccia body, boundary relations, and stratigraphic setting—provide the definitive evidence of origin.

In some depositional environments, sediments develop a certain coherence or rigidity early on, then they break up into angular clasts, still under the influence of the depositional environment. This results in *syndepositional breccias*. Clasts could be produced by desiccation on mudflats or slumping of the banks of tidal channels. Such clasts might show evidence of plastic deformation, having been squeezed together to some extent under their own weight. Clearly, they could not survive much transport; mere contact with flowing water would tend to round them or destroy them.

Some breccias result from short, brief transport of rock materials by gravity acting directly, as in a rock slide. A *talus breccia*, being an accumulation of rock-fall clasts, is created by weathering processes; bedding and tectonic joints in the source rocks control to some extent the sizes and shapes of the clasts. One expects that talus breccias are in steeply inclined contact with their source rocks. Probably all such breccias are oligomictic, containing only one or two or three clast lithotypes. Talus breccias appear not to be very abundant in the rock record, because they form in places of high relief or tectonic activity, such as in mountainous terrains, in canyons, or along sea cliffs, and are likely to be eroded in the longer term. A special kind of talus breccia with much higher preservation potential is the *reef-talus breccia*, which accumulates under water on the foreslope of a reef that is crumbling in the face of heavy wave action.

An *olistostrome* is a chaotic, heterogeneous deposit of rock materials that have slid down a submarine slope (Abbate et al., 1970, p. 522). (It is distinguished from the *melange*, which is formed by tectonic shearing near subduction zones.) Some of the "clasts" of an olistostrome are in fact large blocks or slabs, even as much as 100 or 200 meters thick and a kilometer or two in breadth, that maintained their integrity as they slid down some long muddy slope; these large objects are called *olistoliths*.

Collapse breccias are most characteristically associated with karsts. They are a result of collapse of underground space, such as limestone caverns (see Chap. 9). Other collapse breccias, such as the Mackinac Breccia (Devonian) in Michigan, formed when deeply buried evaporite beds dissolved. Evaporite-dissolution-collapse breccias form after burial, even deep burial, of the soluble formation, and commonly involve substantial thicknesses of overlying rock, including non-carbonate rocks. Many karst breccias, in contrast, form before deposition of overlying sediments and thus involve only the dissolving carbonate rocks; karst breccias generally are closely associated with an unconformity and with the distinctive soils that develop on limestones in humid climates. Some karst breccias contain fragments of *speleothems* (dripstones) and cross-bedded or draped layers of sand or mud deposited by running water. During an ensuing marine transgression, terrigenous sand and mud leak down into a karst, filling underground spaces, including spaces between breccia clasts.

Impact breccias are created when a high-velocity meteoroid strikes the surface. The target is shattered at the point of impact; the resulting clasts move only short distances and hardly at all relative to their immediate neighbors; but some clastic material is excavated from ''ground zero'' and accumulates as an ejecta blanket. The impulse also might cause some finely divided clastic material to be injected into fractures in the target. A special kind of clast, probably exclusive to impact breccias, is the **shatter cone** (Fig. 6-23). Impact generates a shock wave that creates cone-shaped fractures in the target. Apices of the cones tend to point toward ground zero, though some oppositely oriented cones might be created by the dilatation that follows the sudden compression. Impact breccias may contain high-pressure minerals such as coesite, stishovite, or diamond.

Conglomerates and gravels are composed of coarse transported detritus, wherein pebbles have achieved some degree of roundness as a result of transport. Some form subaerially, some in deep water, some under an ice sheet. Gravel deposits are made by debris flows, streams, waves on beaches, and glaciers. Gravel deposits may be cone- or fan-shaped, lobe- or tongue-shaped, wedge or blanket, or elongate piles (as glacial moraines). Some gravel deposits fill a complex underlying topography or conform to the shape of a channel in which they were deposited; some gravel bodies have complex upper surfaces or are shaped like longitudinal or transverse bars. Thin, isolated conglomerates (or breccias) in a stratigraphic succession may be *lag beds* or deflation armors. As a chert-bearing limestone weathers, the chert may accumulate in place as a residuum. Or the wind, or wave action during a transgression, may winnow away the finer grains of a loose deposit, concentrating the coarser grains as a thin

Fig. 6-23. A shatter cone from an impact structure in Indiana.

conglomerate. Some gravels are composed almost entirely of granitic stones, others of limestone clasts, or chert clasts, or quartzite clasts; others comprise a great variety of clasts.

The pebbles of lag beds commonly are blackened. The well-known *desert varnish* is, according to Hunt (1954, p. 183), a dark stain of iron and manganese oxides on rock surfaces. Engel and Sharp (1958, p. 515) pointed to varnished pebbles, in a disused roadbed in the desert, that must have developed their oxide coatings within 25 years duration. But blackening is not limited to desert places. Strasser (1984, p. 1097) concluded that blackened limestone pebbles are a feature of coastal carbonate environments, associated with emergence and subaerial exposure; blackening is caused by organic matter and iron sulfides, a conclusion reached also by Ward et al. (1970, p. 549). Shinn and Lidz (1988, p. 117) suggested that some limestone pebbles at subaerial unconformities are blackened by forest fires.

Walker (1975) identified four properties of transported conglomerates that are useful in diagnosing their origins (Fig. 6-24):

1. Grain size distribution, and whether the deposit is clast-supported or matrix-supported.
2. Long-axis preferred orientation or random orienta-

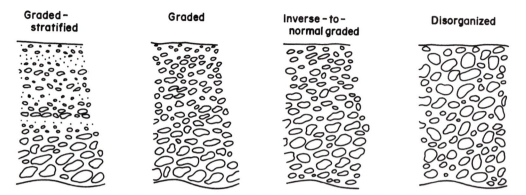

Fig. 6-24. Four kinds of conglomerate. (According to Walker, 1978, fig. 10.)

tions; whether long axis or intermediate axis is imbricated.
3. Stratified or not; stratification horizontal or inclined; layers sharply bounded or intergradational
4. Normal grading, reverse grading, or no size grading at all.

Walker and Mutti (1973) suggested that a distinction be made between *organized* and *disorganized conglomerates*, based on the criteria listed above. As with breccias, the size and shape of a conglomerate body, nature of boundaries and bounding lithotypes, and clast diversity (whether oligo- or polymictic) may be useful criteria of origin. Stream gravels of alluvial fans and alluvial plains typically are bedded or cross-bedded, are clast-supported, and show grading, clast orientation, and imbrication; they are organized gravels and conglomerates. Glacial till typically consists of disorganized gravel, being non-bedded or crudely bedded, very poorly sorted, commonly matrix-supported, and perhaps lacking any kind of grain orientation. A gravel that is matrix-supported or otherwise contains much mud matrix is referred to as a *diamictite*. Conglomerates or gravels deposited from debris flows may be organized or not, depending on the viscosity of the flowing mass, its mud content, and maybe the distance traveled. We examine the characteristics of various gravels and conglomerates at greater length in Chapter 8.

7

Stratigraphy

The study of successions of stratified rocks, and of rocks in their stratigraphic context, is commonly the main event at outcrops of sedimentary rocks. Also, some stratigraphy is conducted on youthful sediments not yet lithified or rendered as fully or readily accessible as an outcrop. In such cases, maybe only the surface of the deposit is examined, or "outcrops" are made, either with pick and shovel or by means of the core sampler or box sampler. The stratigrapher's job is to describe the sedimentary record (or some small part of it), observing the compositions and textures of the stratified sediments and the sedimentary structures and fossils that they contain; to determine thicknesses, shapes, and ages of deposits; to identify and analyze vertical and lateral changes in deposits; to correlate deposits from site to site, thus establishing lateral continuity and synchroneity; to understand deposits in the context of other deposits or other geological events; and to determine the origin of deposits, or the physical, chemical, biological, and tectonic conditions under which they formed. A stratigrapher is an historian, examining the "documents" and "artifacts" of the distant past—commonly incomplete, mangled, ambiguous, or not fully accessible; sometimes suspect, misleading, or of unknown significance; and occasionally available in such great wealth as to require careful selection or culling and summarization based on (one hopes) representative samples.

Much stratigraphy is conducted in the deep subsurface, by examination of rock samples brought up from deep wells, or even without the benefit of rock samples at all, by geophysical methods (seismic stratigraphy), or by so-called petrophysical methods (the well log). Data recovered from deep wells are expensive data and are commonly the only data we have on deeply buried stratigraphic successions. In view of this, governments and commercial concerns strive to preserve this infor-

mation, long after its immediate use to the oil-driller has faded, as libraries of drill cuttings and core samples and as repositories of well logs, typically managed by large computer systems. Without the storehouses maintained by these organizations, much geological knowledge would not exist. But most of the contents of these storehouses have not even begun to be studied adequately.

Some deep wells have been drilled by publicly financed organizations for the express and sole purpose of advancing scientific knowledge. Perhaps the most famous of these was the Mohole, undertaken by the American Miscellaneous Society (AMSOC) in the 1960s and funded by the National Science Foundation (NSF). It was not so much a stratigraphic project as a deep-crustal geophysical exploration. The 1980s found similarly ambitious single-well projects in the Soviet Union. The famous Munsterland-1 well in Germany has been an oracle for the diageneticist. The Deep-Sea Drilling Project (DSDP), launched in 1968 by the Joint Oceanographic Institutions Deep Earth Sampling consortium (JOIDES) and funded by the NSF, has accomplished extensive sampling of sediments of all the oceans' floors. Its great success was probably the inspiration for the Continental Scientific Drilling Program (CSDP), conceived in about 1974 but never adequately funded. "Dry holes," that is, wells drilled in pursuit of the elusive oil pool but that make no financial return on their investment, nonetheless add to our knowledge and are often darkly or resignedly referred to by their promoters as "contributions to science."

Even some surface sections have been officially preserved and dedicated to the stratigrapher and to the public at large—the Morrison and Dakota Formations roadcut in the foothills of the Front Range west of Denver, Colorado; the dinosaur graveyard, underroof, at Dinosaur National Monument in Colorado and Utah; the monumental roadcut in Birmingham, Alabama (Fig. 7-1), touted as the largest in the world, excavated through the Chicamauga Group and Red Mountain Formation and equipped with catwalks and formation markers. One might also view the spectacular stratigraphic section laid bare in the walls of the Grand Canyon as an officially preserved stratigraphic monument and similar (though perhaps more modest) exposures in national parks and reservations everywhere.

Time in Stratigraphy—Correlation and Dating

Stratigraphy keeps a chronological record of events but does not directly record actual dates, and sometimes the chronological record is jumbled. Establishing time and sequence has been the stratigrapher's most consuming task. It seems to rely on five fundamental principles: (1) *superposition*, (2) *cause-and-effect relationships*, (3) *evolving systems effect*, (4) *accumulative effect* of static processes, and (5) *cyclic systems effect*. Let us consider examples of each of these principles:

Superposition

The first is most familiar to the stratigrapher, who knows that younger beds of sediment are deposited on top of older beds and thus are higher in a stratigraphic succession. This is to say that the "arrow of time" in a succession of beds points upward, oppositely to the direction of gravity. It may happen, of course, that a stratigraphic succession comes to be structurally disturbed, so that the direction of the arrow of time is no longer certain, and it is up to the stratigrapher to determine, commonly by observing gravity effects "frozen" into the beds (geopetals in a broad sense), the nature of the misorientation. Superposition also plays a role in situations where the effect of gravity is not significant. In an ooid or a petrified tree trunk, the outer layers are younger than the inner ones; in a pore lined with two or more layers of cement, the innermost layer is the youngest.

Cause-and-Effect Relationships

The principle of cause and effect states that younger events affect older results, and not vice-versa. Cross-cutting relationships are a familiar application. Thus,

Fig. 7-1. The Red Mountain cut in Birmingham, Alabama.

an animal burrow is younger than the bedding laminae that it disrupts. A fracture or vug that intersects a chemical cement is younger than the cement. A younger lunar crater partly disrupts the shape or integrity of an older crater. Certain textural or fabric relationships, observed microscopically, can be useful in establishing sequence of crystallization or recrystallization. Sometimes this entails rather complex cause-and-effect arguments. Dorobek (1987, p. 507) pointed to an interesting relationship between partially silicified pelmatozoan plates (in Devonian limestones in the central Appalachians) and calcite rim cements. The cement overlies only the non-silicified parts, indicating that the silicification occurred first—the silicified parts of a skeletal grain apparently were an unsuitable substrate for precipitation of calcite cement. The principles of superposition and cross-cutting relations permit temporal ordering of events that are millions of years or milliseconds apart.

Evolving Systems Effect

Astronomers determine the age of a star by its intrinsic brightness (luminosity) in relation to its color (or temperature). The position of the star on a plot of luminosity versus temperature (Hertzsprung-Russell diagram) is an indication of the age of the star, because stars evolve, or change in predictable ways, as they get older. We will refer to this as an evolving systems effect. To the earth-oriented scientist, the most significant example of an evolving system is, of course, the biosphere. Organisms have evolved, and the biosphere has become more diverse and complex, and fossil content of sedimentary rocks records the progress of evolution. The fossil content of some particular bed commonly establishes the temporal position of that bed with respect to the entire history of organic evolution.

Accumulative Effect

Cosmologists estimate the age of the universe (time elapsed since the ''Big Bang'') from the present size of the cosmos and its rate of expansion (Hubble constant). We can judge the age of a person by the accumulated ravages of time on face and hands and the age of a horse by the condition of the teeth. The ages or durations of things evidently can be estimated from an accumulative effect. It was the basis for the earliest scientific estimates of the age of the earth—estimates based on the salinity of the oceans or on the extent to which the earth has cooled from an initial high temperature. Today we know that these estimates were founded on certain faulty premises, but modern methods based on accumulative effect are very important and apparently very accurate. We measure the fluorine content in fossil bones, or numbers of fission tracks in certain minerals, or number of mutations in fossil cytochrome c, or degree of surface degradation (of different lunar terrains, for example), and we establish the relative ages of glacial drift plains by the degree of their dissection or the thicknesses of soils developed on them, or the thicknesses of weathering rinds on the clasts that they contain. Certainly the most important, we measure the parent-daughter ratios in radioactive materials. The changes that we see and measure occur at rates that we know, at least roughly; so the magnitude of change tells time elapsed.

Cyclic Systems Effect

Today, we keep accurate time by counting cycles, either the rotations and revolutions of the earth, or the carefully controlled ticks of a mechanical clock, or the resonant vibrations of piezoelectric crystals or of cesium atoms in a magnetic field. Sediments and certain of their components record the ticks of cosmic clocks; certain sedimentary structures or certain mineral grains or fossils contain these records. There is the daily ticking or circadian rhythm, the monthly tick of the tidal cycle, the annual tick of earth's revolution about the sun, and the 11-year tick of the so-called solar maximum, when magnetic storm activity is at its peak. We all are familiar with the use of the annual tree rings in dating events of the past; for western North America, this method has achieved a chronology of nine millennia. Longer cycles, tens or hundreds of thousands of years in length, are related to perturbations of the earth's orbital motions. These so-called Milankovitch cycles (see Chap. 12) may be a cause of climatic variations, including glaciations that bring global sea-level changes. All of these cycles, where recorded stratigraphically, potentially are globally correlatable. The cyclic-systems records, coupled with the various accumulative effects described above, are the basis for *absolute dating*. According to Fischer and Bottjer (1991, p. 1067), ''cyclostratigraphy promises to refine the results of radiometric timing by orders of magnitude.''

Formal Stratigraphic Units

Stratigraphers make careful distinction between rock units and time units (but also recognize a kind of hybrid called the time-rock unit). A *rock unit*, or *lithostratigraphic unit*, is simply a definable body of rock, set up to be in some way "logical" and convenient to the purposes of the stratigrapher, structural geologist, or volcanologist with no other connotation of age, thickness, shape, homogeneity, continuity, induration, or origin. It is defined and distinguished on the basis of its lithology, and it has a stratigraphic "context," a spatial relationship to other units; and it is desirable that the boundaries of the unit coincide with any obvious natural boundaries, such as unconformities or other discordant contacts.

One might consider the total rock fill of some large "basin" as comprising a number of close-fitting distinguishable parts and a rock unit as any one of these parts (Fig. 7-2). Different stratigraphers (or other geologists) might see "parts" differently, and, of course, no one has a detailed, unobscured, all-embracing three-dimensional "x-ray" view of the basin fill when setting out to define and delineate some formal system of rock units for the basin; stratigraphic nomenclatures for basins are protracted joint efforts of numerous sci-

entists. Moreover, a basin fill may be substantially eroded, so that relationships are obscured. Finally, one recognizes that one system might be more logical or more useful or practical or somehow more enlightening than some other system. Thus, any such system is subject to criticism, revision, and refinement as knowledge and understanding of the basin increases and improves. An unfortunate result is a muddled nomenclature of redefined or renamed rock units, mistaken identities, and miscorrelations. Many stratigraphic rock units are not well defined. One need only spend a few minutes with the multivolume *Lexicon of Geologic Names of the United States*, published and updated annually by the United States Geological Survey, to appreciate the severity of the problem.

The stratigrapher is responsible for naming the rock units he or she describes, providing the unit with a proper name in the same sense as a person or an organization has a name. Naming rock units is a formal procedure that today is governed by a set of rules contained in the *International Stratigraphic Guide* (Hedberg, ed., 1976), or the British *Guide to Stratigraphic Procedure* (Holland et al., 1978), or the *North American Stratigraphic Code* (NACSN, 1983). But in spite of the *Code* or the *Guide*, the existing stratigraphic nomenclature has many imperfections. Today, a rock unit in North America is named after a geographic locality, such as a town or mountain range or even an oil well. But many rock units named long ago (before the rules were made) were named for some striking characteristic of the unit—thus, the Mauve Formation and the Redwall Formation of the Grand Canyon in Arizona (and there are many European examples). Too often, names proliferate beyond reason and beyond usefulness. Sometimes this happens because not sufficient work had yet been done to determine whether or not the body of rock at hand is but an extension of a rock unit that already has a name. Rock units correlative with or coextensive with the rocks at hand may be judged by one stratigrapher as one and the same formation and with equal justification by another stratigrapher as distinct formations. Sometimes a rock unit that already has a name is subdivided (or several rock units are combined) into new units, each of which demands a formal name; this is occasionally useful but can lead to unwieldy and burdensome nomenclatures. Economics inspired the proliferated nomenclature in the Pennsylvanian and Permian in the Appalachian coal fields of the eastern United States. There, each coal bed, each sandstone or shale or limestone unit, even if only a foot or two in thickness or occurring

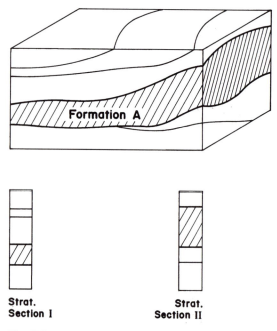

Fig. 7-2. Part of a rock-filled basin divided into formations.

only in the vicinity of a single coal mine, has a formal name. It is very common for rock bodies that transcend political boundaries to undergo an abrupt name change at the border. This is sometimes a result of a language barrier or other failure of communication across borders, but probably chauvinism and provincialism play a role too. Different nomenclatures arise from the study of widely scattered outcrops on the periphery of a large basin; things go wrong when subsurface exploration within the basin finally begins. How should the subsurface stratigraphy be set up? Import one of the surrounding nomenclatures for rocks in the subsurface? Merge the best elements of all surrounding nomenclatures? Or devise yet another nomenclature and hope that it can be related to all of the surrounding ones? Much stratigraphic friction, controversy, and revision occurs when two stratigraphers, having begun their stratigraphic projects in two widely separated locales and working independently, gradually expand their "spheres of influence" until, one day, they come to overlap. Each worker has subdivided the stratigraphic column according to his or her own visions and prejudices and erected his or her own nomenclature, but the two systems, side by side, do not "fit together." Each worker wants, of course, to preserve and defend his or her own nomenclatural system, and each may quite validly claim precedence for having named some rock unit. Ensuing border skirmishes and litigation in the court of scientific opinion may last for years, generating much heat but not much light.

When a new rock unit name is proposed, it is accompanied by a *type section* or *stratotype*, which furnishes the definition of the new formal rock unit. Unfortunately, a type section commonly turns out not to be the best or most typical example of the rock unit, even though it was intended, presumably, to be the most authoritative. Some type sections are contained in deep wells (or at a repository of samples and logs from this well) and are described on the basis of core or cuttings and log profile (the log alone is not sufficient—actual rock samples must be involved in the process). Some rock units, named before the Stratigraphic Code was installed, have no type section; a famous example is the formidable Cambrian-Ordovician Knox Dolomite of the Appalachian region of the United States. After a type section has been set up, it may prove useful to endow certain other sections with some special importance, as *reference sections* or even *principal reference sections*, which supplement the type section, being better exposed or better preserved, more complete, more fossiliferous, or illustra-

tive of some special feature not contained at the type section, or helping to illustrate the geographic variability of the formal rock unit.

Most fundamental of the formal rock units is the *formation*. It is commonly stated that a formation is a *mappable* rock unit, but the term *mappable* is often misconstrued as meaning that the unit be thick enough (or otherwise sufficiently large) that it can be represented on a geologic map of some scale (with much vagueness about *what* scale). Moreover, every formation, no matter how thick, is diminishingly thin near its margins. Is the feather edge of a formation not mappable? The word *mappable should* mean that the unit is distinguishable from other units on the basis of features readily observed by a geologist in the field, and that the boundaries or **contacts** between it and adjacent units are sufficiently well defined that they are portrayable, in a topological sense, on a map of *some* scale. There is nonetheless, a tacit or implicit lower limit on thickness or volume of a formation that geologists in general seem to have consented to, and those who ignore this lower limit risk being pejoratively labeled as "splitters."

The formation commonly yields eventually to subdivision into formal rock units called **members**; these are lithologically distinct in the same sense that one formation is distinct from another. They may be laterally or vertically related to other members in the formation. It is permissible for only some part of a formation to be formally designated as a member, the remainder of the formation not being assigned to any formal member. A **group** is a formal rock unit that comprises two or more vertically contiguous formations. Promotion of formation to group, or member to formation, occasionally becomes desirable as rock units become better known and understood, serving to "make room at the bottom" when it becomes clear that a rock unit would benefit from further subdivision. A group may change laterally to a formation of the same name as formations in the group become very thin, drop out, or lose their identity. Rarely, several contiguous groups or groups and formations are combined into a **supergroup**. (Unfortunately, some have thought of a supergroup as "a particularly large group"; this is wrong. Stratigraphers have even come up with an unneeded stratigraphic entity that they call a "subgroup," apparently filling some unspecified hierarchal level between formation and group.)

Boundaries between rock units typically are quite complicated, demanding special precision in their definitions or descriptions. Shale formation A might grade upward or laterally into sandstone formation B, as the ratio of sand to clay increases over some interval or as thickness and frequency of sandy interbeds increases. The basal boundary of the sand formation might be defined as the lowest occurrence of sandy bed, the highest occurrence of shale bed, the lowest sandy bed of some minimum thickness, or the point where the sandstone-shale ratio is 1:1. Of course, as one defines the base of the sandy formation, one is also defining the upper boundary of the shale formation. One could alternatively consider each shale unit as a ''tongue'' of formation A and each sandy bed as a tongue of formation B. Or the zone of interbedding could be assigned to a separate formal unit.

The *time-rock unit* (or *chronostratigraphic unit*) comprises all rocks deposited during a given time interval. Formal chronostratigraphic units include *systems* (rocks deposited during a geologic period), *series* (corresponding to epoch), and *stage* (corresponding to age). Synchroneity may be established by fossil content (the *biostratigraphic unit*) or remanent magnetic properties (the *magnetostratigraphic unit*), volcanic ashfall beds (bentonite beds) or other *event strata*, or even radiometry or other methods of absolute dating. Boundaries between time-rock units generally coincide with or are parallel to rock-unit boundaries locally, but at some point rock-unit and time-rock-unit boundaries cross or diverge. A time-rock unit contains, in general, parts of many contiguous rock units, each of which is a distinct *facies* or local expression of the deposition that was going on regionally during the given time interval. The lithologic heterogeneity of the time-rock unit emphasizes the obvious (but important) point that sedimentary processes and the types of deposits made differ from place to place at a given point in time. The actual rock contents and the way in which they are arranged in the time-rock unit reflect the geography and physiography at some time in the past—the distribution of land and sea, floodplains and tidal flats, reefs and barrier islands, all coexisting, all contributing simultaneously to construction of the time-rock unit. Thus, recognition of time-rock units is the basis for reconstruction of paleogeography and paleoenvironment and the making of maps to show the disposition of all aspects of the earth's surface at some instant in the distant past.

In their perpetual efforts to clarify and summarize the stratigraphic nomenclature of a region and the lateral, vertical, and temporal relationships between rock units, stratigraphers construct graphical charts, referred to as *correlation charts* (Fig. 7-3). Such charts serve various purposes. Some present the nomenclatural systems devised by different workers, perhaps with an historical perspective, showing how the nomenclature has evolved over many years and how the named rock units of different regions correspond to one another. Other charts show the regional stratigraphic relationships between rocks in different parts of the region, including both spatial and temporal aspects.

Correlation

One of the most important pursuits of a stratigrapher is correlation. There are two kinds of correlation in

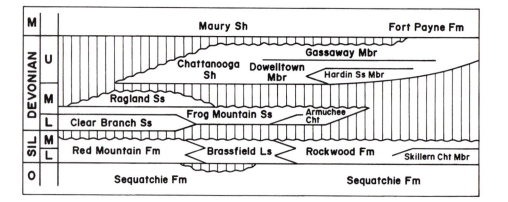

Fig. 7-3. A correlation chart, showing approximate temporal relations among the formal rock units of different regions.

geology, the establishment of (1) simultaneity across localities and (2) lateral continuity of rock units. Some correlations are accomplished by bed tracing, relying on the bedding plane as an isochron (seismic profiling detects these isochrons). Bed tracing establishes lateral continuity, of course, but appears also to establish simultaneity in most cases. A different kind of correlation, which can be applied where bed tracing is impractical or impossible, relies on an internal feature, such as a distinctive heavy mineral suite or pattern of vector properties or some element of a fossil biota that unifies and identifies a deposit. A bed or package of beds that is unique in some way can be identified wherever it occurs on the basis of its unique features.

A certain spatial uniformity of response is assumed also in "fingerprinting" or sequence matching. Here it is *correlated random noise* that indicates a connection between isolated stratigraphic sections. Varved lake deposits in two different localities probably cannot be correlated on the basis of a single varve because one varve looks much like another. Successful correlation relies instead on matching a succession of many varves, where some random pattern of thick and thin appears everywhere, but only once in the history of deposition in the lake. Anderson and Kirkland (1966) correlated lacustrine varves on the basis of their thickness patterns over distances of several kilometers. Hattin (1971) described bedding successions in the Greenhorn Limestone (Cretaceous) that could be identified simply on the basis of bedding patterns (corroborated by bentonite beds) over distances of hundreds of kilometers across Kansas and Colorado. In the correlation of cyclic sections, we look for slower cycles and trends or acyclic (random) events, because in a truly cyclic succession each cycle is indistinguishable from all the others.

Some correlation is accomplished by absolute dating. Two stratigraphic objects are correlative, of course, if they have the same absolute ages, but here we must deal with measurement errors or other uncertainties. It is impractical or impossible to establish the absolute age of every object at every place, and it is not necessary. We build a framework of absolute dates into which other dates can be interpolated or extrapolated by relative dating methods. An absolute date fixes an upper or lower limit on objects or events that are dated *relative* to absolutely dated objects or events. In this way, paleontologically based relative chronologies can become absolute. Thus, absolute dating, relative dating, and correlation interact.

Biostratigraphy is the most universally applicable approach to correlation, capable of establishing synchroneity of rock units even hemispheres apart. It is a complex science and technology that the sedimentologist generally turns over to a specialist. To a first approximation, biostratigraphic correlation is based on presence of a fossil taxon, or an assemblage of taxa, and on comparison of the assemblages of different rocks. If the assemblages are closely similar, then the rocks are of similar age. The best fossils for stratigraphic correlation are fossils having a short time range and wide geographic distribution and that are reasonably common. Some lines of organisms have evolved much more rapidly than others, so that their ranges are relatively short. In general, the lower the taxon, the shorter the range. Each of the several taxa of fossil organisms present in a rock likely has a different temporal range; the several ranges all happen to overlap, of course, in the stratigraphic unit at hand, and the overlap is shorter than the range of any single taxon.

Resolving power of biostratigraphic correlation is directly related to number of taxa that can be compared between two stratigraphic successions; the more taxa, the narrower is the interval of overlapping ranges (Fig. 7-4). Higher resolution is achieved if lower taxa (species rather than genera, for example) are used in the analysis. Further resolution is sometimes achieved if *relative abundances* of individuals are compared between stratigraphic successions that have already been correlated on the basis of range data.

A *biozone* embraces all rocks deposited during the time interval that a particular species lived. A stratigraphic zone based on the overlapping ranges of several species is called an *assemblage (bio)zone*. Of course, not all rocks in a biozone actually contain the species that define the biozone. A taxon may disappear at some stratigraphic level for reasons other than extinction—perhaps a change in the environment caused its disappearance locally or some circumstance denied preservation. It may also happen that some fossils behave as detrital grains and are *recycled* into younger sediments by sedimentary processes, corrupting the relationship between the age and fossil content of the sediments. This is common among the palynomorphs, which are sand-grain-size plant structures of remarkable durability; the palynomorph content of deltaic sediments commonly reflects the ages of rocks in the drainage basin rather than the age of the delta.

The importance of fossils in stratigraphic correla-

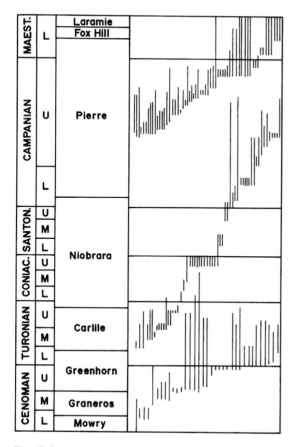

Fig. 7-4. A biostratigraphic zonation based on overlapping ranges of fossil species. Each vertical bar represents the range of occurrence of a certain gastropod species. (After Kauffman, 1970, fig. 6.)

tion and dating is boldly manifest in the contrasts in style and elaboration between Phanerozoic and pre-Phanerozoic stratigraphic nomenclatures currently in use. Also, there may be much uncertainty about the ages of rocks in which stratigraphically useful fossils are rare, such as most aeolianites, most gravelly deposits, and most evaporite deposits.

Biostratigraphy is useful also in establishing sedimentation rates and in locating unconformities or gaps in the rock record.

Unconformities

Unconformities are stratigraphic elements that have no thickness, but they do have ages (in some sense) and

durations, just as rock units do. Unconformities may rise or fall through large thicknesses of strata, merge (or splay), or vanish into bedding planes. An unconformity may truncate older unconformities, erasing a complex history of deposition and erosion (Fig. 7-5). Most unconformities are detrital sources, and most detrital sources are unconformities. If detritus generated in the upland areas of an erosion surface moves downslope into the hollows, then an unconformity gradually buries itself and its active part gradually shrinks to nothing. Marine transgression or regression also causes gradual shrinkage or expansion of an erosion surface. Thus the age and duration of an unconformity varies from place to place. An important distinction between unconformities (or disconformities) and ordinary bedding planes or diastems is that the top of a bed (that is, a bedding plane) is created by the same depositional event that creates the bed itself, and at the same time as the bed. An erosional unconformity, in contrast, forms independently of the depositional event.

The local age of an unconformity could be considered as spanning the interval between the age of the rock just beneath and that of the rock just above. Wheeler (1958, p. 1058) referred to this time interval as a *lacuna*. But time of actual erosion spans only part of this interval (Fig. 7-6). The lacuna has two parts—time represented by erosional loss of rock record, which Wheeler called *vacuity*, and time that passes

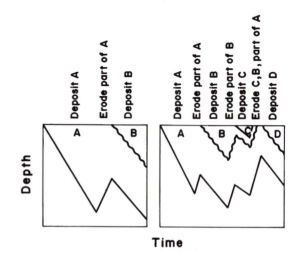

Fig. 7-5. Burial-history (depth versus time) plot for a simple unconformity and a more complicated one.

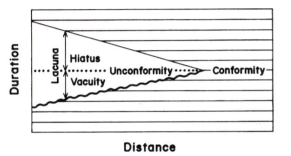

Fig. 7-6. The time structure of an unconformity.

unrecorded due to non-deposition, which Wheeler called **hiatus**. These two parts "accumulate" simultaneously, though commonly at grossly different rates. Duration of the lacuna generally varies geographically, due in part to variation in duration of erosion, certainly, but also in part to variations in the age structure of the underlying rocks—variations that are totally independent of the physical processes giving rise to the unconformity. If, for example, the underlying rocks were deposited very slowly at some locality, then even the gentlest erosion at that place would remove large "chunks" of time and create a large vacuity. If the unconformity intersects and truncates some older unconformity, there is a gross discontinuity in the duration of the younger unconformity at the truncation, as the younger unconformity locally "annexes" the older one. The age of the rock directly overlying an unconformity may be considered, probably without much argument, as representing the time that erosion ceased, and this surely varies geographically. The dominant cause of this variation is topography of the erosion surface, especially if the overlying rocks are marine, because incursion of the sea nearly always abruptly terminates erosion and re-initiates sedimentation, and different regions experience this incursion at different times, depending on their elevations. Thus, topography and rate of relative rise of sea level control the duration of hiatus.

Truncation of sedimentary structures such as stromatolites or burrow fills indicates that a period of mechanical or chemical attrition preceded deposition of the overlying bed. If a surface is subaerially exposed for a long period, it develops a residual soil, which may be preserved in the stratigraphic record as a *paleosol*. Even if a soil is not preserved, there may be discolorations, fractures, degraded minerals, or partly dissolved mineral grains just beneath the unconformity that reflect an episode of weathering. As rock destruction proceeds, certain constituents of the rocks affected may survive and also escape transport away from the region of erosion. For example, certain fossils, hard nodules, or other durable and heavy objects may simply drop out of the eroding bed and contribute to a *lag deposit* that blankets the erosion surface. Commonly there is a concentration of heavy metals such as uranium, phosphorus, or manganese; such a concentrate registers on certain well logs. There may be a biostratigraphic condensation, skeletals of various ages brought together onto the unconformity, having dropped out of the eroding beds above. (A similar feature, occurring in polar or mountain glaciers, is the **ablation surface**, a horizon on which wind-borne dust and other debris has become concentrated during the summer melting and evaporation of part of the previous winter's snow blanket.) The thickness and composition of a lag deposit may indicate the "magnitude" of the unconformity.

But unconformities are not necessarily very evident. Events just prior to deposition over an unconformity may remove paleosols and other evidence of duration. An unconformity representing a billion years might not be physically different from one representing a thousand years. Unconformities within shales are difficult to detect; geologists seldom record them in shale successions. Commonly, unconformity or hiatus is deduced from the absence of some biozone, even though no physical stratigraphic evidence, such as truncations or weathered zones, exists. Unconformity also is indicated if many (or all) of the species contained in some stratigraphic interval should have the same stratigraphic upper or lower bound. Newell (1967) suggested that "It is virtually impossible to demonstrate that a given sequence is free from stratigraphic breaks. . . ." He introduced the term *paraconformity*, which is a biostratigraphic discontinuity within a succession of parallel strata, a stratigraphic break based solely on paleontologic evidence.

Unconformities on carbonate rocks deserve special consideration; here recorded time is dissolved away rather than abraded, and the topography of the dissolved surface may be quite complex. Extensive and well-developed *paleokarst* has been recognized at the top of certain carbonate rock successions. Features of sedimentological interest include red soils (terra rosa); chert lags; irregular topography; and caverns—large or small—ornamented with speleothems (dripstones);

filled with mud, sand, or organic matter washed down from the surface; or collapsed and filled with cave breakdown or collapse breccia. Other less spectacular evidence of emergent carbonate terrains, especially where climate is less humid, is a surface coated or associated with caliche or calcrete, or a hackly surface of rillenkarren, or indication of dedolomitization, calcite recrystallization, or surface silicification (see Chap. 11).

Contrary to expectation, substantial unconformities exist also in the pelagic calcareous oozes deposited in the serenity of the oceanic abyssal plains. These unconformities also are due to dissolution. Below a certain depth in the ocean, seawater aggressively dissolves calcite. If carbonate ooze deposited on a rise or a seamount should be transported, by sea-floor spreading, into the abyssal depths below the so-called *lysocline* (see Chap. 9), or if the lysocline should rise for some reason, then the ooze begins to dissolve, and a dissolutional unconformity appears. Unconformities occur also in deep-basin evaporite beds, again apparently owing to dissolution. Other deep ocean-basin "unconformities" are due to erosion (of soft sediment) by geostrophic currents on the bottom, though only locally do these currents exceed 10 or 15 centimeters per second (Tucholke and Embly, 1984, p. 149).

Some unconformities are considered to be of worldwide scope (though not, of course, continuous worldwide), resulting from eustatic sea-level fluctuations or from global changes in the intensity of tectonism on land. The rock packages that they enclose are called *synthems* (but most stratigraphers refer to them as sequences). At first, only a half-dozen sequences were recognized (Sloss, 1963); today, hundreds of so-called *depositional sequences* (Frazier, 1974) have been identified, each punctuated by hiatus. There is much speculation and debate now about what has caused these episodes and whether or to what extent they are of global significance and correlative from continent to continent (see Chap. 12).

Sloss (1984) made some general observations about unconformities on cratons. Some cratonic unconformities are local and represent syndepositional uplift of local tectonic elements such as fault blocks or shallow intrusive bodies. Regional unconformities are traceable over large-scale cratonic basins or arches and commonly can be correlated with times of orogenesis on the margins of the craton. Other cratonic unconformities are "interregional" in scope, being generally craton-wide and showing little structural discordance between the rocks above and below. Some of the interregional unconformities might be of global proportions. But the eroded rock thicknesses associated with them in cratonic interior regions, and the relief developed on them in some places, are far greater than can reasonably be attributed solely to eustatic sea-level change, and interregional tectonism surely plays a role.

Sedimentation Rates

To determine absolute rate of sedimentation requires that the absolute ages of sediments be known, or at least the absolute length of the time interval represented by a given thickness of sediment. Relative rates of sedimentation can be estimated from certain accumulative effects of time or by comparison of the thicknesses of correlated rock units. Probably thin bedding indicates low average rates of deposition; thick-bedded units likely were deposited more rapidly. Occasionally we find a deposit that is thinner at one locality than at another but comprises the same number of beds in both places. Presumably, each bed becomes thinner in some direction, reflecting a declining sediment discharge and decrease in both instantaneous and average sedimentation rate in that direction.

When two (or more) different materials are being codeposited, the one effectively dilutes the other in the deposit. The magnitude of the *dilution effect* depends, of course, on the relative rates of sedimentation of the two components. One component might be supplied at a constant rate, so that its concentration in a deposit is a measure of the rate of deposition of the other component. For example, one might reasonably assume that over some long term, the organic productivity of ocean surface waters or the rate of infall of cosmic dust is constant. In terrigenous clay muds on the ocean floor below, layer-to-layer variations in the number of diatom frustules, concentration of organic carbon, or extraterrestrial iridium might indicate variations in the rate at which an adjacent continent is providing terrigenous mud to the ocean. Mount and Ward (1986) provided an example and pointed out that it is often difficult to know which component is the constant "background" and which is the variable. Keep in mind that compaction also can increase concentrations. Another index of relative rates of sedimentation is degree of bioturbation; a bed that is thoroughly bioturbated may have been deposited more slowly than an overlying bed of similar sediment that is only lightly or partially bioturbated.

Absolute sedimentation rates are difficult to determine or estimate. One problem is that much sedimentation is intermittent—a bed is deposited rapidly, then there is a comparatively long period during which no sedimentation occurs at all. Even if an instantaneous sedimentation rate is obtained unequivocally by some means, the figure bears little relationship to the longer-term average sedimentation rate. Instantaneous rates are always higher, and generally much higher, than average rates; short-term averages are always higher than long-term averages. This is because long time periods contain erosional intervals or intervals of zero sedimentation rate. Longer hiatuses (or lacunae) have longer recurrence intervals; therefore, the larger stratigraphic intervals have greater numbers of long hiatuses. Long-term averages are referred to as *survival rates*, because they represent the thickness of sediment that has survived to the present rather than thicknesses of sediment actually deposited.

Graphs showing coarse-scale total thickness of sediment accumulation versus geologic time were constructed by von Bubnoff (1963, p. 154), and consisted of two curves (Fig. 7-7), representing history of elevation of top and base of the sedimentary pile (sea-level datum) at some given locality. As the sedimentary pile thickens with time, the curves diverge; during periods of erosion, they converge. Bubnoff referred to these diagrams as *oscillograms* and used them as a characterization of long-term tectonic behavior of a region. Construction of an oscillogram at a given locality requires knowledge of thicknesses and ages (on an absolute time scale) of the layered rocks (including volcanics) in the sedimentary pile at that place. Thickness and age values, ordinarily compiled rock unit by rock unit, are well known or can be estimated within acceptable limits. But elevation of the depositional (or erosional) surface (upper curve) commonly is speculative, and the top of the sedimentary pile, rather than sea level, may be taken as datum. Also, there are compaction effects that influence the shape of the lower curve.

The vertical distance between the depositional surface and a given stratigraphic horizon is the burial depth of that horizon, and the variation of burial depth in time is a *burial history*. The steepest segments of a burial history curve represent the highest rates of sedimentation; reverse slopes represent "negative sedimentation" or erosion. Ordinarily, the magnitudes and rates of negative sedimentation are speculative; a given stratigraphic section does not directly record the magnitude of erosional thinning represented by unconformities, and unconformities do not precisely record their ages, though estimates may be obtained from regional stratigraphic relations. Burial history diagrams commonly reveal that only small portions of geologic time are represented in sedimentary piles. Burial history diagrams have, especially in the past few years, become important tools of the diageneticist (see Chap. 11).

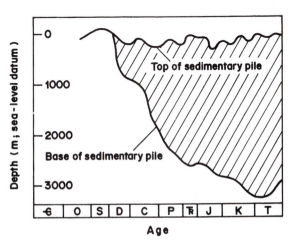

Fig. 7-7. One of von Bubnoff's (1963) oscillograms.

Stratigraphic Cycles

Some depositional systems give rise to stratigraphic cycles, or **cyclothems**, successions of strata that are repeated (at least approximately) many times. Cyclothems are a result of depositional events that have recurred cyclically rather than randomly. But in many stratigraphic successions it is not clear whether repeated stratigraphic motifs are a result of random events or periodic processes. Traditionally, geologists have favored the latter, commonly without much supporting evidence, and probably erroneously in some instances. Below (Chap. 12), we will examine some of the suspected causes of stratigraphic cycles.

The familiar coal cyclothem of Pennsylvanian and Permian Systems of eastern North America and in Europe, wherein several contrasting lithic types are systematically arranged vertically, suggests that deposition followed a plan or pattern. One might view this as a product of a depositional system that always ''knows'' where it is and ''remembers'' where it has been with respect to this plan. This contrasts with certain other stratigraphic successions where it appears that lithic types follow one upon another completely at random. One suspects that stratigraphic successions exist wherein there was a plan, but one that was ''flexible'' and not followed to the letter, or that was vague or noisy, so that the depositional process was not totally deterministic but yet not totally random. The stratigraphic succession that has a memory, however long, short, or fuzzy, displays a *Markov property*. A Markov property exists if the probability of an event A (such as deposition of a bed), given that some other event B has already occurred, is different from the independent-events probability of A as determined by its relative frequency of occurrence in the stratigraphic section (see Harbaugh and Bonham-Carter, 1970, chapter 4; Potter and Blakely, 1968). Thus, existence of a Markov property indicates ''memory,'' or statistical dependence in a time series (or stratigraphic section). This memory might extend only to the immediately preceding event or further back in the series; an event might be statistically dependent to varying degrees on several past events. Hurst et al. (1965) applied a statistical technique that they called *rescaled-range analysis* to various natural systems, such as stream flows, and discovered that many of them exhibit a long-term statistical dependence called *persistence*. Rescaled range is defined as R/S, where R is the difference between maximum and minimum values in a time series of length τ, and S is the standard deviation of the distribution of values in the series. For many natural systems, the rescaled range of measurements of some output variable, such as bed thickness in a stratigraphic succession, closely follows the empirical relation $R/S = (\tau/2)^H$, where H, called the *Hurst exponent*, typically has a value between 0.7 and 0.8. A Hurst exponent greater than 0.5 indicates persistence, meaning that, if a value has been increasing for a certain time period, it is expected to continue to increase for a similar period, and if a value has been decreasing for a given time period, then it is expected to continue to decrease for the same period in the future (see Feder, 1988, p. 181). This is another manifestation of memory. River discharges, ocean wave heights, rainfall abundance, and varve thicknesses are examples of

data that seem to behave in this way (see Mandelbrot and Wallis, 1969, for a stratigraphic example).

Linear feedback is probably responsible for cyclicity in some depositional systems, as for many oscillating mechanical or electronic systems and even some chemical systems. Change comes from within the system; no external cyclic force is required or indicated, though an external ''energy source'' must be present. Such a system is said to be *autocyclic*. In other natural systems, linear or non-linear forcing by external cyclic processes surely is involved—seasonal or longer-term climatic cycles, for example. These are *allocyclic* systems. The stratigraphic character of the deposit depends on the frequency and amplitude of fluctuation of the external force and on the response characteristics (transfer function) of the depositional basin; whereas a low-frequency fluctuation might produce a cyclical deposit, sedimentary processes might be too sluggish to respond to higher-frequency forces.

The *punctuated aggradational cycle* (PAC) of Goodwin and Anderson (1985) seems to be an example of an autocyclic carbonate depositional system. It is a shoaling-upward bed package, 1 to 5 meters thick, with durations estimated to be 30 to 80 thousand years, resulting apparently from an interaction between subsidence or gradually rising sea level and rate of carbonate production. The regressive packages typically are stacked into cyclic successions that show an overall transgressive character. Brett and Baird (1986) observed PACs in Middle Devonian shallow marine shelf sediments of New York State that appear to correlate with more-symmetrical cycles in deeper water. Elrick and Read (1991) described coarsening-upward PACs having thicknesses on the order of a meter in Mississippian limestones of Wyoming and Montana, and Osleger and Read (1991) described similar cycles in Cambrian carbonate rocks in Utah and in Pennsylvania.

Probably all depositional process-response systems are only approximately linear or are grossly non-linear. Non-linear dynamical systems typically appear to be chaotic or completely random, but new mathematical approaches to such systems have revealed certain regularities and predictability, suggesting that many ''chaotic'' systems are just extremely complex (but nonetheless tractable) cyclical systems. When time series of certain properties of ''chaotic'' non-linear systems, including many natural systems, are mapped onto a *phase space*, unexpected patterns emerge called (*strange*) *attractors*. Thickness or lithotype patterns in certain bed successions might prove to behave in this way.

Depositional "Sequences"

Much earlier (Chap. 5) it was suggested that single beds are rapidly deposited, and that bedding planes commonly represent relatively long periods of non-deposition (though some surely record very short breaks). A similar pattern seems to apply to deposition of major bodies of rock. There is a fury of depositional activity for a few millions of years, during which large sediment bodies accumulate. Then there are long periods of dormancy and decay. A depositional episode produces a sediment package that is somehow distinct from adjacent packages, commonly separated from adjacent packages by unconformities; the sediment body typically shows some predictable stratigraphic trend or ordered succession of contrasting lithotypes. The depositional sequence (Sloss, 1963; Vail et al., 1977) is a manifestation of this large-scale episodic sedimentation. The methods of exploration seismology have been particularly useful in the identification and study of depositional sequences.

Seismic Stratigraphy

Seismic stratigraphy is an outgrowth of improvements in exploration seismology made in the 1970s and '80s, as new energy sources and new computer methods of signal processing came into wide use and explorationists realized that seismology could be of value in the search for stratigraphic traps. Seismic surveys were producing large-scale stratigraphic cross sections of lengths, depths, and continuity beyond anything previously available to the stratigrapher, and these cross sections displayed regional patterns of sedimentation never before so clearly and vividly displayed. Much of the seismic work was done on coastal plains and continental shelves. Examples of onlap and offlap, only suspected or presumed before, or inferred from a few isolated boreholes or outcrops, were now clearly revealed in seismic cross sections. Patterns of marine transgression and regression emerged—packages of strata, bounded by unconformities, stacked one upon another in the thick deposits of the continental margin. Other seismic profiles, including some made farther inland, on the cratons, revealed the outlines of ancient reefs, the lagoons and tidal flats behind them, and their

foreslopes into the depths, features that only close drilling, fortuitously located, might otherwise have uncovered. The consequences for stratigraphy and for our understanding of the earth's history are not yet fully realized.

Curiously, seismic profiling reveals *isochrons* (or time surfaces) in a body of sedimentary rock. Similarities in the pattern of bedding from one place to another are correlated in time, and seismic reflections are coherent from trace to trace insofar as these vertical (stratigraphic) patterns are similar from point to point along the profile. Thus, seismic profiling is eminently suited to delineating depositional sequences, which are time-stratigraphic units bounded by unconformities, or by *conformities* that are extrapolations of unconformities along time surfaces. Diachronous boundaries between lithologically contrasting sediments do not ordinarily produce coherent reflections over appreciable distances (Sheriff, 1980, p. 58).

Seismic profiles do nonetheless provide information about lithofacies. A given body of rock has certain mechanical properties that contrast with the properties of adjacent bodies. These properties impose certain characteristics on the "wiggly lines" (traces), such as amplitude, frequency, and phase, that can be of some value in discriminating limestones and shales, and other lithic types, and in distinguishing uniformly bedded limestones from the more heterogeneous reefal limestones, for example. Reflection continuity, abundance of reflections, and their areal configurations are useful criteria for distinguishing among various types of deposits (Sheriff, 1976, p. 537) and for estimating and predicting their geometries and orientations.

On the other hand, even the highest-resolution seismic profile does not "see" the smaller-scale but nonetheless important stratigraphic details that the outcrop or core reveals, and there is much risk of false readings or misinterpretations. For example, certain lateral facies changes might be misinterpreted as depositional or erosional topography or as faults; seismic profiles no doubt miss some unconformities (just as "rock stratigraphers" sometimes do) and even generate false unconformities.

Relative Sea-Level Change and the Stratigraphic Response

The beds contained within stratigraphic sequences show special internal patterns; delineation of these patterns is among the most valuable outcomes of a modern seismic survey. *Onlap* is a progressive updip termination of strata against an underlying inclined surface (Fig. 7-8a). *Offlap* is a progressive basinward shingling of inclined strata (Fig. 7-8b). Onlapping and offlapping geometries are stratigraphic manifestations of transgressions (onlap) and regressions (offlap) of shorelines (of course we knew about offlap and onlap long before seismic stratigraphy). Inclined strata pinch out updip mainly by non-deposition and sedimentary bypassing, and downdip by sediment starvation. Unless one knows the direction of sediment transport or directions of inclination of depositional surfaces at time of deposition, onlap proximal edges and offlap distal edges can easily be confused in seismic profiles.

Let us examine the geometry and age structure of coastal sedimentation and erosion from a conceptual point of view. We will assume, for the sake of simplicity, that sedimentation occurs only in the sea. Thus,

if sea level is falling, then sedimentation ceases on newly exposed tracts and the depositional site shifts seaward; if sea level is rising, then newly submerged tracts immediately begin to deposit sediments.

Consider first the case of relative stillstand. Sedimentation takes place close to the coast, building an offlapping sediment prism. Each new layer slopes seaward and is coarsest and thickest near to its proximal (landward) margin, fining and tapering toward its distal margin in deeper water. As progradation continues, the coarser proximal sediments are deposited over finer-grained, more distal sediments. The ages of both the top and the base of the sediment body increase landward (if we ignore any thin topset or bottomset layers, which a seismic survey probably will not detect anyway). As the sediment body builds seaward into gradually deeper water, the length of its depositional slope gradually increases, and the rate of lateral accretion may gradually decline as the available sediment is spread out over a larger and larger slope area. (Depositional strikeline length also increases.)

The top of such a sediment prism is a level coastal plain lying near to sea level (and base level); it is the *toplap* surface. Sediments "showing" at any given point on the toplap surface are younger than sediments showing at any point more proximal. Thus, at the toplap surface, there is a hiatus that expands landward, because the age of the progradational prism increases landward.

Now if the sea level should begin to fall (Fig. 7-9a), progradation is continually forced down the foreslope or shoreface. Upper shoreface sands might come to rest directly on lower shoreface or offshore muds (see Posamentier et al., 1992, p. 1693). Falling sea level also exposes to erosion the shoreface sands just deposited, and these are continually being reworked basinward. Falling wavebase might also create an ero-

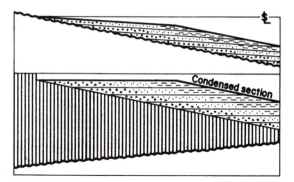

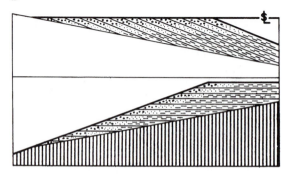

Fig. 7-8. Stratigraphy and Wheeler diagram for **(a)** onlap section and **(b)** offlap section.

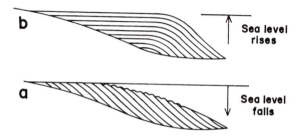

Fig. 7-9. (a) Offlap during sea-level drop. **(b)** Simultaneous offlap and onlap during sea-level rise.

sion surface in the lower shoreface or offshore muds, soon afterward to be buried by prograding upper-shoreface sands. Thus, regression may cause two more or less parallel erosion surfaces, one subaerial and one subaqueous, the subaerial surface lying just a few meters above the subaqueous (wave-cut) surface. Both surfaces, having continually grown areally in the direction of shoreline retreat, are youngest at the coast and age landward. Since the subaerial surface is likely to be topographically rough (owing to valley incision in response to the falling base level), it could locally transect the wave-cut surface below.

It may happen that sedimentation takes place not in the nearshore but farther out to sea, beyond the reach of subsequent subaerial processes. Much of the material in transport might find the heads of submarine canyons and advance quickly onto the surfaces of deep-sea fans, dispatched there by debris flows or turbidity currents. It appears that terrigenous deposition on continental rises is most rapid during marine regressions.

Now consider the case of a sea-level rise following stillstand deposition of an offlapping sediment prism. Here, the site of ongoing coastal sedimentation shifts landward and gradually buries the toplap surface. The age of the toplap surface at its seaward edge marks the time of transition from stillstand or falling sea level to rising sea level. While parts of the surface are now inactive (and preserved in the sedimentary pile beneath transgressive deposits), the duration of the hiatus continues to increase in other parts not yet transgressed.

Transgressive marine deposits are made in the newly available space (accommodation space) between a rising sea surface and the *flooding surface* (Fig. 7-10), that is, the surface undergoing submergence (see Galloway, 1989). Fine-grained estuarine,

lagoonal, and tidal-flat sediments appear first on the flooding surface, but transgression eventually brings a surf zone over these marginal deposits; the surf truncates them, creating a ***ravinement surface***. Thus, transgression may leave two erosion surfaces in the stratigraphic record, just as regression may, but in this case, the subaerial surface lies *below* the subaqeous one. A shoreface sediment prism overlies the ravinement surface. Coarser sediments are deposited close to shore, but finer sediments might be dispersed considerable distances into offshore locales. Each layer is inclined very gently to seaward, tapers gradually seaward, and becomes finer-grained seaward. As transgression proceeds, the whole pattern of sedimentation migrates landward, onlapping the coastal plain. Thus, finer sediments are deposited on top of the coarser sediments, which rest directly on the flooding surface or ravinement surface. The age of the base of the deposit decreases landward; the top of the deposit is approximately isochronous. Hiatus of the unconformity at the base of a transgressive deposit expands landward; this reflects both the landward-increasing age of the underlying offlap deposits and the landward-decreasing age of the overlying onlap deposits. A complication is that transgression causes reworking of regressive deposits. Vacuity also may be present, at least locally, especially if sea-level fall preceded the transgression. Far out to sea, the time interval of transgression is represented only by thin deposits; the stratigraphic section there, in a sediment-starved basin, is said to be ***condensed***. In some locales, condensation may be a result of the gradual shrinking of land masses undergoing submergence; as land masses become smaller, they become weaker sources of detritus. (Condensation also results from erosion by deep marine currents.)

Rate of sediment supply relative to rate of sea-level

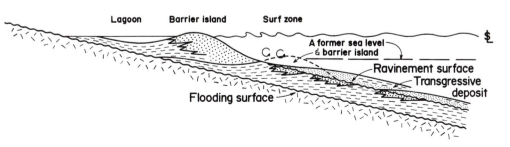

Fig. 7-10. Flooding surface and ravinement created during transgression.

change also controls sedimentation patterns or modifies the patterns described above. If sediment is abundant, coastal progradation may occur even as sea level is rising. The resulting sediment package shows landward onlap and seaward offlap simultaneously (Fig. 7-9b). If sediment supply is weak, complex events might leave very little record.

Carbonate-depositing coastal regions develop sedimentation patterns that are similar in some respects to terrigenous sedimentation patterns and different in other respects. Whereas detrital sedimentation patterns are linked to rates of supply of detritus from the outside, the rate of carbonate sediment production is sensitive to the environment of the depositional basin. Sea-level change or change in seawater temperature or composition can substantially increase or reduce carbonate production. Whereas base-of-slope fan sedimentation in terrigenous depositional systems is accelerated at sea-level lowstand, just the opposite seems to obtain in carbonate systems (Schlager, 1992, p. 44).

This is because the rate of carbonate production on carbonate platforms ordinarily increases during sea-level rise and fore-reef slopes become steeper; thus more carbonate detritus escapes to downslope venues. Carbonate sediment lithifies quickly, and during sea-level lowstands, weathering of carbonate rock produces chemical solutions but not much detritus.

Seismic profiling beneath coastal plains and continental shelves has revealed that the various patterns of sedimentation described above have recurred again and again throughout the world. Offlapping packages, onlapping transgressive packages, and deeper continental slope packages, each wrapped in unconformities or omission surfaces or flooding surfaces, are stacked together in complex ways, recording a complex history of sedimentation and erosion, tectonic and compactive subsidence, uplift, and eustatic sea level change (Fig. 7-11).

The study of the seismic records (made almost en-

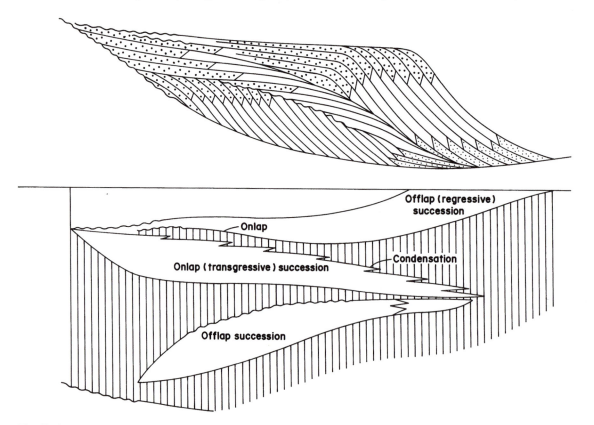

Fig. 7-11. A stack of depositional sequences and its Wheeler diagram.

tirely by the petroleum industry), together with well logs and samples, and biostratigraphy and chronometry, has suggested to some that global (eustatic) rise and fall of sea level is the dominant factor in coastal sedimentation, and that a given era produces a host of more or less identical or comparable sediment packages in coastal and shelfal regions throughout the world. In other words, depositional sequences, and the surfaces that bound them, are globally correlated and synchronous. Others are not convinced, believing that local or regional uplift and subsidence, and variations in rate of sediment supply are at least as important as global sea-level change, and that the evidence of synchroneity is not as strong or abundant as the proponents claim (see Miall, 1986).

Rates of Transgression and Regression

Rates of shoreline advance or retreat depend not only on rate of relative sea-level change but also on the slope of the coastal terrain. The upper surface of a progradational sediment prism, lying generally close to sea level and sloping only very gently toward the sea, would be transgressed very rapidly. Higher on a coastal plain, slopes are somewhat steeper and transgression is slower.

In general, coastal sedimentation is accompanied by subsidence. It is this subsidence that permits the stacking of numerous depositional sequences to thicknesses of thousands of meters. Thus, even if sea level remains constant, there is likely to be some onlap as sediments keep filled the space that subsidence creates. Suppose that this subsidence occurs at a uniform rate but that sea level rises and falls eustatically. One expects transgressions to be more rapid than regressions, because rate of subsidence and rate of sea-level *rise* are *additive*, but rate of subsidence and of sea-level *fall* are *subtractive*. The closer the rate of subsidence is to rate of sea-level change, the greater is the difference between rates of transgression and regression.

Coastal subsidence or uplift affects rate of transgression and regression, but in contrast to eustatic sea-level change, the magnitude and rate of diastrophism vary from place to place along the coast. Thus, there may be a relative rise of sea level in one place and a relative drop in sea level at the same time in another place. This leads to complex stratigraphic relationships along strike that may not be fully appreciated by those who examine exclusively dip-oriented seismic profiles.

Stratigraphic "Events"

Event stratigraphy deals with the stratigraphic manifestations of "rare," short-lasting geological events, but events that, in the long term, occur frequently. A certain randomness or unpredictability of recurrence also is implied. Among such events are severe storms, great floods, turbidity currents, earthquakes, volcanic eruptions, and tsunami. They give rise to single beds, thin packages of a few genetically related beds, or unconformities. The volcanic ashfall bed, or "bentonite bed," is a familiar type of event stratum. Great sea waves have made their mark on coastal lands, depositing coarse gravels even hundreds of meters above the sea (see Moore and Moore, 1988). Another well-known example is the event package produced by a turbidity current; the pattern displayed by the *turbidite* package was carefully described and explained by Bouma (1962), and has come to be known as the *Bouma sequence* (Fig. 7-12a; see also Chap. 8). Pilkey (1988) described very extensive single thick beds on marine basin plains that appear to have been deposited by giant turbidity current events, perhaps triggered by great storms, tsunami, or earthquakes.

It is quite common for an event package to consist of a basal erosion surface created by the event at "full fury" and an overlying organized and predictable succession of textures or sedimentary structures reflecting the waning energy level that follows. By definition, an event stratum is never time-transgressive (except on very short time scales), and for this reason it can be useful in correlation. An event stratum might be reworked, however, if no significant sedimentation goes on between "events."

Events commonly are destructive of life and may even cause mass mortality. A pavement or concentrate of shells may be a result of many organisms suffocating under a volcanic ashfall or storm-generated turbidity, or sustaining an unusual hard freeze. Bone beds associated with alluvial deposits have been attributed to river floods that drowned many animals all at once.

Storms

The *tempestite* (Aigner, 1982) is an event package bearing a distinctive sequence of textures and sedimentary structures caused by a storm on the shallow margin of a sea and probably never more than a few tens of centimeters thick (Fig. 7-12b). In the complete tempestite (Dott and Bourgeois, 1982, p. 665; see also

b

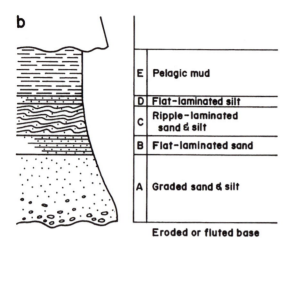

E	Pelagic mud
D	Flat-laminated silt
C	Ripple-laminated sand & silt
B	Flat-laminated sand
A	Graded sand & silt

Eroded or fluted base

a

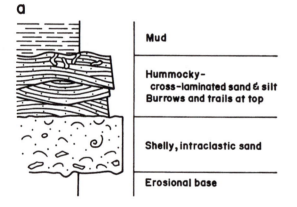

Mud

Hummocky-cross-laminated sand & silt
Burrows and trails at top

Shelly, intraclastic sand

Erosional base

Fig. 7-12. Two common types of event successions: (**a**) the complete Bouma sequence (turbidite); (**b**) sandy, shelly tempestite.

Kumar and Sanders, 1976), there is an erosional base—overlain by a lag gravel composed of pebbles, mud clasts, plant fragments, or shells—passing upward into hummocky cross-laminated and flat-laminated fine-grained sand, thence to ripple-laminated fine sand or silt, and finally a mud layer at the top, deposited after the storm. Bioturbation occurs between storms, muddling the tempestite top-downward. Grain size typically is fine sand and coarse silt; characteristically, mica and macerated plant matter are concentrated on the surfaces of laminae, especially on laminae high in the succession. The thickness of the

succession ranges up to a meter, maybe more; such a succession is generally interpreted to be the deposit of a single storm (or tsunami). The generally fine grain size of the succession suggests suspension transport. Incomplete tempestites are due to erosion during subsequent storms or to unavailability of sedimentary material appropriate to a particular part of the succession. Tempestites change from shallow to deeper waters (Aigner and Reineck, 1982), being coarser and thicker in proximal (nearshore) tracts, and lacking the muddy interbeds, and perhaps lacking the erosional base and the ripple-laminated units of distal tracts.

Consider a shoreface inclined gently away from the coast. At some point the bottom intersects fairweather wave base. Wave base of storms of ever greater intensity intersects the bottom in progressively deeper waters ever farther from shore. It is clear that the parts of the surface that lie in the deeper waters experience wave-generated turbulence only rarely, since the largest storms are also the least frequent. Moreover, this turbulence is due to waves of greater length and height. Bottom tracts in shoaler waters experience the bottom currents generated by these same storms (though here the effects are more intense) as well as the turbulence of more frequent storms of lesser severity. Thus there is a shoreward gradient of long-term accumulated energy expended on the shoreface, and a gradient in storm-event frequency as well. These gradients presumably are reflected in tempestites formed at different depths.

Volcanic Events

Volcanic events can have important sedimentological consequences, forming widespread blankets of air-carried volcanic ash. These layers taper away from their source. Commonly referred to as bentonite beds, they are composed of mixtures of clay minerals derived from the alteration of volcanic ash; zeolites may be present as well. Other components of the original volcanic ash may survive—biotite flakes, apatites and zircons, and feldspars. Some stratigraphic intervals, such as the Ordovician in the central Appalachians of the eastern United States or the Cretaceous of the western interior United States contain large numbers of bentonite beds. Volcanic explosions that occur near to the sea, such as Krakatau's and Tambora's paroxysms of historical times, can raise tsunami that make their sedimentological mark on distant coasts and on the deep ocean floor.

Anoxia Events

Short-term depletion of oxygen in a sedimentary environment may be a local consequence of a volcanic ashfall event or of an algal bloom (the decay of which rapidly consumes oxygen). It may be recorded stratigraphically as a mass mortality, some discontinuity in bioturbation, or an organics-rich layer (black shale). Ocean-floor anoxia, according to Demaison and Moore (1980, p. 1203), is most likely during periods of warm climate, when polar ice caps are absent or much reduced and sea level is high. Some have claimed that ancient marine pelagic sediments bear the record of several *global* anoxia events, but others disagree (see Chap. 12).

Other chemical events are recorded as concretion horizons (see Chap. 11).

Magnetic Events

The occasional reversals of polarity of earth's magnetic field are of global significance and are a basis for worldwide correlation (Fig. 7-13). Remanent magnetism exists in igneous and sedimentary rocks alike, and the record of reversals is a medium by which the absolute ages of igneous rocks can be transferred to sedimentary successions or vice-versa. The longest known polarity interval lasted about 50 million years; 1-million-year intervals are common, and some have lasted less than 100,000 years. There has been some speculation about the effects of field-reversal events on earth's surficial environment. The magnetic field shields earth's surface, and the biosphere, to some degree from cosmic rays and other energetic, mutagenic particles, slowing them down, and channeling them along the magnetic field lines into the polar regions, where they can do little harm. But during a reversal, there is a time of little magnetic field strength that lasts 1000 to 10,000 years (Cox, 1969, p. 238), and there is some evidence that extinctions are more frequent during magnetic reversals (see Chap. 12).

Impacts

Perhaps the most significant and consequential events of all, but not really taken very seriously until about 1980, are the catastrophic *collisions* between earth and large extraterrestrial bodies (meteors and bolides, comets). The record of impacts on the lunar surface and on the surfaces of the rocky planets and their

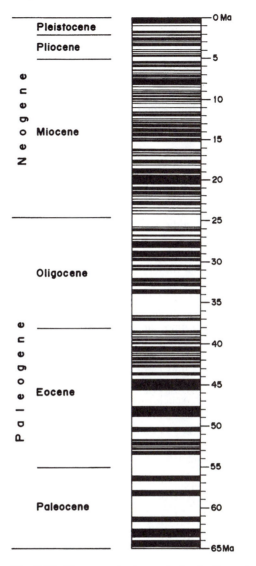

Fig. 7-13. The magnetic time scale for the Cenozoic. (After Harland et al., 1982, fig. 4.6.)

moons is clear and obvious. The record of impacts on earth is less clear and far less dramatic, but it is irrational to imagine that the earth, unlike the other planetary bodies in the solar system, has somehow escaped cosmic bombardment. Shoemaker (1977, p. 1), marveling at the preponderance of impact craters revealed by the recently acquired space-age photographs of the lunar surface and of Mercury, Venus, and Mars, declared that collision is "the most fundamental of all processes that have taken place on the terrestrial [that is, earth-like] planets." Atmosphere burns up many smaller cosmic projectiles and ocean absorbs their energy of flight before they reach earth's rocky surface, but atmosphere and ocean have little effect on the rare, very large objects. Weathering, erosion, and deposition have destroyed or hidden many of the old scars of major impacts, and surely many of those on the ocean floors have been subducted, but a few hundred large ones have survived.

An *impactite* is a blanket of material ejected from an impact site; typically it contains both target and projectile material. Characteristics of impactites, based mainly on the one at the K-T boundary (see Chap. 12), include shocked quartz grains, high-pressure quartz polymorphs, glassy spherules or tektites, soot, and elevated iridium concentrations (and other similar elements, such as platinum, osmium, ruthenium, and rhodium). The thickness of the blanket and size of its clasts decrease away from "ground zero." A bedding surface at the base marks an abrupt change to carbonate-poor, organics-rich sediments above (McLaren and Goodfellow, 1990, p. 159), with decreased $\delta^{13}C$ and $\delta^{18}O$ in carbonate and increased $\delta^{34}S$ in sulfate and sulfide minerals (see Chap. 11); also an abrupt disappearance of many species of organisms. Hildebrand and Boynton (1988) summarized the occurrences of thick deposits, up to 500 meters thick, of coarse breccia at the K-T boundary in and around the Caribbean basin. They consider this to be the deposit of an impact-generated sea-wave, within 1000 kilometers of a tentatively located impact site in the Caribbean Sea.

Shock has various effects on rocks, and these can be useful guides to hypervelocity impact events. Quartz grains show multiple sets of closely spaced microscopic fractures, typically oriented along certain rational crystallographic directions. Such grains also display reduced refractive indexes and reduced birefringence. In polymineralic rocks, the grains of certain minerals selectively undergo phase transitions that do not involve adjacent mineral grains. Hypervelocity impact causes instantaneous melting of target materials, creating glassy spherules of various compositions, which are widely dispersed in an ejecta blanket. Quartz grains in a sandstone might change to silica glass (lechatelierite) without changing shape or reacting with surrounding matrix or cement. Feldspar grains may be transformed to glass but preserve myrmekitic or perthitic microstructure. Chao (1967, p. 212) proposed the term *thetomorph* to describe a glassy phase transformed by shock from a crystalline precursor. High-pressure minerals, such as diamond (C), coesite (SiO_2), and stishovite (SiO_2) occurring in shallow depths indicate impact; the high-pressure quartz polymorphs occur commonly in known impact craters and in their ejecta.

Anthropogenic Events

Several human-caused events have made their way into the stratigraphic record, among these the invention of the Coke bottle; these bottles, with the date and the place of their manufacture molded into them and carelessly discarded, have been incorporated widely into rapidly forming carbonate rock. The "event" in this case is an abrupt *first appearance*. Detonation of the first nuclear bomb created a ^{14}C spike in ooid cortices and karstic dripstones and in lake sediments everywhere. Great fires are recorded in glacial ice, which also tells the tale of lead in automotive fuels. Numerous other examples of human influence on sedimentation could be cited.

8

Terrigenous Depositional Environments

The determination of ancient environments is commonly the intended end result of a sedimentological investigation, bringing together a host of stratigraphic, petrographic, and paleontological observations. It relies heavily on the results of other workers who have probed the anatomies of similar deposits, ancient and modern or whose laboratory experiments or simulations have explicated the workings of complex natural systems.

We have learned already that texture and structure of sediments and sedimentary rocks reflect hydraulics of the environments under which they were deposited. Grain size and sorting in clastic sediments relate to strength of currents and turbulence in the depositional environment. Sedimentary structures make generally clear distinction between the contrasting fluid dynamics of different depositional settings and reveal patterns of current flow. Trace fossils also have proved their value in defining depth zones and hydraulics of depositional environments. In many sedimentary rocks, mineral components reflect environment. Calcite, aragonite, dolomite, gypsum, glauconite, pyrite, and hematite can form in depositional environments, each under a specific set of chemical and physical constraints. State of preservation of organic matter in sediments implies certain conditions at the depositional site.

Paleontology and biostratigraphy are helpful in determining water depth, temperature, salinity, and other physical and chemical properties of the depositional environment. Organisms or their fossils indicate water chemistry, whether saline, brackish, or fresh; there is little overlap between marine and non-marine genera and species. Low biotic diversity in a sediment, or a dwarfed or depauperate fauna, reflect a stressful or unstable environment. Benthonic foraminifers have been especially useful for estimating water depth of ancient

marine environments, because different taxa inhabit different depth zones. Planktonic foraminifers are sensitive to temperature of ocean surface waters. Certain Recent planktonic foraminifers coil to the left in the cold waters of the Arctic Ocean, but in the warmer waters of the North Atlantic they coil to the right. Moreover, the oxygen isotope ratio of their tests depends on water temperature. Occasionally biostratigraphy provides evidence of resedimentation—shallow-water sediments with shallow-water faunas having moved downslope to deeper waters, where they come to be interbedded with sediments containing an indigenous deeper-water fauna.

Geochemistry also is helpful. For example, concentration of boron in a shale is an indication of the salinity of the water in which the shale was deposited (Shaw and Bugry, 1966, p. 49). Boron in solution is more abundant in more saline waters, and certain clay minerals, notably illite, can incorporate dissolved boron in quantities proportional to its concentration. The calcium/strontium ratio in skeletal aragonite seems to indicate temperature of the environment in which the organism grew (Weber, 1973). Ratio of iodine to organic carbon in marine sediments is an index of oxygen level in the environment (Francois, 1987). Iodine exists as iodate in aerobic waters and as iodide in dysaerobic waters; only the iodate is readily adsorbed onto sedimentary organic matter. In modern marine sediments, high ratios of iodine to organic carbon are associated with oxygenated waters; low ratios occur only in anoxic environments.

In this chapter we go beyond the short-term, local effects of sedimentary processes considered in isolation and examine the larger-scale organization brought about by complex natural systems operating in the longer term. To aid us in our quest for depositional environment, we build summaries of the observations

179

and interpretations that we have made; these summaries are called *facies models.*

Facies Models

A *facies* is a body of sediment with specified characteristics, or just the set of characteristics themselves. Our specifications may have broad latitude, so that a given facies embraces a host of sediments that are broadly similar, or our definitions may be quite narrow and severe and fulfilled by a body of particular uniformity. Narrowly defined facies may be set up as subdivisions of broad facies.

Walther's rule, set forth by Johannes Walther in 1894, addresses a relationship between facies patterns, which are "horizontal," and stratigraphy, which is essentially "vertical." Consider two domains side by side; in one, facies A is being deposited, and in the other, facies B (Fig. 8-1). If the two domains, or at least the boundary between them, should in time shift laterally toward A, then facies B comes to overlie A

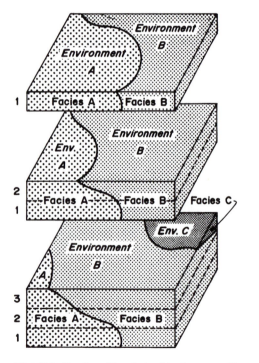

Fig. 8-1. Stratigraphic relationship between adjacent facies with shifting boundary; time-rock units are labeled 1, 2, 3.

stratigraphically; likewise, a shift toward B causes the opposite stratigraphic relationship, that is, A overlying B. Usually, it is sedimentation itself that causes the shift. Some other facies, C, not laterally contiguous with A or B, cannot be in contact with A or B in a vertical section either. We will make two useful corollaries to Walther's rule, first, that vertical (stratigraphic) patterns reflect horizontal (geographic) patterns, and second, that heterogeneity or unpredictability in a vertical profile reflects a similar condition in the geographic distribution of facies. Unconformities disrupt the relationships that Walther's rule prescribes.

Often, depositional slope is a strong ordering principle on facies patterns in marine sediments, reflecting the influence of water depth and distance from shore. On steep depositional slopes, facies patterns tend to be quite regular, typically a series of belts of roughly uniform width and roughly parallel to the shoreline. Transgression or regression over steep slopes results in stratigraphic (vertical) transitions from facies to facies that are gradual. But if depositional slope is very gentle, then stratigraphic contacts tend to be quite sharp and facies patterns haphazard, like Laporte's (1967, p. 91) "facies mosaic" or Potter and Blakely's (1968, p. 158) "crazy quilt." A small sea-level change results in large shoreline displacement, and low-amplitude topographic elements cause large irregularities in coastlines. Also, there may be a broad zone of very shallow water. The nature of deposition itself may determine angle of slope of the depositional surface.

We try to understand the composition and structure, dynamics, and origin of bodies of sedimentary rock by making *facies models.* A facies model is a description of the essential compositional, textural, structural, geometrical, temporal, and biological aspects of a collection of related deposits. It is a summary of the essential elements of a depositional system: a basis for organization of data, an aid to understanding, a guide to prediction of the characteristics of new examples, a standard to which aberrant or varietal examples can be compared (Harms et al., 1975, p. 66). It is typically built from observations made at many isolated localities, commonly by many different geologists, and is subject to alteration or augmentation as more examples become available or as basic understanding improves. The construction of a model might be implemented by statistical synthesis of the characteristics of many examples.

The importance and value of a model is that any

body of sedimentary rock of sufficient scope or delimitation can be assigned to one of a small number of models (maybe only two or three dozen), and, with certain qualifications, a facies or facies model is uniquely associated with a particular sedimentary environment. According to Potter and Pettijohn (1977, p. 314), "A sedimentary model in essence describes a recurring pattern of sedimentation."

It appears that some researchers consider a detailed description of a particularly well-studied single body or even single outcrop to be a kind of model, even though its utility as a standard or guide for other deposits has not been tested and demonstrated. It is in fact a "model" developed for the particular deposit at hand. Maybe it would more appropriately be called an *example*. Rather than a "distillation and synthesis of many examples" (Harms et al., 1975, p. 66), it is but a single case. It is probably true that a single example can serve as a model in the strict sense, but only after this example has been shown to fulfill the requirements and display the characteristics of a model, and this seems to require that many examples be compared. What aspects of an example actually help us to understand the depositional system, and what features in the example make understanding more difficult? What things are essential and what things are spurious? What is the permissible range of characteristics? A single example cannot, in itself, decide the answers to these questions. A model is largely anecdotal at first; generalizations necessarily come later. It is a matter of judgment at what stage of the development of knowledge a model actually emerges, but it seems that a mere one or two examples are not sufficient.

The elements of a complete facies model are as follows:

1. Composition and texture of the constituent materials, including fossils and content of organic matter, isotopic ratios, and trace elements.
2. Size and shape of the deposit; typical thickness or possible range of thickness, length/width/thickness ratios; orientation of the deposit relative to regional depositional strike.
3. Sedimentary structures, including bed thickness; biogenic structures.
4. Internal organization; nature of vertical trends or cyclicity in composition, texture, sedimentary structure; regional patterns of scalar and vector properties; orientation of sedimentary structures relative to shape and orientation of the rock body; arrangement of internal bounding surfaces.
5. Lateral relationships among facies; nature of upper and lower boundaries, whether disconformable, sharp, or gradational; kinds of deposits typically associated or in contact with one another; position of rock bodies within the depositional basin.

Some or all of the model characteristics might be portrayed schematically in vertical profiles, cross sections, or maps. A chronometric framework is important—how are isolated stratigraphic sections connected to one another in time? Commonly we have to guess it, but a plausible stratigraphic interpretation made without such a framework may collapse under the constraints of newly acquired chronometric data. It is important also to understand how diagenesis has affected the sediment body, especially when we attempt to compare modern deposits with ancient ones. Much diagenesis, especially in carbonate sediments, takes place within, or under conditions not far removed from, the depositional environment, so early diagenetic products provide clues to environment of deposition. It is clear that all phases of sedimentology contribute to development of facies models.

But some models are theoretical or conceptual, based not so much on the products of natural depositional systems as on the processes involved. We analyze sedimentary processes and try to guess what the products might be. Sometimes we examine the surface of a modern deposit and build a conceptual stratigraphic profile that would result from progradation or retrogradation of that surface, based on Walther's rule. Sometimes we set up physical or chemical models of sedimentation that are intelligible to a computer. With numerical or symbolic models, initial conditions and boundary conditions can be varied at will, and the consequences of a given set of conditions, as obtained by computation, can be compared with the products of natural depositional systems.

Different facies models might pertain to different scales or different levels of operation of a depositional system. Models have been developed for and applied to rock bodies as small as single beds and as large as whole basins. There may be a physical model of erosion and deposition on a stream bed (and a corresponding facies model that applies to a single kind of stratum) and a completely unrelated model that describes detrital dispersal in a drainage basin and the basin-wide variation in texture and composition of the detritus. At the scale of forces at the stream bed, tectonics and topography are not important; at the larger scale of the drainage basin, dynamics of fluids and of grain motion are not important. Thus, by limiting ourselves to a particular scale, we simplify the analysis. We hope to achieve, ultimately, facies models that embrace elements of all scales.

A model might eventually be resolved into two or more separate models, perhaps representing end points of some range or several closely related "subenvironments" or microenvironments comprising some more comprehensive macroenvironment. On the other hand, several models of limited scope might eventually be found to be related in ways that make a composite model more useful. After all, a whole is greater than the sum of its parts. The larger, more comprehensive the system (or its model), the less likely that it resembles another.

The model concept implies that the *context* of a sedimentary "object" is at least as important to an understanding of its origin or environmental significance as the mere presence of the object or its characteristics considered in isolation. Different environments can make similar products. It may be difficult, for example, to distinguish between ripples formed on a stream bed and those formed on a tidal flat, but if the ripples are associated with bipolar cross-bed sets, *Skolithos* burrows, and oyster-shell lags, then they are more likely of tidal-flat origin. Commonly a certain facies can be diagnosed solely on the basis of its position with respect to other facies, even if its intrinsic properties are equivocal or poorly understood. This contextual aspect of facies arises out of the fact that one depositional environment creates another or follows from another. The two (or the many) are thus closely associated, of course, and tend to be organized geographically and stratigraphically. Thus, a coral reef creates a lagoon; the two environments contrast markedly in their bathymetry, hydraulic regime, biota, and consequently their sedimentary products. Once a lagoonal facies has been identified and its location with respect to the reef established, the location of open-marine facies follows immediately. Context places certain constraints on interpretation. Shallow marine deposits closely interbedded with deep marine deposits, for example, would require drastic and unlikely events or circumstances, and it becomes clear that the interpretation of one or the other of the two facies is defective.

The model might reveal or emphasize some vertical trend or cycle, such as upward decline or increase in grain size, a certain pattern of succession of trace fossils, or a characteristic succession of lithic types. Thus the *vertical profile* of textural properties, bed composition, sedimentary structure types and dimensions, fossil occurrences, and so on reflects *change*, and the manner in which a depositional environment changes or evolves at a given locality, as reflected in the vertical profile, is one of the most important and useful elements of a facies model. Visher (1965b) was among the first to use the vertical profile as a model of a (three-dimensional) depositional system. We like to have models with the vertical format because our stratigraphic section descriptions, whether from outcrops or from wells, are like this. That vertical profiles are useful or adequate to the task points up once again the significance of Walther's rule.

The composition of the atmosphere and chemistry of sea water influence depositional systems; these things have changed or evolved. The most ancient sedimentary rocks arose under reducing conditions, before free oxygen became abundant; weathering may have occurred under nearly abiotic conditions. Throughout most of the earth's history, including the first third of Phanerozoic time, weathering, erosion, and deposition on land took place in the absence or near absence of vegetal ground cover. Lands in humid regions may have closely resembled the desert lands. Tectonism, driven by internal heat sources more intense than at present and operating on a thinner crust (possibly) may have influenced depositional systems differently. In the distant past the moon might have been closer to earth than it is today, and tidal ranges greater. Solar output was less and may have created unfamiliar climates long since replaced by more modern ones. The oldest preserved sedimentary rocks were deposited at a time when the whole universe was only about half the size it is today; physicists argue whether this has influenced the gravitational "constant" and the force of earth's gravity.

The sizes and shapes of organic reefs are determined in large part by the peculiarities of organisms. Reefs could not form until the appropriate organisms had evolved. The distribution of reefs today is largely determined by the tolerance of reef-building organisms to climate, water depth, turbulence and turbidity, and water composition. Marsh grasses seem to be a rather recent evolutionary development; before their appearance, certain coastal environments surely were very different from those of today. Planktonic calcareous foraminifers and diatoms did not become abundant until Mesozoic time, when oceanic pelagic oozes of Jurassic and Cretaceous and younger deep basins supplanted the starved-basin facies of earlier times.

Depositional models developed for, or based upon, the more recent deposits—and implicitly assuming "normal" rates of denudation, "normal" sea water composition, and "normal" tidal range—might not apply to the very old rocks. Thus, it may be necessary to develop special models for very old depositional systems or to make certain allowances or disclaimers when "modern" models are applied to ancient systems. One of the special problems facing those involved with Recent near-sea-level environments such as deltas, barrier islands, marine shelves, and organic reefs is that they formed or changed during a time of deglaciation and rapid sea-level rise, so that modern systems represent only one side of the transgression-regression "coin."

The great value of a facies model based on modern deposits is that the physical, chemical, and biological processes creating the deposit are operating in the present; they can be monitored and measured and sediment responses to these processes directly observed. Also, we are dealing unequivocally with an isochronous surface. An important drawback is that a deposit is not really complete until it gets buried, and there are components of the modern deposit that will not be preserved. Diagenetic effects adulterate the sedimentary fill and modify its thickness and shape. Moreover, many modern environments, paradoxically, are not as accessible or convenient as the ancient ones. An ancient deltaic deposit in western Kansas is criss-crossed with good roads; cross sections of it stand in full view, ready for direct scientific observation, measurement, and sampling. Obtaining a cross section of a modern delta is another matter, requiring special equipment and expensive procedures. Much of the system is under water; samples are soupy and sedimentary structures mercurial.

More and more today, we expand our view of the depositional system to include its interactions with adjacent systems and its long-term evolution. On the very large scale, we see the depositional products of a system in its youth, overlain by or laterally adjacent to the products of the system in maturity and on the wane, or evolving into some completely different system. We see very different transportational and depositional processes all operating in turn, perhaps on the same sediment, or in concert on different components of the total sedimentary input. At this large scale it is the *depositional architecture* that we seek to describe and understand. We might find numerous fluvial channel sandstone bodies embedded in a very large elongate body of mudstone, the sandstone bodies recording a river that changed course many times during its long life, and that, as a result of thousands of seasonal floods, gradually built a thick body of overbank mudstones. The channel deposits might exhibit certain regional trends in texture, sedimentary structure, and geometry that reflect local distance from the mouth of a river system of great areal and temporal extent. Large-scale vertical trends might reflect gradual decline of topographic relief of a drainage basin, maybe even of a whole continent, or some major change in tectonic behavior of the continent. In a bajada or delta we might observe certain cyclical patterns that record tectonic pulses or climate changes.

Characteristics of Depositional Systems

A depositional system is a conflation of external influences: geography or physiography, climate, astronomical forces, tectonics, the aptitudes and ingenuity of the biosphere. In studying a deposit and making a depositional model, we hope to learn not only the nature of the depositional system but also the external conditions that created or influenced the system.

Physical, chemical, and biological properties of environments, listed below, are measured routinely in modern, active, vital depositional systems, and we make valid attempts at estimating them in ancient environments as well. Not only is the magnitude of the property important but also the amplitude and frequency of its fluctuation:

Ambient fluid
    Temperature, heat capacity
    Density, viscosity, pressure or depth
    Climate: humidity, rainfall
    Currents: direction and strength; turbulence
    Turbidity
    Intensity and quality of light
    Salinity, acidity, dissolved gases (especially O_2, CO_2)
    Nutrient content, trace elements
Substrate
    Orientation, depth or elevation
    Composition and texture
    Firmness and stability
    Rate of sedimentation/erosion
    Nutrient content, oxygen content
Biology
    Organic productivity: biomass, populations, rate of overturn
    Structure of food chain: predator-prey relations, symbioses

Distribution of organisms and their influence on substrate

These properties are interdependent. For example, intensity of light in the ocean depends in part on turbidity, which depends on turbulence and quantity of plankton, which are controlled in turn by yet other factors. Light is dimmer in cold climates—surficial heat and light are quantities that depend on mean sun angle. Snow and ice are more reflective of solar energy than ground free of snow or a sea surface not frozen, so the cold is amplified and perpetuated. Biological communities are more diverse and complex in warmer climates than in cold. Some interdependencies form closed loops, leading to delicate balancing acts or to oscillations or instabilities. Mound- or reef-building marine organisms grow and thrive in places that happen to be hydrodynamically favorable, but the community itself alters the hydrodynamic environment. Organisms are interdependent in many ways; one organism creates opportunities for others. In the Niobrara Formation (Cretaceous) in the western interior United States, for example, large benthonic bivalves created islands of "terra firma" in a sea of soupy pelagic mud and attracted hard-substrate-loving fauna that could not have tolerated the muddy places.

Depositional systems differ because of differences in the sediment available to them. Though many depositional systems produce organized or complex sedimentary patterns, others make apparently chaotic or homogeneous products. Perhaps the available sediment was so uniform that the environment could not differentiate it, or it was supplied too rapidly for efficient processing. A depositional system might have the potential, by virtue of its hydraulics, to produce ripples or cross beds, but if only mud is being provided to the system, no ripples or cross beds can appear. The output of a natural system, such as a river, depends not only on the physical mechanisms operating there but also on the quantity and quality of the input (sediment load). To a large degree, the sediment load controls the structure and dynamics of the river, but the properties of the river, in turn, control and modify the characteristics of the load. Different rivers have contrasting physiographies and different behaviors because the textures of their loads are different.

The quantity of or rate at which sediment is provided to a depositional system is largely a factor over which the system has no control. It is determined by external conditions—the extent and physiography of the surrounding lands, the vigor of the weathering processes, and the competence and longevity of delivery systems. Thus we observe starved basins ready to receive sediments that never arrive; in other places we find a supply that seemingly overwhelms the depositional basin, sediment so bountiful that depositional processes cannot efficiently organize it into a sensible pattern. A marine transgressive depositional system can be transformed into a very different regressive one simply by an increase in sediment supply. In carbonate depositional environments, where sediments are produced internally, we find shoaling-upward or deepening-upward successions, depending on the rate of carbonate production relative to rate of basin subsidence or sea-level change, rate of production interacting with relative sea level. The arrangement of the various sedimentary facies in a basin typically is grossly asymmetrical, and the pattern of sediment dispersal defines a certain *basin polarity*, as suggested by Potter and Pettijohn (1977, p. 341). In most major basins, sediment enters from one side or one end and flows toward the other; this gives many basin fills a wedge geometry, and a regional trend in sediment texture and composition. It is clearly expressed in such diverse depositional systems as deltas and carbonate ramps.

The very process of sedimentation evidently can bring about gradual change in the conditions under which sedimentation itself takes place. It is the cause of vertical trends in texture and sedimentary structure in shoreface deposits. As sediment gradually fills a basin, the size and shape of remaining basin space changes, of course. This leads not only to changes in topographic or bathymetric relief but also to changes in circulation patterns of currents, distance of detrital sediment transport and patterns of sediment dispersal, locations of depocenters, even changes in climate and hydrology, and, in many basins, change from marine to transitional to terrestrial environments. Lakes and swamps, because they are sedimenting systems, are self-destructive; a lake makes its deposit in the geological short term, then vanishes. The wind can build great fields of dunes that conceal and cover a grassland, a playa lake, or a coastal marsh. The stratigraphic effect is abrupt and profound. Just so, the wind can locally uncover a water table that is everywhere else concealed beneath 100 meters of the driest sand. The contrasts between desert dune and oasis again are clearly recorded in the rocks.

Classification of Environments

Depositional environments can be classified, rather broadly, as terrestrial, transitional, and marine. The surface of the sea is obviously an important environmental boundary or barrier. A simple listing divides the world environment into parcels of manageable size that can be considered as separate and to some degree independent and distinguishes them on the basis of major depositional process or agent or physiographic unit:

Terrestrial	Transitional	Marine
Alluvial fan and fan delta	Deltaic	Sublittoral (marine shelf, including a host of carbonate-depositing environments)
Fluvial	Littoral	
Lacustrine	Tidal	
Glacial	Estuarine	
Aeolian		Bathyal marine fan
		Abyssal plain

But some glaciers and alluvial fans are partly marine or transitional; a deltaic or littoral environment may be associated with a lake that is wholly terrestrial. It is difficult to know where boundaries should be placed. Separate models have been developed for different kinds of rivers (fluvial environments), such as braided and meandering types; to different kinds of deltas, such as elongate (or birdfoot) and lobate; to different parts of littoral or sublittoral environments, such as lagoons, barrier islands or bars, tidal deltas, and organic reefs. Lacustrine environments of humid, temperate regions are different from lacustrine environments of arid, tropical regions; each demands its own models. The apparent simplicity of deep-sea deposits may be only a figment of our limited experience and diminutive sample of these remote deposits. As access to the major deep oceanic regions improves, we shall undoubtedly make new, more specific models pertinent to bathyal and abyssal depositional processes and products. Some scientists are engaged in the making of new models applicable to other planetary surfaces.

In the following pages, we examine the major features of each of the major environments of detrital deposition and their sedimentary products as determined from study of both ancient and modern examples. We will explore carbonate depositional systems later (Chap. 9).

Alluvial Fan and Fan-Delta Environments and Facies

The front end of the sedimentary cycle; detritus not far removed from its source.

Alluvial fans are conical piles of detritus, built by running water flowing out of an elevated land onto an adjacent lower land. The water and its load of young, locally derived detritus issues from a highland valley as essentially a point source at the apex of the fan. Deposition occurs because an abrupt decrease in gradient, or expansion of the flow, or infiltration of the running water reduces the competence and capacity of the flow. Some fans are composed almost entirely of coarse sand and gravel, others mostly of mud; the latter generally are steeper than the former. Climate affects the style of sedimentation and rate of growth. Some fans form from outwash gravels at the terminus of a glacier.

Fan Physiography

Fans characteristically form along steep scarps (Fig. 8-2), as in rift zones and block-faulted terrains. A long mountain front builds many alluvial fans, side by side, and these eventually coalesce into a generalized wedge of detritus called a *bajada*. Some fans of arid regions (dry fans) are completely devoid of vegetation; modern wet fans—that is, fans of humid regions—commonly are forested. Fans of arid regions are much smaller and steeper than those of humid climates.

Streams emerging from a mountain valley onto the apex of the fan commonly are incised there; a so-called *fanhead trench* may be as much as 12 meters deep. Probably such a trench forms because the fan is aggradational and the canyon floor is degradational; the fan to some extent onlaps the mountain valley, and the valley stream cuts downward into the *proximal fan*. A fanhead trench bifurcates downslope into a radial pattern of many small, shallow, braided channels of the *midfan*. The beds of these channels may be quite permeable, so that water flows in them only during and shortly after flood surges originating in the highland watershed or during heavy or prolonged rainfall on the fan surface itself. The latter may result in a generalized sheet flood on the fan. In the long term, channelized flow switches from one ephemeral stream to another, because the channels are aggrading; a chan-

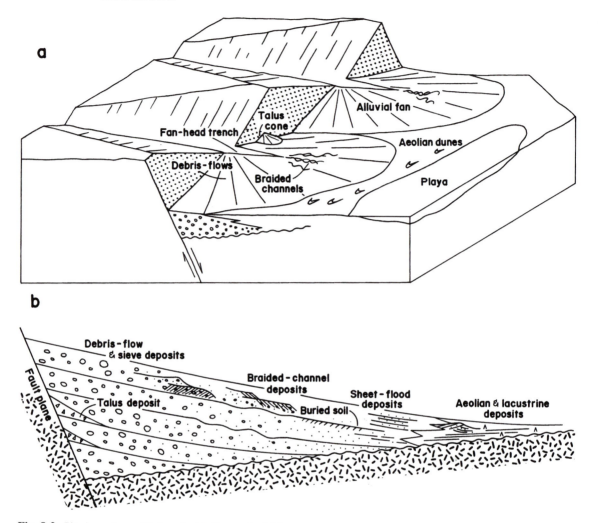

Fig. 8-2. Physiography and facies of arid-climate alluvial fans.

nel bar or debris flow plugs an active channel and di-
verts the flow to another channel. Channels deteriorate
downfan, and ***distal fan*** deposits are made largely by
sheet flows on very gentle slopes. Deposits of the distal fan merge almost imperceptibly with the fine-
grained, flat-lying deposits of the basin floor.

    In arid regions where fans are being built, detritus
typically enters elongate basins or graben from the
flanks, then turns and follows the basin axis. As fans
expand, they may eventually merge distally; up until
that stage, detritus from each flank remains quite separate. The floors of some tectonically made intracontinental basins are below sea level, and no drainage to

the sea can be established until the basin is filled with
detritus. Much debris-laden water must flow over the
long term before this can happen. In the meantime
large lakes may form on the basin floor. When a new
external base level is established, fans may begin to
erode.

Fan Sediments and Sedimentation

Fan sediments include braided stream channel deposits, sheetflood, debris flow, and sieve deposits.
Whereas debris flow (or mudflow) deposits are the
dominant component of dry fans, streamflow deposits

are the major component of wet fans. Perennial streams drain the wet fan, but occasional flood surges seem to be the dominant agent of transport and sedimentation on such fans.

Stream channel deposits are lenticular, cross-bedded sands and gravels, typically bearing oriented and imbricated clasts and devoid of mud matrix. Sheet flood is a fast, shallow flow created by lateral expansion of a flow that has left its channel or by rapid rainfall on non-channelized fan surfaces; *sheet-flood deposits* are broad, lobate sheets of flat-bedded sands and fine gravels, closely interbedded, and dotted with isolated outsize clasts. *Debris flow deposits* typically are narrow lobes of poorly sorted, clast-supported or matrix-supported gravels, with some preferred orientation of the larger clasts and reverse size-grading at the base, or with chaotic fabrics, depending on the viscosity (water content) of the flow. They are volumetrically more important on dry fans than on wet fans and require that a certain amount of mud be present; some debris flows consist entirely of mud (and may be referred to as mudflows rather than debris flows). According to Gloppen and Steel (1981, p. 51), maximum particle size in a debris flow is correlated with thickness of the flow. Some of the largest clasts may protrude from the top of the flow deposit. There may be a drape or thin cap of finer-grained, water-washed material overlying the flow deposit.

Sieve deposits (Hooke, 1967, p. 454) are waterlaid small hummocky lobate bodies of largely mud-free, clast-supported, commonly angular openwork gravels; rapid percolation of the water out of the permeable mass has caused an almost instantaneous freezing-up of the material in transport. Typically the material at the distal (downslope) end of the lobe seizes up first, forming a rampart against which the material just behind is then brought to rest. Later, a matrix may be added as fine-grained sediment illuviates into the gravel during emplacement of an overlying sediment layer.

Nilsen (1982, p. 65 ff; see also McPherson et al., 1987, p. 335) listed many criteria for recognizing ancient alluvial fan deposits, among them the following: sediments polymict and compositionally immature or with composition closely reflecting that of the catchment basin; sediments poorly sorted and angular; fossils lacking except possibly for some vertebrate bones, plant fragments, and certain animal burrows; mixture of streamflow and debris flow deposits; a limited suite of sedimentary structures; radial paleocurrent patterns

and rapid downfan decline in maximum and mean clast size; soils and calcretes; distal interfingering with alluvial plain, lacustrine, marine, or aeolian facies; and wedge or lenticular geometry. One occasionally encounters rip-up mud clasts, armored mudballs, and antidune bedding, convolute bedding, and root casts. Arid fans commonly bear aeolian deposits, deflation surfaces armored with faceted, varnished pebbles, and caliche soils and crusts. Humid fans may contain much plant matter and soils and other weathering products characteristic of humid climate, and of course they lack the distinctive sedimentological features of arid environments. Vegetation was lacking, of course, on the older fans, such as the Precambrian system of humid fans that McGowan and Groat (1971) described.

A fan deposit as a whole is wedge- or fan-shaped, 100 to 1000 meters thick on its proximal edge, thinning distally, and grading into non-fan fluvial deposits, or inland sabkha deposits, or lacustrine or marine fan-delta facies. Many fan deposits are bounded by a fault on the proximal side, and contemporaneous faulting offsets beds in the proximal fan; ancient fans commonly are bounded above and below by unconformity, possibly with substantial relief. Vertical trends or cycles in grain size reflect tectonic and climatic trends and cycles during fan construction. Along seismically active mountain fronts, the proximal fan may incorporate talus deposits, slope wash (colluvium), and landslide masses. The proximal fan contains much coarse gravel in thick deposits seemingly lacking structure. Mid-fan channels characteristically have longitudinal bars; distal-fan channels have transverse bars.

Fan Deltas

Some fans descend distally into a subaqueous environment, either lacustrine or marine; the subaqueously deposited part of a fan is called a *fan delta*. Sands and gravels prograde into standing water as simple delta foresets, whose set thickness, even as much as 30 meters, reflects water depth. Ordinarily, an abrupt change to steeper slope marks the water line. Stream channel, sheetflood, sieve deposits, and aeolian features are not present here, of course. Instead, debris-flow deposits are common, and grain-flow deposits might be present. There may be evidence of wave action and tides; there may be lacustrine or marine fossils; fan-delta deposits may interfinger with lake marls, marine skeletal carbonate sands, or turbidites.

Fluvial Environments and Facies

Detritus farther removed from the source makes its way to the sea, carried along by the great freshwater vascular system that drains the lands.

Fresh water returns to the sea via the rivers, and rivers do most of the work of transport of terrestrial sediment to the sea. A stream or river is essentially a unidirectional flow confined to a channel; some streams are less than a meter in width and discharge a fraction of a cubic meter per second or even nothing at all for extended periods; some rivers are thousands of times larger. A typical channel changes along its length; size, gradient, channel pattern, discharge, and nature of sediment load change more or less progressively; a channel may divide into several channels, then recombine downstream. The channel may, even in the short term, be quite flexible or deformable, locally changing its width, depth, and course from day to day, though most dramatically during flood.

A river constantly changes, arranging and rearranging, adjusting and exchanging its sedimentary load, making do with whatever materials happen to come down from the countless, nameless sources upstream, and in whatever quantities. The channel is at one place or another cutting rock materials and carrying them away, and at other places depositing rock materials. Ordinarily there is a net erosion in the upper reaches of the stream and a net deposition in the downstream reaches. Those reaches where sedimentation is going on were in some earlier time reaches of net erosion, and to some extent, the deposition that a river does in its old age reverses the work that it did in its youth. A river's year's work organizing its bedload vanishes in a minute with the passage of the new season's flood crest. Rebuilding or repair of the point bars and braid bars of normal stage are subject to time and chance, the whims of storm and drought.

Physiography

There are two styles of river, *meandering*, and *braided* or *anastomosing*. These two types of river produce very different kinds of sedimentary deposit. Many rivers fall into a middle ground or are braided over some part of their length and meandering over others. There are various degrees of meandering or braiding in a river, of course. The former is quantified by *sinuosity*,

being the ratio of channel length to valley length (Fig. 8-3). Some channels have sinuosity less than 1.5 and are said to be *straight* (Leopold et al., 1964, p. 281). Many high mountain streams with steep gradients are straight; some distributary channels of large river deltas (such as the Mississippi delta) are straight. The most sinuous rivers have values of 4 or more. Braiding also has been quantified; Brice (1964, p. D27) introduced the *braiding index*, defined as twice the total length of islands and braid bars divided by valley length, and being an approximation of the sum of island or bar perimeters in a reach. (Note that sinuosity of a simple straight channel is unity; braiding index of the same channel is zero. Note also that the gradient of a channel is the valley gradient divided by sinuosity.) Rust (1981, table 1) suggested that the term *braided* be reserved to multi-channel streams of low sinuosity and "anastomosing" to multi-channel high-sinuosity streams.

Deposits of Meandering Streams

The deposits made by rivers in their valleys are broadly divided into those made within the active channel or channels of the river and those made be-

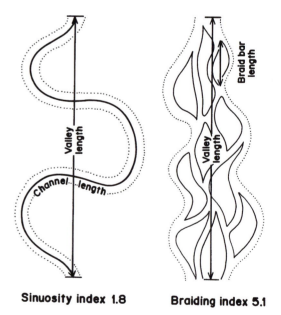

Sinuosity index 1.8   **Braiding index 5.1**

Fig. 8-3. Sinuosity of a sinuous channel; braiding index of a braided channel.

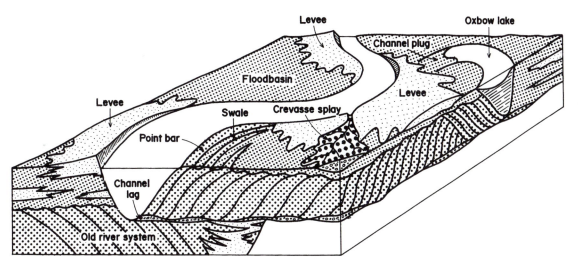

Fig. 8-4. The major physiographic and sedimentological features of a meandering stream.

yond the channel during flood; the former are called *channel deposits*, and the latter are called *overbank deposits*. Allen (1965d) described two kinds of channel deposit of meandering rivers, the channel lag deposit and the point bar deposit. A *channel lag deposit* is a lenticular sheet composed of the coarsest or heaviest materials available to the river; these materials— including gravel, mud clasts, pieces of wood, bone, and shells—accumulate in the deepest parts of the channel, along the *thalweg*. Channel lag deposits thinly and discontinuously mantle an erosion surface representing the bottom of a channel that has migrated laterally and are overlain by the finer-grained point bar deposits. A *point bar deposit* is an arcuate or crescentic body of sand and maybe some gravel that forms by lateral accretion on the convex bank of a river meander (Fig. 8-4). It is the major deposit, volumetrically, of many meandering streams. Meander loops are continually lengthening as the river cuts away at its concave bank and deposits bedload on the convex bank (or point). Grain size decreases upward, and downstream, in a point bar deposit (Fig. 8-5); fining-upward is considered to be one of the most distinctive characteristics of a fluvial sandstone. Ideally, the lower parts of the point bar sheet display flat laminasets; these are overlain by trough-style cross bedding, created by migration of dune bedforms; higher in the sheet, tabular cross-bed sets and ripple bedding occur. Such a succession reflects lower flow regimes at channel mar-

gins than exist near the thalweg. Continued accretion of the point bar results in a wide ribbon of fairly uniform thickness that corresponds approximately to channel depth. Puigdefabregas and van Vliet (1978, p. 473) described accretionary bedding in meandering

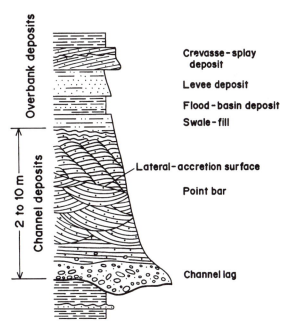

Fig. 8-5. Vertical profile of the deposit of a meandering stream.

stream sandstones of Tertiary age, being inclined units (15 to 20° dip) tens of centimeters thick, delineated by drapes of mud or silt; individual units contain other, smaller-scale cross lamination. The top of a point bar deposit typically has a corrugated "accretional" topography, a system of arcuate swells, called *scroll bars*. These features reflect an episodic lateral accretion of the point bar sheet owing to variations in discharge and stage of the river. Occasionally, part of a river flow finds the shorter path across a point bar, cutting a minor channel called a *chute* (Fig. 8-6); a *chute bar* forms near the downstream end of the point bar, where the chute reenters the main flow (McGowan and Garner, 1970, p. 85). As river meanders grow, move, and change their shapes, they are continually cutting into old point bar deposits, reworking some sand over and over again. A major channel avulsion or aggradation of the valley floor over the long term inactivates point bar deposits and leads to their burial as elongate or broad tabular sand bodies embedded in or interbedded with the much finer-grained overbank deposits.

Overbank deposits are made by river floodwaters outside the channel (Fig. 8-4); they are, on the whole, finer-grained than channel deposits. *Swale-fill deposits* are mud and plant debris laid down in time of flood in the furrows or swales between scroll bars. Closest to the channel and commonly overlying swale fills are *levee deposits*. A natural levee is a ridge of sediment that borders the channel, deposited by floodwaters out of the suspended load. It is thickest, or highest (perhaps a few meters), at the channel bank and tapers onto the floodplain; it is better developed along the concave banks of meanders than along the convex banks, where the point bars are. Grain size in a levee deposit decreases distally, away from the channel. Sandy beds

display ripple cross lamination; these are draped or overlain by laminated muds. Beyond the levees and near the margins of the floodplain, *floodbasin deposits* are made, being thin, muddy layers that settle out of the broad, shallow, ponded floodwaters. They are the finest-grained of all overbank deposits. The muds are laminated or bioturbated; there may be root molds, desiccation cracks, films of evaporitic salts; organic matter may be abundant. Levee and floodbasin deposits are overlain locally by *crevasse-splay deposits*, formed where floodwaters have breached the levee, and spilled out onto its gentle backslope and onto the floodplain. They are palmate or lobate bodies of sandy river sediment, perhaps a few meters thick, but generally thinner than a meter, locally incised into levee deposits. Typically, they are somewhat coarser than the sediments that make up the levees. They commonly bear thin tabular cross bedding, flat lamination, and ripple lamination.

Channel-fill deposits are overbank deposits made in channel segments rendered inactive by avulsion, chute cut-off, or meander-neck cut-off (Fig. 8-4). The abandoned channel segment that remains after a meander cut-off typically contains standing water and is called an *oxbow lake*. It eventually fills with muddy, organics-rich lacustrine or river-flood sediments, making a curvate "plug" that partially circumscribes point bar deposits. The muds may be laminated in part and bioturbated in part; roots may penetrate the top layer of the fill. Channel-fill deposits tend to be more resistant to erosion than point bar deposits, and constrain the meander belt to some degree.

Deposits of Braided Streams

Braided streams also make both channel deposits and overbank deposits, but the latter are thin and discontinuous. Typically, bedload is coarser and more voluminous than in meandering streams. Braiding results when a channel deposits the coarser components of its bedload as a medial braid bar that splits the flow. *Longitudinal bars* are thin sheets, of rhomboidal plan, and elongate parallel to the main flow (Fig. 8-7). Just as in point bars, the coarsest grains (pebbles or coarse sand) are at the upstream end of the bar. The downstream end typically is a slip face. Even as deposition is occurring on some parts of the bar, other parts are being eroded, mainly by strong channelized flows on the margins. Internally, braid bars show fining upward, clast orientation and imbrication, tabular cross bed-

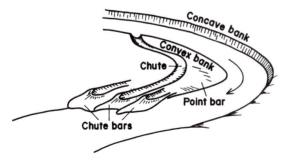

Fig. 8-6. Chute and its bar deposits.

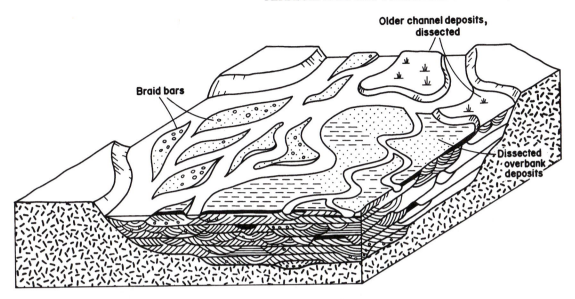

Fig. 8-7. Major physiographic and sedimentological features of a braided stream.

ding with reactivation surfaces, and horizontal bedding that overlies cross-bed sets if they are present.

Besides the longitudinal bars of braided streams, there are *transverse bars* with sinuous crests and slip faces arrayed across the channel. They grow and migrate in the downstream direction and produce extensive tabular cross-bed cosets. Transverse bars occur in reaches downstream of the longitudinal bars, where bedload is finer-grained.

Discharge of braided streams tends to be rather erratic. During times of low flow, braid bars emerge, and the flow skirts them, guided along gently sinuous, shallow channels between the bars and commonly cutting into the edges of the bars. Eynon and Walker (1974, p. 53) observed, in a Pleistocene glacial outwash stream, that ''side-channels'' flanking the gravel bars commonly are transporting sand; thus gravel is transported in the higher levels of the system, sand in the lower levels. One suspects that low or normal flow can move only whatever sand and other fine-grained sediment is available; transport and deposition take place only between the bars, whose tops are emergent. Gravel is moved mainly during flood stage, when sheets of water pour over the surfaces of the bars; material that the river could not move at low stage now rolls and slides along the bar tops, and the bars grow and migrate downstream. Braided streams occasion-

ally deposit mud and sand beyond the active channel during flood, the mud typically accumulating in abandoned channels. Grass and other plant growth help to stabilize this material during long intervals of normal stage. Wind-generated ripples may decorate the abandoned parts of braided stream deposits.

In the aggregate, channel deposits of braided streams are broad ribbons or sheets of sand and gravel with flat, scoured base. They are an amalgamation of many separate depositional units in erosional contact with one another and are structurally complex and heterogeneous. There may be thin fining-upward units and coarsening-upward units as well (Fig. 8-8). Fining-upward units typically are lenticular, erosionally cross-cutting other similar units. Cross-bedded gravel and coarse sand at the base of a channel fill gives way upward to ripple-laminated fine- and medium-grained sand, thence to flat-laminated silt and clay mud of overbank deposits. Tabular sets are the dominant cross-bedding style, formed by braid-bar accretion and migration; commonly they are overlain by thin flat-bed sets. Directional variability is generally quite low, but in channel-fill deposits in the Donjek River in Canada's Yukon Territory, variation within stream channels was, peculiarly, found to be much greater than variation between channels (see Williams and Rust, 1969, fig. 28). Trough cross-bed sets may be rare or

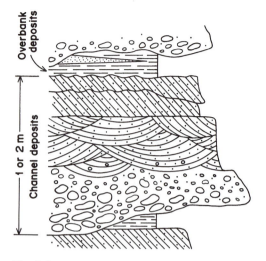

Fig. 8-8. Vertical profile of the deposit of a braided stream.

abundant, occurring mainly in the lower parts of channel deposits (Cant and Walker, 1976, p. 104 and 106).

Large-Scale Structure

Alluvial architecture, that is, the large-scale distribution of channel deposits and overbank deposits within a large alluvial deposit (Allen, 1965d, figs. 35, 36), is controlled by ratio of coarse to fine sediment, channel pattern, channel mobility, and rate of valley aggradation. Coarse-member units are broad and relatively thin or narrow and deep. Spacing of channel sandstone bodies is related to their size and shape, sinuosity, and proportion relative to the muddy overbank deposits in which they are embedded. Some coarse-member units are isolated within an embracing fine-member body (Fig. 8-9). In other large alluvial deposits, one coarse-member unit erosionally transects another. Stacked channel-sandstone bodies are said to be *multistorey*; channel sandstone bodies that are in lateral contact are said to be *multilateral*.

The alluvial body as a whole may be fan- or delta-shaped or may conform to the shape of a long, broad alluvial valley. Allen (1965c) described large-scale cyclicity in a famous alluvial deposit, the Old Red Sandstone (Devonian) of England and Wales, each cyclothem consisting of a fluvial channel sandstone on an erosional base, grading upward to or sharply overlain by overbank mudstones. Similar cyclicities occur in the "Catskill facies" (Devonian) in the northern Appalachian region of the United States (Allen and Friend, 1968, p. 48 ff).

Allen (1983) and Miall (1988) defined fluvial channel facies architecture in terms of a hierarchy of internal bounding surfaces. Low-order boundaries separate sets or cosets of cross strata and define the deposits left by a single migrating bedform or cluster of mutually similar and coeval bedforms perhaps related to bar construction. Higher-order boundaries divide a sandstone body into internally organized units representing discrete channels, each with a fairly predictable succession of textures and sedimentary structures, as in multistorey and multilateral channel deposits. The low-order boundaries commonly are truncated by higher-order boundaries.

Lacustrine Environments and Facies

Progress to the sea delayed; detritus and chemical solutions intercepted by little basins and held back temporarily from the "big basin" of the ocean.

Most of the land slopes to the sea; from where you are now standing, no matter that you are a thousand kilometers from the ocean, you could probably reach the coast without ever taking an upward step. It is the long-continued action of running water that has made it so, filling in the hollows, bringing down the heights, and making monotonic downward paths to the sea. But there are places on the land where this has not been accomplished because the topography is youthful or

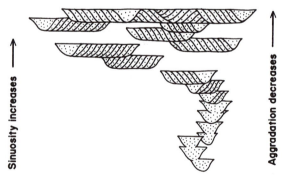

Fig. 8-9. Large-scale depositional architecture of alluvial deposits; here, a low- sinuosity, rapidly aggrading stream changes over time to high-sinuosity, non-aggrading stream.

arid and running water has not yet completed an efficient and comprehensive drainage system or certain events have created temporary barriers to runoff or deranged or rejuvenated a mature drainage system. Thus, lakes have formed behind ice dams or beaver dams, landslide debris, lava flows, and glacial moraines, in volcanic craters and caldera, impact craters, ice kettles, cirques, and the riegels of glacial staircases, on the surfaces of glacial ice sheets, between aeolian dunes, in dolines, on river flood plains, and on continental rifts and downwarps. Some lakes were once part of the sea, landlocked now in the incompletely sutured boundaries between collided continental plates, or disconnected from the sea by uplift or by coastal sedimentation. Human enterprise creates lakes also by erecting barriers to drainage, making excavations, or diverting water into basins otherwise lacking water.

Lakes are ephemeral, dynamic, and diverse—they evolve rapidly and are self-destructive; small lakes change more rapidly than large lakes; large lakes last longer than small lakes, taking longer to eutrophy and fill with sediment. Many lakes change size with the seasons or even completely vanish for part of the year. Lakes in different climates or lakes of very different size are very different in their chemistry and organic productivity, their physical structure and dynamics, and the nature of their deposits. According to Fouch and Dean (1982, p. 87), ''... to propose a set of physical and biological criteria that uniquely identify specific lacustrine depositional environments is difficult and potentially misleading.''

Physiography

Lakes occur in arid and humid climates alike, in cold climates and tropical. Some lakes are small; a person of modest means might even own one or two of them. There are very large ones shared among nations; one's view of such a lake might be comparable to an ocean view. Some very large lakes are quite shallow; some rather small lakes are very deep. Lake Baikal in Russia, at 1740 meters, the world's deepest lake, has about the same surface area as Lake Erie in North America, which is only about 65 meters deep.

Water enters lakes via inflowing streams, general runoff, groundwater flux, and rainfall onto the lake surface. Water exits through an outflowing stream, infiltration to groundwater, and by evaporation. If most of the water loss is by evaporation, then the lake acts as a distiller, concentrating and precipitating dissolved salts. Lakes on rivers intercept and filter particulate matter, so that the river that flows out of the lake is cleaner and clearer than the river that flowed in. Lakes also regulate discharge of through-flowing rivers, attenuating flood peaks downstream and maintaining flow during drought.

Lakes may develop high salinity, far exceeding that of the ocean, if evaporation is a dominant item in the hydrological budget of the lake. Ordinarily, streams and rivers contain small quantities of dissolved solids that are carried to the ocean, but in places of *internal drainage*, runoff and its solution load are intercepted by local, closed basins having no outlet. Saline lakes and their chemical precipitates are, by and large, a feature of arid climates, where inflow of water to the lake basin typically is of low volume and seasonal or flashy. Sediments of such lakes are mineralogically and texturally different from most other sediments.

Physical and Chemical Properties of Lake Waters; Circulation of Lake Waters

The dissolved ion content of most lakes is dominated by sodium, potassium, calcium, and magnesium, chloride, carbonate, sulfate, and nitrate. Lakes of humid climates are mildly acidic, and their waters typically contain calcium, magnesium, and carbonate. Lakes of arid climates, in contrast, typically are alkaline, with elevated levels of sodium, potassium, chloride, and sulfate.

The water of humid, temperate-climate lakes is divided into three layers of contrasting physical and chemical characteristics (Fig. 8-10). In the uppermost layer, called the *epilimnion*, water temperature fluctuates with the seasons. The epilimnion is in contact with the atmosphere, exchanges gases and heat with

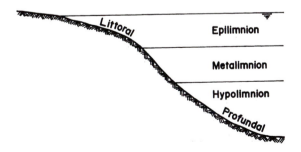

Fig. 8-10. Section through a small temperate-climate lake showing stratified water column.

the atmosphere, and receives abundant light for plant photosynthesis. Typically this layer is saturated with oxygen. The *metalimnion* is an intermediate layer in which temperature characteristically declines with depth; the downward decline in temperature is called the *thermocline*. The bottom layer, called the *hypolimnion*, is a zone of constant cold. The hypolimnion is in contact with the sediment at the bottom of the lake. In the hypolimnion, organic matter decays; this depletes oxygen and creates plant nutrients; pH generally is lower than in the overlying waters. At some depth, the rate of oxygen consumption balances the rate of oxygen production; this so-called *oxygen compensation depth* lies usually within the metalimnion but fluctuates diurnally and seasonally. Some lakes are too shallow to have a meta- or hypolimnion.

Stream water entering a lake may be more or less dense than the lake water. If the stream water is very cold or laden with fine suspended sediment, it hugs the bottom of the lake, flowing down the slope of the lake basin as a density current. Stream water that is warmer or fresher than the lake water forms a plume that spreads over the surface of the lake. Or the riverine influx may form an intermediate water layer in the lake, finding its place within the density gradient of the metalimnion.

The thermal structure of a lake lends stability; as long as warmer water overlies cold, dense water, circulation or interchange of water between layers does not occur. But freshwater lakes of temperate climates experience, each year, a convective overturn. It is quite abrupt, and in a small lake might take only a few hours. Lake waters are warmed in the summer; the surface water, in contact with warm air, warms first, and because it is less dense than the cooler waters at depth, it remains at the surface. The deeper water warms more slowly and generally remains cooler than the surface water. In the autumn, the surface water of a lake gradually cools, and when it becomes colder than underlying water, it sinks to the bottom, and bottom water rises to the top; this circulation continues until all of the water in the lake has cooled to about 4°C, at which temperature fresh water attains its maximum density (Fig. 8-11), and heat-driven circulation (convection) ceases. Surface water that cools below 4° remains at the surface because it is *less* dense than the underlying water of 4°C; when the surface water cools to about 0°C, it freezes, of course. In the spring, a weaker overturn may occur, as near-freezing water warms toward its density maximum.

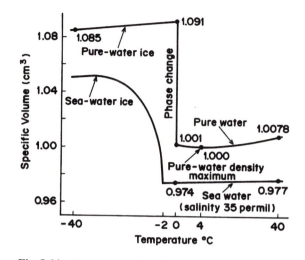

Fig. 8-11. Thermal properties of water and its ice. Temperature scale is distorted for clarity. Specific volume is volume of 1 gram.

Overturn renews a lake, restocking the bottom waters with oxygen and the surface waters with nutrients. If the surface of a lake remains frozen throughout the winter, the water beneath the ice may become quite anoxic. This water is isolated from the atmosphere and photosynthetic production of oxygen is curtailed. Bacterial decay continues, however, and consumes whatever oxygen the lake water can offer. Fish and other aquatic animals may suffocate. In northern parts of the United States and in Canada, sports-fishing lakes commonly are mixed and oxygenated artificially during the long winter to ameliorate the winter kill.

In some lakes, especially the deeper ones, there is a dense, deep layer that does not participate in convective overturn; such lakes are said to be *meromictic*, in contrast to completely circulated *holomictic* lakes. Lakes in the tropics and subtropics are warmest at the top year round and do not undergo convective overturn; they are said to be *amictic*. Coefficient of thermal expansion of water varies considerably with temperature (see Fig. 8-11). Whereas it is very low near the freezing point, at the higher temperatures it is much greater, and small temperature differences cause relatively large density differences. Thus, even small temperature gradients in warm tropical lakes are sufficient to maintain a stable stratification. The bottom water, because it does not mix, becomes anoxic and acidic. Typically, deep lakes in polar regions also have a stagnant bottom water layer. Lakes with density stratification due not to temperature gradients, but rather to salinity contrasts behave anomalously in the face of seasonal temperature change.

Besides the convective overturn, there is also a wind-driven circulation. Wind blowing steadily over the surface of a lake pushes surface water toward the downwind edge of the lake, where lake level rises as a result. Excess water sinks and feeds a countercurrent of deeper lake waters back toward the windward margin, where these cooler, deeper waters well up to the surface. When wind shear is relieved the tilted water surface of the lake gives rise to a seiche—a periodic sloshing of the lake from bank to bank for a time. Wind-driven changes in lake level have been a problem in Lake Michigan in the United States (the second largest of the Great Lakes); high water on the downwind (eastern) shore has caused erosion of high banks of weakly consolidated glacial sediments and loss of lakefront properties. Because of upwelling on the west margin of the lake, surface waters there are about 10°C cooler than water on the east side.

Spores and pollens, sticks, and leaves may be carried in from surrounding lands, but this organic detritus, according to Dean (1981, p. 224), is usually only a minor component of a temperate-climate lake's total organic carbon. Terrestrial vegetation contributes a substantial portion of the lake's *dissolved* organic matter, however. The lake itself generates both particulate and dissolved organic matter. This indigenous, or autochthonous, organic matter can be distinguished chemically from allochthonous organic matter, derived from the terrestrial plants outside; the ratio of chlorophyll derivatives to carotenoids is higher in allochthonous organic matter, and pigment diversity is less than that of aquatic vegetation.

Water chemistry strongly influences the composition of the biota. Organic diversity decreases with increasing salinity. In freshwater lakes, macrophytes may be the major producers, but in very saline or alkaline lakes, macrophytes may be absent, and phytoplankton and photosynthetic bacteria are the major producers. In alkaline lakes, green algae are important; in lakes where dissolved silica is available, diatoms may be important producers.

Organic productivity of a lake is controlled in part by the rate at which nutrients are carried in and by rate of internal recycling of nutrients. Most of the organic carbon contained in a lake is continuously recycling; only minor quantities of it leak out of the cycle and become incorporated in lake sediments. Recycling of nutrients is controlled by frequency of overturn, which, in turn, is controlled by climate and water temperature and by size and shape of the lake. Light levels, temperature, and salinity directly affect the vigor and well-being of a lake's biota and thus influence productivity.

The development or evolution of a lake's biota and chemistry is divided into three stages. The early **oligotrophic** stage is characterized by low nutrient content; thus it supports only a small plant crop; oligotrophic lakes characteristically are well oxygenated. As a lake ages, plant nutrients leached from surrounding lands and carried into the lake stimulate increasing organic productivity. The lake progresses through a **mesotrophic** stage and finally becomes **eutrophic**, at which stage it is characterized by high organic productivity and low oxygen content. Anoxia in the hypolimnion of eutrophic lakes is due largely to bacterial decay of the large quantities of plant matter that accumulate there. Human activities can greatly accelerate the evolution of a lake toward eutrophism—agricultural runoff and sewage effluents contain large quantities of nutrients, especially nitrate and phosphate.

Lake Sediments

A lake is an island of deposition in a sea of erosion. It is a local site of sedimentation strongly influenced by its surroundings; it records events in the surrounding drainage basin, where rock records are, by and large, being destroyed. Lakes and their sediments change rapidly in response to rapid change in the drainage basin; even seasonal changes may be recorded. We monitor lakes and their sediments today so that we may know the severity of pollution and the effectiveness (or ineffectiveness) of our remedial efforts and so that we may detect trends in climate (perhaps also human-caused). In short, lakes furnish a detailed record of terrestrial environments where otherwise there is very little record.

Beaches and wave-built bars may form at the margin of a lake, in what is called the **littoral zone**. The beaches of lakes are simpler, in general, than the beaches of oceans, because lakes experience only the weakest of tides and gentler and less complex wave climates. Pelagic (open water) sedimentation of terrigenous silt and clay dominate the **profundal zone**—that is, the central or deeper parts of the lake basin—or thin turbidite layers may be deposited from density currents originating on delta slopes. Lake deposits commonly are interbedded with or otherwise closely associated with fluvial and alluvial deposits. Deltas or alluvial fans may prograde into a lake basin and eventually fill it completely.

Sediments of moderate-sized temperate-climate lakes are of several different types, distributed in a lake basin in an organized manner. Certain sediment types form early in a lake's evolution; other sediment types form near the end of a lake's life cycle. The first, basal lake sediment is detrital sand and mud eroded from the drainage basin surrounding the lake, and largely undiluted with sedimentary materials made by and within the lake itself. There may be a residual soil at the bottom, which was there before the lake appeared. Overlying the basal clastic blanket and commonly the most abundant sediment in a temperate-zone lake is *marl*, microcrystalline calcite and other carbonate minerals, mixed with clay and other fine-grained terrigenous sediment. The carbonate minerals are chemically precipitated out of lake water that has become saturated with carbonate introduced from the surrounding drainage basin. Precipitation is hastened by release (to the atmosphere) of carbon dioxide from solution as lake water warms in the spring; photosynthesis by aquatic plants also consumes carbon dioxide.

Most of the carbonate sediment in lakes is precipitated inorganically, though photosynthesis induces part of this precipitation. Generally, carbonate production is more rapid in the littoral zone than in pelagic waters. In freshwater lakes, low-magnesium calcite is the major carbonate mineral precipitate, but in lakes where the magnesium/calcium ratio is high, high-magnesium calcite or even dolomite may be precipitated, and some ancient lake sediments contain the iron-rich carbonates ankerite and siderite. Most of the lake marl of profundal zones precipitates as very fine grained calcium carbonate in the surface waters, where phytoplankton photosynthesis keeps carbon dioxide levels low and pH high. As this carbonate sifts down through the hypolimnion, where carbon dioxide concentration is high and pH is low, some or even all of it redissolves and never reaches the bottom. There may be a substantial skeletal carbonate component—the shells of bivalves, gastropods, and ostracods, and the tiny calcitized oogonia of charaphytes, but only locally is skeletal carbonate abundant.

The phytoplankton is abundant in open (pelagic) water and rains down lipid-rich organic matter into the profundal zone (the bottom environment overlain by a hypolimnion), making a layer of brown or black organic sediment called *gyttja* (pronounced "YIT-cha"). At the edges of the lake (littoral zone), rooted vascular plants grow and proliferate—sedges, water lilies, and cattail rushes—and *sedge peat* accumulates; it grades laterally into the gyttja of lake center and in the long term prograges over lake-center sediments. Bacterial decay of this plant matter reduces the oxygen content of the lake; methane and hydrogen sulfide may form in the bottom waters. As the lake basin fills over the long term, the lake contracts and shoals into a *bog* (Fig. 8-12), forest cover advances over the sedge peat, and the lake eventually vanishes as a physiographic entity. The distinctive deposit that it makes records the lake's evolution from an oligotrophic, carbonate-precipitating body of water of low organic productivity to a eutrophic peat bog.

Stromatolites and onkoids occur in some lakes. Schafer and Stapf (1978) illustrated various complex forms from (Recent) Lake Constance in Germany and from Permian lacustrine beds a few kilometers away. Among the Recent algal structures, they found both hard, smooth, laminated onkoids and soft, spongy onkoids enclosing little algal "bushes." Littoral carbonate, either soft mud or hardened algal bioherm, may build a bench or shelf with a lakeward margin that falls away steeply and abruptly into deeper waters.

Seasonal changes in lakes are commonly reflected in the lake sediments, summer layers distinguished from winter layers in composition and texture. Many lakes precipitate carbonate only in the summer, when vigorous photosynthesis keeps carbon dioxide levels low and pH high. In the winter, the lake deposits a dark layer of fine-grained organic matter, especially if the surface of the lake is frozen and the water is chemically reducing. Such an annual sediment couplet is called a *varve*; it is typically a millimeter to a centimeter thick. If bottom waters are stagnant and anoxic, varves may be protected against subsequent bioturbation or stirring by currents. Glacial varves are alternating coarse and fine detrital sediment layers that reflect seasonal changes in rate of meltwater production. Varves in the ancient deposits of large lakes commonly are continuous and correlatable over distances of 10 kilometers or more. Varve counting gives precise elapsed time (in years), durations of cycles, and sedimentation rates.

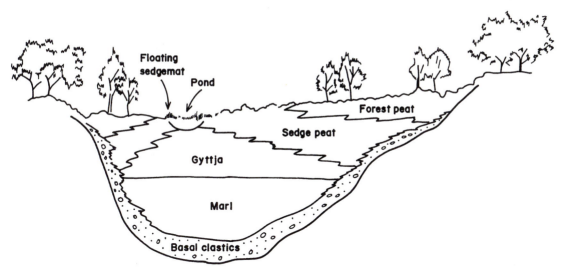

Fig. 8-12. The sediment fill of a lake that has evolved into a bog. (After Dean and Fouch, 1983, fig. 1, with permission of AAPG.)

Among the distinguishing characteristics of ancient lacustrine deposits are a freshwater biota and various mineralogical and isotopic peculiarities (not many lakes precipitate aragonite); presence of varves; association with fluvial and alluvial deposits, little deltas, and aeolian dunes; but lack of any tidal facies. A lake makes a lenticular deposit, not too thick, with sharp edges. Because lakes shoal with time and ultimately fill with sediment, their deposits are fundamentally regressive (Picard and High, 1981, p. 247), and detrital lacustrine fills typically coarsen upward and toward the margins (Galloway and Hobday, 1983, p. 198). Seasonal or longer-term fluctuations of lake level give rise to fluvial cut-and-fill structures, wave-cut terraces and associated beach gravels, local soils, mud cracks, or algal mats in the peripheral sediments and certain cyclicities in sediments of the deeper basin. Berner (1984, p. 608) noted large differences between modern marine and freshwater lacustrine organics-rich sediments in the ratio of carbon to sulfur. Decay of organic matter creates anaerobic conditions in the sediment. Anaerobic bacterial reduction of dissolved sulfate, which is abundant in sea water but not in fresh, precipitates reduced sulfur as certain ferrous sulfides, which eventually become pyrite (see Chap. 11). Organics-rich sediments of freshwater environments do not contain much pyrite.

Lakes of arid regions are very shallow or experience wide fluctuations in depth, responding rapidly to rainfall or snowmelt in the catchment basin, then completely evaporating during dry spells. Shallow, ephemeral, evaporitic lakes are called *playas*, or inland sabkhas. In the sandy and muddy deposits there may be shallow-water sedimentary structures, desiccation features, and structures due to displacive crystallization of evaporite minerals. Very saline lakes may precipitate, in addition to the calcium carbonate minerals of fresher lakes, the sodium carbonates trona ($NaHCO_3 \cdot Na_2CO_3 \cdot 2H_2O$), nahcolite ($NaHCO_3$), and natron ($Na_2CO_3 \cdot 10H_2O$); sulfates, such as gypsum ($CaSO_4 \cdot 2H_2O$); various halides, such as halite (NaCl) and sylvite (KCl); nitrates, such as soda nitre ($NaNO_3$); and borate minerals, such as borax ($Na_2B_4O_7 \cdot 10H_2O$). Some alkaline lakes associated with volcanism in the east African rift contain substantial silica in solution that precipitates as hydrous sodium silicate. The distinctive deposits of arid-climate lakes are described further in Chap. 10.

Deltaic Environments and Facies

The rivers complete their work and relinquish their burden to the waves and tides. Detritus crosses the line and enters the marine realm or poises on the brink.

Deltas are progradational deposits made where the channel-confined flow of a river expands and decelerates into standing water and drops its load thereby. Deltas tell us that rivers move sediment more effectively than ocean-marginal currents do. Though the delta is named for its "triangular" shape, deltas in fact have many shapes; each shape is a balance or interplay between riverine processes and coastal processes. The visible part of any modern delta is but an inkling of its true proportions; most of the depositional surface is hidden beneath the waves. Since world sea level rose substantially following the last ice age, the subaerial parts and upper layers of all modern marine-margin deltas have formed in only the last few thousand years. The architecture of a delta—that is, its size and shape and the nature and manner of arrangement of its constituent facies and its progradational lobes—is determined by sub-delta bathymetry, that is, the configuration of the bottom over which the delta progrades, and by rate of influx of riverine sediment and its texture, wave climate at the delta margin, tidal regime, and the nature of permanent ocean currents (especially the wind-driven geostrophic circulation). In many or most deltas, basin subsidence and sea-level changes are influential. There are some indirect effects as well—climate determines the quantity and variability of river discharge, whether the river and receiving basin freeze in winter, the kinds and amounts of sediment to be disposed of, and the kinds and amounts of vegetation on the delta. Density contrast between river water and the standing water into which it debouches determines how the two waters merge and mix and how the river drops its load.

Physiography

Many workers consider that a delta has three physiographically contrasting parts, the delta platform (or delta plain), delta slope (or delta front), and prodelta. The *prodelta* is the deep-water region where fine-grained fluvial sediments are deposited slowly from suspension (Fig. 8-13). The *delta front* is the site of most abundant sedimentation and progrades over the prodelta; sediments are coarser than those of the prodelta. The *delta platform* is the emergent or nearly emergent part of the delta, where heterogeneous sedimentation occurs in distributary channels, on levees and splays, in lakes and marshes, beaches and barrier islands, lagoons and tidal flats. In some deltas it is useful to refer to deposits of the prodelta, delta front, and delta platform as *bottomset*, *foreset*, and *topset* deposits, respectively, as Gilbert did in 1885 in his study of the simple deltas of lake margins. The foreset or slope deposit is of greatest volume. Foreset slopes of most deltas are much gentler than the 10 to 25° slopes that Gilbert described. There is a general coarsening upward, though the topset deposit may contain much fine-grained sediment in addition to its coarser complement.

The areal plan of a delta platform is governed by the relative importance of fluvial, wave, and tidal processes (Fig. 8-14). River-dominated deltas consist of deposits made by distributaries, including deposits within the channel, in overbank sites, and at the channel mouth. These deltas have elongate or birdfoot plans. Wave-dominated deltas tend toward a cuspate plan resulting from continuous reworking of riverine

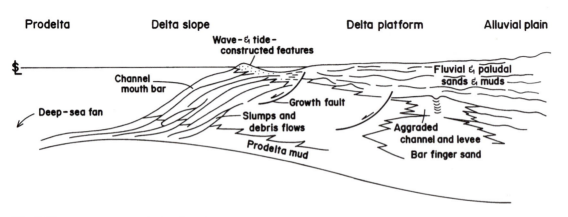

Fig. 8-13. Physiography and facies of a delta.

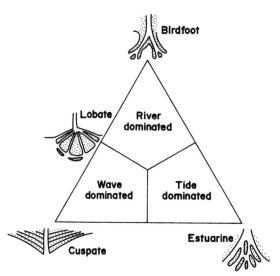

Fig. 8-14. Classification of deltas and dependence of their physiography on the relative influence of riverine, tidal, and wave processes. (After Galloway and Hobday, 1983, fig. 5-8.)

sediments by ocean waves and associated longshore currents. Tide-dominated deltas have irregular, estuarine plans; distributaries flare onto tide-swept platforms. Tidal currents flow up and down the river channel(s) on a regular basis and modify riverine sediment into shore-perpendicular tidal ridges.

Deltaic Sediments and Sedimentation

As a river flow issues from the channel mouth, it deposits its bedload and much of its suspended load on the lateral margins of the flow as levees and just beyond the orifice as channel-mouth bars. The lateral deposits are underwater extensions of channel levees and are termed **subaqueous levees**. **Channel-mouth bars** (Fig. 8-15) are the site of most rapid sedimentation. They are fan-shaped or lobate caps of sand on subaqueous muddy "bulges" that form beyond the mouth of a channel (Fisk et al., 1954, p. 81). The surface of a channel-mouth bar ascends gently to a crest, then slopes more steeply downward toward the prodelta. The back-bar surface, facing the river flow, receives the coarsest sediment.

Position and shape of the mouth bar depend on buoyancy and momentum of the river water that issues from the channel mouth, on bathymetry of the delta front, and on texture of the sediment in transport. Also,

ocean waves and tidal currents affect mouth-bar sedimentation in marine deltas.

Where the effects of waves and tides can be neglected, water that issues from a river mouth is subject to three forces (Wright, 1977, p. 857)—inertia of the flow, buoyancy of the flow, and friction at the bed. River water may emerge from the channel mouth with sufficient momentum to form a flat plume called a **plane jet** (Fig. 8-16), which maintains its integrity for some distance beyond the channel mouth. As the jet decelerates and expands laterally, it gradually drops its suspended sediment load. Coarser sediment falls out nearer the mouth, finer sediment in the deeper wa-

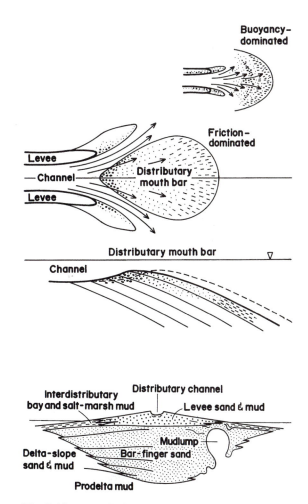

Fig. 8-15. Channel-mouth bars and bar fingers in a river-dominated delta. (After Wright, 1977, figs. 5, 6.)

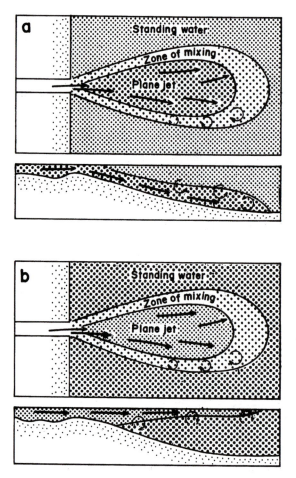

Fig. 8-16. Dynamics of river flow entering the sea: (**a**) effluent of higher density than standing water in the receiving basin, (**b**) effluent of lower density. (After Wright, 1977, fig. 3.)

ter beyond. If the river enters the sea, then the plane jet keeps to the surface, because river water is less dense than sea water. An arcuate bar (Fig. 8-15) receives the coarsest sediments. Finer suspended sediment is carried well out over the delta slope before it finally settles out. If the channel mouth is quite shallow, perhaps as a direct result of channel-mouth sedimentation, then bottom friction hastens plane-jet deceleration, and sedimentation takes place nearer to the channel mouth. Over time, a mouth bar centrally located at the mouth of the channel gradually rises into the flow and splits and deflects the flow toward the marginal levees; the bar may emerge ultimately, leading to bifurcation of the channel.

As channel-mouth sedimentation continues, the channel and its levees (both aerial and subaerial) lengthen and advance over the mouth-bar deposits. Over the long term, channel-mouth sedimentation and channel elongation result in a *bar finger* (Fisk et al., 1954, p. 89), an elongate body of sand, tens of meters thick, doubly convex in transverse section. The channel itself, flanked by levees, runs the length of its upper surface, like a backbone. A bar finger coarsens upward, because the lower parts are deposited farthest from the channel mouth, downslope from the mouth-bar crest, and the upper parts are deposited as coarser proximal bar sediments that prograde over the distal bar deposits. Distal bar sediments commonly are laminated or contain thin graded beds or slumped sediments. The higher, proximal bar sediments display trough-style cross-bedding in thick sets. All parts of a bar finger may contain dish structure and other evidence of dewatering of rapidly deposited fine sediment. Typically, plant macerals are abundant. Bar fingers thin laterally into delta-slope and interdistributary muds. They are overlain by levee deposits that grade laterally into interdistributary marsh and bay muds, and they rest upon fine muds of the prodelta, which commonly contain marine fossils and burrows.

Bar fingers are well developed in the Mississippi delta, forming the "toes" of the birdfoot. Fisk (1961, p. 30) described the gulfward motion of the Southwest Pass mouth bar of the Mississippi delta during the nineteenth century (before construction of artificial levees). Advancing at a rate of some 75 meters per year, the bar maintained a lunate form, though its crest was transected by minor submerged channels radiating from the mouth of the pass (distributary channel). Bar-finger deposits of the Mississippi delta are substantially thicker than might be expected on basis of typical maximum depths of sand deposition on the mouth-bar front, and this apparently is due to settlement of the bar into the encompassing muds, even as it is being built; post-glacial sea-level rise also has played a role.

Levees are small and poorly defined near channel mouths, but farther upstream, where they are older, they are higher and wider; vegetation stabilizes them and traps sediment borne by flood waters. Continuing compaction and subsidence of the deltaic sediment prism, which affects thickness of bar-finger deposits, also increase the thickness of a levee deposit to values considerably in excess of the height of the levee. Distributary levees may break during times of flooding, and *crevasse splays* form on the surface of the delta, much like those that form in meander belts upstream. Crevasse-splay deposits are lobate wedges of sand and mud that are channeled proximally and thicken distally; grain size decreases distally. Typically, they are interbedded with other *interdistributary deposits*, including organics-rich muds and peats of marshes and shallow lakes. A crevasse splay on a delta may remain active for a considerable period, making sheets of sand and mud that spread over interdistributary areas, displacing and filling marshes and shallow lakes and embayments of the sea. The Mississippi delta has built several major splays in the past hundred years. Some of the crevasses have drawn off much of the flow of the main channel, even causing abandonment of old lobes and establishment of new ones. This has created stacked coarsening-upward cycles in deltaic successions (Coleman and Gagliano, 1964). When a distributary channel becomes inactive, it receives a plug of organics-rich clay, much like a channel plug deposit upstream in the alluvial valley.

Prodelta mud forms the foundation of the delta; though thin, it is the most continuous of all deposits of the delta. Deposited in the deeper waters beyond the delta front, it eventually is buried as the delta progrades over it. Much of the mud is deposited out of suspension, and turbidites may be present.

Burke (1972) described submarine canyons and fans that form an integral part of the Niger delta. Marine longshore currents flow toward the "corners" of this lobate delta, spilling sediments into the heads of submarine canyons there. These canyons nourish fans on the sea floor, which are eventually buried beneath prograding delta-slope sands and muds.

Ancient deltas in the rock record are typically a coarsening-upward succession of mudstones and sandstones, fully marine at the bottom, non-marine or marginal-marine at the top. Thickness of the succession reflects depth of the body of water into which the delta prograded; thickness might also be influenced, of course, by tectonic subsidence or sea-level rise. Lineback (1966) described a Mississippian-age delta (Borden Siltstone) and associated deposits, in southern Illinois that is interesting because variations in thickness of its foreset prism helped to define subdelta depositional topography. Individual beds (subsurface) could be traced on well logs from the topset portion of the delta southward through the foreset portion into the bottomset beds, a vertical distance of about 200 meters. This delta encroached upon, and eventually bur-

ied, a crinoid bank on the west (Burlington and Keokuk Limestones) that stood 100 meters above the floor of the basin. Many delta slopes are very gentle, and in ancient deltaic deposits, the dip directions and displacement directions of slumps and growth faults may be the best, most obvious evidence of direction of depositional dip and progradation. Gentle foreset slopes also are clearly revealed on seismic profiles.

Widespread prodelta mudstone at the bottom of an ancient deltaic succession is laminated or bioturbated; fragments of terrestrial plants are abundant in deltas younger than mid-Paleozoic. Prodelta mudstones grade upward into the mudstones, siltstones, and fine-grained sandstones of the delta slope. There may be graded turbidite layers in the lower parts of the slope deposit. At some higher stratigraphic level there may be evidence of passage through wave base, above which sandstone beds, commonly with rippled surfaces, become more abundant. The upper part of a deltaic succession contains the coarsest deposits, the sandstones of distributary channels, channel-mouth bars, and perhaps wave- or tide-built bars. Crevasse-splay and levee deposits, coal beds and paleosols, add to the heterogeneity of upper parts of deltaic successions. Distributary channel deposits are fining-upward sandstone bodies scoured into underlying beds. There may be lag pebbles and coal "spars" (coal pebbles) at the base; trough-cross-bedded sandstone gives way to flat-laminated and ripple-laminated beds above.

Fisher and McGowan (1969) found the delta front facies of the Wilcox Group (Eocene) of the Texas gulf coast to be mostly sandstone, which they interpreted as distributary mouth bars reworked to various degrees by waves and currents of the marine environment. In the delta platform facies, they found thick, narrow sandstone bodies with symmetrical cross sections. These bodies are not very sinuous, have distributary areal patterns, and are embedded in interdistributary muds and lignites; sandstone constitutes 40 percent of rock volume. Fisher and McGowan mapped 16 distinct lobes, stacked in a way that suggests three distinct phases of delta progradation.

Horne et al. (1978), who made detailed observations of the coal-bearing Carboniferous strata of the Appalachian region of the eastern United States, found coarsening-upward, trough-cross-bedded sandstone bodies, 15 to 25 meters thick, that they interpreted as distributary mouth bars. Slump structures and contorted or load-casted bedding occur on the flanks and seaward margins of these bodies. Levee deposits are thin and poorly developed, being poorly sorted silt-stone and sandstone, beds inclined at about 10° away from the associated channel. Thin, cross-bedded and ripple-laminated crevasse-splay sandstones coarsen upward and distally, away from levee crevasses. These various sandstone bodies are closely associated with very fine grained deposits—mudstones of interdistri-butary bays or intertidal flats—and they grade seaward into mudstones that have marine faunas. Horne et al. found interdistributary bay-fill successions containing black shales with some lenticular limestones and sid-erite concretion horizons, locally overlain by thin, fine-grained, ripple-bedded crevasse-splay sandstones. Coal beds seem to be laterally adjacent to levee de-posits. Higher on the delta plain (upstream), sandstone bodies are more sinuous than those downslope, and commonly are multistorey (that is, contain more than one fining-upward cycle) or multilateral. They are flanked by levee deposits much thicker than those downstream.

Delta-Margin Processes

The edges of delta platforms are coasts, and thus the sites of coastal depositional and erosional processes. These processes can exert major influence on delta ar-chitecture. Where coastal reworking of the delta mar-gin is substantial, sandbodies characteristically are sheets rather than fingers. Wave action, combined, per-haps, with permanent ocean currents, reworks deltaic sediments and redistributes sediments laterally, par-allel to depositional strike. If wave action is particu-larly effective, it creates a lobate or cuspate delta; much of the sand delivered to the delta margin is trans-ported alongshore, and a series of shore-parallel beach ridges accretes onto delta margin segments between distributary mouths. Waves also assemble barrier is-lands or bars around the margins of active delta lobes. There may be aeolian dune fields atop the beaches and barrier islands. The sands are finer grained and better sorted than channel-mouth deposits that have not been reworked; the sands coarsen upward and grade down-ward into prodelta muds. There is a vertical succession of sedimentary structures characteristic of the shore-face (described later).

Strong tides, on the other hand, transport deltaic sediments *transverse* to regional depositional strike. Tide-modified distributary channels are broad and shallow and have mouths that flare, as in the Mahakam delta of Kalimantan, Indonesia. Strong tidal currents rework channel-mouth bars into tidal bars (or "ridges") that are elongate parallel to the tidal cur-rents. Such a bar deposit rests on a basal scour with shelly lag and fines upward. Cross-bedding may be the dominant sedimentary structure; it may have a bipolar aspect. There may be broad, muddy tidal flats and salt marshes flanking the bars and the fluvial channels. Tidal flat deposits, consisting of flaser-bedded sands and muds and muddy tidal creek fills, may eventually prograde over tidal-ridge sands. Tidal reworking cre-ates sand sheets in the deltaic system, just as wave reworking does.

Different segments of the delta margin undergo the destructive effects of waves and tides at different times and to different degrees. Riverine constructional pro-cesses may dominate and outpace any destructional processes on an active delta lobe, but if crevassing or distributary avulsion displaces major delta building to some other lobe, the newly inactive lobe succumbs to coastal degradation. The important factor thus seems to be rate of sediment input to the delta margin (rela-tive to rate of reprocessing at the margin), but rapid subsidence of deltaic sediment also helps to protect it from subsequent modification by coastal processes. Deltas may evolve from one kind to another. As a delta progrades into deeper water and as the delta margin lengthens, sediments delivered to the delta come to be spread ever thinner, so that coastal processes gradually become more and more important. Sea-level change or tectonism or changes in rate of input of water and sediment to the delta also alter its regime.

Coastal deposits of the delta may eventually be dis-sected by distributary channels and buried beneath ag-gradational deposits of the delta platform as the system subsides and progrades. Many ancient delta deposits are overlain by a sheet of marine sandstone that was deposited during delta abandonment and marine trans-gression over the delta platform.

Delta Subsidence and Other Deformations

Subsidence is an important influence on the thickness, shape, and depositional architecture of large deltaic sediment bodies. Some subsidence is due to loading of the crust, some is due to early compaction and de-watering of rapidly deposited sediment. Deltas com-posed largely of mud are subject to considerable com-paction and subsidence. Localized rapid deposition, as on mouth bars, causes instabilities that give rise to slumps and debris flows, and turbidity currents that flow basinward onto the prodelta of deeper waters. Even the very gentlest of delta slopes may show the effects of submarine slope failure—scarps of rotational slumps, listric growth faults (both of which cause a disconcerting dip reversal), lobate mudflow deposits and the *gullies* cut by them, and mudlumps. Slope instability apparently is due largely to syndepositional pore-fluid overpressure, including the pressure of gases evolved during decay of incorporated organic matter. Doust and Omatsola (1989) documented the complex interplay between subsidence and sediment supply in the Niger delta. The delta has prograded as a series of growth-fault-bounded units; thin transgressive shale units within normal regressive successions mark times of particularly rapid subsidence.

Bar fingers of the Mississippi delta are disrupted and broken up internally by mud diapirs. These structural features, called *mudlumps* (Morgan et al., 1968, p. 145), were described by Fisk (1961, p. 48) in the Mississippi delta, and Burke (1972, p. 1977) noticed ''shale diapirs'' in the Niger delta. They are masses of mud that have intruded the bar fingers, even rising to the surface and forming small mounds or islands. Commonly they are elongate in plan and may contain highly folded and faulted delta-slope and prodelta muds.

Glacial Environments and Facies

Special handling of detritus by a special transporting medium.

A glacier is a mass of ice that forms from snow and flows or spreads under its own weight. Glaciers are sedimentologically important and interesting because they create, entrain, transport, and deposit detrital rock materials. Moreover, the glacial ice itself is a sedimentary deposit, but one that is continually in flux. Like other sediment bodies, it is stratified and subject to the attention of stratigraphers; in some respects it yields the same kinds of information that lake sediments provide, such as year-to-year variations in temperature and precipitation. Flow displays regionally organized patterns. A glacier has the greatest competence and capacity of any discrete flow, but at the same time it has the lowest rate of sediment discharge. Glaciers leave both erosional and depositional features on the lands over which they flow, and the sizes, shapes, orientations, distribution, and textures and fabrics of these features relate closely to size, shape, flow direction, etc., of the glacier that made them. When glaciers melt away, their meltwaters leave behind various lacustrine and fluvial deposits reworked out of deposits that had been made directly from the ice; there may also be some aeolian reworking of the fines. These various *proglacial* deposits may or may not, in themselves, bear the mark of glacial origin.

Physiography

Glaciers are classified broadly into *ice caps*, which spread laterally over more or less flat topographies, and *ice streams*, which are constrained to channels (stream valleys), so that flow is generally unidirectional. Some glaciers change downcurrent from one type to the other.

That a glacier flows (or has flowed) implies a certain minimum volume and thickness. Flow of glacial ice is laminar (viscous). Flow velocity is greatest at the top of the ice, of course, being the sum (integral) of all incremental flow in the layers of ice beneath. As in streams of water, shear stress, and thus rate of strain, is greatest at the base of the flow and decreases approximately linearly to zero at the top. A glacier also slides over its bed (at least some of the time), so there is an additional component of shear stress and flow velocity.

A glacier is an open system with respect to ice, entrained rock materials, and heat content. Some or all parts of the surface of a glacier are being nourished intermittently by new snowfall; new snow becomes ice over a period of a few years, and this ice flows through the system, ultimately reaching some point where it ablates—that is, leaves the system through melting, sublimation, or calving. The time that any given parcel of ice and its sediment load spends in a glacier is far less than the life of the glacier. Rate of *nourishment* and *ablation* vary with time and determine the *mass balance*, or budget, of the glacial system. If net nourishment exceeds ablation, then the glacier expands or advances; if ablation exceeds nourishment, then the glacier contracts or recedes. All glaciers begin, obviously, with positive mass balance, and end in ''red ink''; in midlife they are fairly steady-state systems. Glaciers melt by absorbing heat; some glaciers lose heat to cold surroundings. Glaciers of polar regions surely are colder than glaciers of temperate regions; the temperature of a glacier affects the way it moves and the way it transports its sediment load. A very thick ice sheet develops a more or less linear thermal gradient below a certain depth. In the upper half of the Antarctic ice sheet, temperature is between −25 and −30°C; but in the deeper ice, temperature increases with depth at the rate of 3.25°C per 100 meters (Gow et al., 1968, p. 1011). Temperature at the base of the ice sheet is a little above 0°, and there is a thin layer of pressure-melt water between the ice and the underlying bedrock.

A mass of glacial ice is divided into three zones, a basal, or subglacial zone, a supraglacial zone at the surface of the ice, and an internal, or englacial zone. Ice of the *subglacial zone* is influenced by the bed of the glacier. In a cold glacier, the subglacial zone may be frozen to the substrate, so that basal sliding does not occur and subglacial erosion and deposition are insignificant. A warmer glacier may ride over a thin layer of pressure-melt water, and the substrate may or may not be frozen. Commonly the base of a glacier is in the process of melting or freezing. The *supraglacial zone* is influenced by seasonal changes; glacial nourishment and ablation take place here. Detached masses of ice that is no longer flowing are included in this zone. The *englacial zone* embraces the large core region of a glacier; long-term, long-distance transport takes place here.

Sediment Transport and Sedimentation

Sediment load may account for a substantial proportion of the total volume of a glacier. Because glacial ice is so viscous and so competent, it entrains and transports very large blocks as readily as it moves silt and clay. Large blocks move downglacier about as fast as the fine-grained sediment does, and there is little or no sorting of the material in transport. The rock material that has been transported and deposited by a glacier is called *drift*, and this is divided into unsorted and unstratified drift, called *till*, and sorted, stratified drift, called *outwash*, which has been reworked by glacial meltwater.

Entrainment of bed material into a glacier is accomplished in two ways, abrasion and plucking. Glacial ice itself is not hard enough to abrade or gouge bedrock or frozen soil—rock detritus embedded in the ice and pressed tightly against the bed does this work. Striae and crescentic marks appear in the bed, recording not only the fact of glacial abrasion but also the local direction of glacial motion. The abrasion of rock

against rock produces angular silt called *rock flour*, which, when released from an ablating glacier, accumulates on the floors of glacial lakes or is blown by the wind into blankets of loess. Plucking, or the quarrying of blocks of bedrock, typically occurs on the downstream flanks of bedrock knolls, ice apparently catching the edge of a joint-bounded block and rotating it out of its recess. Probably there is some pressure melting and refreezing, owing to contraction and expansion of the flow in the immediate vicinity of the knoll. Basal melting in a glacier causes deposition of entrained rock material; basal freezing causes substrate materials to be entrained and eroded.

Material deposited at the base of a glacier is called *lodgement till*. Deposits of lodgement till characterize all glaciations and are the critical sedimentological sign of ancient glaciations. These direct deposits of glacial ice are blanket deposits, up to several meters thick, thinner over topographic highs, and in sharp erosional contact with underlying bedrock or residual soils; a glacier bed composed of soft sediment may be sheared and folded. Tills are compositionally heterogenous, commonly reflecting the local composition of bed material but also containing erratic clasts of distant provenance. Tills are texturally diamict, typically bimodal or multimodal, and contain faceted and striated clasts; elongate clasts may show a preferred orientation parallel to glacial flow or across the flow in some instances, and clasts commonly are imbricated. Tills are essentially unstratified, but some contain lenses of stratified drift, have a shear-induced lamination (see Kruger, 1979, p. 331), or are banded owing to incomplete mixing of rock debris from different sources.

The supraglacial zone carries rock materials that have fallen onto the surface of the glacier (as from the walls of a valley that constrains the flow) or englacial or subglacial materials that have been exhumed to the surface of the ice by ablation. Supraglacial materials clearly are most abundant near the terminus (downstream margin) of the glacier, where ablation rates are greatest. As a glacier ablates, supraglacial tills are let down as a blanket of *ablation till* over generalized areas of ice stagnation and disintegration and also as discrete bodies at the active margins of glaciers. Ablation till forms deposits of distinctive geometry, such as kames, and lateral, medial, and end moraines. It may show some reworking by glacial meltwater, debris-flowage, or slumping where lateral ice support has melted away. It typically overlies lodgement till.

Drumlins are large longitudinal bedforms composed mainly of lodgement till; in some, the till is quite fine grained (essentially locally derived mud or sand). They are ten meters to a few tens of meters high and a kilometer or more in length (Fig. 8-17). They occur in swarms of even hundreds of individuals; hardly ever do they occur alone. Drumlin fields typically are broad bands located some distance behind the glacial terminus and parallel to it, where glacial ice flowed radially toward the margins of a lobe or where ice flowed upslope. Lundqvist (1969), working in Sweden, described large *transverse* bedforms called **rib moraines** or **Rogen-type moraines**, also composed of lodgement till. These bedforms are transitional with drumlins. (It is reminiscent of the transition between the longitudinal and transverse aeolian dunes.)

Kames and *eskers* are outwash deposits made beneath or in close contact with glacial ice. These deposits are distinguished by their shapes and locations relative to the margins of the ice or the terminal moraines. Kames are steep-sided hills, conical or elongate, of meltwater-deposited crevasse-fill gravels let down onto the drift plain as the ice melted completely away. Eskers are steep-sided, long, sinuous hills of outwash deposited on the beds of subglacial meltwater streams. The banks of these streams were the walls of ice tunnels; one suspects that pipe-full conditions obtained occasionally, as happens also occasionally in karst passages.

The **proglacial** environment embraces glacier-margin conditions, where deposition is effected by flowing or standing meltwaters. Essentially, it is a place where glacial till, deposited directly from ice, is remobilized, reorganized, and redistributed into alluvial fans or aprons called **outwash plains** or **sandars**, and deposits made in lakes that are ponded against the ice margin or held in kettles, where buried ice blocks have melted. Outwash fans or aprons resemble small alluvial fans and fan deltas, and their sediments bear textures, sedimentary structure, and internal organization similar to those of fans and fan deltas. The composition of the detritus resembles that of the glacial till from which it was derived; there may be some faceted and striated clasts. Glacilacustrine deposits include varved (or otherwise rhythmic) mudstones, commonly with multipartite summer layers and studded with dropstones that have deformed underlying laminae and are draped by overlying laminae. Some glaciers flow to the sea or into large lakes, the terminus spreading and floating over the surface of the water as an ice shelf. Ablation of such a glacier results in subaqueous outwash aprons that grade basinward into laminated debris flow, grainflow, and turbidite deposits, commonly containing dropstones released from melting icebergs. Glacimarine laminites (seasonal or not) deposited from meltwater density currents and intercalated with ice-rafted debris may blanket large areas of a continental shelf and slope.

The structure of a glacial facies association is determined by elevation, relief, and extent of the glaciated terrain; duration of glaciation and shorter-term cyclicities of climate; and bathymetry and sea-level changes for glaciers close to the sea. Elevation and relief determine whether a glacier is confined between valley walls of a highland region or spreads radially over a lowland. Elevation (together with latitude) determines local rates of nourishment and ablation; in mountainous regions, even the compass orientation of the slope may be important. Topography and physiography influence the size and shape of a glacier and

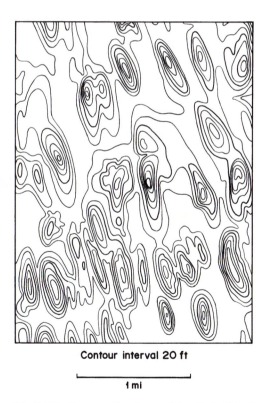

Contour interval 20 ft

|—————————————|
1 mi

Fig. 8-17. Physiography of part of drumlin field in New York State.

the direction of its flow as well as the size, shape, and distribution of ice-contact and ice-marginal deposits. Climatic cycles of a certain order cause halting or intermittent advance and retreat of glacial ice, adding complexity to glacial deposits. Eustatic sea-level change, combined with the isostatic effects of glacial loading and unloading, is an integral part of continental-scale glaciation. As a glacier melts away, part of the terrain it covered may become part of the sea, so that the glacial deposits come to be modified by marine processes.

Terrestrial glacial depositional successions are of three types (Edwards, 1986, p. 464): There is a *lodgement till facies zone* (Fig. 8-18a) consisting of a layer

of lodgement till, perhaps several meters thick, in sharp contact with an eroded or glacitectonically deformed substrate and commonly overlain by a thin layer of glacifluvial outwash sands or gravels or glacilacustrine varved muds. This is the most proximal of the three facies zones. Downstream is the *supraglacial facies zone*, containing a basal lodgement till that is overlain by end-moraine or ice-stagnation deposits, perhaps intercalated with lacustrine mudstones. The *proglacial facies zone* embraces ice-marginal alluvial-fan gravels and sands, lacustrine muds, and aeolian loess. This generally fines upward and fines and thins away from the ice margin.

Marine glacial facies are formed below sea level and

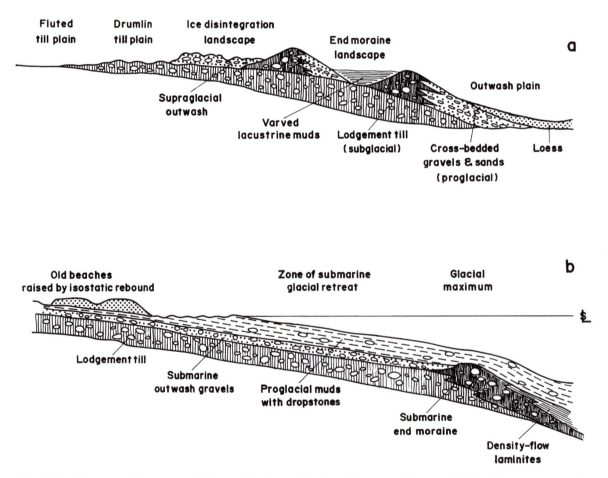

Fig. 8-18. Facies of glacial deposits. **(a)** Facies of land-based glaciers. (Edwards, 1986, fig. 13.11.) **(b)** Facies of marine-based glaciers. (After Edwards, 1986, fig. 13.12; Sugden and John, 1976, fig. 11.9.)

are either in contact with grounded ice, or under an ice shelf, or in deeper waters beyond the glacial maximum (Sugden and John, 1976, p. 224; Edwards, 1986, p. 467). The *marginal glacimarine facies zone* contains end-moraine and outwash deposits made generally near the glacial maximum and perhaps a thin basal lodgement till. In the lower part of the succession there may be beds of ripple-laminated sand made by melt-water density underflows. The upper part of the succession may contain ice-rafted diamictites and lamin-ites with dropstones, these deposits made during the glacial retreat. Boulton (1990, p. 19) observed that un-derwater sediment fans that lean against the ice margin are pushed seaward and smeared as a result of ice-margin advance. Moreover, glacial ice that is advanc-ing over muddy sediments causes these sediments to intrude crevasses in the ice from below; later, when the ice has melted, these intrusions are preserved as reticulate patterns of till ridges on the sea floor. The *proglacial marine facies zone* (Fig. 8-18b) contains laminated mudstones deposited by meltwater density currents, ice-rafted diamictites, and subaqueous de-bris-flow diamictites. Boulton noted that, though melt-water may form underflows retaining contact with the bottom, most meltwater rises to the sea surface as tur-bid, brackish-water plumes; sedimentation of the sus-pended mud takes place even 10 or 20 kilometers sea-ward of the ice margin.

Aeolian Environments and Facies

The winds disperse the sand and dust and loft the spores and spiders from land to land or to the sea.

Air is an efficient medium of transport of the smaller sedimentary particles as long as they are not bound together by vegetation, moisture, or mineral cements. Ordinary winds move sand grains up to about 1 mil-limeter in size; larger particles in a wind-swept bed are left behind. Grains smaller than about 0.1 millimeter move on quickly into regions where aeolian processes are not particularly evident. In desert places, where sediment-binding vegetation is sparse or even lacking altogether, the wind can be a dominating erosive agent. But humid places too can experience the erosive effects of the wind—we squint at the drifting snow, smell the clouds of sand and dust rising from the plowed field, and cringe from the searing sand-laden gusts at the beach before the storm. Before vegetation establishes itself on the outwash and moraine of re-ceding glaciers, the wind lifts out the fine rock flour and spreads it as a blanket of loess over large areas. The wind transports also the finer components of vol-canic ash great distances. High-speed (250 kilometers per hour) stratospheric airflow (the jet stream) can dis-perse fine particles globally. Wind is the fastest me-dium of particle transport, but discharge rates gener-ally are low. Flight paths readily transcend the boundaries between land and sea—windblown dust seems to be an important component of sediments of the deep ocean.

Areas that are deserts now, once were not—ocean currents change their courses and alter the patterns of rainfall on the lands; mountain ranges rise up into the paths of humid winds and intercept their moisture. To-day the great deserts of the subtropical latitudes are expanding at rates that human beings can perceive in their lifetimes. Dry air in motion is a virtually limitless sink for water; the reservoir behind the Aswan High Dam in Egypt never filled, in part because so much of its water is carried off by the hot winds. Vegetation slows and baffles the winds, anchors and shelters the sand and dust against the aeolian forces, and filters or strains the aeolian loads out of the wind. In earlier times, before terrestrial vegetation made its appear-ance, aeolian processes might have been far more im-portant and effective than now.

Physiography

The great sand seas, or *ergs*, cover areas of tens of thousands of square kilometers in sand hundreds of meters thick. The erg shows a hierarchy of bedforms or landforms; the largest forms are the *draa*, which are large mound-clusters of smaller forms called *dunes* (not the same as the fluvial bedforms described in Chap. 5); some of the larger complex forms appear to be composed of smaller, simpler dunes. Dunes, in turn, are covered over with ripples. All these forms move with the wind, the smaller ones much faster than the larger ones. The smaller forms can respond to short-term changes in wind direction; the larger ones re-spond far more sluggishly.

Dunes are the forms of intermediate scale; they have wavelengths of hundreds or thousands of meters and heights of tens or hundreds of meters. The *barchan* (Fig. 8-19) is an elegant form with a crescent-shaped plan and a well-defined slipface that extends to the tips

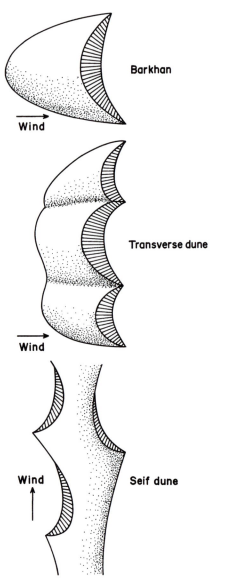

Fig. 8-19. Various types of aeolian dunes.

rium relationship between the maximum (that is, stable) size of a barchan, wind speed, and grain size. Barchan formation requires unidirectional winds. Under excessive sand supply, these short-crested forms come together into long-crested transverse dunes.

The long-crested **transverse dune** bears a more or less continuous but scalloped crest that is oriented transversely to the wind; the dune advances slowly downwind. It is a large dune of regions where sand cover is complete, due either to a large supply of sand upwind or to some local condition that decreases the transporting capacity of the wind. By virtue of their size and orientation, transverse dunes are especially disruptive of the wind-flow patterns that give rise to them, and this leads to certain internal structural complexities that the barchan or the longitudinal dune types lack.

The **longitudinal** or **seif dune** is a wind-parallel ridge of sand, commonly 200 to 300 meters high and tens of kilometers long (though there are much smaller ones only a meter or two in height and a kilometer or two in length) occurring on otherwise smooth, flat, wind-swept topographies where limited sand volumes are transported quickly by steady winds. The large seif dune has a sinuous or scalloped crest, with small slipfaces arrayed alternately on either side of the crest. One suspects that transverse and longitudinal forms are responses mainly to different average wind speeds; as in the case of the subaqueous bedforms, grain size may be a factor as well. Other large, complex forms with branching or star-burst crests are called **star draa** or **rhourd**. Commonly they are covered over with smaller dune forms. Draa and other large dunes are more complex than smaller dunes; this, again, is because the larger forms substantially disrupt the flow of the wind and because they respond so slowly to changing wind conditions.

Internal Features of Aeolian Deposits

Various workers have tried to find relationships between the internal sedimentary structure of dunes and dune type. Glennie (1970, fig. 68; 1972), for example, has suggested that different types of dunes display different angular distributions of cross beds, as manifested in histograms or stereographic plots of large numbers of cross-bed measurements. Studies of this kind have not met with much success; a major obstacle is the difficulty of dissecting a modern dune. Cross-bed sets in aeolian dunes typically are a meter or sev-

of the horns, which point downwind. Barchans (pronounced "BAR-can") are among the smaller dune types; they occur in swarms but nonetheless are separate and distinct from one another. They lose sand from their horns and spawn new barchans; little barchans race ahead and collide and merge with the larger, slower ones. Probably there is some equilib-

eral meters thick. Bounding surfaces between sets are planar in sections parallel to the wind and dip gently downwind; in sections across the wind, sets are broadly trough-shaped. According to McKee (1966, p. 57), wedge-shaped sets form if the wind changes direction. Tabular types in which set boundaries dip moderately or steeply represent slipface deposits formed on the lower part of a dune, largely through avalanching of sand. Tabular-planar sets with roughly horizontal set boundaries mostly form high in the dune. Trough sets, never a dominant type, seem to form mainly along dune crests. These relationships were discovered in deep cuts made by bulldozer in aeolian dunes at White Sands National Monument in New Mexico.

There is a higher-order boundary—extensive, gently inclined surfaces that separate a thick aeolian sand sheet into cross-bed cosets of tabular form (Fig. 8-20). Stokes (1968, p. 511) thought that multiple parallel truncations in Jurassic aeolian sandstones in Arizona and Utah might be due to regional deflation to a water table contained within the aeolian blanket. Brookfield (1977, p. 310) objected to Stokes's hypothesis on grounds that a surface deflated to water table ought to show some of the features of inland sabkhas. Brookfield (1977, p. 316) and Kocurek (1981, p. 760) have suggested alternatively that the coset-bounding surfaces demarcate draa-scale bedforms.

Laminae within a cross-bed set have certain features that aid in distinguishing ancient aeolian sandstones from those deposited in aqueous environments. Angle of inclination is high, even in excess of 30°, but not consistently so. Near the base of a set there is commonly a rather abrupt change to very gentle dips (Fig.

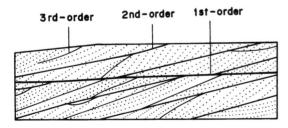

Fig. 8-20. First-, second-, and third-order surfaces between tabular cosets in an ancient aeolian dune sandstone. (After Brookfield, 1977, fig. 2, with permission of Blackwell Scientific Publications, Ltd.)

8-21) and discontinuities or low-angle truncations of laminae. An inclined lamina, especially one that is somewhat thicker and a bit more steeply inclined than the average, may end rather abruptly in an *avalanche toe* (or *sand-flow toe*) against the curving underlying lamina; it is due to a thin grainflow on a slipface steepened beyond angle of repose. Truncations higher in a set are probably reactivation surfaces, made during lulls or transients in direction of the wind. Deformed laminae indicate slumping during a rainfall or melting of buried snow. Oversteepening, and associated slumping or flow, can probably occur only in damp sand; slumping may be of small scale, involving only a few laminae, or it may involve a substantial part of a cross-bed set several meters in thickness. Aeolian ripples typically bear long straight crests that have rounded cross sections. Ripple index, that is, the ratio of wavelength to height (see Chap. 5), is high, greater than 15. Coarser grains tend to be concentrated on the crests. Ripples may be preserved on inclined laminae, their long, low, rounded crests running parallel to dip of slip faces of cross beds (McKee and Bigarella, 1979a, p. 192). The peculiar translatent stratification that Hunter (1977a) described seems to be diagnostic of aeolian sand deposits (see Chap. 5). Little "buttons" with raised rims are made by raindrops on dry sand; they are considered to be good evidence of aeolian deposition. Bioturbation is not uncommon in aeolian sands and sandstones (see Ahlbrandt et al., 1978), and their lebensspuren are quite different from those of subaqueous deposits. Glennie (1970, p. 81, 167) found that a sand dune fines upward and that sand at the top of a dune is better sorted and more angular. Eschner and Kocurek (1986) described large-scale contortions that occur when unconsolidated aeolian sands are transgressed by marine waters.

Though most aeolian sands are composed dominantly of quartz, there may be substantial quantities of rock fragments; soft materials such as calcite cement fragments and glauconite peloids recycled from older rocks also may be present. Some aeolianites are composed *mostly* of carbonate grains (see Chap. 9) or even gypsum, among the softest of minerals.

Many aeolian systems include distinctive *interdune deposits*, generally isolated, lenticular bodies of flat-bedded or ripple-laminated or adhesion-rippled sandstone, perhaps with muddy drapes and desiccation-cracked muds, evaporitic crusts, bioturbation, plant root molds, and vertebrate and invertebrate fossils. Some interdune surfaces are quite dry, others intersect

Fig. 8-21. Sharp break in angle of dip of aeolian cross bed. Avalanche toe in an aeolian cross-bed set.

a water table. Ultimately the surrounding dunes encroach upon the interdune hollows and bury their special sedimentary products. Fryberger et al. (1988) described some modern examples of interdune surfaces caused by deflation to a water table (Stokes surfaces). The water table limits aeolian scour because of the cohesion of damp sand; early cements and algal mats also may bind the surface. Stokes surfaces may be diachronous, owing to encroachment of adjacent dunes and lateral shift of the deflated region where scour occurs (this is analogous to the diachronous scour surface developed by lateral migration of a river channel).

Aeolian sand sheets are areally extensive but thin wind-laid deposits that seem to be formed where vegetation plays an important role, acting as filter and baffle to the sand-laden winds and to some extent stabilizing the drifting sands. Ahlbrandt and Fryberger (1982, p. 26) listed eleven characteristics of modern sand-sheet deposits, among them high-index ripples in coarse sand; horizontal or gently inclined cross-laminae (up to 20°); curved or irregular internal erosion surfaces; and abundant root casts and bioturbation traces. Sand sheets are in most cases marginal to ergs. Fryberger et al. (1983, pp. 302, 310) suggested that aeolian sand sheets of Pennsylvanian-Permian age in

Wyoming are deposits of an aeolian sand sea that prograded into coastal waters.

Ancient Aeolianites

Ancient aeolian deposits are laterally extensive, homogeneous bodies of fine-grained, well-sorted sandstone with large, sweeping cross-bed sets; thick cross-bed cosets may be set apart by lenticular interdune deposits. Pebbles and granules are lacking except maybe at the base of some cosets and on the perimeters of the aeolian deposit. (Do not confuse pebbles with concretions.) Body fossils are not abundant except perhaps in interdune deposits, but trace fossils are distinctive. Aeolian sandstone bodies may pass laterally into playa or alluvial fan deposits or contrasting marine beds. Aeolian deposits probably are not transitional with anything; aeolian sands may *interfinger* with lacustrine or alluvial fan or marine sediments, but do not *grade* into such sediments.

Among the ancient sandstones considered to be aeolian are the Coconino (Permian), Wingate (Triassic), and Navajo Sandstones (Jurassic) of the Colorado Plateau in the southwestern United States (McKee, 1933, 1945; Sanderson, 1974; McKee and Bigarella, 1979a); part of the New Red Sandstone (Permian) in England

(Laming, 1966), and Barun Goyot Formation (Cretaceous) of Mongolia (Gradzinski and Jerzykiewicz, 1974). Walker and Harms (1972) interpreted exposures of the Lyons Sandstone (Permian) near Boulder, Colorado, as aeolian, though they had earlier been described as littoral. Sedimentary structures ruled the interpretation—there are thick cross laminasets with low ripples whose crests are invariably oriented parallel to *dip* of the inclined laminae; also, there are small avalanche (or sand flow) features, rainprint "buttons," and small reptile or amphibian tracks.

Littoral Environments and Facies

Detritus on the tumultuous edge of the sea.

The complex boundary where land and water meet is shaped by the waves and the tides. There is probably more mechanical energy expended here than in any other environment. Different wave climates and different tidal regimes make different kinds of beaches. Some coasts are shaped mainly by storm waves, others by fair-weather waves, still others mainly by tidal processes. Other influential factors are textural qualities of the detrital sediment and rate of delivery of detrital sediment to the coast; climate, insofar as it determines frequency and intensity of storms, nature of vegetation on the coast, and whether evaporite deposition occurs on tidal flats; and trends or fluctuations in relative sea level. An important problem and goal of the diagnostician of an ancient coastal deposit, besides inferring wave climate and tidal regime, is to determine whether the coastal system was transgressive, stationary, or regressive in the long term.

Physiography

A typical wave-constructed beach (Fig. 8-22) comprises three zones, called shoreface, foreshore, and backshore. The *shoreface*, which is a seaward slope that extends from mean low tide level to mean fair-weather wave base, is the zone of shoaling and transforming waves, that is, waves that are expending mechanical energy against the bottom and slowing down, steepening, and breaking. To seaward of the shoreface is a zone of transition to the offshore, between mean fair-weather wave base and mean storm wave base. Landward of the shoreface is the *foreshore*, which lies between mean low tide and mean high tide levels. It is the zone of wave bores, that is, swells that have collapsed into solitary waves or surges, and it includes a wave-swash zone, or *beach face*. The *backshore* lies behind the foreshore, above the mean high tide level; commonly it slopes gently landward. The backshore is bounded landward by an aeolian dune ridge, sea cliff, or lagoon or salt marsh.

A beach is a place of complex currents. Waves produce nearly symmetrical oscillatory currents on the lower shoreface, currents that are roughly perpendicular to the shoreline. Higher on the shoreface, the os-

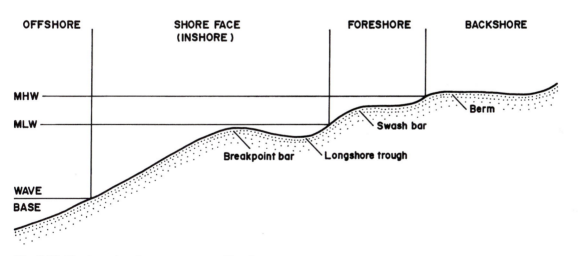

Fig. 8-22. Physiography of a wave-constructed beach.

cillations are asymmetrical, with a strong landward surge. Transforming waves generate unidirectional currents near the shore, some flowing parallel to the strand (*longshore current*), others flowing seaward (*rip current*). These currents are superimposed on tidal currents, which are especially strong and consequential in estuaries and on the coasts of broad marine shelves. Storms are disruptive; they erode the sediment bodies made during fair weather and transport coastal sediments both landward onto the backshore and seaward into deeper water of the lower shoreface or the offshore. Storm wind stress transports surface water to the coast, and causes sea level to rise, a so-called *storm tide* or *storm surge*. Part of the sea-level rise is due to reduced atmospheric pressure associated with storm cells. Much of this excess water returns to the open sea along the bottom, eroding and steepening the upper shoreface. After the storm passes, fair-weather waves will gradually restore the upper shoreface sediment prism and profile.

The shoreface commonly contains a longshore bar constructed by the surf. The position of a so-called *breakpoint bar* (Fig. 8-22) on a shore profile is determined by the point at which waves break and vice-versa; there may be more than one breakpoint bar, corresponding to different wave conditions or different tide levels. On the foreshore, a *swash bar* is formed by wave bores surging up the beach face, then retreating. Just to seaward of the swash bar, there may be a *ridge-and-runnel* system, created by a complex interaction between tide- and wave-generated shoreward and longshore currents. A runnel or *longshore trough* is a shore-parallel trough or furrow that carries a longshore current; during the ebb tide, water drains back to the sea in part via the runnel. The *ridge*, on the seaward flank of the runnel, is formed by the waves (it may actually be a breakpoint bar) but modified by the tides. Typically it is asymmetrical, being much steeper to landward, and decorated with shore-parallel ripples. Runnels connect with *rip channels*, which transport water and sediment to the offshore. The backshore may contain a ridge or berm of sand that advances landward during the highest tides or during times of heavy swells.

Beach Deposits

Ancient progradational wave-dominated coastal successions (Fig. 8-23) ordinarily coarsen upward from dark, fossiliferous, bioturbated offshore mudstones to

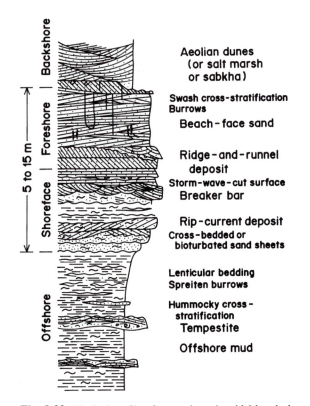

Fig. 8-23. Vertical profile of regressive microtidal beach deposit. There is much variation in beach deposits, reflecting wave climate, the variable pattern of wave-generated currents, and the texture of the sediments.

shoreface and foreshore sandstones (and imbricated gravels in some beach deposits). They range in thickness from about 10 meters to 50 or 60 meters, presumably reflecting depth to wave base. There is a succession of sedimentary structures. Lower parts reflect storm wave conditions; upper parts reflect fair-weather conditions. Sediment eroded from the upper shoreface during storms is deposited on the lower shoreface, near storm wave base, where it is shaped and molded by storm waves but will be largely unaffected by subsequent fair-weather wave conditions. The upper shoreface, in contrast, typically consists of deposits made by fair-weather waves, though scour surfaces created by storms are preserved in the upper shoreface as well.

Near storm wave base, in the zone of transition between shoreface and the offshore, fine sands and silts alternate with mud. The coarser sediments, deposited during storms, extend well offshore. They characteristically display hummocky cross-stratification (HCS) or landward-oriented tabular cross-bed sets. Muddy sediment settles out of suspension during fair weather and drapes the HCS. Strong rip currents, which move coastal sediments seaward, make lobate deposits with seaward-oriented cross-bed sets on a scoured base on the lower shoreface or offshore. Bioturbation by an infauna active during the intervals between storms may largely obliterate these structures.

Sandstones in the overlying shoreface succession typically display wave-ripple lamination, reflecting the gentle currents made on the bottom by the persistent but ordinary waves; transforming waves also create dune bedforms. The observations of Clifton et al. (1971) suggest that landward-oriented trough cross-bedding is present in lower shoreface sandstones, overlain by flat-laminasets and seaward-oriented trough cross-bedding in higher sandstones of the shoreface. Wave-generated currents of the inner surf zone and flood tides transport sand shoreward over the crest of the ridge, or longshore bar, and dump it into the runnel, where a longshore current takes over. Thus, the landward flank of the ridge is a slipface, and under certain conditions the ridge can advance over the runnel and weld itself to the swash bar. (Also, the swash bar may prograde seaward over the runnel.) In this way, the beach progrades during fair weather. Migration of ridge-and-runnel systems, longshore bars, or rip channels creates erosion surfaces within the coastal succession (Davidson-Arnott and Greenwood, 1974, p. 700; Hunter et al., 1979, pp. 720 and 724; Clifton, 1981, p. 177). There may be a lag gravel on the bed of a runnel, or a shell pavement, or trains of ripples.

Beachface sandstones display swash cross-stratification, being thin wedge-shaped sets of planar laminae gently inclined toward the sea. One might be lucky enough to find swash marks, rill marks, or adhesion ripples in some of these laminated sandstones of ancient coastal successions. There is generally not much bioturbation. A backshore facies may be preserved at the top of the coastal succession, sandstones with landward-dipping tabular cross-bed sets or flat-laminasets representing accreted ridges, aeolian cross beds and ripples, or adhesion ripples; wind deflation creates shell lags. Driftwood may be deposited there during storms or high tides. Backshore sandstone might be profusely bioturbated locally.

The lower part of the Gallup Sandstone (Cretaceous) in New Mexico (Campbell, 1971) comprises a set of accretionary units inclined gently to seaward,

each unit bearing backshore facies at its upper, landward edge, replaced successively downdip by foreshore, shoreface, and offshore facies. This is clear evidence of coastal progradation. The abundant sediment supply suggests that this coast was part of a cuspate (wave-dominated) delta. But progradation was episodic; surfaces that separate the imbricate units are gently erosional and apparently mark brief degradational events (stormy times). Locally, and especially to landward, the coastal sandstones are overlain by coals or aeolian dune sandstones. McCubbin (1982, p. 259 ff) found the coarsest sand in the middle shoreface part of the Gallup coastal deposit, so that only the lower part of the succession coarsens upward, the upper part fining upward. This feature, together with a total thickness and overall grain size that are somewhat greater than usual, suggest a beach experiencing heavy surf. Clifton et al. (1971, pp. 663, 669) had earlier described a modern gravelly beach of a stormy coast, pointing out that the coarsest sediment is well down the beach face, where storm deposition is important.

Coastal Systems with Barrier Islands

Under certain conditions, a beach forms offshore, as a *barrier island* (Fig. 8-24). It seems that some barrier

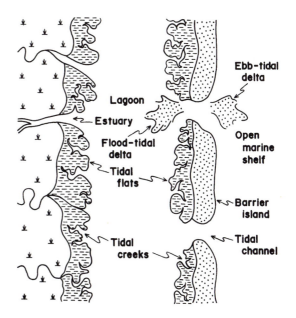

Fig. 8-24. Physiography of tide-influenced coast with barrier island.

islands form from longshore bars that grow upward and breach, or emerge during drop in sea level; others grow by longshore accretion of spits; still others form as a result of sea-level rise and flooding of lowlands surrounding or flanking beach ridges or cheniers. As sea level changes, barrier islands migrate landward or seaward or else vanish beneath the waves or amalgamate with the mainland. The backshore of a barrier island borders on a *lagoon*, a place of shallow, perhaps stagnant water insulated from ocean swells. The restricted waters become hypersaline if climate is arid, but in a humid climate, lagoonal waters are diluted by fresh waters from the mainland. Unlike the sands of the barrier island, sediments of the lagoon (Fig. 8-25) may be muddy, organics-rich, and extensively bioturbated. Muddy tidal flats may occupy the margins of the lagoon. Storm waves may carry barrier-island sands into the lagoon, forming *storm washover fans*. These are lobate sheets with erosional base, channeled proximally and gently tapering distally. They consist of flat-laminated sand; a basal shell-lag may be present; bioturbation may overprint the current-generated structures. Washover fans owe in part to elevated sea level of the storm surge. McLane (1982; 1983) found lagoonal sediments in the Codell Sandstone (Cretaceous) in south-central Colorado behind a barrier island; they consist of strongly bioturbated, weakly muddy, fine-grained sandstones. Spreiten burrows obliterate virtually all bedding. Punctuating this otherwise homogeneous body of bioturbated sandstone are a few isolated thin beds or lenses of somewhat coarser, well-washed sandstone, each having an erosional base; some are flat-laminated; others comprise a single landward-oriented cross-bed set; some beds are perforated with *Ophiomorpha* burrows and contain numerous molluscs that apparently were transported with the sand. Each isolated sand layer appears to have been deposited by an abrupt, short-lived, competent current, most likely a storm-washover current.

Tidal channels (or tidal inlets) transect the barrier island at intervals, kept open by a strong tidal ebb and flow. Ebb currents dominate the deeper axis of the channel; flood currents sweep the shallow banks. Flanking the axial *ebb channel* are linear bars that resemble levees; these bars separate the ebb channel from the much shallower marginal *flood channels*, bearing flood-oriented bedforms (Hayes, 1980, p. 141). Perhaps only the flood currents or only the ebb currents are manifested in the cross bedding of a given tidal channel deposit. A tidal channel is like a very short river with bidirectional flow and a delta at each end. On the lagoonward end, a tidal channel shoals onto a *flood-tidal delta*, a thin, lobate sand body that matures into a shield shape (Fig. 8-26); flood-tide waters fan out over the gently seaward-sloping surface, or *flood ramp*, of the shield, constrained to some extent within small distributary channels; then they spill over its marginal slipface and flow around the margin of the shield back toward the tidal inlet during the ebb, building *ebb spits* on the shield margin.

The flood ramp may bear transverse bars (Boothroyd, 1978, p. 326), which are in turn mantled with flood-oriented sand waves and ripples (Hayes, 1980, p. 146). Sands of the flood-tidal delta display flat-laminasets and flood-oriented tabular cross-bed sets. Cross laminasets commonly give way to overlying ripple laminasets that are draped with films of clay deposited out of suspension during the slack time between tidal flow and ebb. McLane (1982, p. 78) found a tabular body of well-washed sandstone about a meter thick, locally overlying lagoon-fill sandstones of the Codell, and apparently connected to a tidal inlet. It has an erosional base and comprises flat-laminasets and tabular cross-bed sets, and is gently burrowed. He interpreted this tabular body as a thin flood-tidal delta shield. Its complementary ebb-tidal delta also is preserved.

An *ebb-tidal delta* guards the seaward end of a tidal channel, though it may be totally submerged, and di-

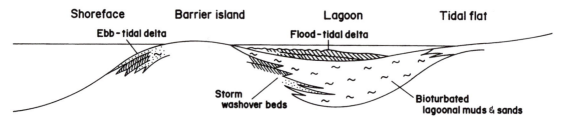

Fig. 8-25. Facies of lagoonal and tidal deposits associated with barrier island.

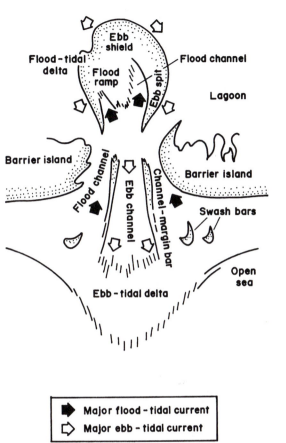

Fig. 8-26. Physiography of flood-tidal delta and ebb-tidal delta. (After Hayes, 1980, figs. 1, 10.)

the longshore drift. A recurved spit on the updrift flank of the channel grows laterally into the channel, leaving in its wake a series of curved beach ridges with marshy swales between (Hayes, 1980, p. 149). Tidal-channel migration gives rise to a lateral accretion deposit with an erosional base mantled with lag gravel (Kumar and Sanders, 1974, fig. 6).

Chenier Plains

A **chenier** is a sandy beach ridge embedded in muddy marsh sediment. On some coasts there is a series of these ridges, made of sand, gravel, and shells, roughly parallel, alternating with narrow mudflats or marshes and creating an extensive coastal lowland called a chenier plain (Fig. 8-27). Chenier plains are progradational deposits apparently made on coasts receiving an abundant but intermittent or fluctuating supply of sand and mud. They probably originated as barrier islands or beach ridges nourished by longshore drift. Allen (1965a) pointed out that a system of barrier islands on the margin of the Niger delta is but the latest

minished and modified by ocean swells and longshore currents. Lying in deeper water and on a steeper slope than the flood-tidal delta, it is thicker and areally more compact. Under energetic wave climate the ebb-tidal delta lobe lies close in on the terminus of the tidal inlet; in milder wave climates the lobe extends farther seaward. On some ebb-tidal deltas, there are short swash bars built by waves on the shoals between ebb and flood channels of the tidal inlet. Typically, the ebb-tidal delta deposit coarsens upward and bears a succession of sedimentary structures from bioturbation below to cross bedding, flat lamination, and ripple lamination in upper layers, where wave and tidal currents exert their major influence. Parts of some ebb-delta sands are interbedded with mud.

A tidal channel migrates slowly in the direction of

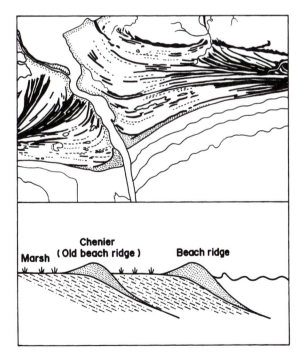

Fig. 8-27. Cheniers of the Louisiana Gulf Coast. (After Byrne, et al., 1959, plt. 1.)

of a series of shore-parallel beach ridges, the older ones stacked in behind and separated by shallow ponds, lagoons, or creeks. The best-known chenier plain lies downdrift, just to the west of the Mississippi delta, on the Louisiana and Texas coast of the Gulf of Mexico. Individual cheniers there are about 3 meters high and as much as 50 kilometers in length.

Rhodes (1982) presented a description of the Carpentaria chenier plain of northern Australia. Here, cheniers are composed mostly of shelly material. Organics-rich mud derived from the mouths of nearby rivers builds upward to low-tide level and progrades seaward, but during times of short mud supply, shell debris accumulates as beach ridges over the low-tide muds.

Transgressive Systems

In many cases transgression gives rise to a simple fining-upward succession, comprising cross-bedded coastal sandstones that grade upward into ripple-laminated fine-grained sandstones, thence to offshore shales, perhaps containing isolated storm-deposited sandstone beds.

Bourgeois (1980) described what she considered to be a transgressive coastal succession, the 200-meter-thick Cape Sebastian Sandstone (Cretaceous) in southwest Oregon. There is an erosional base, overlain by a basal conglomerate facies containing rounded sandstone boulders up to 0.5 meter in diameter, thought to be concretions eroded out of the underlying sands. The conglomerate grades upward into trough-cross-bedded pebbly coarse sandstone and flat-laminated shelly sandstone with escape traces. This facies gives way to hummocky cross-bedded, flat-laminated, and burrowed siltstone and sandstone. The exceptional thickness of the deposit suggests that it was the product of a stormy sea, and that the transgressed surface was rather steeply inclined; probably the coarse lower part of the succession was influenced by a sea cliff.

One of the most famous marine transgressive deposits is the one at the base of the Paleozoic in western North America, comprising Tapeats, Flathead, and Deadwood Sandstones (among others) and coextensive over half the continent. The age of the sandstones decreases from Early Cambrian near the western margin of the craton to Early Ordovician toward craton center. The cross-bedded sandstones, resting unconformably on Archean metamorphics and Proterozoic sedimentary rocks, show local variations of thickness and grain size that reflect topography and lithology of the transgressed surface. Typically the sandstone succession fines upward and grades into overlying shales.

Transgressive successions are more complicated if a barrier island and associated lagoon are present. During transgression, the surf zone advances over the barrier superstructure (Nummedal and Swift, 1987, p. 245) and truncates it. Meanwhile, the site of barrier island and lagoonal sedimentation moves landward. An erosion surface produced by scour in the surf zone comes to separate any surviving marginal marine deposits below from reworked barrier island sands above—this is the *ravinement surface* of Swift (1968, p. 444; see also Swift et al., 1971, p. 241). The ravinement surface is diachronous, lying stratigraphically above the flooding surface and more or less parallel to it. Sediments between the two surfaces are marginal marine muds (lagoonal or tidal-flat or estuarine) and washover sands (Kraft et al., 1987); sediments above the ravinement are shoreface sands grading upward into offshore muds.

The slope of a transgressed surface, and the rate of sediment supply to the coast relative to rate of sea-level rise (or rate of subsidence) influence the nature of an ancient transgressive coastal deposit, determining in particular what is preserved and what is destroyed. Transgressive coastal deposits likely are thinner than regressive successions and in some cases represented only by a thin lag bed on an unconformity (Kraft, 1971, p. 2183). Thickness asymmetry of transgression-regression cyclothems is due in part to sea-level changes superimposed on longer-term subsidence, so that transgressions are more rapid than regressions. Gross stratigraphic relationships in the Cretaceous of the western interior of North America suggest that rapid subsidence provided ''room'' in which coastal successions, whether transgressive or regressive, could be and were fully recorded and substantially preserved.

Tidal-Flat Environments and Facies

Detritus on the fuzzy edge, experiencing conditions that vary regularly with the time of day.

Commonly a mixture of sand and mud, the tidal flat is a place where flooding and drying alternate rapidly. How can an environment apparently so unstable lead to any sort of well-defined sedimentary succession that is stable over the long term, and that recurs in time and place? What sedimentological results get preserved—those created by ebb or those created by flood? What sedimentary responses have periods equal to the tidal period? What responses are long-term averages?

Some coasts are strongly influenced by the astronomical tides, others are hardly affected at all. A tide regime is *microtidal* if tide range is less than 2 meters, *mesotidal* if it is between 2 and 4 meters, and *macrotidal* if tide range exceeds 4 meters. Coasts differ from one another depending on the relative influence of waves and tides. On microtidal coasts, barrier islands (made by waves) may be well developed; numerous tidal inlets transect the barrier islands of mesotidal coasts. On macrotidal coasts, tidal ridges and broad tidal flats develop, but no barrier islands.

Across the tidal flat there is an essentially unbroken transition from nearly continuous submergence to nearly continuous emergence (Fig. 8-28). Typically there is a pronounced asymmetry; that the ebb current is stronger and shorter-lived than the flood causes mud to be carried landward and sand to be carried seaward. Thus, the higher tracts of a tidal flat, close to high-water level, are muddy, and the lower tracts, close to low-tide level, are sandy. Sandy beds or lenses are draped with muddy films (flaser bedding), and muddy beds contain lenses of sand (lenticular or linsen bedding). These bedding types are characteristic of environments where there is an alternation of flow and stagnation. The sand waves and large ripples made on the lower tracts by the flood tide deflect and constrain the return flow, especially during the latter stages of

the ebb, when the troughs act like so many little parallel channels, each carrying its own rivulet. This return flow is at a right angle, of course, to the direction of the flood current that made the sand waves. Small ripples are thus impressed onto the troughs of the larger ripples or sand waves, giving rise to the ladderback ripples (see Chap. 5). There also may be double-crested or flat-topped ripples. Sand waves develop reactivation surfaces. Net sedimentation rate is low, and intense bioturbation may affect all deposits of the tidal flat. Mud cracks due to desiccation of clay mud or carbonate mud commonly form on the tidal flat during low tide, but those formed in the supratidal belt, above the mean high tide, are more likely to be preserved. Here, wetting occurs only during the spring tides (a monthly cycle rather than a diurnal or semidiurnal cycle) and during storms. Mud polygons become quite dry and hard and are scattered as large, platy clasts into the intertidal zone by the infrequent high-water events. Sediment laminae formed in the supratidal zone are less likely to be muddled by burrowing organisms, though they may be penetrated by plant stems and roots.

Tidal creeks are small, sinuous channels with numerous tributaries that carry tide waters onto the tidal flat during the flood and drain the flat during the ebb. (We shall take care here to distinguish between these and *tidal channels*, which are tide-swept inlets through barrier islands.) Particularly distinctive of these creeks is the rapid increase in channel width from source to mouth (Fig. 8-24). The creeks meander, cutting away at concave banks and building up the convex banks by lateral accretion, much as ordinary meandering rivers do. The point bars are composed of thin beds of sand

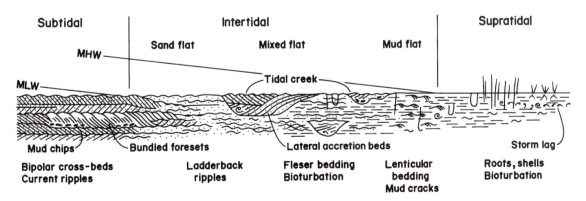

Fig. 8-28. Physiography and facies of a tidal flat.

and mud that dip gently into the creek channel and prograde over an erosion surface marking the creek floor, which may be paved with shells or mud clasts. Slump structures are formed where an oversteepened concave bank has collapsed into the channel at low tide.

One suspects that the thickness of a tidal flat deposit is comparable to tidal range, and that tidal flat deposits thicker than a few meters are made during subsidence or long-term sea-level rise. Since the tidal flat is muddy on the landward edge and sandy on the seaward edge, progradation (presumably during relative sea-level rise) results in a fining-upward succession (as Evans, 1965, p. 230, found in the Wash on the coast of England, for example), the whole succession complexly dissected by tidal creek deposits. The opposite trend, of course, suggests a sea level that is rising over the tidal flat and displacing it landward; one would expect these tidal flat deposits to be overlain by more open marine, subtidal facies, such as Harrison (1975, p. 37) envisioned for Recent deposits on the coast of Virginia.

Evidence of tidal-flat origin of ancient deposits, or at least tidal influence, includes bipolar (herringbone) cross-bed sets, reactivation surfaces (see the excellent illustrations of deMowbray and Visser, 1984), flaser and lenticular bedding, desiccation cracks, mud chips, other sedimentary structures indicating intermittent emergence, and small channels. Swett et al. (1971) found current ripples on some cross-bed slip faces, their crests parallel to dip, and noted also the common occurrence of *Skolithos* burrows and escape traces in the Eriboll Sandstone (Cambrian) in Scotland. In the Eureka Quartzite Ordovician) in Nevada and eastern California, Klein (1975) found ripples oriented by currents 90 or 180° to those that made the associated cross-bed sets and noted also shallow channels and scours lined with mud-chip conglomerates, mud cracks, and deep burrows. In the Curtis Formation (Jurassic) in Utah, Kreisa and Moiola (1986) found *sigmoidal tidal bundles*, each embracing all laminae in a tide-generated laminaset formed during a single tidal ebb or flow. Boundaries between bundles, called *pause planes*, represent slack-water intervals. Systematic variability within a tidal bundle records waxing and waning flow velocity during the tidal episode; variability between bundles presumably records monthly modulation of tidal amplitude.

Many ancient deposits considered to be peritidal are cyclic, but it is not the tides that were responsible.

Deposition occurring at the very edge of the sea is sensitive to longer-term changes in sea level; even minor changes, and peritidal sediments, it seems, record readily the gentle fluctuations in sea level. These changes might be periodic in some cases, possibly driven by Milankovitch processes (see Chap. 12).

Mazzullo (1978) described a rapidly prograding tidal flat deposit, the Winchell Creek Member of the Great Meadows Formation (Lower Ordovician) in eastern New York. The most complete sections comprise three parts: A lower part is siltstone and silty sandstone in medium to thick cross-bed cosets capped by ripple beds and flat-laminated beds. Cross-bed sets are predominantly trough-style, and sets higher in the section are thinner than those below; they show bipolar orientations and reactivation surfaces. Ripple crest lines deviate from cross-bed strike by as much as 60°. The middle part of the Winchell Creek Member contains interbedded siltstone and shale; the siltstones are flat-laminated, with parting lineation, or bear ripples with flattened crests; the shales are riddled with desiccation cracks. There are large channels, up to 4.5 meters wide and 1.5 meters deep; those higher in the section are smaller. Channels contain skeletal and siltstone-clast conglomerates (thalweg deposits) and silty sandstone in cross laminasets overlain by flat-laminated and rippled siltstone; the trace fossil *Skolithos* is locally abundant. Finally, the upper part of the Winchell Creek comprises shale and silty shale, either interlaminated or lenticular bedded.

Mazzullo interpreted the three parts of the Winchell Creek as a progression from high-subtidal to high-intertidal depositional setting, reflecting a rapid progradation. Subsequent transgression terminated tidal-flat deposition and brought on clear-water carbonate deposition.

One expects to find tidal deposits mainly at the interfaces between marine and terrestrial deposits. Thin coal beds may be closely associated with tidal sandstone-and-shale deposits of humid climates; in arid climates, nodular evaporites (gypsum or anhydrite) and algal-laminated carbonates suggest close association with an intertidal zone.

Tidal flats of arid regions typically, though not necessarily, are composed of carbonate and evaporite materials; arid flats are called *sabkhas* (see Chap. 9). They share many of the characteristics of siliciclastic tidal flats, but the different climate and biota of the sabkha give rise to special features such as algal mats, evaporite crusts, and displacive nodules.

Estuaries

The shallow coastal sea invades the river mouth; waters mix and river flows compete with ocean waves and tides.

Dalrymple et al. (1992, p. 1132) defined **estuary** as a marine-flooded part of a river valley that receives sediment from both fluvial and marine sources and whose sedimentary facies are influenced by fluvial, tidal, and ocean wave processes. Estuaries are characteristically associated with marine transgressions; regression or sediment progradation may cause them to evolve into strand plain, tidal flat, or delta.

Either waves or tides may dominate as a sediment-transporting agent. If waves are dominant, a bar, barrier island, or spit develops at the mouth of the estuary (Fig. 8-29a). This feature prevents much of the energy of ocean waves from entering the estuary. Sand (or gravel) is deposited also at the upstream end as a bay-head delta. The basin or ''lagoon'' between the coastal bar or barrier island and the delta typically receives organics-rich muds. A lagoon may develop into an extensive salt marsh as it fills.

Where tides are important, longitudinal bars or sand ridges form at the mouth of the estuary (Fig. 8-29b). Harris (1988, p. 282) reported that, if sand is abundant, then fields of sand waves form instead. At intermediate rates of sand supply, *linear sand banks* form, oriented 10 or 15° to the direction of peak tidal flow. Sand waves adorn these banks, flood-oriented on one flank, ebb-oriented on the other. Though sand ridges dissipate wave energy, they let the tide waters through, and the converging estuarine shores accelerate the flood-tide currents toward the head. Tidal currents tend to diminish or otherwise modify the bay-head delta. Mud accumulates on tidal flats along the banks of the estuary, but the longitudinal sand-mud-sand zonation of wave-dominated estuaries is not so evident in tide-dominated estuaries.

Because estuaries are features developed during transgressions, they migrate landward. Facies zones move landward as well, but sand bodies at head and mouth tend to expand and merge, shrinking the central basin. Wave action may remove or rework the upper parts of the estuarine succession (forming a ravinement surface); lateral migration of tidal channels could have a similar effect. The whole succession is bounded below by an unconformity (the old valley floor), this

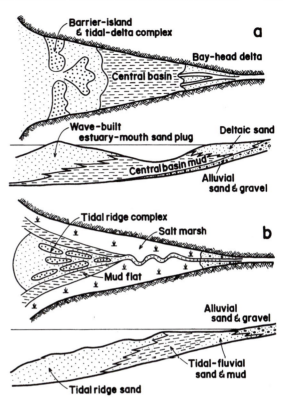

Fig. 8-29. Estuarine sediment facies patterns: **(a)** wave-influenced estuary, **(b)** tide-influenced estuary. (After Dalrymple et al., 1992, figs. 4, 7.)

surface perhaps mantled with fluvial deposits. Riverine processes gradually become dominant, especially as transgression slows, and both wave- and tide-dominated estuaries evolve into deltas.

Epicontinental Shelves

Fine detritus dispersed by storm waves and tides beyond the direct influence of the coast.

Shallow epicontinental, or epeiric, seas and the shallow seas of pericontinental shelves make deposits that typically are muddy (comprising either clay mud or lime mud). *Epeiric seas*, so important in the past, hardly exist today in their past splendor, and we do not understand them very well. Modern continental shelves are not entirely analogous, and what's more, *they* are not particularly well understood either, though Brenner (1980) used them as partial analogs of ancient epicontinental seaways. Laterally extensive blanket shales typically show only the most subtle and gradual lithological variations, but important distinctions can be made between bioturbated, fossiliferous shales and black, laminated, bituminous shales. There may be shell beds or banks, isolated silty or sandy layers, and volcanic ash beds. Sandstone bars also may be present, many kilometers from any shore. Commonly it is the coarser deposits in a shelfal mudstone that tell much of the story of shelf environments and shelf sedimentation.

On the modern continental shelves, patterns of sedimentation are controlled by the distribution and magnitude of sediment sources and by bottom currents. Sediment is supplied to the shelf by rivers or by the wind; on some shelves, volcanism is a significant sediment source; there may be reefal or other organically produced carbonate. Transgressions and regressions alter the conditions under which sediment is being supplied to the shelf. The deposits of many modern shelves are a product of Pleistocene glacial events. Some shelf sediment is actually relict non-shelf sediment that was reworked after a transgression.

Coriolis-modified wind-driven flow, called the *geostrophic circulation*, sweeps the broad shelves and causes long-term regional sediment transport. There is a Coriolis deflection to the right in the Northern Hemisphere, so that surface water piles up against the shore of a land mass to the right of the wind. The hydraulic head of excess water at the coast causes a return flow along the bottom, also under Coriolis influence, of course, and in addition affected by bottom friction. Similarly, *upwelling* currents can be created along coasts that lie to the *left* of the wind. (In the Southern Hemisphere, reverse right and left.) Shelf topography influences current patterns, and submarine canyons and other hollows intercept sediment being swept across a shelf.

Various periodic or episodic flows on a shelf may be superimposed on a simpler, regional flow. Storm winds blowing parallel to a coast can induce substantial currents in shelf waters, not only at the surface but in the bottom waters also. Tidal currents also may be important and regionally complex. Where tidal currents sweep the shelves, the flood currents generally are geographically separated from the ebb currents, so that certain parts of the shelf are flood-dominated and other parts are ebb-dominated.

Among the more prominent depositional features of shelves are ridges or bars that are elongate parallel to regional shelfal currents. These large constructional bedforms are as much as a few tens of kilometers in length, a few kilometers wide, and up to about 10 meters high. Their surfaces commonly are covered with ripples or dunes. They have been mapped on modern continental shelves, and certain ancient sandstones embedded in regionally extensive shales have been interpreted as similar large longitudinal bars of marine shelves. The tidal currents organize sand into large transverse bodies and longitudinal bodies. McCave (1971) described sand waves off the coast of Holland in the North Sea, with heights up to 7 meters and wavelengths of 200 to 500 meters, which he attributed to tidal currents. Kenyon (1970) described current-parallel *tidal sand ribbons* up to 15 kilometers long and 200 meters wide on the sea floor around the British Isles. Some sand ridges on the floor of the southern North Sea, in water depths of 30 to 40 meters, rise within 2 or 3 meters of the surface of the sea (Caston, 1972). In transverse bodies (sand waves), there are large-scale cross-bed sets, commonly bounded by reactivation surfaces, mud drapes, or pavements of mud clasts and developed into cosets several meters thick. The more complex cosets, possibly even bearing bipolar foreset orientations, are considered to reflect varying velocity asymmetries of tidal currents (Allen, 1980, p. 303).

Storm- and tide-generated currents seem to have been important also in the epeiric seas (see Klein and Ryer, 1978). Ancient sandstone bodies deposited in shallow epicontinental seas (Fig. 8-30) are lenticular, a few meters thick, and encased in shelf mudstones, commonly making sharp contact with underlying shale. They are coarsening-upward successions with a fine-grained, thin-bedded, rippled and burrowed lower part and a fine- or medium-grained and cross-bedded upper part. Mud drapes and mud clasts may be abundant, also shelly sheets resembling tempestites. Sedimentary structures indicate that they were affected by storm-generated currents, tidal currents, or generalized ocean currents attributable to geostrophic circulation. Offshore-bar sandstone lenses associated with the lower Gallup Sandstone in New Mexico (Campbell, 1971) are about 3 meters thick, 3 kilometers wide, and 60 kilometers long, embedded in offshore siltstones and mudstones. Sandstone beds in the bars are sea-

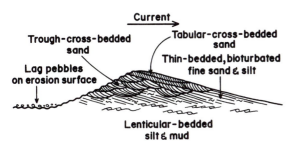

Fig. 8-30. Section through a marine-shelf sandstone body. (After Gaynor and Swift, 1988, fig. 11.)

ward-imbricated, abutting the flat upper surface of the bar and grading seaward into muddy beds. Many of the beds are bioturbated, but here and there cross laminasets show through, dipping uniformly in a direction parallel to the length of the bar. The bars formed on a shallow marine shelf perhaps 150 kilometers wide, deposited by shore-parallel currents carrying sand from some delta or river mouth perhaps 300 kilometers distant.

In many cases, the source of shelf sand appears to be some distant delta, and storms are perceived as the agents of movement of this sand to the shelf. The Shannon Sandstone (Cretaceous) in Wyoming is laterally equivalent to coastal sandstones 100 kilometers or more to the west (Spearing, 1974). It comprises coarsening-upward elongate lenticular bodies up to 20 meters thick, embedded in shale. Paleocurrent directions, as indicated by ripples and cross beds, are parallel to elongation of the bars and roughly parallel to the distant shorelines; they reflect some dominating regional shelf current probably related to the tides and to passage of swells (Gaynor and Swift, 1988, p. 872), though some bedforms and sedimentary structures suggest the influence of transverse storm currents. Tillman and Martinsen (1984, p. 141) suggested water depths of 20 to 100 meters. Another Cretaceous shelf sandstone, the Duffy Mountain sandstone body in the Mancos Shale in northwestern Colorado (Boyles and Scott, 1982), also is elongate parallel to a distant shore; bounding surfaces of cross-bed sets are inclined toward that coast. The sandstone body climbs stratigraphically shoreward over bioturbated mudstones, and it appears that the body grew or migrated slowly in that direction. Boyles and Scott suggested that two substantially different energy regimes operated—intermittent storm-generated currents that caused the

shoreward migration and weaker fairweather currents transporting sand along the bar crest.

Asquith (1974) elucidated the large-scale patterns of shale deposition on a marine shelf, based on the Pierre Shale (Upper Cretaceous) of Wyoming and associated sandstones. The Pierre (rhymes with *fear*) seems to consist of several cycles comprising nearly flat-lying shale and siltstone of a shelf environment, siltstone and shale of a slope environment (inclined 2 or 3°), and shale and marlstone of a basin environment. Shelf shales and offshore bar sands merge landward with coastal deposits, including barrier island sandstones and paludal and fluvial mudstones and sandstones. Depositional relief may be a hundred or several hundred meters (determined in the case of the Pierre from variations in stratigraphic distance between correlated bentonite beds). Depositional environments shifted with marine transgression and regression, but the only significant sedimentation was regressive, depositional episodes bringing basinal shales into abrupt contact with underlying thin marginal marine or nonmarine deposits. Asquith envisioned Late Cretaceous sedimentation in Wyoming as a complex of asynchronous progradations of several widely separated deltas.

Deep Basin-Margin Environments and Facies

The end of the line for detrital sediments: they come to rest in the mysterious dark depths and await drastic fates.

The waters of the ocean and the ocean floor are divided into distinct parts, based on depth (Fig. 8-31). The marine environments of the continental shelves, extending to depths of about 200 meters, are referred to the **neritic realm**; environments beyond the shelves are of the **oceanic realm**. The bottom, or *benthonic*, environments are classified as **littoral** between the high and low tides; **sublittoral** between low tide and the edge of the continental shelf (200 meters depth); **bathyal** on the continental slope, between 200 meters and 4000 meters depth; **abyssal** on the plains that lie at 4000 to 5000 meters depth; and **hadal** below 5000 meters, in the trenches. Environments of the overlying oceanic waters are the **pelagic** environments, comprising **epipelagic** (waters down to 200 meters depth, but excluding the neritic waters), **mesopelagic** (200 to

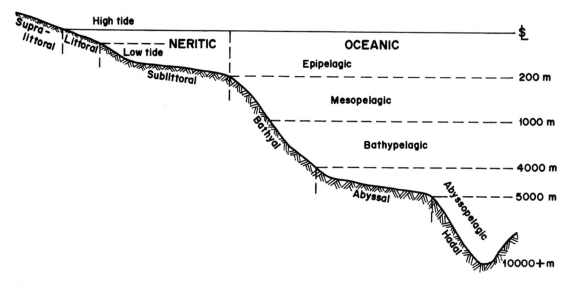

Fig. 8-31. Partition of the ocean environment.

1000 meters), **bathypelagic** (1000 to 4000 meters), and **abyssopelagic** (below 4000 meters).

Most of our knowledge of the deep ocean floor is gained today by the ship-borne acoustic profilers and side-scan devices, the dredge and coring equipment, and the deep-diving submersible vehicles. The work is expensive and dependent on special technologies such as satellite positioning and remote manipulation. But some ancient deep-sea deposits have been elevated and accreted onto land masses and are thus accessible to ordinary stratigraphic methods.

Submarine Fans and Turbidite Deposits

The **submarine fan** is a gently sloping fan-shaped or lobate body of mostly sand and mud deposited at the base of a continental slope and advancing over an abyssal plain. A fan deposit may be 1000 meters or more in thickness. Many fans are nourished by a submarine canyon that incises the continental shelf, intercepts deltaic or other coastal sediment, and funnels it to the shelf break and continental slope. The largest modern fan is the Bengal fan in the Indian Ocean, fed by the Ganges and Bramaputra rivers. It has a radius of 3000 kilometers, and covers an ocean floor area of some 3 million square kilometers (Emmel and Curray, 1985, p. 107). A single submarine channel at the apex of this and other modern fans splits into distributaries

that course downslope onto **suprafan lobes** (Fig. 8-32). Channels are cut by turbidity currents and modified grain flows; they are typically flanked by levees formed by "flooding" of the muddy components of these flows onto the general interdistributary surface of the fan. Coriolis deflection of these currents causes the levee on the right side (Northern Hemisphere) to be higher than the one on the left. Hesse (1986) described Yazoo-type channels on the Northwest Atlantic Mid-Ocean Channel in the Labrador Sea, being tributary channels with deferred entrance to the main channel because of the main-channel levee.

Like the subaerial fan, the submarine fan divides naturally into **proximal fan**, **midfan**, and **distal fan** (Fig. 8-32); these are physiographically and sedimentologically distinct (Normark, 1970, p. 2189). The proximal, or inner fan is marked typically by a single, leveed, straight or meandering channel, perhaps several kilometers wide and a hundred meters or more deep. Suprafan lobes mark the midfan surface. They are broad, subdued, overlapping bulges of sandy sediment that moved down the channels of the upper fan, then spread over large lobate areas. The suprafan is scored by sinuous, radiating distributary channels that do not have levees. The largest of these channels may be a few hundred meters wide and a few meters deep. The distal (or outer) fan is composed typically of muddy turbidites; the surface is smooth, lacks chan-

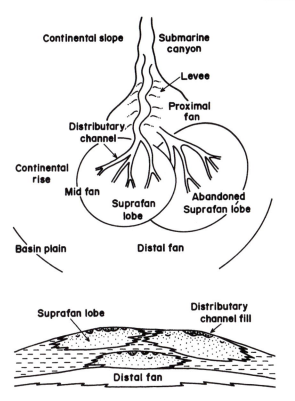

Fig. 8-32. Physiography and facies of a deep-sea fan. (After Walker, 1978, figs. 13, 18.)

nels, slopes very gently, and merges imperceptibly onto the abyssal plain.

Different parts of a submarine fan comprise different lithofacies. Debris-flow deposits occur mainly on the inner fan, commonly in association with slumps and confined between the walls of a fan channel. Also there are modified grain-flow deposits (see Chap. 4) on submarine fans, typically composed of medium or coarse sand and occurring mainly on the inner and middle fan. Grain-flow deposits grade downfan into turbidity-current deposits, comprising distinctive, laterally persistent graded units up to a meter or two in thickness, occurring principally on the outer fan.

Debris-flow deposits are conglomerates, clast-supported or matrix-supported, in graded, reverse-graded, or reverse-to-normal graded beds a meter or more in thickness, with aligned-clast, imbricated-clast, or disorganized-clast fabric. Davies and Walker (1974) described such conglomerates of the Cap Enrage For-

mation (Cambrian and Ordovician) in the Gaspe of Quebec, Canada, qualifying them as *resedimented*, that is, shallow-water deposits remobilized to a deeper venue. Grain-flow beds are either reverse-graded or not size-graded. There may be outsize clasts scattered throughout the bed, or "coherent lumps" enveloped in eddy-like swirls of finer-grained material derived from the lump. In grain-flow deposits, dish structure is common. The sharp base of a grain-flow bed may bear drag marks or slide marks, frondescent marks, or load structures.

The turbidite unit was formalized by Bouma (1962), who subdivided it into five parts, traditionally labeled A through E: (A) a sandy, graded, basal part with a grooved and fluted erosional base; (B) a fine-sandy, flat-laminated part; (C) a silty, ripple-laminated part; (D) another flat-laminated part, finer-grained than B; and (E) a muddy part without much structure, though perhaps graded. This succession has come to be known as a *Bouma sequence*. It represents a single depositional event—namely, the passage of a single turbidity current. There is an abrupt erosive first stage, followed closely by deposition from suspension in a hydrodynamic milieu that decays gradually from an initial vigorous high regime to a final standstill. The erosional base is made by the head of the turbidity current (see Chap. 4). The lower two parts of the Bouma cycle (A and B) are deposited in upper flow regime as the body of the turbidity current is passing; antidunes are preserved occasionally. The units of finer-grained sediment just above (C and D) are a product of lower flow regime of the tail of the turbidity current. C is rippled, but D is apparently too fine-grained for ripples to appear (see Chap. 5). The topmost muddy layer of the cycle (E) is deposited from still water; it is the silt and clay that settles out after the turbidity current has passed and dissipated. This grades upward into ordinary pelagic mud such as mantles the abyssal plains. Typically a turbidite bed is overlain directly by another; commonly a scour surface separates the two, locally decapitating the bed cycle beneath.

Many workers have observed certain variations on the theme of the complete Bouma cycle, apparently reflecting turbidity currents that develop gradually in the inner or middle fan from grain flows or debris flows, achieving full maturity in midfan, then gently collapsing on the outer fan or abyssal plain beyond. Thus, a distinction can be made between a *proximal turbidite* and a *distal turbidite*. Distal turbidite units are thinner and finer-grained than proximal deposits, generally lack signs of basal scour, and are laterally more uniform and persistent; the lower Bouma divisions commonly are missing from distal turbidite units. R.G. Walker (1967, p. 27) suggested that the proportion of Bouma cycles that begin with an A unit compared to cycles that begin with a C unit is an index of proximality (the so-called *ABC index*, based on a moving average of 25 or more units). Other stratigraphic characteristics that indicate proximality are bed thickness, frequency of occurrence of erosion surfaces between Bouma units, and sand/shale ratio; these quantities all decline downfan. Based on ancient fans in the circum-Mediterranean region, Mutti (1974) suggested that inner fan deposits consist of large, lenticular sandstone bodies enclosed by mudstone units that are essentially devoid of sandstone beds; that middle fan deposits are coarse, thick, channel-fill sandstones enclosed within finer-grained, thin-bedded deposits; and that outer-fan deposits consist of coarse-grained and fine-grained laterally persistent turbidite beds. Thinning-upward successions (*positive megasequences*) of turbidite units, involving 3 to 6 or more sandstone units, appear to result from the filling of channels (Ricci-Lucchi, 1975, p. 8; Ricci-Lucchi and Valmori, 1980, p. 264). Thickening-upward successions (*negative megasequences*) seem to be due to progradation of lobes (Fig. 8-33). *Monotonous* successions (Ricci-Lucchi, 1975, p. 18) show vertically uniform bed thicknesses or erratic variations and are products of overbank deposition. External conditions (quality and quantity of the sediment supply) contribute to bed thickness variations as well.

Submarine fans may be 5000 meters thick or more in their proximal parts, thinning gradually toward the distal margins. Typically the ancient ones are strongly deformed as a result of continental margin tectonics. Proximal sections of ancient fans are dominated by channel-form sandstones a few meters thick; many Bouma sequences comprise only A and E divisions. Channel-form units change downfan into lobate sheets, orientations of flutes and grooves reflecting the change from confined, channelized sediment transport to fan-wise dispersal. The more distal sections, in contrast, are dominantly shale, with interbedded thin, flat-laminated or convolute-laminated siltstones and graded, fine-grained quartz sandstones. Bouma sequences typically lack the lower divisions. Number of sandstone beds and aggregate sandstone thickness decline in a systematic fan-wise manner away from the apex of the fan. Paleocurrents, as determined from

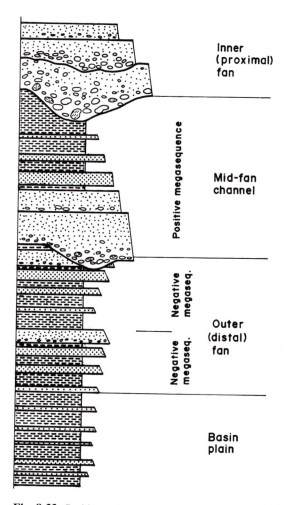

Fig. 8-33. Positive and negative megasequencees in turbidite. (After Walker, 1978, fig. 14.)

sole-mark orientations, radiate from the apex. Finally, on many fans there is an overall *upward* increase in thickness of sandstone beds, reflecting gradual change from more distal to more proximal deposition due to progradation of the fan (see MacDonald, 1986, p. 255).

Trough-shaped basins of the ocean floor typically fill axially. McBride (1962, pp. 84–85) noticed this in the Martinsburg flysch (Ordovician) in the Appalachian region of the eastern United States, explaining that turbidity currents flowed down the submarine slopes onto the basin floors, where many of them changed course and flowed longitudinally for some distance. Other examples are described by Dzulynski et al. (1959, p. 1096) in the Carpathian flysch in Poland and Hsu et al. (1980, p. 1044) in the Ventura basin in California.

The submarine fan is a place where the methods of seismic stratigraphy excel (see Weimer, 1990). This is because thickness is so great, depositional slopes are so gentle, downslope thinning is so gradual, and channels and their levees are so broad that these features are difficult to discern by ordinary stratigraphic methods; but seismic profiling reveals them clearly. Moreover, all the modern or recent ones are located in very deep water, and are virtually inaccessible.

Contourites

Contour currents, where the geostrophic circulation impinges against a continental slope, attain speeds of a meter per second and transport much fine-grained sediment along the contours of the slope. Heezen et al. (1966) considered them to be important in shaping the continental rise, pointing out (p. 508) that "the thickest sediments in the ocean are found beneath or very near the axes of deep geostrophic contour currents." Contour currents, it appears, also rework turbidites (Heezen and Hollister, 1971, p. 335 ff). Beneath the Gulf Stream, a north-flowing so-called western boundary current of the North Atlantic, there is a deep countercurrent that flows southward. Currents in the Northern Hemisphere are deflected toward the right (Coriolis effect), which, for the deep countercurrent is *landward* (in contrast to the *seaward* deflection of the Gulf Stream itself). Western boundary undercurrents *in general* are deflected landward and hug the upper continental rise or lower continental slope; they may be sufficiently competent, at least locally, to transport silt and fine sand along the contours. Bouma (1972, p. 919) enumerated the characteristics of **contourites**, or deposits made by contour currents: they are composed of very fine sand and silt, well sorted and nearly mud-free, with no pebbles or intraclasts; beds are sharply defined, less than 5 centimeters thick, commonly laminated or cross-laminated. Stowe and Lovell (1979) differed in some details, describing the contourites as fine sand and mud, laminated or bioturbated, and commonly associated laterally (upcurrent) with lag gravels, presumably marking the provenance of the fine-grained sediment. Bein and Weiler (1976) described a Cretaceous contourite in Israel,

some 3000 meters thick, composed of calcareous detritus.

Tsunami Deposits

Kastens and Cita (1981) described what they considered to be a Recent tsunami-laid deposit on the floor of the Mediterranean Sea and suggested the historical event that created the wave. Acoustic profiling on the Mediterranean abyssal plain revealed a very homogeneous unit in certain areas, near the top of the sediment column, filling local depressions. Several cores were taken. The curious deposit turned out to be one big (several meters) graded bed of silt and clay; not even x-radiography could detect any other sedimentary structure. Apparently the fine sediment was washed off the basin walls and resedimented on the basin floors. Radiometric dating (of certain associated beds) suggested an age between 3100 and 4500 years. Archaeologists have determined that the volcanic island Santorin exploded in about 1500 B.C. (some consider this to be the landmass that Plato referred to as Atlantis; see Vitaliano, 1973, chaps. 9 and 10). The explosion, Kastens and Cita suggested, created a tsunami that disturbed substantial quantities of fine-grained bottom sediment, putting it in suspension. The turbid layer slipped sideways into the basins, where, in the ensuing calm, it gradually dropped its suspended silt and clay back to the bottom.

Deep-Ocean-Basin Sediments

Much fine-grained terrigenous material is deposited on the abyssal plains. Most of this material is delivered to the oceans via river deltas and by the wind; then it is transported by the generally weak thermohaline circulation or is deposited from the thin, stagnated residues of turbidity currents (the nepheloid layers). These typically are blanket deposits. Some land-derived detritus is mixed with marine-volcanogenic muds and sands or is substantially diluted by biogenous materials—the multitudinous tests of planktonic calcareous foraminifers and coccolithophores and the siliceous diatoms and radiolarians (see Chap. 9).

The red muds of the abyssal plains are composed mainly of particles of various clay minerals derived from continental soils and from intrabasinal volcanics. There is also detrital silt of quartz, feldspars, and ferromagnesian minerals. The composition and distribution of terrigenous detritus on the ocean floor reflects provenance and the rates and patterns of dispersal by ocean currents. Kaolinite is the characteristic clay mineral of tropical soils, and this is the dominant clay mineral of equatorial pelagic muds. Illite is more abundant in the mid-latitude oceanic sediments. Chlorite is largely a higher-latitude clay mineral. The distribution of smectite suggests that this clay mineral is derived mainly from oceanic volcanic centers. The reddish muds get their color probably in much the same way as continental redbeds do (see Chap. 11). Sedimentation is very slow, and the muds are exposed for long periods to the oxygen-rich bottom waters, which alter the iron-bearing silicate minerals and create ferric oxides and hydroxides in small amounts.

There are currents on the deep ocean floor; gentle but persistent currents are related to the thermohaline circulation, that is, the pattern of density currents that geographic variations in temperature and salinity bring. The very cold, dense water of the Antarctic convergence, for example, flows northward over the floor of the Atlantic Ocean, far into the Northern Hemisphere. Geostrophic circulation, comprising the Coriolis-modified wind-driven currents of the ocean, sweep the continental shelves and slopes. Geostrophic

and thermohaline currents are locally intensified by ocean-bottom topography. Tidal currents also sweep the deep ocean floors, and make ripples in the bottom sediment. Lonsdale and Malfait (1974) described "abyssal barchans" and asymmetrical transverse dunes, composed of foraminiferal sand, on the Carnegie Ridge (eastward from the Galapagos Islands in the eastern Pacific), probably made by episodic dense currents spilling over sills in the ridge. Lonsdale and Smith (1980) described another location in the Pacific, near the Marshall Islands, where muddy sediment is being deposited in large waves. At a depth of 5000 meters or more, this surface is locally covered with regular waves 50 meters in height, with wavelengths of 5 kilometers (apparent, since the acoustic profile might not be strictly perpendicular to the crests of the waves). Some waves make internal reflections on acoustic profiles that resemble foreset layers. Large waves of muddy sediment at abyssal depths generally are considered to have been deposited out of suspension from a strong, steady bottom current. A major source of fine sediment in this case may be occasional turbidity currents flowing off the Marshall-Gilberts Ridge.

9

Composition and Classification of the Carbonate Rocks; Depositional Environments

The chemical rocks comprise only about 10 or 12 percent of all sediments, but they are unusually varied in their textures and compositions. Some are composed of chemically or biologically formed grains of great variety; others are crystalline aggregates formed by chemical precipitation from complex solutions, commonly under rather unusual environmental conditions. Most of the chemical rocks are limestones and dolomites, rocks composed mainly of the carbonate minerals calcite and dolomite, though there are other chemical rocks composed of silica, apatite, hematite, gypsum and anhydrite or halite and sylvite. Examination of the non-carbonate rocks is reserved to a separate chapter (Chap. 10).

Much attention was directed toward an understanding of carbonate rocks in the late 1950s and early '60s, as the petroleum industry, until then fixated on sandstones (terrigenous rocks) as the prime reservoirs of oil and gas, began more and more to find oil and gas in carbonate rock reservoirs and to target these rocks specifically in their exploration programs. It soon became clear to the petroleum geologists that they knew little about these rocks. The remedy for this knowledge vacuum involved the pleasant prospect of visiting the sites of modern or recent carbonate sedimentation, such as the Bahamas, Florida Bay, the Persian Gulf, the Great Barrier Reef, and the coral atolls of the Pacific. Much of our knowledge of carbonate rocks has come to us through studies in such locales, and many of these studies were sponsored by the petroleum industry.

Carbonate sediments can form in almost any aqueous environment, marine or lacustrine, brackish or hypersaline, arid or humid, warm or cold. The shallow waters of the ocean and the waters of most lakes are essentially saturated with calcium carbonate ($CaCO_3$) and can readily precipitate calcite or other carbonate minerals. Almost no carbonate forms in the *deep* waters of the oceans, however, which are *undersaturated* with respect to calcium carbonate. But the calcitic skeletals of pelagic plankton may settle slowly onto the deep ocean floor, or carbonate sediment of the shallow waters may move rapidly downslope to great depths as turbidity currents.

Most $CaCO_3$ precipitation in the ocean is mediated by organisms in shallow water. Life-forms discovered $CaCO_3$ as a useful and readily available structural material at least as far back as the beginning of the Cambrian Period, and nearly all carbonate sediment since then has formed through the intercession of plants and animals. Significant volumes of older carbonate rock are a by-product of algal or bacterial photosynthesis, which has been going on for at least 3 billion years.

Diagenesis plays a very important role in carbonate rocks, and it is difficult to understand these rocks without addressing the diagenetic effects. (Nonetheless, much of our analysis of these matters is deferred until Chap. 11.) Significant diagenesis of carbonate sediment begins right from the start, and probably no carbonate rock is beyond some further modification of its composition or its texture under diagenetic conditions.

The Carbonate Cycle

Atmospheric carbon dioxide (CO_2) participates in many important earth-surficial chemical reactions, both organic and inorganic. Because CO_2 dissolves readily in water, it is abundantly available in many aqueous environments. While some natural processes consume CO_2, other reactions produce it. The flow of CO_2 through systems of importance to us here may be described in summary form as

$$H_2O + CO_2 \rightleftharpoons CH_2O + O_2$$
$$\updownarrow \qquad\qquad\qquad (9\text{-}1)$$
$$CaCO_3 + H_2CO_3 \rightleftharpoons Ca^{2+} + 2HCO_3^-$$

This is the net result of a set of concurrent or simultaneous reactions involving exchange of CO_2 gas between water and atmosphere, plant photosynthesis, animal respiration, dissociation of carbonic acid (H_2CO_3), and precipitation and dissolution of $CaCO_3$ which collectively comprise the **carbonate cycle**. The most important of these reactions are

$$CO_2 + H_2O \rightleftharpoons H_2CO_3 \qquad (9\text{-}2)$$
(formation of carbonic acid)

$$H_2CO_3 \rightleftharpoons H^+ + HCO_3^- \qquad (9\text{-}3)$$
(first dissociation of carbonic acid)

$$HCO_3^- \rightleftharpoons H^+ + CO_3^{2-} \qquad (9\text{-}4)$$
(second dissociation of carbonic acid)

$$Ca^{2+} + CO_3^{2-} \rightarrow CaCO_3 \qquad (9\text{-}5)$$
(precipitation of calcium carbonate)

$$CO_2 + 2H_2O^* \rightarrow CH_2O + H_2O + O_2^* \qquad (9\text{-}6)$$
(photosynthesis reaction)

$$CH_2O + O_2 \rightarrow CO_2 + H_2O \qquad (9\text{-}7)$$
(respiration reaction)

$$CaCO_3 + H_2CO_3 \rightarrow Ca^{2+} + 2HCO_3^- \qquad (9\text{-}8)$$
(dissolution of calcium carbonate)

The fictitious compound CH_2O of Eq. (9-6) and (9-7) is to be understood as the building block of the sugars, such as glucose ($C_6H_{12}O_6$), and their polymers, the starches and celluloses. The asterisk in the photosynthesis reaction of Eq. (9-6) indicates that the free oxygen resulting from photosynthesis derives from the splitting of water molecules (and not CO_2 molecules). Photosynthesis reduces and fixes atmospheric carbon and can lead or contribute to production of coal or petroleum. Note that, in terms of reactants and products, respiration (Eq. 9-7) is the reverse of photosynthesis. Also, the oxidation of organic matter in weathering environments is similar to respiration. In the dissolution reaction of Eq. (9-8), one of the bicarbonate ions comes from the reaction of hydrogen (H^+)

with $CaCO_3$ and the other comes from the dissociation of H_2CO_3, as shown in Eq. (9-3).

It is clear (from LeChatelier's principle of chemical equilibrium) that any process that removes CO_2 from aqueous solution causes the reactions of Eqs. (9-2), (9-3), and (9-4) to proceed to the left and encourages precipitation of calcium carbonate, that is, the reaction of Eq. (9-5). Among the mechanisms for this are increase in temperature, evaporation, upwelling of deep marine waters to (shallower) regions of lower pressure and higher temperature, photosynthesis, and the bacterial decay that produces ammonia. On the other hand, any process that increases the availability of CO_2 causes dissolution of $CaCO_3$ or reduces the rate of its production; such processes include volcanism, decreased photosynthesis or increased respiration or organic decay, great wildfires, and the human consumption of fossil fuels.

The concentrations of carbonic acid (H_2CO_3), bicarbonate (HCO_3^-), and carbonate (CO_3^{2-}) are related to pH of the system (Fig. 9-1). At low (acidic) pH, most of the carbonate exists as H_2CO_3; at neutral and mildly basic pH, most is in the form of HCO_3^-; and at high (basic) pH, the dominant ion is carbonic acid (CO_3^{2-}). Dissociation of H_2CO_3 is a two-step process, described by Eqs. (9-3) and (9-4), which yields hydrogen ions, HCO_3^-, and CO_3^{2-}. The equilibrium constant for Eq. (9-3) has been determined to be

$$K_1 = [H^+][HCO_3^-]/[H_2CO_3]$$
$$= 10^{-6.4} = 4.2 \times 10^{-7} \qquad (9\text{-}9)$$

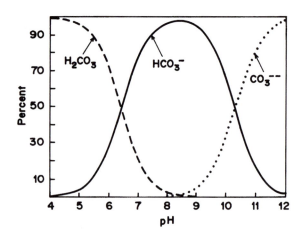

Fig. 9-1. Concentration of H_2CO_3, HCO_3^-, and CO_3^{2-}, as a function of pH.

at 25°C; and for reaction (9-4)

$$K_2 = [H^+][CO_3^{2-}]/[HCO_3^-]$$
$$= 10^{-10.3} = 5.0 \times 10^{-11} \qquad (9\text{-}10)$$

The ion product for water

$$K_w = [H^+][OH^-] = 10^{-14} \qquad (9\text{-}11)$$

also is useful here. Suppose that a carbonate system described by reactions (9-2), (9-3), and (9-4) has a neutral pH, for example. Then, from Eq. (9-11), $[H^+] = [OH^-] = 10^{-7}$, and we obtain, from Eqs. (9-9) and (9-10),

$$[HCO_3^-]/[H_2CO_3] = 10^{0.6}$$
$$[CO_3^{2-}]/[HCO_3^-] = 10^{-3.3}$$
$$[H_2CO_3]/[CO_3^{2-}] = 10^{2.7}$$

This indicates that bicarbonate is the dominant ion at neutral pH (Fig. 9-1), and carbonate ion is the least abundant ionic species; concentration (or rather, *activity*) of undissociated carbonic acid is intermediate. The system attains highest bicarbonate-ion concentration at pH of about 8.3, which is roughly the pH of ordinary sea water.

Let us find the equilibrium constant for the reaction (9-8), the dissolution of $CaCO_3$:

$$K = [Ca^{2+}][HCO_3^-]^2/[H_2CO_3] \qquad (9\text{-}12)$$

K may be considered as a product of K_1, $1/K_2$, and K_s:

$$K = [H^+][HCO_3^-]/[H_2CO_3]$$
$$\times HCO_3^-]/[H^+][CO_3^{2-}] \times [Ca^{2+}][CO_3^{2-}]$$

Calcite ($CaCO_3$) is a solid in this chemical system and its "concentration" (or activity) is taken as unity. The solubility product of calcite at 25°C has been found experimentally to be

$$K_s = [Ca^{2+}][CO_3^{2-}] = 10^{-8.3} = 4.5 \times 10^{-9}$$

Therefore

$$K = 10^{-6.4} \times 10^{10.3} \times 10^{-8.3} = 10^{-4.4}$$

Now let us find the concentration of dissolved Ca^{2+}. We know from Eq. (9-8) that the concentration of HCO_3^- is twice that of Ca^{2+}, so we may replace $[HCO_3^-]$ in Eq. (9-12) with $2[Ca^{2+}]$:

$$K = [Ca^{2+}][HCO_3^-]^2/[H_2CO_3]$$
$$= [Ca^{2+}]\{2[Ca^{2+}]\}^2/[H_2CO_3]$$

The concentration of H_2CO_3 in water that is in equilibrium with air is about 10^{-5} molar. Thus we have

$$K = 10^{-4.4} = 4 \times [Ca^{2+}]^3/10^{-5}$$

or

$$[Ca^{2+}] = 4.7 \times 10^{-4} \text{ molar}$$

Suppose that CO_2 content of the air increases; then the concentration of H_2CO_3 in water increases, which leads to dissolution of $CaCO_3$ and increase in $[Ca^{2+}]$.

Temperature also controls precipitation and dissolution of carbonate. Cold, carbonate-undersaturated sea water that wells up from the depths onto a platform precipitates calcium carbonate when it warms up. When the carbonate skeletals of planktonic organisms fall through the warmer surface waters of the ocean into the cold, deep layers, they dissolve. One of the reasons that carbonate is less soluble in warm water than in cold is that CO_2 (like all gases) is less soluble in warm water than in cold.

Fresh water (as in lakes) precipitates calcite as dissolved CO_2 escapes to the atmosphere or is consumed by photosynthesis. Fresh water does not greatly supersaturate with respect to calcite, and certainly not with respect to aragonite, whose solubility product is even higher than that of calcite. Thus, calcite, inorganically precipitated, is by far the dominant calcium carbonate phase in lake sediments.

Sea water has a mildly alkaline pH of about 8 and has the ability to resist to a considerable degree any externally imposed change in its pH. This is to say that sea water is *pH-buffered*. Simple buffers are solutions of a weak acid (or weak base) and one of its salts in water. Sea water contains a weak acid (H_2CO_3), which dissociates partially to HCO_3^-, and there is additional HCO_3^- in the solution, which may be viewed as the anions of a dissociated (dissolved) bicarbonate salt. The addition of acid to sea water drives the reactions

$$H^+ + CO_3^{2-} \rightarrow HCO_3^-$$

and

$$H^+ + HCO_3^- \rightarrow H_2CO_3$$

which consume most of the added acid (substituting strong acids with weak acids). Addition of a base to sea water is countered by

$$OH^- + H_2CO_3 \rightarrow HCO_3^- + H_2O$$

and

$$OH^- + HCO_3^- \rightarrow CO_3^{2-} + H_2O$$

But sea water is a more complex buffer if it is in contact with solid $CaCO_3$, as it commonly is; in this case acid is consumed also by dissolution of $CaCO_3$, and base is consumed by precipitation of $CaCO_3$:

$$H^+ + CaCO_3 \rightarrow Ca^{2+} + HCO_3^-$$

and

$$OH^- + Ca^{2+} + HCO_3^- \rightarrow CaCO_3 + H_2O$$

Sillen (1967, p. 1189) pointed out that the capacity of the buffering systems ordinarily credited with maintaining oceanic pH is in fact "pitifully small, compared with the amounts of acids and bases that have passed through the ocean system in the course of time." While the carbonate system accomplishes rapid local pH buffering, it seems that a different chemical system is responsible for long-term, large-scale buffering. Garrels (1965) and Mackenzie and Garrels (1965) advocated silicate minerals as major oceanic buffers. Clay minerals and other finely divided or "degraded" silicates react with sea water (though not as rapidly as the carbonate system does), and may represent the major long-term influence on oceanic pH.

The fresh waters of lakes and streams typically have salinities (that is, total concentration of dissolved salts) of 0.1 to 1 permil (g per kg water) and Mg/Ca of 1/10 to 1/3. Evaporated lakes have wide-ranging salinities and Mg/Ca ratios. Normal sea water has a salinity of about 35 permil and Mg/Ca of about 3/1. Sea water that is evaporating, as on an arid tidal flat, may have salinity as high as 350 permil and Mg/Ca of 5/1 or 10/1 or more (owing to early precipitation of Ca-bearing minerals). These hypersaline waters may mix with normal sea water or with meteoric water to give a wide range of total salinity and Mg/Ca ratios generally higher than that of normal sea water.

Composition of Rock-Forming Carbonate Minerals

There are three major sedimentary carbonate minerals, aragonite (orthorhombic $CaCO_3$), calcite (rhombohedral $CaCO_3$), and dolomite [rhombohedral $CaMg(CO_3)_2$]. Much carbonate has a composition that is intermediate between calcite and dolomite; that is, it contains magnesium, but not as much magnesium as true dolomite contains.

The calcite of sedimentary environments generally contains a certain amount of Mg, randomly substituting for Ca; if the amount of $MgCO_3$ molecule is less than about 4 mole percent, the mineral is referred to as *low-Mg calcite*, or simply as calcite; if there is more magnesium, typically 10 to 18 mole percent $MgCO_3$, then the mineral is called *high-Mg calcite*, or just Mg-calcite. (Ca-Mg carbonates intermediate between dolomite and magnesite, $MgCO_3$, are practically non-existent.) Stehli and Hower (1961, p. 367) measured the magnesium content of high-Mg calcites, and found a statistical mode at 12.5 mole percent. Some cement precipitated from sea water is high-Mg calcite, and certain organisms also produce it as skeletal material.

In the mineral dolomite, layers of $CaCO_3$ "molecules" alternate with layers of $MgCO_3$, and Ca and Mg are in stoichiometric one-to-one ratio, but again, the proportion of the two cations varies over some range in sedimentary dolomites. In some dolomite, the magnesium content is substantially below its stoichiometric proportion; magnesium ions are more or less randomly substituted for calcium ions, and the term *Ca-rich disordered dolomite* is used. Much disordered dolomite forms not by precipitation but by chemical alteration of calcite.

Aragonite is a polymorph of calcite that forms in sedimentary environments by inorganic precipitation or through biological intermediaries; it is less stable in diagenetic environments than calcite and typically dissolves or recrystallizes to calcite.

Sedimentologists have learned—through fieldwork, petrography, chemical analysis, and laboratory experiments—that the composition of the chemically precipitated calcites, dolomites, and aragonite reflects the chemical and physical environments under which these minerals form and controls their crystal habits and crystal sizes (see Chap. 11). The most important controls on carbonate composition and crystal form seem to be total salinity of the solutions—including concentration of Na^+, SO_4^{2-}, Ca^{2+} and Mg^{2+}, HCO_3^-, and CO_3^{2-}—and ratio of magnesium to calcium in the carbonate-precipitating solutions. The presence of organic films on the surfaces of mineral crystals seems to suppress crystal growth in some cases and to catalyze growth in other cases. Rapid crystal growth results from high total concentration of dissolved ions (salinity) or rapid loss of CO_2 from the water. Slow crystal growth results from either low total salinity, chemical and physical conditions that are not changing rapidly, or possibly from presence of organic matter in solution or coating the surfaces of crystals. Diagenetic conditions lead to dissolution of carbonate minerals, some more rapidly than others, or to alteration of their compositions or crystal sizes and shapes.

Other cations besides Ca^{2+} and Mg^{2+} may occur in the sedimentary carbonate minerals, notably Fe^{2+}, Mn^{2+}, and Sr^{2+}, as substituents for Ca^{2+}. The contents of iron and manganese in calcite or dolomite are determined largely by oxidation state (redox potential) at time of carbonate precipitation (see Chap. 11). The redox potential (Eh) of natural waters varies from about -0.6 to 0.6 v and is controlled largely by relative rates of organic respiration, bacterial decay, and photosynthesis. Strontium occurs mostly in (skeletal) aragonite, and its concentration is, according to Odum (1957), phylogenetically determined, though environmental temperature also is influential (Smith et al., 1979; Beck et al., 1992). When aragonite inverts diagenetically to calcite, most of the strontium, which is not readily accommodated in the calcite structure, moves to the pore water. Recrystallizations in general expel impurities and lattice substituents to crystal surfaces, where they are readily removed by pore fluids in contact.

Sodium also may be present in carbonate minerals, but it seems that most of it is contained in liquid-filled vacuoles rather than as part of the crystal structure. The presence of sodium in calcite or dolomite is considered by many workers to be an indication that the mineral precipitated from saline water.

The major skeletal materials of organisms are calcite, aragonite, opal, apatite, chitin, and lignin. Different organisms form skeletons or excretions of one or another of the carbonate minerals, depending on their taxonomy (Table 9-1). Calcite, either low-magnesium or high-magnesium, and aragonite are by far the most important skeletal minerals—all skeleton-building phyla exploit calcium carbonate, and calcium carbonate constitutes the greatest skeletal biomass of all skeletal minerals. Organic evolution, according to Wilkinson (1979) has caused the composition of carbonate rocks to change through time, owing to changing dominance of the various groups of carbonate-precipitating organisms (this might, in turn, be related to changes in the composition of sea water). Paleozoic organisms made low-Mg and high-Mg calcite; Mesozoic organisms made high-Mg calcite and aragonite, and Cenozoic organisms are dominantly aragonite makers. Amorphous silica (opal) and certain phosphate minerals also are important, silica mainly in the diatoms, radiolarians, and certain sponges and phosphate mostly in the vertebrates; these materials contribute substantially to certain sedimentary rocks.

Allochems

Sedimentary particles having more or less well-defined shapes and internal fabrics and having been formed by processes taking place at the depositional site are called **allochems**. Several important types of allochems (Fig. 9-2) are widely distributed among limestones:

Ooid—consists of a nucleus surrounded by a cortex of one or more concentric layers or coats of calcite or other mineral; some have a radial-fibrous cortex; diameter generally less than 1 millimeter.

Peloid—a small, ovoid body of mud; peloids are of dubious origin, though probably most are fecal. "Grain constructed of an aggregate of cryptocrystalline carbonate, irrespective of origin" (McKee and Gutschick, 1969, p. 101). A *pellet* is a grain of proven or presumed fecal origin. Most are smaller than 0.1 millimeter.

Intraclast—"pieces of penecontemporaneous, usually weakly consolidated carbonate sediment that have been torn up and redeposited by currents" (Folk, 1962, p. 63). Intraclasts are generally larger than peloids, even attaining dimensions of several centimeters, and their sizes and shapes and internal textures are much more variable. A *grapestone* is a type of intraclast in which "carbonate sand grains of various kinds are clumped together to form compound grains" (Bathurst, 1975, p. 87) or a small number of "grains become firmly cemented together" (Taylor and Illing, 1969, p. 80). These objects are also called *lumps*. A *steinkern* is an internal cast, that is, hardened mud or cement filling an internal cavity of a skeletal, which has been released from its mold in the depositional environment.

Skeletal—a piece of the hard, mineralized tissue of an organism; a structural element such as a shell or a tooth.

Table 9-1. Skeletal Materials of Organisms

Red algae	High-magnesium calcite, opal
Green algae	Aragonite
Protozoans	Aragonite, calcite, opal
Sponges	Calcite, opal
Corals	Aragonite, calcite
Bryozoans	Aragonite, low-magnesium calcite
Brachiopods	Low-magnesium calcite, apatite
Annelids	Aragonite, calcite
Molluscs	Aragonite, low-magnesium calcite, apatite
Arthropods	Aragonite, calcite, apatite
Echinoderms	Calcite
Vertebrates	Aragonite, apatite

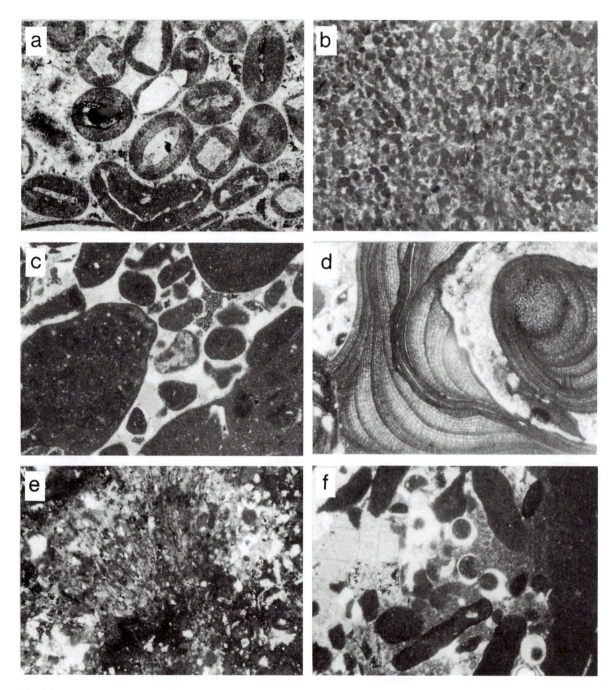

Fig. 9-2. Various limestones and their allochems: (**a**) ooid grainstone; (**b**) peloid grainstone; (**c**) rounded lumps or intraclasts, calcite cemented; (**d**) coralline alga ball (rhodolith); (**e**) interior of spongy onkoid with algal-filament molds; (**f**) *Umbellina* and intraclasts;

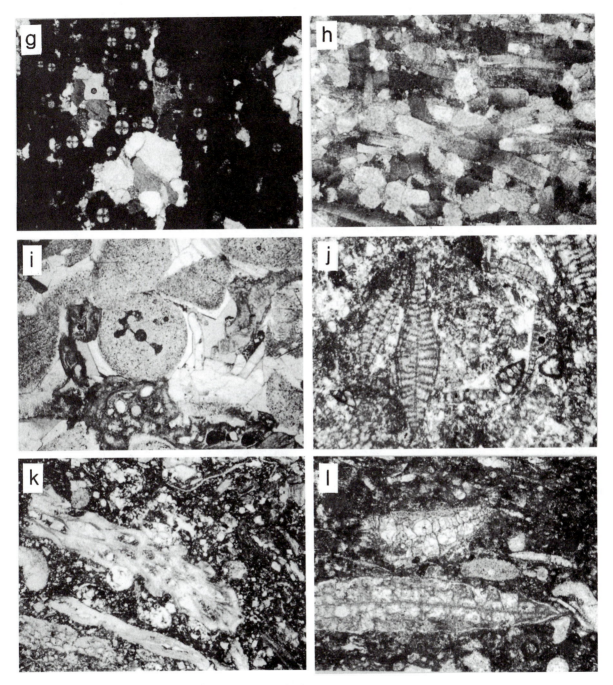

Fig. 9-2. *(Continued)* **(g)** calcispheres; **(h)** grainstone composed entirely of fragments from the prismatic layer of large bivalves *Inoceramus*; **(i)** skeletal grainstone composed of echinoderm plates and bryozoan fragments; **(j)** skeletal packstone with large foraminifers *Leptocyclina*; **(k)** skeletal packstone; **(l)** skeletal wackestone with bryozoans.

Some skeletals of carbonate rocks are, of course, not carbonate.

Onkoid—algal-coated grain; a spheroidal, unattached stromatolite; Aitken (1967, p. 1164) distinguished "simple" onkoids that have concentric coats and "spongy" onkoids bearing more or less radial openings, which are the molds of algal filaments.

Thromboid—term derived here from Aitken's (1967, p. 1164) thrombolite; a structureless clot of lime mud agglutinated or bound together by algal slimes.

Ooids

Ooids are so distinctive in shape and internal structure (though the structure is variable) that their presence draws immediate attention from the curious amateur and the jaded professional alike. It is surprising that Nature produces such objects apparently without the intercession of organisms (they resemble little eggs, hence the name). The environmental conditions that produce them give rise to great numbers of them; they are hardly ever formed in isolation.

The origin of ooids has been a subject of observation, experimentation, speculation, and controversy for at least 150 years. Modern marine carbonate ooids are composed of layers (or coats) of aragonite built up concentrically around some nucleus, such as a carbonate peloid or skeletal, but occasionally of some completely different material, such as a quartz sand grain. Some ooids are composed not of carbonate but of hematite or apatite.

Many modern and ancient ooids display a radial-fibrous fabric, but layers with radial structure commonly are interlayered with coats that have tangential or micritic structure. Newell et al. (1960) described ooids of the Bahaman oolite shoals, and these have become a kind of standard for ooids that *do not* have the radial fabric. In the Bahaman ooids, aragonite crystallites are tangentially oriented in some coats and randomly oriented in others. Bathurst (1975, p. 308) concluded that the aragonite needles are c-axis tangential to the surface of the ooid originally and that radial fabrics develop during subsequent diagenesis. Loreau and Purser (1973, p. 432) suggested just the reverse, that radial fabric is primary and tangential orientation is a secondary effect of grain-to-grain collisions during ooid formation.

Kahle (1974) and Sandberg (1975) determined that aragonite ooids of the Great Salt Lake in Utah are radially structured right from the start. Sandberg's ex-

cellent scanning electron microscope (SEM) photomicrographs of Great Salt Lake ooids show radial crystals traversed by narrow, well-defined accretionary bands; tiny disconformities transect and truncate these accretionary layers. Sandberg suggested that ooid growth involves the accumulation or precipitation of a thin layer of very fine, randomly oriented aragonite prisms on the surface of a nuclear grain. Much larger, nearly radial crystals grow out of the substrate of widely divergent crystallites, culminating in a tight palisade structure. One suspects that unfavorably oriented crystallites get buried beneath the crystals that happen to have the radial orientation; it is analogous to the growth of columnar cement crystals in a pore. Apparently there are periods of abrasion or otherwise interrupted growth, after which accretion begins again from a new random layer. Experiments by Weyl (1958) suggest that the growth of ooids requires periods of transport alternating with periods of (shallow) burial, such as occurs in migrating ripples or wavebuilt bars or tidal bars. Agitation seems to cause the c-axis tangential fabric, and a balance between growth and attrition may be responsible for limiting ooids to diameters not much greater than 1 millimeter. Davies et al. (1978) performed laboratory experiments suggesting that ooids created in quiet water exhibit radial fabric but that agitated water favors tangential fabrics. Ooids of the hypersaline waters of Baffin Bay on the Texas Gulf Coast, according to Land et al. (1979, p. 1276), have tangential aragonite coats, forming under the more agitated conditions, and radial coats of high-magnesium calcite, forming under quieter conditions. Heller et al. (1980, p. 945), noted that ooids of the Warrior Formation (Cambrian) in Pennsylvania have radial centers and, if the ooid exceeds 0.6 millimeters diameter, tangential outer layers. They suggested that during normal ooid growth, radial layers form initially, but when an ooid has grown to greater than 0.6 millimeter in diameter, tangential layers begin to form in response to increased abrasion of the now heavier particle. In marine ooids, coats with randomly oriented needles are made by boring algae or other organisms, according to Margolis and Rex (1971, p. 848), and are analogous to micrite envelopes (described below). Clay particles inhibit development of the radial structure.

Halley (1977, p. 1113) suggested that radial structure is characteristic of freshwater and hypersaline ooids. But Land et al. (1979, p. 1275) could find no salinity control (or magnesium/calcium control) on ooid structure. Sandberg (1975, p. 524) also had dismissed salinity as a control on ooid structure.

Though modern ooids seem to be composed of aragonite in general, ancient ones are of calcite. This has generally been attributed to recrystallization (neomorphism) of the unstable aragonite, but preservation of fine structure in ancient ooids suggested to Sandberg (1975, p. 511) that they were of calcite originally; he argued that neomorphic inversion would have destroyed the fine structure. More recent results suggest otherwise—that fine structure *can* be preserved under neomorphism. Mazzullo (1980, p. 415), for example, described detailed preservation of microfabric of botryoidal aragonite cement that had neomorphosed to calcite (see Chap. 11).

Some objects produced by organisms might occasionally be misconstrued as ooids—for example, the peculiar objects (Fig. 9-2f) tentatively assigned to the nodosinellid foraminifers and named *Umbellina* (Loeblich and Tappan, 1961; nicely illustrated by Rich, 1965). Also the calcispheres (Fig. 9-2g), tiny hollow spheres of calcite typically filled with a radial cement. These are abundant in some Devonian and younger limestones and are attributed to *Eovolvox*-like colonies (see Brazier, 1980, p. 52). Cross sections of productid brachiopod spines faintly resemble ooids. Fossil spore cases have sometimes been misidentified as ooid molds (pores left when an ooid dissolves diagenetically).

Peloids

Peloids are probably the most abundant and ubiquitous of the allochems. Surely some objects that are not fecal are called peloids, but, by and large, peloids are of fecal origin. The fecal matter of some animals, more or less continuously excreted throughout the animals' lifetimes, can survive in large quantities as sedimentary particles. Most fecal pellets are initially quite soft, and under overburden pressures are readily mashed into what commonly comes to be labeled as matrix (see Shinn and Robbin, 1983, p. 612). *Much* carbonate mud (matrix) in limestones, under close petrographic examination, proves to be pelletized. Some carbonate rocks are composed almost entirely of peloids, and these may all be essentially the same size, as though produced by large numbers of individuals of a single species. Some peloids (or pellets) are internally struc-

tured, reflecting certain complex protuberances in the lower gut of the animal that produced them. There are hard fecal pellets, too. Certain gastropods make hard pellets that do not disintegrate readily and contaminate their feeding sites thereby (Schreiber, 1986, p. 197). A *coprolite* is a larger defecated mass, more likely to be preserved in non-marine environments.

Hattin (1975) discussed the possibility of pelletization-mediated sedimentation of coccolithophorids (certain algae, with skeletal parts similar in size to clay particles) by *planktonic* herbivores, such as copepods. He described Cretaceous carbonate rocks consisting largely of peloids composed of the tiny coccolith skeletals. Schrader (1971) described the membrane-encapsulated fecal pellets of certain living planktonic herbivores. These pellets contain the empty shells of diatoms and silicoflagellates and occur in deep-water sediments; not only did pelletization (presumably) hasten sedimentation of tiny skeletons but the fecal membranes also protected the contents from dissolution during the descent to the bottom.

Marshall (1983, p. 1136) observed certain peloids (10- to 80-micrometer spheroidal bodies of microcrystalline high-magnesium calcite) and "micropeloids" (5 to 10 micrometers) filling or partially filling skeletal pores, borings, and other cavities in the southern part of Australia's Great Barrier Reef, which he suggested are formed by chemical precipitation. Small peloids like this occur also in ancient rocks. Chafetz (1986) suggested that some such peloids are bacterial precipitates.

Intraclasts

Intraclasts are of various origins. Some are rip-up clasts, a result of some thin layer of soft but coherent sediment at the surface being ripped apart by a vigorous current. Perhaps this layer was desiccated and divided into thin polygonal plates with upturned edges; subsequent tidal flow or storm surge scattered them, as a gust of wind disperses dry leaves. Beds composed of imbricated, rounded, platy intraclasts of lime mud, even as much as 10 or 15 centimeters across but only a centimeter in thickness, are numerous in early Paleozoic carbonate formations. Commonly referred to as "edgewise conglomerates," these rocks seem to be far less numerous in younger formations (Sepkoski, 1982a, p. 371).

Other intraclasts are formed when the oversteepened bank of a tidal creek gives way and the mud breaks into angular blocks; when a layer of halite dissolves and an overlying semiconsolidated mud layer collapses and breaks; or when a cemented surface layer founders into unconsolidated sediment beneath. Grapestones and other objects formed by local cementation within the depositional environment, such as pieces of beachrock, must also be considered as intraclasts. Intraclasts composed of ooid grainstone often are reported, embedded in similar ooid grainstone. For example, Clark (1980a, p. 133, 136) described them in Zechstein limestones (Permian) in the Netherlands and suggested that they are lumps of beachrock. Some grapestones or lumps are held together by algal filaments. Commonly they are intensely micritized by boring algae. Winland and Matthews (1974, p. 922) observed that in grapestones of the Bahamas, grains are bound together by encrusting foraminifers or by calcified cyanobacterial filaments. Bathurst (1975, p. 316) advised that grapestones have not been widely recognized in ancient limestones, but that their distribution in modern carbonate depositing environments suggests that they are a major component of ancient limestones and a valuable index of environment of deposition.

A steinkern forms when a shell, filled with hardened mud, breaks open or is otherwise stripped away in the depositional environment and lays bare the mud fill, which then behaves as a discrete sedimentary particle. These objects do not seem to be very common, though probably some are misidentified as the more ordinary kinds of intraclasts. Dissolution of shells can occur diagenetically, giving rise to a "false steinkern," surrounded by a skeletal-moldic pore. A burrow-fill that has survived some syndepositional erosion might be referred to as a steinkern.

Skeletals (Fossils)

Skeletal grains come in endless variety of size and shape, though single limestone samples generally contain only a few kinds (Fig. 9-2). A single organism may give rise to a single sedimentary particle, such as a shell; to a large collection of grains of various sizes and shapes, as a multi-membered skeleton disarticulates; or to some small part of a contiguous and coherent large structure such as a reef. A single diatom, ostracod, or trilobite typically forms and moults several complete frustules or carapaces during its lifetime,

just as a single tree forms millions of leaves, seeds, or spores. A shark has multiple ranks of deciduous teeth; the teeth of a single shark may be scattered over large areas of the sea floor. One might consider a shark, a trilobite, or a bird as a collection of skeletal grains and fecal pellets temporarily equipped with fins or legs or wings. Skeleton-bearing or pellet-making organisms that are vagile for at least part of their lifetime are dispersing sedimentary particles (or potential sedimentary particles) just by moving about.

Identification of the various skeletal grains is a task ordinarily reserved to the paleontologist, but the petrographer can (and must) make at least broad taxonomic assignments for skeletal grains (see Horowitz and Potter, 1971; Bathurst, 1975; Scholle, 1978; Flugel, 1982; Adams et al., 1984). The calcareous algae may be the most difficult; for a detailed treatment see Wray (1977) and the volume edited by Flugel (1977). Johnson (1961) gives a binomial taxonomic classification of algal structures preserved in ancient carbonate rocks, and offers many excellent photographs and photomicrographs. B.R. Pratt (1984) described the diverse lower and middle Paleozoic algal structures, also with excellent photomicrographs.

Limestones differ markedly insofar as the kinds of skeletals contained in them differ. Thus, trilobite packstones are a common occurrence in the lower Paleozoic, but pelmatozoan grainstones are a hallmark of the upper Paleozoic. Their different skeletal contents give them very different appearances. Any reasonably knowledgeable geologist can distinguish them at a glance and correctly rank them as older and younger. The distinctive stromatoporoid limestones and dolomites of the Devonian, or the very different foraminiferal limestones of the Cretaceous and Tertiary, can never be confused with carbonate rocks anywhere else in the stratigraphic succession.

Wilson (1975) contrasted the limestones of different ages, and Heckel (1974) described the differences between carbonate buildups of different ages, reflecting evolution of the biota (Fig. 9-3). Early Paleozoic skeletal grainstones, packstones, and wackestones contain brachiopods, bryozoans, and trilobites; reefs and mounds are stromatolitic, or composed of calcareous algae and sponges. Middle Paleozoic limestones contain pelmatozoans, bryozoans, and rugose corals; reefs are constructed out of stromatoporoids and rugose and tabulate corals. Pelmatozoans and bryozoans are characteristic of the late Paleozoic limestones; corals and

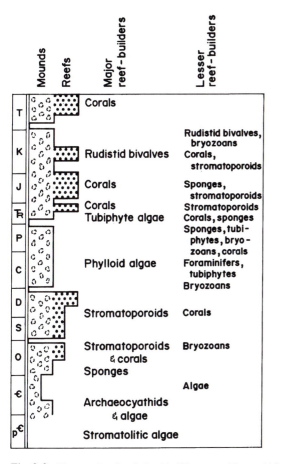

Fig. 9-3. History of major skeletal buildup assemblages. (After James, 1983, fig. 61.)

Onkoids

Onkoids come in various sizes and internal structures (see Chap. 5), probably determined in part by the kinds of algae or bacteria that made them. Many of them form around a nucleus, such as a shell. They are probably much more abundant in sedimentary rocks than generally thought; some of them may be confused with a diagenetic product, called the vadose pisoid. Bosellini and Ginsburg (1971) recommended that one kind of algal nodule, the **rhodolith**, made by precipitation of carbonate by coralline algae (*Rhodophyta*), be distinguished from the onkoids, which are products of sediment trapping and binding, mostly by green algae (*Chlorophyta*) or blue-green algae (now called *Cyanobacteria*). Rhodoliths of differing internal structures appear to reflect different sedimentary environments.

Thromboids

Some onkoids or other carbonate masses have a "spongy" structure, rather than laminae; they are composed of little formless clots of lime mud, some containing thin, tangled or branching tubes that are the molds of algal filaments. The clots, here called thromboids, apparently are made by green algae. Probably many of these algal clots have been construed as peloids. Kennard and James (1986, p. 494) attributed the clots to in situ calcification of coccoid-dominated microbial communities, and Chafetz and Buczynski (1992, p. 289) suggested that they are a result of bacterially induced precipitation (cf. cement peloids, described above). In some environments, thromboids form large mounds several meters thick.

Carbonate Mud and Carbonate Cement

Crystal size of non-allochem carbonate in limestones or limy sediments varies from less than a micrometer to several millimeters. **Lime mud** is the very fine calcite or aragonite material of carbonate sediments produced in depositional environments. As carbonate mud ages and lithifies, it becomes a somewhat coarser material called **micrite**. Coarser crystals in carbonate rocks are produced either by chemical precipitation in pores or by recrystallization of finer-grained carbonate. Some chemically precipitated carbonate is fibrous or acicular; some is blocky or equant and is called

stromatoporoids declined and reefs were small until the phylloid green algae made their appearance in late Paleozoic time. Fusulinid foraminifers also appeared and made large contributions to latest Paleozoic limestones. At the close of the Paleozoic Era, sponges, hydrozoans, and encrusting algae were important reef builders. Mesozoic limestones contain echinoids and certain molluscs in abundance; rudistid bivalves contributed to extensive biostromes. With the Cenozoic Era came large and varied foraminifers, a great variety of bivalves and gastropods, and new kinds of reef-building corals and algae. Petrographic contrasts in limestones of all (Phanerozoic) ages are amply displayed by Horowitz and Potter (1971, Pls. 64–100).

spar. Recrystallization yields aggregates of generally equant crystals intermediate in size between micrite and spar.

Lime Mud

Lime mud has several origins—chemical precipitation from evaporating solutions; biochemical precipitation from sea water, occasioned by photosynthesis; biological construction of very small skeletal components; and mechanical breakdown, commonly by organisms, of coarser carbonate materials. Some carbonate skeletals, such as coccoliths, are of mud size without any attrition. Apparently, certain bacteria precipitate carbonate minerals as mud-size crystals. The carbonate skeletons of many organisms, such as certain green algae, consist of very small platy or acicular crystallites bound together by an organic matrix. When the organism dies and the organic matrix decays, the skeleton collapses into a heap of loose crystallites of very small size.

Mechanical abrasion or attrition of carbonate skeletals by boring algae or by predators or grazing animals such as certain gastropods and starfishes produces lime mud. Many allochems, especially skeletals, undergo *micritization* in shallow marine depositional environments. It is not entirely clear why this happens, but it is usually attributed to the activities of certain epiphytic or endolithic algae that coat the grains and become calcitized (Bathurst, 1975, p. 388; Kobluk and Risk, 1977) or that bore into the surface of the grain, "grinding" the surface layer into micrite or creating little pockets into which lime mud from the outside accumulates. Micritization may progress to the center of a grain, and cause the grain to disintegrate completely.

Micrite

An abundant constituent of most carbonate sediment, micrite is lime mud that as a result of early and gentle recrystallization, consists of equant particles or crystals of calcite ranging in size up to about 4 micrometers. Folk (1965, p. 32) determined that the "normal" size of lithified micrite of marine origin is about 1.3 to 2.5 micrometers, but that micrite of many freshwater limestones is smaller, ranging between 0.5 and 1.5 micrometers. Aragonite needle mud (of certain green algae) readily changes into equant calcite micrite upon early diagenesis. The average aragonite needle, according to Folk, is so small that some ten to a thousand needles must come together to make one average micrite unit. Wilkinson (1979, p. 525) has pointed out that organisms that make most of the aragonite needles in modern environments did not proliferate until Mesozoic or Cenozoic times, so older micrite is necessarily of some other origin. Micrite can continue to grow, through recrystallization (neomorphism), into coarser aggregates of microspar and pseudospar (see Chap. 11).

Fibrous Carbonate

Some chemically precipitated carbonate is fibrous. It includes thin crusts and botryoidal masses of parallel or radially arranged crystal fibers composed of aragonite or high-magnesium calcite. Aragonite needles have square ends, in contrast to the steep, sharp terminations of calcite needles. These cements form mostly in normal marine environments by chemical precipitation and are the earliest of cements in a pore. They invert or recrystallize diagenetically to coarser material, and in the process exsolve magnesium or strontium or other ions to tiny inclusions or to the pores.

Spar

Coarser, equant or blocky, pore-filling, chemically precipitated carbonate is called *spar*. Spar is composed of low-magnesium calcite or ordered dolomite, but never of high-magnesium calcite, calcium-rich dolomite, or aragonite. Spar ranges in size from 10 micrometers to several millimeters and is typically anhedral, though both calcite and dolomite spar may bear crystal faces. Petrography of carbonate rocks, and cathodoluminescence and isotope studies (see Chap. 11), indicate that much calcite and dolomite spar cementation occurs very early in the diagenetic history of a carbonate rock, while it is still under the influence of marine or near-surface meteoric waters, but some precipitates from saline formation waters under deep burial.

Calcite and dolomite spar are easily distinguished in thin section; they both may be polysynthetically twinned, but the twin composition planes are different: calcite lamellar twins are $\{01\bar{1}2\}$, so that the traces of twin planes are parallel to cleavage traces or to long diagonal of cleavage rhombs; dolomite twins are $\{02\bar{2}1\}$, and twin plane traces are parallel to both long and short diagonals of cleavage rhombs but never parallel to cleavage traces. The rhombohedron is by far the dominant crystal form in dolomite but quite rare in calcite, where the scalenohedron is generally the most important form. Dolomite characteristically shows curved cleavages and crystal faces and corresponding sweeping extinction; this is the so-called **saddle dolomite** or **baroque dolomite** (*baroque* from old French, "a misshapen pearl"). Alizarin red-S, a staining solution routinely applied to a carbonate rock thin section before the cover slip is affixed, stains calcite but not dolomite.

Some calcite spar (pseudospar) is a product of neomorphism of micrite or of fibrous carbonate cements. Petrographers have developed various criteria for distinguishing between the chemically precipitated cements and the neomorphisms (see Chap. 11), but some problems and uncertainties yet remain.

Caliche

That caliche or similar calcareous materials originating in arid-climate soils or occurring on the surfaces of carbonate sediments in a vadose zone could be preserved and recognized in ancient carbonate rocks was proposed by Dunham (1969a and 1969b) and Thomas (1965), based on stratigraphic and petrographic studies of the Permian Capitan reef complex of southern New Mexico and west Texas. Until that time the pisoids (or pisoliths) of the so-called Carlsbad facies of the reef complex were widely regarded as algal onkoids, formed in a shallow lagoon behind the reef rock of the Capitan facies. Dunham's evidence that they are diagenetic consisted of several observations about fabric of these pisoids and pisoid-bearing rock: downward elongation, that is, growth layers thicker or more numerous on the undersides of the pisoids; fitted polygonal geometries of the pisoids, and laminae shared among adjacent pisoids; inclusions of crystal silt perched on upward-facing parts of laminae; laminae truncated by leaching, or fractured; reverse size-grading of pisoids; lack of admixed sediment; and lack of sedimentary structures characteristic of coarse, current-transported deposits. Thomas, apparently working independently, cited lack of stratification or lack of interbedding of pisoid deposits with stromatolites or other deposits of unquestioned algal origin. Other characteristics of vadose pisoids (see Gerhard et al., 1978, table 1; or Gerhard, 1985, p. 201) are internal shrinkage cracks, nuclei of broken pisoids, and nu-

merous concentric bands with radial-fibrous fabric (Fig 9-4). Onkoids, in contrast, typically have skeletal or intraclast nuclei, wrinkled or pustular muddy laminae, and algal filaments.

Gill (1985) described subaerial and vadose deposits capping Middle Silurian pinnacle reefs in the Michigan basin. Wardlaw and Reinson (1971) found pisoids in the Middle Devonian Winnipegosis Formation of south-central Saskatchewan, Canada. Gerhard (1985) and Gerhard et al. (1978) documented their occurrence in the Mississippian Mission Canyon Formation of western North Dakota, making distinction between vadose pisoids and small, structureless, mud-supported *weathering pisoids* (Fig. 9-5) commonly associated with caliche crusts.

In spite of these and many other studies of both

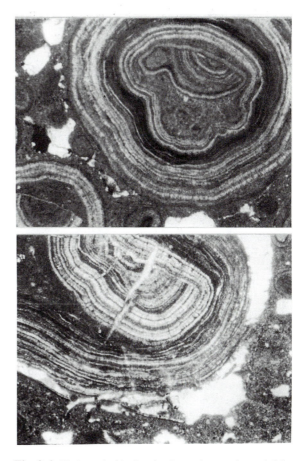

Fig. 9-4. Vadose pisoids showing internal truncations, shrinkage cracks.

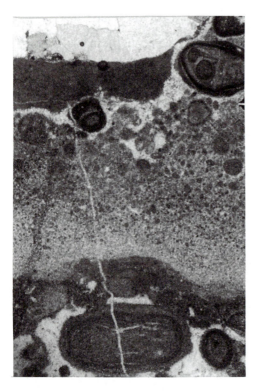

Fig. 9-5. Reverse-graded "weathering pisoids."

ancient and modern carbonate deposits containing (presumably) similar pisoid bodies and associated textures and structures, there is still some debate and considerable confusion about them. Aissaoui, Coniglio, et al. (1986, p. 120) noted radial-fibrous layers and reverse grading in "pisoids" of a Miocene reef talus in Egypt; nonetheless they considered these objects to be algal onkoids. Land and Goreau (1970, p. 461) described micrite crusts in Jamaican reef rocks that they considered to be marine cement, but they cautioned that these crusts closely resemble caliche crusts. Moreover, Land and Moore (1980, p. 367) referred to submarine internal sediment in this same reefal mass that resembles vadose silt (see Chap. 11). The distinction between pisolite (that is, pisoid-bearing rock) and onkoidal rocks is probably most difficult when pisolite development is in its early stages.

Opinion on Terminology

As originally defined, *pisoid* (or *pisolith*) is a concentrically coated grain larger than 2 millimeters; thus, an ooid larger than 2 millimeters would be called a pisoid. But now that we have come to recognize the genetic differences among coated grains, this distinction seems specious. No vadose pisoid and no onkoid should be called an ooid because it happens to be smaller than 2 millimeters. By the same logic, an object that is considered to be an ooid on genetic grounds should be called an ooid, even if it is large. The term *pisoid* should be reserved for the coated objects formed by vadose processes.

Beachrocks

Carbonate sand beaches—composed of skeletals and skeletal fragments, peloids, and ooids—occur on many tropical and subtropical coasts. Skeletal grains typically are abraded and well rounded, and they may be heavily micritized. Rapid, penecontemporaneous cementation of this sand creates *beachrock*, typically only a surface layer or crust underlain by uncemented sand. Such a layer may be broken and reworked by the waves into pebbles and slabs, and some may be encrusted or bored by marine organisms; the slabs come to be incorporated into the loose sand of the beach, forming a breccia of sand-supported clasts composed of the same kind of sand. Beachrock may contain the little vugs or keystone pores caused by entrapment of air pumped into the sand by waves; grains in general are loosely packed. Gently inclined lamination typical of the shoreface environment commonly is preserved. Some modern beachrocks form within a few years (Meyers, 1987, p. 558). The cement is precipitated out of sea water as acicular or platy aragonite or fine-granular high-magnesium calcite, forming very thin isopachous fringes on the grains or showing meniscus and pendant configurations. In ancient rocks, this cement has altered to low-magnesium calcite, and additional cementation likely has taken place.

Beachrock surely is more common in the ancient record than the dearth of documented examples suggests. As with carbonate aeolianites (described below), the beachrock concept probably does not come to mind as often as it should. Bathurst (1975, p. 369) declared that recognition of ancient beachrocks is "likely to be difficult," but that the tell-tale signs of original aragonite or high-magnesium calcite cement, coupled with lack of early compaction, are useful criteria. Donaldson and Ricketts (1979) described what they considered to be a Precambrian beachrock (see Chap. 11).

Classification

As with terrigenous rock classification, we owe to Folk (1959 and 1962) a popular classification of carbonate rocks (Fig. 9-6a). Carbonate rocks are considered to consist essentially of three components, allochems, micrite (or lime mud), and carbonate cement (usually referred to simply as spar). Carbonate rocks in which allochems are abundant and in which the quantity of spar exceeds quantity of micrite are called *sparry allochemical limestone*, or *type I* limestone. If allochems are abundant and micrite exceeds spar, then the rock is *micritic allochemical limestone*, or *type II*. If there is less than 10 percent of allochems, then the rock consists mostly of lime mud and is called *micrite*, or *type III*; there is little possibility of (spar) cement in such rocks. Finally, there are special carbonate rocks that are the rigid skeletons or frameworks assembled by corals, algae, and other organisms; these are called *biolithite*, or *type IV* limestone. One may require that the allochem pole be resolved so as to specify types of allochems, perhaps according to the scheme also suggested by Folk (Fig. 9-6b).

The ternary presentation of type I, II, and III limestones makes clear that certain proportions of the end members (allochem, mud, spar) are not possible. In the first place, allochems cannot be made to occupy space fully; there is always some intergranular space. If there is little or no mud, then the allochems must be in mutual contact, of course. How tightly packed they are and the total quantity of intergranular space available for precipitation of spar depends on their sorting (as in sandstones), their shapes, and their plasticity. Loose packings with porosities as high as 40 or 50 percent (or even higher) are possible, especially where the allochems are skeletals of very irregular shape. Tight packings, with porosities less than 10 percent, can be achieved through mashing together of soft peloids. Thus, spar, which occupies intergranular space, exists in quantities between about 10 percent and 40 or 50 percent in limestones that lack mud. Of course, some or all of the intergranular space may be filled with micrite instead. But mud, unlike spar, is a rock-forming material by itself, and the quantity of mud or micrite in a rock may go beyond the limit imposed by the requirement of grain support; allochems may be floating (that is, embedded in the mud), or altogether absent. For purposes of classification, empty primary pore space is construed as spar (in contrast to Folk's sandstone classification, wherein pore space, cemented or not, is ignored). Spar that fills secondary pore space should be ignored; any "spar" formed by recrystallization of micrite should be construed as micrite.

Folk (1962) also offered a "spectral" classification that includes, but goes beyond his ternary scheme (Fig. 9-7a). Limestones containing mostly spar in the spaces between the allochems are divided into *rounded biosparite*, *sorted biosparite*, and *unsorted biosparite*.

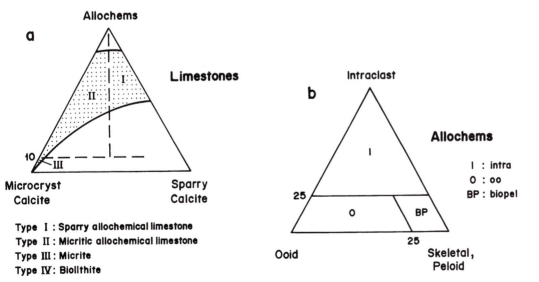

Type I : Sparry allochemical limestone
Type II : Micritic allochemical limestone
Type III : Micrite
Type IV : Biolithite

Fig. 9-6. Ternary classification of limestones (according to Folk, 1962): **(a)** allochem-mud-spar triangle, **(b)** allochem triangle.

Folk (1962)   **a**

matrix grains	Mostly lime mud				subequal mud and cement	Mostly cement			
	0-1%	1-10%	10-50%	> 50%		poorly sorted	well sorted	well rounded	
	Micrite Dismicrite	V Sparse (Allo) -micrite	Sparse (Allo) - micrite	Packed (Allo) -micrite	Poorly- washed (Allo) - sparite	Unsorted (Allo) -sparite	Sorted (Allo) -sparite	Rounded Allo -sparite	Biolithite
	Type III	Type II				Type I			Type IV

Dunham (1962)   **b**

contains mud			lacks mud		Bound- stone
mud-supported		grain-supported			
Mudstone	Wackestone	Packstone	Grainstone		

Fig. 9-7. Classification of carbonate rocks: **(a)** according to Folk (1962) and **(b)** according to Dunham (1962). Note the correspondences between the two.

This assumes that the allochems are skeletals, for the most part; if they are peloids, then the limestones would be called pelsparite; if there are both peloids and skeletals, then the rock would be called a bio-pelsparite or pel-biosparite (put the *least abundant* allochem *first*). Limestones containing about equal amounts of spar and lime mud between the allochems are called ***poorly washed biosparite***. Those containing larger quantities of mud are called ***packed biomicrite***, ***sparse biomicrite***, and ***fossiliferous micrite***. A limestone composed almost entirely of lime mud is called ***micrite***; if the micrite contains small openings (called fenestrae; they are fenestrae even if cemented) it is called ***dismicrite***.

Another classification in wide use, and especially in the petroleum industry, is that of Dunham (1962). Here, carbonate rocks are divided into those that lack mud, those that contain mud and are grain-supported, and those that are mud-supported (Fig. 9-7b). In ***grain-supported*** rocks, grains (allochems) are in mutual contact; in ***mud-supported*** rocks, grains are embedded in mud and are not in mutual contact. Limestones that lack mud are called ***grainstone***; spaces between the grains are either empty or filled with chemical cement. If lime mud is present, but in quantities less than about 25 or 35 percent, then this mud fills (or partially fills) the spaces between grains, but the rock is still grain-supported; such rocks are called ***packstone***. Some in-

tergranular cement may be present in these rocks, occupying space not filled with mud. If mud is more abundant, then the grains on the whole can no longer be in mutual contact (they are said to be *floating* in the matrix); if allochems still constitute more than 10 percent of a mud-supported rock, then it is a ***wackestone***; otherwise it is a ***(lime) mudstone***. An important distinction can be made in the wackestones: In some, skeletals are disarticulated and worn or broken and apparently were transported into a muddy environment. In other wackestones, skeletals are unbroken and may even be in growth orientation; these are the so-called *whole-fossils wackestones*. Rocks corresponding to Folk's biolithite are called ***boundstone***.

Embry and Klovan (1971, p. 734), viewing Folk's and Dunham's classifications as addressing reef limestones inadequately, subdivided Dunham's boundstone into three new categories: ***bafflestone***, a muddy limestone containing abundant stalked or dendroid skeletals interpreted to have baffled the currents, causing these currents to drop their load of suspended mud; ***bindstone***, composed of mud and grains that have been bound together into a rigid mass by encrusting or filamentous organisms; and ***framestone***, consisting of a more or less continuous rigid structure of frame-building organisms, commonly enclosing spaces that may be filled with mud or cement. Closely associated with most reefs are flanking deposits of coarse reef talus that Embry and Klovan called *floatstone*, essentially a coarse wackestone containing fragments of reef rock, and *rudstone*, being a coarse packstone of reef-rock fragments.

Since most carbonate rocks are composed overwhelmingly of two mineral species, calcite and dolomite, and commonly contain terrigenous materials as well, it is useful to some purposes to classify them according to the relative proportions of these components, as Leighton and Pendexter (1962) did (Fig. 9-8).

In at least one important respect, the classification of carbonate rocks parallels that of detrital rocks. The relative proportion of framework and matrix, which distinguishes terrigenous arenites, wackes, and mudstones, also distinguishes carbonate grainstones, wackestones, and mudstones. This parallelism recognizes the contrasting hydrodynamic behavior of grains and mud, no matter what their composition, and the importance of grain-to-matrix ratio in distinguishing among depositional environments. A carbonate rock that contains abundant lime mud probably was not deposited in an environment where waves and currents stirred the bottom; agitation, such as occurs on beaches or tidal shoals, would tend to remove the mud. Lime mud comes to rest in quiet places, such as the deeper waters or the shallow lagoons. There are exceptions—lime mud that has become entrapped in the crannies between coral skeletons on a reef or mud that is bound into a tough mat by algae.

What controls the sizes and shapes of allochems in carbonate rocks? Certainly the controls are at least as varied as the kinds of allochems. Of course, the sizes and shapes of skeletal grains and of fecal pellets are given by biological factors—the anatomy of the organism that released the grains. Some skeletal grains have a particular shape or structure that causes them to break readily and in a certain way. The extent to which skeletals have been disarticulated or abraded bears on the intensity of mechanical energy in the environment, but predation, scavenging, and bacterial decay also wear down skeletal grains. Some skeletal structures simply collapse under their own weight or fall apart as organic connective tissue decays postmortem. Whereas large clasts in a detrital deposit indicate that some competent transporting agent was involved, a large skeletal in a carbonate sediment may have grown at its present location, essentially out of nothing. Good sorting in allochems ordinarily is accomplished hydrodynamically, but this is not always the case. Many peloid grainstones or packstones are very well sorted, even though they were deposited in very quiet water; the grains are all about the same size because they were made that way. Thus, the sizes and shapes of carbonate grains, their sorting, degree of disarticulation or breakage or surface damage, orientation, and grain-to-matrix ratios may have little relationship to hydraulics or may have much to do with hydraulics. This is an important distinction between carbonate rocks and terrigenous rocks and bears on the ways by which they are studied and interpreted.

Dolomite and Dolomitization

Dolomitization is a diagenetic process, a chemical alteration of calcite or magnesian calcite. Much dolomitization occurs within or closely associated with depositional environments and is controlled by characteristics of these environments. It seems appropriate to examine this subject here and now, rather than later, in Chap. 11, where dolomitizing processes that have nothing to do with depositional environments are treated.

The Dolomite Problem

The occurrence of the mineral dolomite in sedimentary rocks has for many years been referred to among sedimentologists as "the dolomite problem." It is widely perceived that the mineral does not precipitate directly

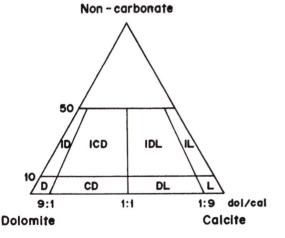

Fig. 9-8. Classification of carbonate rocks according to proportions of mineral components. IDL, for example, is the abbreviation for impure dolomitic limestone. (Leighton and Pendexter, 1962, fig. 2.)

from the usual natural surface waters under "ordinary" environmental conditions, as calcite does. Though sea water is at least twenty times supersaturated with respect to dolomite, the nucleation and growth of dolomite in ordinary sea water is very slow. Animals and plants, though many of them mediate the precipitation of calcite or aragonite, do not make dolomite. Thus, it has been generally accepted that the dolomite that we observe in certain modern depositional environments, and most of the vast volumes of dolomite contained in ancient rocks, was formed by chemical alteration and recrystallization of calcite (that is, of limestone). This process, called *dolomitization*, is carried out by magnesium-rich fluids moving through pores in sediments and sedimentary rocks.

Dolomitization seems to be a dissolution-precipitation reaction rather than a cation-exchange reaction. On a very small scale, calcite or aragonite dissolves in the presence of a dolomitizing fluid, and dolomite precipitates in its place. There is a transfer of magnesium ions from the solution to the solid phase and of calcium ions from the solid phase to the solution. Thus, as dolomitizing brines are bringing in large quantities of magnesium ions, they also must be flushing away equally large quantities of calcium ions. An initially high magnesium/calcium ratio in the fluid, which seems to be important in promoting dolomitization, no doubt declines rapidly. Moreover, much of the magnesium that the fluid brings to a dolomitizing system can never participate in the reaction (it is the magnesium still contained in a fluid that is no longer saturated with respect to dolomite). Perhaps dolomitizing fluids gradually change into calcite-cementing fluids.

Morrow and Ricketts (1988, p. 35) observed experimentally that dolomitization is promoted by high carbonate and bicarbonate concentrations. In other experiments, Baker and Kastner (1981) observed that sulfate in solution inhibits dolomitization, even sulfate concentrations less than 5% of that of normal sea water. It is not so much a high magnesium/calcium ratio that promotes dolomitization, according to Baker and Kastner, but low sulfate concentration. Morrow and Ricketts (1988, p. 31) concurred but noted that high sulfate concentration, while inhibiting dolomitization, does not necessarily inhibit direct precipitation of dolomite. In coastal sabkhas of Baja California, Mexico, studied by Pierre et al. (1984), dolomite and gypsum precipitate together in equilibrium from the pore water. One suspects that other variables, heretofore unsuspected, play a role. Certain organic substances occurring in natural environments are known to inhibit or promote dolomitization, but much more work is needed in this matter.

Origins of Early Dolomite

Dolomitization of limestone seems to require two conditions; it might be said, therefore, that the "dolomite problem" has two parts: (1) How does Nature come up with a more or less alkaline solution having a high magnesium/calcium ratio or high carbonate and bicarbonate concentrations? (2) How can substantial quantities of this fluid be made to pass through limy sediment and exchange some of its magnesium ions for calcium ions in the sediment? Various workers since about 1960 have proposed that high magnesium/calcium ratios in a pore fluid might be achieved through precipitation of calcium-rich minerals from the fluid, which thus becomes relatively enriched in magnesium; through mixing of marine water with fresh meteoric water; or through release of magnesium ions into pore water from chlorophyll or other plant-derived organic compounds. Some early-dolomitizing solutions apparently are derived from volcanic terrains. To get the pore water to move through the sediment, various workers have proposed that dense dolomitizing brines sink through a porous sediment, displacing less dense pore waters; that pore water moves upward and replaces water evaporated from pore space near the surface; or that movement of dolomitizing fluids is associated with sea-level changes, wave pumping, seasonal change in water-table elevation, or thermal convection.

The first model of syndepositional dolomitization to achieve wide popularity was formulated by Adams and Rhodes (1960). Called dolomitization by *seepage refluxion*, it seemed to explain the frequently noted association between dolomite and evaporite deposits in both ancient rocks and present-day arid tidal flats known as sabkhas. On the restricted carbonate shelf or broad lagoon in arid climates (Fig. 9-9), sea water evaporates and becomes quite hot, dense, alkaline, and saline. At first it loses calcium to organism-mediated precipitation of carbonate mud and skeletal calcite and aragonite (some of this destined shortly to undergo dolomitization). On the landward shoals and supratidal flats, this water precipitates even more calcium as evaporitic gypsum or anhydrite. In the process, a hot, hypersaline brine is formed; magnesium/calcium ratio increases to many times that of normal sea water (also

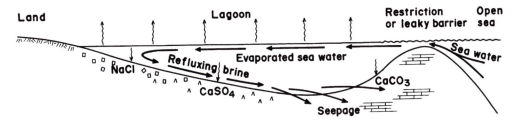

Fig. 9-9. The seepage-reflux mechanism in a lagoon. Refluxing brines seep into limy sediments and dolomitize them. (After Adams and Rhodes, 1960, fig. 3.)

sulfate concentration is reduced to low levels). This brine is considered to be an aggressively dolomitizing fluid. It sinks to the bottom of the lagoon, according to Adams and Rhodes's model, forming an underlayer that tends to spread and flow toward any depressions and to flow generally back toward the sea, this seaward flow being called refluxion. Some of the water seeps into underlying carbonate sediments, displacing any less dense pore waters that they contain, and dolomitizes these sediments. But as it performs its office, it gives up much of its magnesium, gains calcium, and eventually loses its dolomitizing powers. Normal sea water continues to flow *into* the lagoon from the open sea, riding over the dense, saline underlayer and replacing water lost to evaporation and seepage.

Patterson and Kinsman (1982) described brine formation and dolomitization on the coastal sabkha along the Persian Gulf, but the mechanism and pattern of circulation is different. Sea water is propelled onto the supratidal surface of the sabkha by strong onshore winds, where it puddles in shallow depressions. As this water evaporates, it precipitates gypsum, and the magnesium/calcium ratio increases to values as high as 100 to 1. Some of the saline water seeps into the calcareous sediment and dolomitizes it. Patterson and Kinsman noticed that dolomitization seems to take place only under reducing conditions but could not explain why; probably it is related to bacterial reduction of sulfate, which could locally reduce the inhibiting effect that sulfate has on dolomitization.

One would expect that dolomites formed by seepage reflux (or its variants) would be closely associated with gypsum or anhydrite deposits, and that they might bear a cryptalgalaminate structure, desiccation cracks and tepees, and other indicators of arid, shallow, restricted, marginal marine environments. Beales and Hardy (1980) suggested that "occult" gypsum or anhydrite,

being tiny crystal inclusions in dolomite crystals, might indicate that the dolomite is a product of a hypersaline environment, even if other evidence of such an environment is lacking. Such inclusions are so small that they are easily overlooked in thin sections, but they may be detected in insoluble residues. Length-slow chalcedony in carbonate rocks also seems to be an indicator of the former presence of evaporite (see Chap. 11).

Hsu and Siegenthaler (1969) doubted that seepage reflux is adequate to explain dolomitization of large masses of sediment of low permeability and offered, as an alternative, a mechanism that they called *evaporative pumping* (anticipated by Deffeyes et al., 1965). Here, evaporation of pore waters in sediment at or near the surface of an emergent carbonate flat provides a hydraulic head that pulls phreatic sea water upward into the pores of the desiccated surface layer. Continued flow toward the surface enriches the dissolved ion content of the pore fluid in the surficial sediment, and the concentrated fluid is supposedly responsible for dolomitization. This might account for the dolomite crusts of certain Recent arid supratidal flats, but it is not clear how the process could achieve the dolomitization of large sedimentary volumes that refluxion presumably could not. Hsu (1967) had already discounted the need for a high magnesium/calcium activity ratio, which the evaporative-pumping mechanism does not seem to provide. It is noteworthy that the path of pore-water movement is opposite to that of seepage refluxion, but, again, dolomites formed by this process ought to be associated with evaporite deposits and sedimentary structures indicative of slightly emergent, arid coastal environments.

Runnels (1969) examined the geochemical consequences of the *mixing* of various natural waters; there are implications in all of chemical diagenesis. Two waters that are saturated or undersaturated with respect to some mineral substance may, if mixed together, yield a solution that is supersaturated in that mineral; similarly, two waters that are saturated may mix to form an undersaturated solution. Such conditions result when the solubility of a substance is a non-linear function of the chemical and physical parameters that are influencing the system. For example, solubility of calcite as a function of partial pressure of carbon dioxide is not linear (Fig. 9-10a). Suppose that water of composition A mixes with water of composition B. Since points A and B lie on the curve, they both represent saturated solutions. But mixtures of the two solutions lie on the *straight line* between A and B, such as the 1:1 mix represented by C. Point C lies below the saturation curve, in the region of undersaturated solutions. Thus, a solution of composition C would tend to dissolve more calcite. Plummer (1975, p. 224) presented saturation curves for calcite in seawater-freshwater mixtures at various temperatures and carbon dioxide partial pressures.

The solubility of calcite also varies non-linearly with the concentration of other ions in solution, such as chloride or sulfate. Two waters of different total salinity, but both saturated with respect to calcite, become undersaturated when mixed together (Fig. 9-10b). Similarly, two waters saturated with respect to gypsum or anhydrite, but with different concentrations of NaCl, yield mixtures that are undersaturated with respect to the calcium sulfate minerals, and hence tend to dissolve these minerals (Fig. 9-10c). Some solubility curves are concave, and mixing creates supersaturation. Concave curves are ordinarily a consequence of the common ion effect.

A natural fluid of relatively low-salinity and high magnesium/calcium ratio can result from the mixing of normal sea water with fresh water. In marine coastal regions, sea water saturates the pores of rocks or rock materials. But on the land, meteoric water seeps down through the rock materials, displacing and flushing away the saline pore waters. By virtue of its lower density, the fresh pore water floats on the sea water, and there is a boundary between the two waters, this boundary surfacing at or near the coastline and sloping landward (Fig. 9-11). The equilibrium profile of the interface is such that its local depth (below sea level) is roughly forty times local elevation of the (fresh) water table above sea level, though this relationship may be modified by various chemical or physical fac-

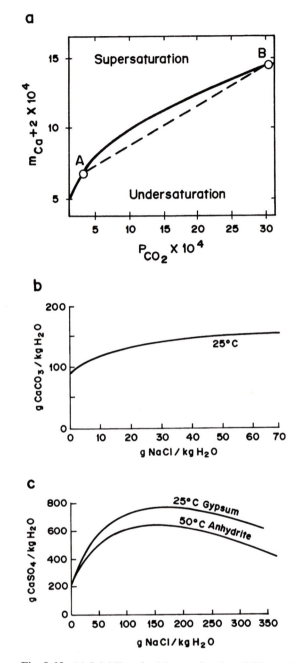

Fig. 9-10. (a) Solubility of calcite as a function of CO_2 partial pressure, (b) solubility of calcite as a function of salinity, (c) solubility of gypsum and anhydrite as a function of NaCl concentration. (After Runnels, 1969, figs. 1, 2, 3, with permission of SEPM.)

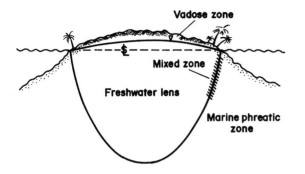

Fig. 9-11. Mixing-zone dolomitizing environment.

tors. There is a certain amount of mixing at the interface, due in part to diffusion and in part to vertical fluctuation of the interface; the interface moves as a result of seasonal change in rainfall (which changes the thickness of the freshwater saturated zone) or to changes in sea level. As this mixing zone moves across a body of carbonate sediment, the mixed water supposedly dolomitizes the sediment.

Hanshaw et al. (1971, p. 721) and Badiozamani (1973) suggested that mixing of sea water and meteoric water creates solutions that are undersaturated with respect to calcite and oversaturated with respect to dolomite—that is, fluids which might cause dolomitization. Badiozamani's so-called *dorag* (a Farsi word meaning ''mixed blood''), or mixing zone mechanism, is an early diagenetic process that requires emergent (subaerially exposed) terrains and establishment of a phreatic lens or pillow of fresh (meteoric) water. Instead of arid climate, a humid condition seems more conducive to dolomitization.

Choquette and Steinen (1980) studied the locally dolomitized Ste. Genevieve Limestone along the LaSalle anticline in southeastern Illinois; they accepted a mixed-water dolomitization mechanism, mainly on the basis of facies relationships and lack of any evidence of evaporites. Various textural and chemical evidence suggested that dolomitization was early, and that it was precipitated from a diluted sea water. Ward and Halley (1985) likewise presented petrographic and isotopic evidence for mixing-zone dolomitization in late Pleistocene reef limestones on the Yucatan peninsula.

Purser and Aissaoui (1985) and Aissaoui, Buigues, et al. (1986) found dolomite in the Mururoa Atoll of French Polynesia, forming a lenticular mass up to 100 meters thick beneath the island reefs (Aissaoui et al., p. 36), thinning toward the central lagoon and toward the peripheral island flanks. There are very fine-crystalline (5 to 50 micrometers) chalky dolomites formed by replacement of limestone; coarser, pore-lining dolomite cement; and even coarser dolomite (0.1 to 0.25 millimeters) apparently formed by recrystallization of earlier, finer dolomite. Studies by SEM revealed that dolomite began as a *cement* within the microscopic pores of a fine-grained limestone substrate, the initial very small cement rhombs then growing, engulfing and digesting surrounding calcite, rather like amoebas engulfing food particles. Dolomitization of the coarser limestones proceeded selectively, attacking biogenic aragonite most vigorously; the aragonite first dissolves, then dolomite spar cement precipitates immediately in the voids. Clearly, this dissolution of calcite or aragonite and contemporaneous precipitation of dolomite is accomplished by a fluid that is undersaturated in calcite (and aragonite) and saturated with dolomite.

Dolomitization in the Mururoa reef, according to Aissaoui et al., is related to mixing of marine and meteoric waters. A hydrodynamic system driven by oceanic currents and by thermal convection within the atoll mass caused sea water to move inward toward the center of the reef and upward toward the floor of the lagoon. Sea water was the source of the magnesium; this is supported by isotope studies. That meteoric waters participated in the dolomitizing process is suggested by the intimate association of dolomite with dissolution of aragonite and calcite skeletons (see Chap. 11). That dolomite does not persist into the oceanic flanks of the atoll implies that undiluted sea water was not able to bring about dolomitization.

von der Borch and his associates (1975, 1976, 1979) described modern dolomite forming in certain ephemeral alkaline lakes in the Coorong region of South Australia. Apparently there is an early disordered phase that gradually becomes true dolomite. It forms out of evaporitically modified meteoric ground water that has picked up magnesium from nearby basic volcanic terrains, and that flows through aragonite and high-magnesium calcite precursor muds, dolomitizing them. Although the dolomite-forming site is near to the sea, and there is a complex and diffuse boundary between the continental and oceanic waters (von der Borch, 1976, p. 959), the mixing zone seems to have nothing to do with the creation of this lacustrine dolomite.

There is some evidence that deep ocean waters have intrinsic dolomitizing properties. Saller (1984) observed that dolomite is forming near the base of the Eniwetok atoll in the Pacific at depths of more than 900 meters below sea level. Apparently, ordinary sea water is the dolomitizing fluid. Sea water at such depths is calcite-undersaturated but greatly supersaturated with respect to dolomite. Saller suggested that thermal convection drives the flow—cold sea water that enters the base of the carbonate mass is gently heated by the underlying volcanic basement. Baker and Burns (1985) also described dolomite formation in deep water, at ocean depths of 1000 to more than 3000 meters, in organics-rich carbonate sediments. They explained that pore waters always have reduced salinities relative to sea water, and magnesium/calcium ratios near 5 to 1. Dolomitization occurs under these conditions, but only in the upper few meters of sediment, and especially in the zone of microbial sulphate reduction.

Gebelein and Hoffman (1973) proposed an algal origin for certain dolomites, namely the dolomite laminae that alternate with limestone laminae in many cryptalgalaminites. Recent algal mat deposits in Florida, Bermuda, the Bahamas, Western Australia, and the Persian Gulf are composed of thin layers of particulate carbonate sediment alternating with layers of algal filaments. Gebelein and Hoffman observed that partially decomposed algal layers are studded with large numbers of minute rhombs of high-magnesium calcite attached to the gelatinous sheaths that envelop algal filaments. Experimental algal cultures were observed to concentrate magnesium in the sheath material. The sheath material is quite stable, but when it does eventually decompose, it liberates magnesium, which is then available for direct precipitation of dolomite within the decomposing organic layer, or for early dolomitization of a carbonate lamina in direct contact with the organic layer.

It is paradoxical that one class of dolomitization models demands a hypersaline fluid and another a hyposaline fluid; each accomplishes what ordinary sea water, in the middle, does not readily accomplish. It is paradoxical also that a mineral that ostensibly is so reluctant to form appears to form under a broad range of chemical and physical conditions, including conditions that we have not yet looked into (see Chap. 11). In spite of the numerous studies of the process and its sedimentological results, dolomite is still a problem.

Texture of Early Dolomite

Early dolomitization typically begins in the carbonate muds or in the mud matrix of grainy carbonate sediments. Probably this is because mud is initially quite porous, commonly greater than 70 percent (Enos and Sawatsky, 1981, p. 917), and has large internal surface area. In limestones only lightly dolomitized, dolomite occurs as scattered, small (typically <50 micrometers long diagonal) rhombohedra in mud matrix. At higher levels of dolomitization, rhombs are more numerous; presumably those that nucleated first are larger than the younger rhombs. Possibly the larger rhombs form at the margins of a plume of dolomitizing fluid, then smaller rhombs add to the population as the plume expands. As dolomitization continues, rhombs become ever more numerous, and crowded, and neighbors begin to interfere as they grow, and to grow together. Peloids, though composed of mud, nonetheless seem to resist dolomitization even as dolomite rhombs are forming in an enclosing matrix. No doubt, peloids are more compact than mud, and they probably are permeated or coated with organic slimes that inhibit entry of or close contact with early dolomitizing fluids. Lucia (1962, p. 856) observed that, in Devonian crinoidal carbonate rocks in Texas, matrix was dolomitized first, then crinoid plates. Ward and Halley (1985, p. 408) observed similarly that matrix is preferentially replaced, leaving skeletals generally undolomitized, in Recent carbonate sediments in Yucatan, Mexico.

In ancient carbonate rocks it is clear that certain skeletal grains are more resistant to dolomitization than others; the petrographer may find that brachiopods or rugose corals are dolomitized, but that other skeletals are not. Mattes and Mountjoy (1980, p. 275) noted that stromatoporoids in Devonian dolomites in Alberta, Canada, were strongly resistant to dolomitization. Sibley (1982, p. 1087) indicated that high-magnesium calcite and aragonite are more susceptible to dolomitization than low-magnesium calcite.

Dolomites commonly have higher porosities than associated limestones. If dolomitization occurs early, in a porous lime mud, then the dolomite will inherit much of the pre-dolomite porosity and inhibit subsequent porosity-destructive compaction. If dolomitization is accompanied by dissolution of calcite or aragonite, then escape of dissolved calcium carbonate from the system might cause development of high porosity.

Shallow Marine Carbonate Environments

Animal, vegetable, and mineral—partners in the crime. Perhaps the most complex of depositional systems, the carbonate shelf is a product of life and of death. Reefs are Nature's cities, where populations are large, dense, and diverse, and engaged in manufacturing, construction, and maintenance, and disposed toward self-perpetuation.

Carbonate production occurs most vigorously, and with greatest sedimentological consequence, in warm, shallow, clear waters of tropical and subtropical seas. Most modern reefs and mounds are forming in tropical or subtropical marine waters, on the western sides of oceans, where they are bathed by the warm, strong *western boundary currents*. In such environments carbonate production is intrinsically rapid enough that it can keep up with almost any credible tectonic subsidence or sea-level rise and can therefore maintain and perpetuate the shallow marine conditions that are conducive to continued carbonate production. On the other hand, where conditions are less than ideal, subsidence or sea-level rise might shut down a carbonate factory. Change in climate, or restriction of circulation, causing anoxia or hypersalinity, can bring carbonate production to an abrupt end. High nutrient contents generally do not favor reef growth; where nutrients are abundant, non-reef-building organisms become dominant and compete with the reef-builders (Hallock and Schlager, 1986, p. 390). Contamination of the environment by terrigenous muds or abundant fresh waters from large rivers also is detrimental to reef growth. Wilson (1974, p. 819) pointed out that carbonate buildups generally are buried not by normal marine or deep-water facies indicative of transgression but by evaporites or shales, indicating emergence or terrigenous influx.

Rapid shallow-marine carbonate production commonly gives rise to a deposit of major proportions called a ***carbonate platform***. The carbonate platform manifests the "all-or-nothing" sensitivity to water depth of carbonate-producing mechanisms. The Bahama Banks is an example (Fig. 9-12); its flat, shallow surface is a place of vigorous carbonate production, both organic and inorganic, and its margin is a steep declivity into adjacent deep water. Shoal waters produce nearly all the carbonate and maintain the flat-topped platform. At least 5000 meters of carbonate rock underlie this surface, indicating that the platform has maintained itself as a shallow-marine carbonate factory in spite of long-term subsidence. In the deeper waters surrounding the platform, hardly any carbonate is produced at all. The regional subsidence has caused the deeps, which receive only whatever carbonate happens to tumble off the edge of the platform.

Where subsidence or sea-level change is not happening, carbonate bodies build laterally. Many carbonate deposits of the geological past formed broad shelves that prograded seaward from the shallow margins of land masses into deep water.

Read (1985) distinguished among carbonate ramps, rimmed shelves, isolated platforms and oceanic atolls, and drowned platforms (Fig. 9-13), each the product of a particular rate of subsidence or sea-level rise, wave-and-current regime, and rate of carbonate production. The ***carbonate ramp*** is a very gently sloping surface on which shallow, wave-agitated facies pass gradually downslope into deeper-water facies; reefs or other rimlike bodies are generally lacking, though there may be some locally developed grainstone banks or muddy bioherms defining the seaward margin of a tidal flat or lagoon, or isolated carbonate buildups well downslope. The ***rimmed carbonate shelf*** is a shallow platform with well-defined margin and steep foreslope into surrounding deep water. The shelf margin, agitated by waves, is marked by more or less continuous reef (barrier reef) or shoals and islands of skeletal or ooid sands. The steep slopes seaward of the rim contain reef breccias or carbonate sands that have spilled off the edge of the shelf. ***Isolated platforms and atolls*** are like rimmed shelves, but the rim forms a circle that faces the open sea in all directions. Rim facies and slope facies of leeward margins differ from those of windward margins; carbonate sands originating to windward are swept across the platform to the leeward margin, where they cascade down the leeward slope. On atolls, windward reefs are broader than leeward reefs and have different biota; windward foreslopes are steeper than leeward slopes.

A ***drowned platform*** is a carbonate platform that has failed to maintain upward growth during subsidence or sea-level rise. On drowned platforms, shallow platform facies are overlain by nodular and thin-bedded whole-fossils wackestones; there may be numerous hardgrounds in the succession. Isolated buildups (patch reefs) may develop on a drowning ramp or platform. Drowning ramps develop transgressive successions.

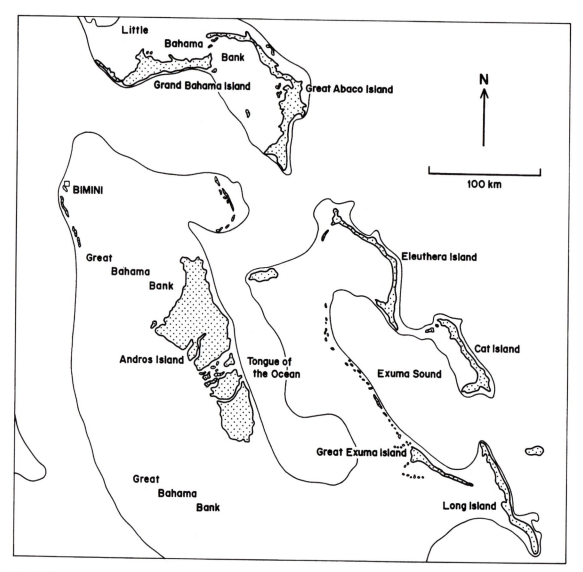

Fig. 9-12. Physiography of part of the Bahama bank; thin contour line at a depth of 180 meters. (After Sellwood, 1986, fig. 10.8.)

Marine Carbonate Facies

In terrigenous sediments, depositional environment is reflected in texture, structure, and architecture of the deposit; this is also true of carbonate sediments, but here the grains themselves also reflect the depositional environment, because they are, by and large, created in the same environment in which they come to be deposited. Carbonate lithofacies commonly corre- spond closely with biofacies. Thus, petrography plays a greater role in analysis of ancient carbonate depo- sitional environments than in analysis of terrigenous environments, where the larger-scale stratigraphic ob- servations are the more useful and diagnostic.

Based on his long and varied experience in ancient carbonate deposits, Wilson (1975, pp. 25, 350 ff.) de- fined nine standard carbonate facies belts, distinguish-

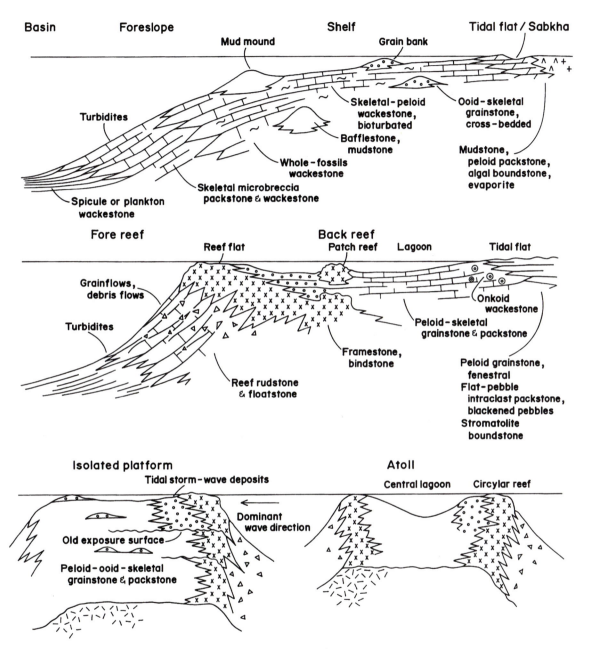

Basin   **Foreslope**   **Shelf**   **Tidal flat / Sabkha**

Mud mound   Grain bank

Turbidites

Skeletal-peloid
wackestone,
bioturbated

Ooid-skeletal
grainstone,
cross-bedded

Bafflestone,
mudstone

Whole-fossils
wackestone

Mudstone,
peloid packstone,
algal boundstone,
evaporite

Skeletal microbreccia
packstone & wackestone

Spicule or plankton
wackestone

Fore reef   **Back reef**

Reef flat   Patch reef   Lagoon   **Tidal flat**

Grainflows,
debris flows

Turbidites

Onkoid
wackestone

Peloid-skeletal
grainstone & packstone

Framestone,
bindstone

Reef rudstone
& floatstone

Peloid grainstone,
fenestral
Flat-pebble
intraclast packstone,
blackened pebbles
Stromatolite
boundstone

Isolated platform   **Atoll**

Tidal storm-wave deposits   Central lagoon   Circylar reef

Dominant
wave direction

Old exposure surface

Peloid-ooid-skeletal
grainstone & packstone

Fig. 9-13. Profile of carbonate ramp, rimmed shelf, isolated platform, and atoll.

251

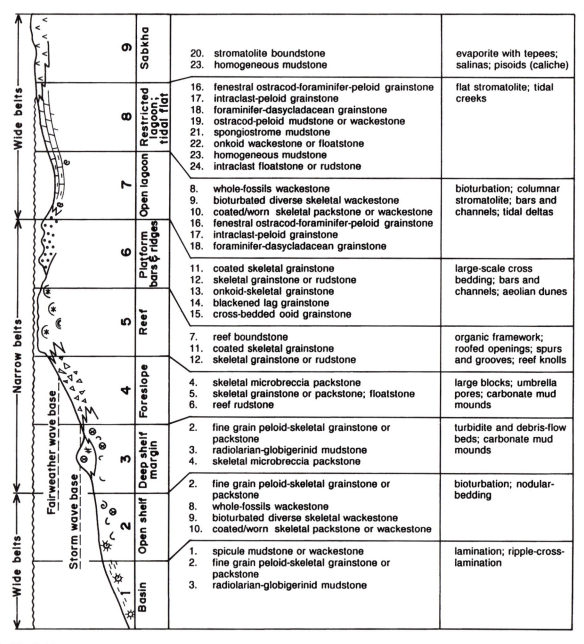

Fig. 9-14. Standard carbonate facies belts. (According to Wilson, 1975.)

able on lithological and biological grounds (Fig. 9-14). They are, from farthest offshore to farthest onshore, as follows:

1. *Basin facies*, where water is too dark and cold for abundant benthonic production of carbonate. Fine-grained suspended carbonate, or siliceous or organic matter, produced elsewhere, settles here, forming dark, laminated carbonate and clay mudstones.

2. *Open marine shelf facies*, comprising sediments deposited in circulating, oxygenated waters of normal salinity but below fair-weather wave base. Sediments are whole-fossils wackestones with diverse fauna, bioturbated, wavy- or nodular-bedded. Beds typically are tabular or broadly lenticular, traceable for many meters; thicker beds and thinner beds are "shuffled" together; bed surfaces commonly are knobby, with calcareous shale partings or interbedded thicker shale units.

3. *Basin-margin (base-of-slope) facies*, similar to the shelf facies, but formed at the margins of a shallow-water carbonate platform. Thick-bedded, dark- or light-colored carbonate mudstones, commonly cherty, containing skeletal material and exotic blocks derived from upslope.

4. *Platform foreslope facies*, deposited on the steep slopes that flank a carbonate platform. Sediments are skeletal wackestones and packstones, perhaps with exotic blocks; there may be some slumping or sediment-flow structures; there may be some muddy bioherms.

5. *Platform-margin organic reef or mound* in shallow, wave-agitated waters. Boundstone masses with pockets of grainstone and packstone are formed here. There is not much bedding, but exposure surfaces can develop as a result of sea-level change or halting subsidence.

6. *Winnowed platform-edge facies*, comprising carbonate sands of current-scoured channels, wave-swept bars and beaches, and aeolian dunes. Light-colored skeletal and ooid grainstones typically are cross-bedded, diverse skeletal grains broken and abraded.

7. *Open platform facies* of shallow, moderately circulating waters of normal or near-normal salinity. Grainstones, packstones, wackestones, and mudstones reflect environmental diversity and variability; typically they are bioturbated or cross-bedded. A diverse biota is characteristic of open shelf environments.

8. *Restricted platform facies* of various shallow, commonly hypersaline, perhaps oxygen-poor waters of lagoons and tidal flats. Sediments here are dolomitized peloid packstones, stromatolite boundstones, skeletal wackestones, and mudstones. Some sediments are cross-bedded, some are bioturbated. There may be fenestral fabrics. Faunas typically are limited, depauperate, comprising certain foraminifers, ostracods and small gastropods; some individuals may be stunted (dwarfed) or deformed, reflecting stressful environ-

ment. Restriction may be due to a barrier or to great width of a shallow shelf.

9. *Platform evaporite facies* of arid supratidal zones. Sediments mostly are laminated dolomite and nodular anhydrite or gypsum. Hypersalinity or anoxia excludes a burrowing fauna; flat lamination without burrows is characteristic of sediments of such environments. Spring tides or wind tides can bring sea water onto the dry flats, making little ponds that precipitate halite or anhydrite or the finest carbonate mud. There may be some windblown terrigenous material.

Wilson further subdivided carbonate facies into 24 *standard microfacies* (SMF), each facies belt comprising some combination of three or four (more or less) of these SMFs. In addition, he listed 79 kinds of sedimentary structures that can occur in carbonate successions.

Carbonate Reefs and Mounds

A *reef* is a hard mass of calcium carbonate erected by organisms in place. Reefs consist of large, contiguous skeletons, built collectively by large numbers of colonial sessile organisms, and of smaller skeletons and other carbonate particles bound together by organisms and commonly by early marine carbonate cements. Much of the fine material of a reef is the product of associated boring, rasping, and grazing organisms, which erode and degrade the coarser skeletal materials of reefs. Skeletons of the living reef surface continually build upon the old, dead skeletal materials of earlier generations, and reefs grow upward and outward. Because reefs are products of shallow marine environments, they are affected by the currents generated by waves and tides and the turbulence of the surf. This controls their directions and rates of growth, and thus their sizes and shapes. Heavy surf locally breaks down the hard reef structure, creating fragments large and small that are dispersed by the currents over the shallow surface of the reef and into surrounding deeper waters. Thus, reefs are a major source of carbonate detritus in both shallow- and deep-water sediments.

Most reefs are asymmetrical, reflecting a hydrodynamic gradient, one side being upcurrent or facing heavy surf, the other side downcurrent or facing quiet lagoon waters. Some reefs are elongate, forming a barrier or fringe parallel to a mainland, or a full circle that fortifies an island or lagoon pool; other reefs are isolated knobs, pinnacles, or patches. Many reef complexes divide naturally into several physiographic zones. There is a *reef crest*, typically composed of bindstone or framestone that sustains the brunt of the wave action. This gives way to a current-swept *reef flat* behind, where the solid core of the reef is paved with grainstones made of small or disarticulated skeletals. The reef flat slopes gently into the *back reef*, where bioturbated lagoonal mudstones and peloid packstones accumulate. Patch reefs may dot the lagoon. Seaward of the reef crest is the *reef front*, which slopes, commonly quite steeply, into deep, open marine waters; it is a place where numerous framebuilders thrive, and where bafflestones and framestones are formed. Downslope, reef talus accumulates on the *fore reef* as rudstones and grainstones. Many elongate reefs are dissected by *grooves* or surge channels that divide the reef crest and reef front into a series of *spurs*, and that transport wave surge and tide waters between the back reef and fore reef and carry reef talus seaward onto the fore reef slope.

A solid reef "core" ordinarily is only a small part of any mature carbonate reef complex; it forms the reef crest. Many reef cores are actually composed of loose or soft skeletal grains that the shallow, agitated marine waters have quickly cemented. Many reef frameworks contain large pockets into which mud or skeletal sand has fallen; these internal sediments also are apt to be quickly lithified. Active reef cores are continually degraded by storm waves and bioerosion; this creates much of the reef talus that accumulates seaward on the fore reef. As a reef progrades, old reef core comes to be mantled with grainstones of the reef flat, then buried beneath the wackestones and mudstones of the back reef.

There is a striking zonation of organisms on reefs, reflecting reef-transverse contrasts in water depth and turbulence. A low-diversity community of robust wave-resistant corals and encrusting red algae flourish in the surf zone. Certain corals of modern reefs, such as *Acropora palmata* and *A. cervicornis*, make branches that reach toward the incoming waves. Coral atolls of the Pacific are wider on the windward segment of the ring, where wave action is more vigorous, and the distribution of organisms varies between windward and leeward reef. Diversity is much greater on the reef flat, where the environment is less stressful; corals with bulbous or cabbage-head form, benthonic foraminifers, molluscs, delicate green algae (codiaceans), and various burrowers, grazers, scavengers, and predators reside there. Lagoons may be populated with ostracods and certain molluscs, annelids, and green algae (dasycladaceans). Down the slope of the

reef front turbulence gradually declines; so does light level. Living among the rubble of dead corals of the fore-reef slope are a diverse community of pioneers—flexible ahermatypic corals, crinoids, bryozoans, basket-shaped sponges, and red algae that can tolerate low light levels. The reef talus eventually serves as a foundation on which the main reef builders erect their edifices as the reef grows and expands seaward.

In the Precambrian there are stromatolite reefs of major dimensions. A spectacular example occurs in the early Proterozoic Pethei Group in the vicinity of Great Slave Lake in the Northwest Territories of Canada (Hoffman, 1974). Platform facies consist of laminar and columnar stromatolites and ripple-laminated ooid grainstones. Stromatolite columns typically are strongly elongate in plan, parallel to tidal currents that swept the shelf. Their morphologies are in other ways wonderfully diverse, apparently reflecting local water depth and gradual temporal changes in water depth.

Another kind of carbonate buildup, more properly called a *mound*, is composed of fine-grained calcium carbonate (mostly mud) that has been chemically precipitated by certain organisms or entrapped and bound together by them. Any reef-like or mound-like buildup of biologically produced carbonate is also called a *bioherm*. A laterally extensive, relatively thin body built by organisms in place is called a *biostrome*.

Thrombolites are examples of (soft) carbonate mounds; they are composed mainly of clotted mud (thromboids). The loaf-shaped bodies with dimensions of meters or tens of meters that Aitken (1967) studied and described were of early Paleozoic age; rather similar examples in Mississippian rocks (Charles Formation, subsurface in Williston basin, eastern Montana) contain a very conservative biota of calcispheres and gracile ostracods, even ostracod "nests" or "nurseries" in the fenestrae between thromboids (Fig. 9-15). Pratt and James (1982) described Ordovician thrombolite mounds in Newfoundland, being elongate banks up to 50 meters in length, comprising closely spaced domal bodies composed of thromboids and a few corals and sponges. Kah and Grotzinger (1992) reported on early Proterozoic examples in Canada's Northwest Territories.

Waulsortian mounds are a peculiar and not well-understood kind of carbonate mud mound that appears to have been, in the main, a Paleozoic phenomenon. Examples have been carefully described not only around the village of Waulsort in Belgium (LeCompte, 1937 and 1970; Sandberg et al., 1992), but also in

Fig. 9-15. Ostracod nursery in fenestra of a thrombolite.

England (Parkinson, 1957), Ireland (Lees, 1961 and 1964; Schwarzacher, 1961), Canada (Bourque and Gignac, 1983), United States (Pray, 1958; Cotter, 1965; Ross et al., 1975; MacQuown and Perkins, 1982; King, 1986), and Australia (Wallace, 1987). The mounds are steep-sided (even as steep as 50°) "haystacks" or cones rising as much as 100 meters above their substrate. They are composed of lime mud with pelmatozoan plates and fenestellid bryozoans, many of which, though delicate, are quite intact and undamaged; other fossils are not abundant. Also, there are the numerous enigmatic calcite-cemented lenticular openings known as stromatactis (see below). The mounds prospered, it appears, on the deeper parts of carbonate ramps or beyond the margins of carbonate platforms. They are thought to have formed where thickets of pelmatozoans and fenestellid bryozoans baffled the gentle currents wafting by and filtered the suspended mud, new generations of the creatures continually repopulating the surface of the growing mud mound. Schwarzacher (1961) presented evidence that the mud knolls had to compete continuously with the currents, which eroded and undermined their flanks, and shaped them into elongate mounds. Parkinson (1957, p. 520) found limestone breccias on the flanks of some of the mounds that he examined, indicating that the mounds lithified early.

Bathurst (1982) examined the stromatactis problem in detail, referring to examples in the Ordovician of Sweden, Devonian of Western Australia and Belgium, and the Carboniferous of England. In all occurrences, according to Bathurst, *stromatactis* is a system of cav-

ities in a marine carbonate mud mound, the cavities receiving, under marine phreatic conditions, internal sediments and carbonate cements. The lenticular openings are approximately parallel to successive contemporaneous surfaces of the carbonate buildup. The host sediment may be homogeneous mud, or a pelleted or clotted mud, or mud with various skeletals (fossils). The smooth or undulating base of a stromatactoid cavity often can be shown to be the surface of an internal sediment, in some places interlaminated with cement. The digitate roof may locally show skeletal support, as though the cavity were a shelter pore, but such support is incidental, not essential. The cavities are cemented with inclusion-rich radiaxial calcite (see Chap. 11) and lesser amounts of calcite blocky spar. Bathurst proposed that lightly cemented crusts form episodically on the mound surface, modulated by varying sedimentation rate. Sea water circulating in the pores erodes uncemented muds from under the crusts, then emplaces carbonate cement in the spaces. Crusts commonly are ruptured and mutually displaced a bit. Cavity floors are the surfaces of old crusts (commonly covered over with internal sediment); cavity roofs are the uneven undersides of crusts. Bathurst suggested that certain "zebra rocks" are a less deformed, more regular kind of stromatactis. Ross et al. (1975) described an example of zebra rock in the base of an Ordovician calcareous mud mound on Meiklejohn Peak in southern Nevada. The quite regular zebra stripes, being an alternation of lime mudstone, pelleted internal sediment, and radiaxial calcite, grade upward to the less regular, typical stromatactis in the mound core. Stromatactis has also been reported in thrombolite mudmounds (see Pratt, 1982a).

Comparable to Waulsortian mounds in some ways are the muddy, sandy carbonate mounds in 600 or 700 meters of water in the Florida Strait that Neumann et al. (1977) described. These modern mounds, steep-sided and elongate parallel to a gentle bottom current, are composed of partly cemented packstone and wackestone of crinoids and certain "soft corals," such as alcyonarians and sea pens.

Modern deep-water carbonate mud mounds have been described also by Roberts et al. (1990), associated with gas seeps in the northern Gulf of Mexico. Some exceed 20 meters in height. Carbonate mud is precipitated possibly by bacterial mats that have been observed near some seeps, then cemented by calcite or dolomite. The hard, cemented surface of the mound commonly is colonized by sponges, crinoids, coralline algae, and other organisms or, on mounds in water depths exceeding 300 meters, by certain chemosynthetic mussels and tube worms sustaining themselves by oxidizing methane that is seeping out through the cracks. Ancient examples have been recently discovered and described by Beauchamp and Savard (1992), Campbell (1992), Gaillard et al. (1992), and von Bitter et al. (1992), among others.

Carbonate-Platform and Ramp Depositional Architecture

On the Bahama Banks (Fig. 9-12), well-washed skeletal and ooid sands on the margins are swept by waves and tidal currents. Some of the bank-margin sands, such as those at the south end of Tongue of the Ocean on the Great Bahama Bank and at the north end of Exuma Sound, are tidal bars arrayed perpendicular to the margin. Individual tidal bars range up to 20 kilometers in length and 8 or 9 meters in height. The crests are decorated with large bedforms oblique to the bar axis, and bar flanks typically are rippled. Other marginal bars are parallel to the bank margin. Like barrier islands on other coasts, these bars commonly are dissected by tidal channels with ebb- and flood-tidal deltas and storm spillover lobes. Large and small bedforms adorn the surfaces of these deposits. Sand of windward bank margins is swept over the platform by tide- and wave-generated currents and carried ultimately to the leeward edge, augmented by skeletals and peloids created on the platform interior itself. Sediments of platform interiors are peloid sands and muds. Locally, grains and mud are stabilized into small mounds by marine vegetation, such as *Thalassia* grass.

Ancient thick carbonate successions typically reflect progradation of carbonate deposits from basin margin toward basin center. But basal or proximal successions may contain transgressive facies; depending on climate, proximal successions may contain evaporite facies or karst features. Transgression over gently sloping terrain provides a broad zone of shoal waters in which carbonate production can become firmly established. Basal transgressive carbonate rocks grade basinward and upward into progradational ramp facies, which, in turn, grade upward to shelf facies. Shelf facies may end abruptly to seaward in a marginal reef or rim facies, beyond which the depositional surface descends steeply to deep-basinal facies.

The Mississippian System (Madison Formation or Group and its equivalents) in the western United States is a vast, integrated carbonate complex comprising a transgressive, deepening-upward, lower facies and upper, shoaling-upward, regressive facies deposited on a shelf with sharply defined margin and steep, west-facing slope to a rapidly subsiding deep starved basin. The basal transgressive phase of eastern sections (central Montana) is a discontinuous unit, up to 10 or 20 meters thick, of medium- to thick-bedded, laminated or cross-laminated or bioturbated, glauconitic grainstone, packstone, and wackestone. Greater water depth is indicated for the overlying unit, which consists of 30 to 70 meters of thin bedded, dark, argillaceous lime mudstone or wackestone with partings of more argillaceous carbonate or replacive chert. Many of the limestone beds contain unbroken fenestellid bryozoan fronds and small, thin-shell brachiopods; some beds are bioturbated. Large-scale intraformational truncation surfaces were probably created by submarine mass movements. Locally there are Waulsortian-type lime mud mounds. This unit represents transition from the transgressive phase to a much more consequential regressive phase, which lies above. The regressive succession is the eastern proximal part of an extensive, westward-thickening and westward-prograding foreslope deposit.

The foreslope facies shoals upward. Gradationally overlying and interfingering landward with this facies are diverse limestones of shoal and intershoal environments deposited near wave base. They are medium- to thick-bedded, cross-laminated skeletal-peloid packstone and grainstone; in addition to pelmatozoan plates and fenestellid and ramose bryozoans, there are rugose corals, and the colonial coral *Syringopora* in growth orientation. These beds, in turn, grade upward and farther landward (eastward) into laminated dolomites and evaporites of arid-climate intertidal and supratidal environments; locally these rocks are altered to vadose pisolite.

Far to the west is the shelf margin, marked by a narrow, curvilinear ridge in the shelf facies (Rose, 1976, p. 548), with an abrupt westward thinning trend just beyond, accompanied by profound change of foreslope facies to deep slope and basinal facies.

In the (Madison-equivalent) Deseret Limetone in Utah, Gutschick and Sandberg (1983) discriminated between shelf, slope, and basinal rocks on the basis of lithotype and bed thickness, color and organic-carbon content, coral, algal, and foraminiferal biofacies, and ichnofacies (Fig. 9-16). Benthonic green and red algae

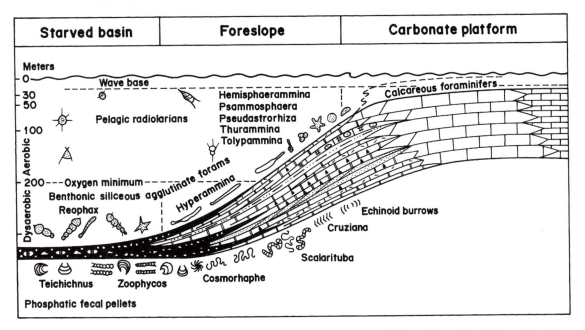

Fig. 9-16. Depositional environments and facies of the Deseret Limestone in Utah. (According to Gutschick and Sandberg, 1983, fig. 7.)

occur on the carbonate platform, but only the red algae persist into deeper waters of the foreslope. Corals of the foreslope are small, solitary forms without dissepiments. Certain conodonts seem to be characteristic of the shelf margin. Planktonic foraminifers had not yet evolved, so calcareous forms generally are not present on the slope; but arenaceous (agglutinated) foraminifers existed, and they were adapted to slope and basin floor, including dysaerobic habitats. Trace fossils on the foreslope include *Zoophycos*, *Cosmorhaphe* or *Helminthoida*, and *Scalarituba*. Basinal facies include spicular cherts and phosphatic shales. Gutschick and Sandberg noted that, as a rule, platform rocks, with <1 percent total organic carbon (TOC), are light gray, that rocks of the foreslope (TOC 1 to 3 percent) above the oxygen minimum layer are olive gray to black, and that basinal rocks below the oxygen minimum layer are brownish gray to brownish black.

The remarkable Middle Ordovician carbonate succession of the Appalachians, eastern United States, was deposited on the interior margin and slope of a long northeast-trending foreland basin of the Taconic highland (Fig. 9-17). This is an example of a carbonate ramp rather than a rimmed shelf (Read, 1980), there

being no abrupt shelf margin; facies at the landward margin indicate humid climte. Closest to the craton are peritidal lime mudstones and fenestral peloid packstones deposited during marine transgression of a regional karst surface; hardly any of these rocks are dolomitized. They are overlain by shallow subtidal cherty onkoid-skeletal wackestones, packstones, and grainstones. Shallow ramp biostromes overlie, consisting of pelmatozoan-and-bryozoan grainstones and wackestones. Farther downslope are buildups comprising lime mudstone, wackestone, and algal boundstone, flanked by draping beds of pelmatozoan-bryozoan grainstone and packstone. The lower slope deposits are black, shaly limestones and nodular limestones, some with slump breccias and turbidite bedding. At basin axis, the lower-slope limestones interfinger with siliceous turbidites and terrigenous sandstones and conglomerates poured into the trough from the other side, derived from the Taconic orogen.

In Devonian reefs of the Canning basin, Western Australia, long forereef slopes of 30 to 35° (Playford and Lowry, 1966, pp. 32, 63) suggest basin depths of about 200 meters. Forereef floatstones contain talus blocks as much as 100 meters across. In the Jurassic

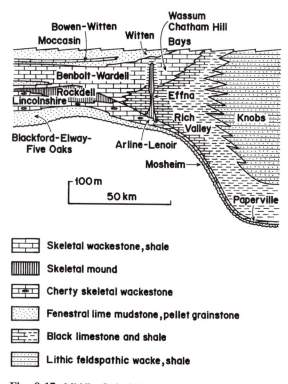

Skeletal wackestone, shale

Skeletal mound

Cherty skeletal wackestone

Fenestral lime mudstone, pellet grainstone

Black limestone and shale

Lithic feldspathic wacke, shale

Fig. 9-17. Middle Ordovician ramp-to-basin facies pattern, Virginia. (From Read, 1980, fig. 5, with permission of AAPG.)

of central Greece, there are carbonate breccias and megabreccias on the slopes below a carbonate platform margin. Johns (1978, p. 571) described small (2-meter diameter) ponds of laminated, fine-grained carbonate turbidite in the hollows between the large blocks of reef talus. Meissner (1972, p. 213), referring to the spectacular Capitan reef complex (Permian) in Texas and New Mexico, noted that much terrigenous detritus bypassed the carbonate platform and accumulated to substantial thicknesses in the deeper waters beyond the forereef margin (Fig. 9-18). In the Pethei Group in Canada (Fig. 9-19), clast size decreases rapidly from carbonate intraclast grainstones and conglomerates of the stromatolitic shelf margin to lime muds and clay muds of the deep basin. Basin and slope facies contain carbonate rhythmites consisting of very even-bedded light gray limestones alternating with thinner layers or partings of dark gray shale. Units of carbonate rhythmites, 25 or 30 meters thick, alternate with siliciclastic turbidite units of the basin floor.

Transgressive and Cyclic Carbonate Deposits

Transgressive carbonates typically are thin blanket deposits with numerous facies changes. Laporte (1967) discriminated shallow-marine carbonate rocks in the Manlius Formation (Lower Devonian) of New York as supratidal, intertidal, and subtidal facies. The supratidal facies consists of algal stromatolitic dolomite with desiccation cracks and fenestral fabric. Many of the laminae contain peloids, but there are few fossils and little bioturbation. Intertidal facies comprises peloid-skeletal grainstones and packstones and ooid-intraclast grainstones, generally not dolomitized. Subtidal facies consists of thick-bedded, bioturbated peloid wackestones with encrusting and globular stromatoporoids. Some stromatoporoids, up to 30 centimeters in diameter, are toppled and abraded. The stratigraphic and geographic arrangement of the three facies indicates irregular lateral shifts of environments, with an underlying transgressive trend. Manlius facies pass eastward into coarse skeletal grainstones of the Coeymans Formation in eastern New York. The Coey-

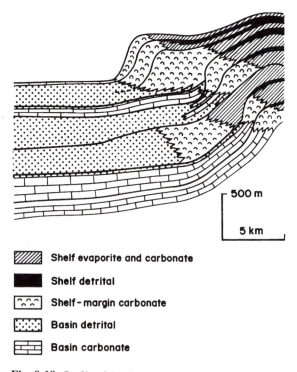

Shelf evaporite and carbonate

Shelf detrital

Shelf-margin carbonate

Basin detrital

Basin carbonate

Fig. 9-18. Profile of the Capitan reef. (After Meissner, 1972, fig. 2.)

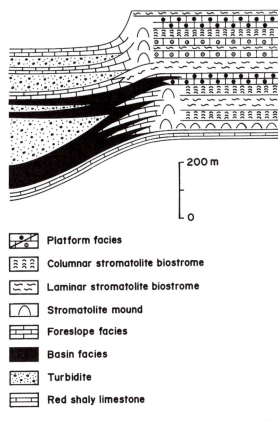

200 m

0

- Platform facies
- Columnar stromatolite biostrome
- Laminar stromatolite biostrome
- Stromatolite mound
- Foreslope facies
- Basin facies
- Turbidite
- Red shaly limestone

Fig. 9-19. Facies in the Pethei Group. (According to Hoffman, 1974, fig. 3.)

mans seems to be a coastal bar or barrier island facies behind which, in a broad lagoon, the Manlius was deposited.

Many transgressive successions consist of a stack of thin regressive cycles, and many shallow marine carbonate successions appear to be cyclic. Probably most are adequately described in terms of Goodwin and Anderson's (1985) *punctuated aggradational cycles* (see Chap. 7). Cyclic marginal-marine successions typically reflect repeated rapid submergence followed by gradual shoaling due to sedimentation, as Read (1973a) described in the Pillara Formation (Devonian) of the Canning basin, Western Australia; Hoffman (1975) in the Rocknest Formation (Proterozoic) in the Northwest Territories of Canada; and Wilson (1967) in the Duperow Formation (Upper Devonian) of the Williston basin in North Dakota, Montana, and Saskatchewan. These seem to be examples of

punctuated aggradational cycles (PACs). Commonly the transgressive phase is represented by a thin lag bed or intraclast packstone layer or a weathered surface coated with clay mud. The stratigraphic record of the regressive phase is thicker and typically contains skeletal-peloid packstones or onkoid wackestones passing upward into intertidal fenestral peloid grainstones or stromatolitic beds. Arid-climate cycles may be capped by laminated and chickenwire anhydrites and dolomite microbreccias. In the cyclic Helderberg Group (Devonian) in the Appalachians, lobes of cross-bedded calcareous quartz sandstone were deposited during regressions and early phases of transgression, according to Dorobek and Read (1986, p. 605). The thickness of a cycle, which can be 2 meters to as much as 20 meters, presumably indicates magnitude of the sea-level rise that initiated the cycle. The cycles that Wilson described in the Duperow Formation are traceable over areas of 30,000 square kilometers (subsurface), though each is typically only 3 to 5 meters thick.

Mixed Carbonate-Terrigenous Successions

Interbedded or mixed quartz sand and carbonate successions are in some cases a result of aeolian sand advancing onto a carbonate shelf or of a marine transgression over a coastal dune field. Storms and tidal currents can mix quartz sand and carbonate sand that were originally deposited separately.

The Helderberg Group (Devonian) in the Appalachians is among the best examples of mixed quartz sand and carbonate sedimentation. Driese and Dott (1984, p. 264) concluded that it was deposited in intertidal and open shelf settings experiencing marine transgressions and regressions. Carbonate sediments prograded rapidly in response to regressions; sheets of sand spread basinward more slowly and also migrated landward during transgressions.

Mixing and interbedding of volcanogenic detritus with carbonate sediments (Fig. 9-20) occurs in certain tectonically active tropical and subtropical locales, either around isolated volcanic islands or along zones of plate convergence. The volcanic particles may be formed by weathering of volcanic rock (epiclastic), then transported by streams and ocean currents into carbonate depositional environments. Or the particles may be formed directly by volcanic processes (pyroclastic), perhaps falling from the air onto carbonate-depositing environments. Much mixing of carbonate and terrigenous material is accomplished during re-

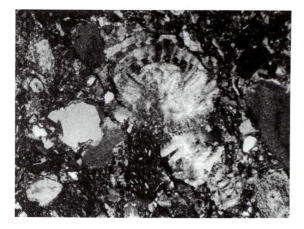

Fig. 9-20. The rock is a mixture of carbonate and volcanic materials; the skeletals are camerinid foraminifers.

sedimentation of shallow-marine sediments at the base of long, steep submarine slopes.

The Sabkha

Marginal marine carbonate depositional systems in arid climates differ in many respects from those of humid climates. An arid coastal flat subject to periodic or sporadic tidal inundation or storm surge is called a *sabkha*. Much of our knowledge of sabkha environments is based on the modern ones of the Trucial Coast of the Persian Gulf, an extremely arid region. Persian Gulf waters are only slightly more saline than normal sea water, but barrier islands along the coast create lagoons in which salinities are about twice normal. Lagoons are floored with muddy skeletal sand that grades landward to peloid sands and algal mats; tidal inlets are floored with ooid sands. A broad shelf extends seaward, mantled with coral reefs and skeletal sand and gravel in the shoal areas and carbonate muds in the hollows.

On the margin of the gulf are the salt-encrusted supratidal flats known as *coastal sabkhas*. They are 8 to 10 kilometers wide and grade landward to wind-swept *continental sabkhas*. Only slightly above sea level, a coastal sabkha is a product of coastal offlap; beneath the surface of the sabkha is a wedge of intertidal and marine-shelf or lagoon sediments, capped with supratidal evaporitic crusts and a thin aeolian layer in some cases. The sabkha environment itself contributes little

in the way of additional sediment thickness; its main effect is that it imposes a characteristic diagenetic overprint on sediments deposited outside this environment. Occasional inundation of the sabkha leaves shallow, ephemeral puddles or ponds of saline water called *salinas*; some of the standing water evaporates, some of it leaks into the sediments. The water table is 1 to 2 meters below the surface of the coastal sabkha; the pore waters are as much as ten times more saline than normal sea water; brine temperatures average 34°C; sabkha surface temperature may range up to a scorching 60°. The hot brines replace some original sediment components and precipitate evaporite minerals within the original sediment as poikilotopic cement and as displacive nodules. Gypsum crystal mushes form in a belt nearest the strand (Fig. 9-21); farther inland, nodular anhydrites predominate, some a product of dehydration of gypsum, some precipitated directly as anhydrite. Magnesium/calcium ratios in the brine increase as a result of calcium sulfate precipitation, commonly to values even exceeding 12, compared to about 5 for normal sea water. High magnesium/calcium brines induce dolomitization of the lime muds.

On the continental sabkha, aeolian processes are more evident. Aeolian dunes of quartz sand locally dominate; there are carbonate sands as well, blown in from the coast. Anhydrite that formed earlier under coastal sabkha conditions hydrates to gypsum, or gypsum dehydrates to bassanite ($CaSO_4 \cdot \frac{1}{2}H_2O$ or anhydrite. Additional calcium sulfate precipitates from inland-derived groundwaters. Halite precipitates in wind-deflated areas.

Ancient coastal sabkha deposits typically are preserved in the upper parts and landward margins of prograded marine carbonate successions formed in arid climates (see Chap. 10). Fryberger et al. (1983, pp. 302, 310) suggested that the Pennsylvanian-Permian Minnelusa Formation and correlative units in the Powder River basin in Wyoming are an ancient example of a continental sabkha on which aeolian processes were important.

Carbonate Aeolianites

Wind-laid carbonate deposits are widely distributed in modern or Recent warm coastal regions and even hundreds of kilometers inland. Hardly ever have they been recognized in the ancient rock record, however. They probably exist there, but the idea of carbonate aeolian-

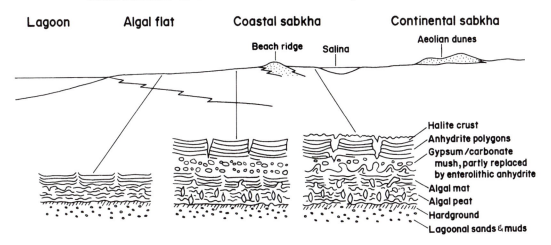

Fig. 9-21. Physiography and facies of the sabkha. (After Warren and Kendall, 1985, fig. 2.)

ite does not often occur to the geologist, and these deposits can be difficult to distinguish from certain carbonate deposits formed in the sea. Most carbonate aeolianites contain the skeletal remains of marine organisms, which belie the terrestrial environment in which these sediments were deposited.

The grains of carbonate aeolianites typically are small, abraded skeletals or skeletal fragments, pellets, and ooids (Glennie, 1970, figs. 127 and 128, presented excellent thin-section photomicrographs of carbonate aeolianite). In many respects other than composition, carbonate aeolianites are similar to their more common siliceous counterparts (see Chap. 8). They consist of well-sorted fine to medium sand-size particles; dunes contain the same kinds of cross beds and ripples as the siliceous dunes; plant-root structures such as dikaka and rhizoconcretions and terrestrial skeletals may be present. Paleosols, caliche crusts, or karst surfaces form when these aeolian deposits are no longer active. Carbonate aeolianites typically are cemented early but incompletely; needle-fiber cements are common. Carbonate cements have vadose features, such as meniscus and pendent arrangements.

Most carbonate aeolianites are deposited as dunes adjacent to beaches of mud-free carbonate sand. There are some in the Great Salt Lake Desert in Utah, in the western United States, containing calcite ooids, skeletals, and selenite crystals. In the Thar Desert of western India, 400 kilometers from the nearest coast, aeolian dunes are composed of foraminiferal tests (Goudie and Sperling, 1977). Many aeolian deposits

of quartz sand contain a carbonate sand and dust component; much of the carbonate contained in loess is detrital.

Karst

In the dark underground, an environment of carbonate dissolution and precipitation carried out by flowing groundwaters.

Karsts are forming today, at and close to the surface of the ground in equable climates, so they are readily accessible, commonly even their interior parts. Some karsts are considered to be quite scenic; "spectacular" or "exotic" are attributes justifiably attached to many of the caverns of currently or recently active karsts. Some karsts have been imposed on limestones just deposited, others on limestones exhumed to the surface after long residence in the deep subsurface. Some ancient buried karsts are of special economic importance as places of rich lead and zinc mineralization and as petroleum reservoirs of unrivaled productivity.

Karst-Forming Processes

A karst is the quintessential carbonate dissolution-precipitation product. Karst is to humid climate what caliches are to semiarid climate. Both phenomena are a kind of hybrid of weathering and diagenesis that takes place in vadose and upper phreatic groundwater zones. In both caliches and karsts, dissolution and precipitation of calcium carbonate are the essential processes, and in general there is a downward transport of carbonate. Soils form in each case, and a soil seems important insofar as it is a reservoir of CO_2, the major chemical agent of the limestone-karst-forming and caliche-forming processes. Whereas the content of CO_2 in air is 0.03%, the gas in soil contains about 0.6%, or 20 times more (Hunt, 1972, p. 259), and an upper limit of CO_2 concentration in soil gas seems to be about 7% (Ford, 1988, p. 28). This gas is derived from metabolism of microorganisms and decay of organic matter; also, plant roots exhale CO_2 to the soil. Some workers think that there are fewer examples of caliche and karst terrains in pre-Mississippian times, before the proliferation of land plants and development of true soils. The capacity of meteoric water to dissolve calcium carbonate ranges from 50 to 400 mg/l, and dissolution rates range up to 100 or 150 $m^3/km^2/yr$ in some rain forests. In some cavern systems, such as the great Carlsbad Caverns of New Mexico, hydrogen sulfide (H_2S) is an important chemical agent; CO_2 and H_2S are supplied by nearby leaky gas reservoirs (see Hill, 1990). In karst waters, H_2S converts limestone into gypsum, which is even more water-soluble than carbonate.

Caverns form in both vadose and phreatic zones; many caves reach their maximum development at the boundary between the two zones, that is, at the water table. Flowstones, dripstones, and other *speleothems* are products of the vadose zone. Cave sediments of vadose zones differ from sediments of phreatic zones in the kinds of sedimentary structures and cements that they have. Water flows faster, and flows are "flashier" in vadose zones; underground streams in vadose zones commonly have a free surface. After a hard rainfall, subterranean streamflows are very energetic and competent, capable of lifting cobbles and rolling large boulders. Below water table, in contrast, flows are more sluggish and have no free surface. Sediments from above may fall into the phreatic regions, but only the finest sediment particles are transported within the phreatic zone. Of course, water tables are very irregular and fragmented in karst systems, and water tables rise and fall over the long term, so a vadose karst may later become a phreatic karst, or vice versa. Some caverns, close to the sea, form in a zone of calcium carbonate undersaturation where fresh water and sea water mix and may eventually be drowned by the sea or otherwise influenced by the marine environment. During transgressions karst cavities may fill completely with marine sands and muds.

Cave **breakdown** is a special kind of cave deposit, created by collapse of walls or ceilings (the term also refers to the process that creates the deposit). Breakdown is caused by overwidening of cave passages or rooms, dissolution along the fractures and bedding planes that surround an unsupported or weakly supported rock mass, or removal of the buoyant support of a cave roof when water drains from a passage or room. Breakdown is coarse and chaotic; some blocks may be several meters across.

There is abundant finer-grained sediment in caves. Red mud, consisting of clay and quartz silt eluviated from the soils above, is a common cave sediment, occurring as fluvial deposits on the floors of passages, forming laminated lacustrine deposits, or coating the walls and ceilings. Where sand and pebbles are available, rippled or cross-laminated fluvial deposits are possible. Poorly sorted, chaotic breccia deposits are common in some caves, apparently created by *pipe-full conveyance*—that is, transport by a totally confined flow, a flow without a free surface.

Ancient Karsts

The paleokarst that makes the top of the Mississippian System in much of the western United States has received much attention. Sando (1988) summarized observations made over many years of part of the Mississippian karst, that which lies at the top of the Madison Limestone in southern Montana and central and southeastern Wyoming. The enlarged joints, the sinkholes, "tubes," and "pipes" extending downward as much as 30 meters below the top of the Madison, contain red muds (the residuum of limestone dissolution) and breccia clasts formed by roof collapse, but the dominant fill is quartz sand that leaked into the fissures and cavities during a post-karst marine transgression. The quartz sand is part of a fluvial and marine sandstone (the Darwin Member of the Amsden Formation) that blankets a large area of the Madison karst. Sando suggested that most of the Madison caves formed below the water table, hence the general lack of flowstones and dripstones.

The Madison karst appears to have been influenced by presence of one or two zones of evaporite, and in some places features created by large-scale *deep* intrastratal dissolution of evaporite beds might be confused with surface karst. Intraformational breccias in the Madison Limestone near Livingston, Montana, and in the nearby Bridger Range near Bozeman appear to have resulted from collapse of space created by deep dissolution of evaporite (Roberts, 1966), rather than by surface karst processes. But any dissolution cavities containing sediments leaked from above during deposition of the Darwin or other directly overlying post-Madison deposits are clearly pre-Darwin, and *not* deep-burial dissolution-and-collapse features. Budai et al. (1987, p. 915) noted collapse breccias in the upper part of the Madison of western Wyoming with features suggesting a multiple collapse history, including events that occurred after deposition and lithification of some part of the overlying Amsden Formation.

Another famous paleokarst is the unconformity at the top of the Knox Group (Cambrian and Ordovician) in Virginia and Tennessee. Mussman et al. (1988) have recently summarized the salient features of this paleokarst, which locally hosts important lead and zinc sulfide mineralization (see Kyle, 1983). Chert-dolomite breccias and red mudstones locally overlie the unconformity, and there are filled caverns and intraformational breccias below. Mussman et al. suggested that the caves formed at the water table, on a coastal wedge of fresh water overlying a marine phreatic zone. Meteoric waters drained generally seaward and gradually opened up the subsurface conduits. Kyle observed both early and late breccias in the Knox in central Tennessee. The early ones contain a fine detrital matrix and breccia clasts with bleached (weathered) rims and are directly related to karst processes. The later breccias typically have fitted clasts (mosaic or crackle breccia fabric) and cements rather than matrix; they are due to collapse of cavities created by intrastratal dissolution at depth and commonly are overprinted on early breccia.

There is a special kind of carbonate dissolution that takes place in carbonate sands soon after their deposition and before they have become fully lithified. Meyers (1988) documented a Mississippian example (at the top of the Lake Valley Formation in New Mexico); most of the dissolution took place between carbonate grains rather than along joints and bedding planes; it is a karst mainly at the petrographic scale rather than the physiographic scale. The process occurred entirely in a vadose zone.

Carbonate Grainflows, Turbidites, and Olistoliths

Carbonate sediment formed in the halcyon shallows moves downslope.

Some debris flows and turbidites are of carbonate composition but otherwise similar to the silicic ones described in Chapter 8. Most of the resedimented carbonate particles, such as ooids and coral and algal skeletal fragments commonly contained in such deposits, could only have formed in shallow water. But sedimentary structure and stratigraphic relationships and admixing of pelagic fossils indicates that deposition took place in deep water. Many examples, ancient and modern, have been described.

In certain Cambrian and Ordovician carbonate slope facies in Nevada (Cook, 1979), there are numerous spectacular glide or slump masses (olistoliths). On the upper slopes, these bodies occupy deep, wide channels; original beds are deformed into big, tight folds or broken into large blocks and boulders. Alvarez et al. (1985) described some of the many slides in ancient pelagic limestones in Italy, noting that the rear of the slide mass characteristically is stretched and thinned and that the forward part is shortened and thickened, folded and thrusted.

Besides the slides and slumps, there are debris-flow deposits consisting of large and small clasts embedded in lime mud, and (modified) grain-flow or high-density turbidity-current deposits consisting of poorly sorted grainstones and packstones. Flow deposits of the lower slopes are finer-grained and display clast orientation, imbrication, and grading. Hiscott and James (1985) found single carbonate debris flow beds up to 100 meters thick in Cambrian-Ordovician rocks in Newfoundland. Hesse and Butt (1976) described carbonate turbidites in the Alps and Apennines and suggested ways of distinguishing between those deposited above the carbonate compensation depth (see below) and those deposited below. Mullins and Cook (1986) distinguished line-source **carbonate aprons** from the point-source turbidite fans more common in siliciclastic settings. Recent carbonate turbidites exist in Exuma Sound in the Bahamas; grain flows and debris flows also are abundantly represented there (Crevello and Schlager, 1980).

It appears that base-of-slope carbonate sedimentation increases during sea-level rise or sea-level highstand (see Chap. 7). Moreover, compositional differences between highstand and lowstand input have been identified. Everts (1991, p. 239) reported that, in Tertiary rocks of Spain, highstand turbidites contain more ooids, pellets, and grapestones than lowstand turbidites. Crevello and Schlager (1980, p. 1134) considered that large quantities of lithoclasts in Bahaman slope deposits are due to rim collapse during Pleistocene interglacial highstand. During lowstands sedimentation rates on the slopes are reduced.

Pelagic Environments and Facies

The calcareous and siliceous oozes of the oceans' Great Plains, tilled by the languid, lingering creatures of the dark underworld.

The pelagic plankton oozes reflect two environments, the surface waters, typically warm and well lighted and bandied by the breezes, where countless calcareous and opaline particles come into being; and the cold, sullen darkness of the bottom waters, where the particles come to rest. Clays and fine silts from distant lands also make their way to the abyss, carried by the winds and ocean currents; oceanic volcanoes create their special kinds of sands and muds; and meteoric dust drifts down steadily but sparsely on all the oceans' floors.

Sedimentary Processes

Sedimentation on the abyssal plain is mainly the gentle pelagic rain of terrigenous clay and skeletons of the plankton. The nature of the plankton crop of ocean surface waters is determined by climate and surface currents. Foraminiferans are quite cosmopolitan, raining their calcareous (low-magnesium calcite) tests onto the depths at all latitudes and making foraminiferal oozes on those parts of the floors of all the oceans that lie generally above 5000 meters depth. Radiolarians enjoy the warm surface waters of the tropics and contribute to siliceous radiolarian oozes, especially in the equatorial Pacific. Diatoms prefer the cold waters of polar or subpolar regions; their siliceous oozes lie along the northern margin of the Pacific and encircle Antarctica. Coccolithophores have made important carbonate contributions to warm- and cool-climate marine sediments. Pteropods are widely distributed little pelagic snails with very thin aragonite shells. They contribute aragonite to pelagic sediments above 1000 meters depth.

The rate of sedimentation of the skeletons of planktonic organisms is determined by the rate of organic production at the surface relative to the rate and extent of dissolution of the skeletons as they fall to the bottom. Productivity is controlled mainly by nutrient content. The deep, cold waters of the ocean, including the waters of the *oxygen minimum layer*, are rich in nutrients. Where these waters well up to the surface, generally near to continental margins, the planktonic biomass increases. Productivity is least in the centers of geostrophic gyres, as in the so-called ''Sargasso Sea,'' being the eye of the great North Atlantic gyre. (Sargassum weed floating in these waters has collected there from faraway coastal provenances.)

Surface waters are saturated or supersaturated with calcium carbonate, but the deeper waters are undersaturated. In the oceans there is a zone called the **lysocline**, in which there is a rapid decrease in carbonate saturation with depth. Calcareous skeletals formed in the surface waters dissolve gradually as they sink through the lysocline into deeper waters. The smaller skeletals sink very slowly, perhaps only a few centimeters per day, and may dissolve completely before they ever reach the ocean floor. Others survive their passage through the lysocline, then dissolve during long exposure on the bottom. Many skeletals drop to the ocean floor as components of fecal pellets, some with organic sheaths or coatings that increase their potential for survival as pelagic sediment. The depth level below which sea water is undersaturated with respect to calcite and below which calcite is generally not preserved as ocean-floor sediment is called the **calcite compensation depth** (CCD), between 4000 and 6000 meters in modern oceans. The CCD is not a constant but varies geographically and temporally, depending on the CO_2 content of the water column, which in turn depends on the water temperature, rate of decay of organic matter in the water, and rate of infall of carbonate detritus from the surface. Some carbonate skeletal matter is rapidly inserted into ocean floor regions below the CCD by turbidity currents; quickly buried, this carbonate survives dissolution. Moreover, carbonate sediments deposited above the CCD on the oceanic ridges or rises are slowly transported into deep sub-CCD waters of the flanking basins by the tectonic spreading of the sea floor away from the rifts; some of this carbonate escapes dissolution. Thus there *are* carbonate sediments below the

CCD, but special processes were involved in their deposition and preservation. Aragonite is even more soluble than calcite in the deep waters, the compensation depth for aragonite (ACD) in modern oceans being 2000 meters or less. The aragonite shells of Mesozoic ammonites commonly dissolved completely during or after their descent to the bottom, but the calcite aptychus or operculum survived. This seems to be a consequence of the seawater chemical environment that exists between the ACD and CCD. In the bivalves *Inoceramus* in Cretaceous chalks of western Kansas and eastern Colorado, Hattin (1981, p. 843) observed that the calcite prismatic layers are abundant, but the aragonite nacreous layers are nowhere preserved. But in this case, selective dissolution of aragonite might have occurred diagenetically under burial (see Chap. 11).

Siliceous skeletons (opal-A) also dissolve in sea water, but with respect to silica dissolution, the deeper waters are *less* aggressive than the shallow waters. Probably less than 5 percent of total planktonic silica production actually reaches the bottom, but if a siliceous skeleton does manage to do so, it contributes permanently to the sediment. The siliceous oozes of the ocean floors lie at depths generally below the CCD, where carbonate diluents ordinarily cannot be. According to Berger (1976, p. 322), the siliceous skeletons of silicoflagellates of modern oceans dissolve most readily, followed by diatoms, radiolarians, and sponge spicules.

Pelagic Sediments

Pelagic oozes typically display a multimodal size distribution, reflecting the characteristic sizes of the various skeletal components; this multimodality commonly survives diagenetic recrystallization. The degree to which the skeletals have undergone dissolution is evidence of depth of the depositional surface with respect to the lysocline, but in this matter, diagenesis obfuscates the evidence. In all pelagic sediment there is a certain amount of fine-grained terrigenous detritus. In the deep basins, where it is not much diluted by pelagic skeletals, it accumulates very slowly as red mud, in some places studded with the sluggishly precipitated ferromanganese nodules (see Chap. 10). Other silicate rock detritus—mainly fine ash, glass fragments, and plagioclase—derives from oceanic volcanism; quartz silt probably is of continental provenance. Lateral change between these terrigenous

muds and the calcareous and siliceous oozes is very gradual.

Secular fluctuations of the lysocline and CCD, related to eustatic sea-level fluctuations or to physical changes in seawater, can cause stratigraphic variations in carbonate content. Sea-floor spreading imposes a certain stratigraphic pattern on the pelagic sediment pile and causes lateral and vertical facies changes in pelagic sediments by transporting the depositional site across different environments. At the crests of active oceanic ridges or rises, hot, acidic, metalliferous sulfide exhalations alter the top of the oceanic basaltic crust (ophiolite) and darken the earliest pelagic sediments accumulating directly on this crust; the so-called "basal ferruginous layer" contains iron oxyhydroxide or manganese oxide or both. As ocean crust spreads away from the ridge, younger carbonate ooze accumulates over the basal layer. Further plate-spreading transport carries the crust into deeper waters farther from the crest and below the CCD, where the carbonate layer partially or fully dissolves and where pelagic clays and siliceous oozes accumulate. Remanent magnetic orientations can provide evidence for geographic locations at which deposition took place. Because oceanic crust is impermanent, and eventually subducts, there are almost no pre-Mesozoic pelagic sediments on the ocean floor. Some Paleozoic oceanic rocks *have* escaped subduction, however, and occur in melanges on the margins of the continents.

Ancient deep-sea pelagic sediments that are now exposed on land include red, nodular limestones and colored cherts; black or dark brown ferromanganese nodules may be present. Some pelagic successions can be shown to be resting on or interbedded with ophiolites. Pelagic cherts are red, green, yellow, or brown, making uniform beds a centimeter in thickness (ribbon bedded), with partings or beds of siliceous mudstone, also colored. Red nodular limestones consisting of centimeter-size nodules closely set in red marl are deposited in shallower water than are the radiolarites or red cherty limestones. Mesozoic and Cenozoic pelagic limestones commonly are composed largely of planktonic foraminifers, such as globigerinids (Fig. 9-22). Paleozoic pelagic and hemipelagic rocks were deposited before the important calcareous plankters (the calcareous foraminifers and coccolithophores) evolved; (siliceous) radiolarians were the only abundant plankton, though pteropod-like animals (*Hyolithes*), and the (chitinous) graptolites made minor contributions to early Paleozoic sediments. The association of radiolarians and graptolites with black, organics-rich shales of the Paleozoic suggests that there were other plankters that had no preservable skeleton.

Many ancient chalks are pelagic deposits of epicontinental seas much shallower than the bathyal and abyssal depths of the oceans. Late Cretaceous epeiric pelagic chalks include the nodular chalks of Normandy and the South of England (Kennedy and Garrison, 1975). In addition to the planktonic calcareous microfauna, these rocks contain a diverse benthos of sponges, echinoderms, bryozoans, brachiopods, and molluscs. Locally, the chalks are siliceous or phosphatic. In the western interior United States are the Greenhorn Limestone (Upper Cretaceous), a bioturbated deposit of pelleted coccoliths (Hattin, 1975), and the Niobrara Formation, a little higher in the Upper Cretaceous Series, being bioturbated foraminifer-and-coccolith chalk and argillaceous chalk with numerous *Inoceramus* and other molluscs (Frey, 1972; Hattin, 1982). A benthonic macroinvertebrate fauna of low diversity (only three or four species) in the Greenhorn added thin, fragile shells to a much more diverse and prolific planktonic and nektonic fauna. For the Greenhorn, Hattin (1971, p. 421) suggested a water depth of 120 to 150 meters, based on stratigraphic relationships. In the Niobrara, bottom-dwelling large inoceramids furnished about the only hard substrate in the soupy oozes, and their shells played host to local, complex communities of other molluscs, bryozoans, barnacles, serpulid worms, and sponges. Finely disseminated organic matter and pyrite are abundant in many layers. Volcanic ash layers (bentonite beds) are numerous. That these chalks are not much diluted by

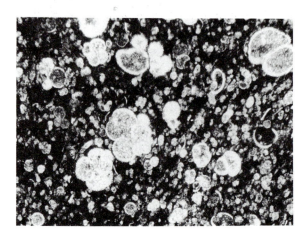

Fig. 9-22. Pelagic limestone with globigerinid foraminifers.

terrigenous detritus is mainly a reflection of the remoteness of emergent terrains. There *are* thin beds or partings of argillaceous chalk, thought to record times of somewhat depressed sea level (and expanded detrital source areas), and there are alternations between darker and lighter-colored layers or between heavily and lightly bioturbated layers that reflect fluctuations in oxygen level in the bottom water, which may, in turn, reflect cyclical changes in the structure of overlying water layers. Many individual beds can be recognized in sections many kilometers, even hundreds of kilometers, apart. The patterns of bedding breaks and of degree of bioturbation and content of organic matter have been tentatively attributed to Milankovitch cycles.

10

Other Sedimentary Rocks

Besides the sandstones, mudstones, and limestones that comprise the bulk of the sedimentary rocks, there are some other sediments that are very important economically and have great scientific importance but exist in much smaller volumes. Many of these rocks are chemically "peculiar," and their origins are in various respects problematical. Nature has found special or unusual ways of disposing of certain substances in sedimentary environments, precipitating them out of saline ponds or volcanic vents or taking advantage of the special chemical environments created by or within living systems. Chemical rocks keep a significant portion of the total crustal quantity of a substance in storage, in highly concentrated form. Existing evaporite rocks contain about twice the total quantity of salts currently in solution in the ocean. The marine-deposited Phosphoria Formation in the Northern Rocky Mountains region contains more than five times the amount of phosphorus dissolved in the world ocean today. Some beds, with 30% or more of P_2O_5, represent a 2-million-fold enrichment with respect to P_2O_5 in ordinary seawater.

Evaporite Rocks

Evaporites are rocks formed by precipitation from solutions supersaturated due to evaporation of the solvent. They are the soluble residues of the chemical weathering of rocks; most are halides, sulfates, borates, and nitrates (Table 10-1). Some are deposited in restricted extensions of deep ocean basins, others in shallow marine-marginal lagoons, still others in broad, shallow intracratonic sags or in small, fault-bounded troughs of continental interiors far from the sea. The accumulation of substantial thicknesses of evaporite minerals implies the evaporation of much water, with or without ultimate desiccation of the basin. Evaporite sedimentation can be very rapid, greatly exceeding net sedimentation rates of most other sediments, and evaporites can accumulate to thicknesses of 1000 meters or more.

There are thick, ancient evaporite deposits in the Michigan, Williston, and Paradox basins of the United States; Elk Point basin in western Canada; Zechstein basin in northern Germany; Moscow basin in Russia; and in Siberia and in the Middle East. These are Phanerozoic deposits, but bedded evaporites have been found in Proterozoic successions also. In fact, the abundance of ancient evaporites is something to be marveled at in view of the softness and solubility of these rocks. Many evaporites have survived because they were deposited in cratonic basins and then buried under a thick protective layer of younger and far more durable sediments. The generally very low permeability of evaporite deposits protects them to a large degree from intrastratal dissolution (percolating water is rarely a problem in salt mines). Old sedimentary rocks, evaporites included, at the bottoms of cratonic basins can last, it seems, as long as the continents can last. However, under burial in a deeply subsiding basin, tectonic compression of a basin, or regional metamorphism, evaporite deposits are squeezed out of the sedimentary pile and destroyed. Dissolution of evaporite occurs locally in the deep subsurface, and evaporite-bearing formations in the subsurface lose their evaporite mineral components quickly on uplift and exposure.

Evaporites accumulate today in internally drained arid continental basins, as in the Great Salt Lake region (Bonneville basin) of the United States, where there has not been sufficient long-term runoff to establish (or require) drainage to the sea and where rainfall does not reverse the work of evaporation. Marine coastal

Table 10-1. Some Evaporite Minerals of Sediments

Chlorides	
Halite	NaCl
Sylvite	KCl
Bischofite	$MgCl_2 \cdot 6H_2O$
Carnallite	$KMgCL_3 \cdot 6H_2O$
Sulfates	
Anhydrite	$CaSO_4$
Gypsum	$CaSO_4 \cdot 2H_2O$
Polyhalite	$K_2MgCa_2(SO_4)_4 \cdot 2H_2O$
Kieserite	$MgSO_4 \cdot H_2O$
Kainite	$KMgSO_4Cl \cdot 3H_2O$
Epsomite	$MgSO_4 \cdot 7H_2O$
Langbeinite	$K_2Mg_2(SO_4)_3$
Glauberite	$Na_2Ca(SO_4)_2$
Carbonates, bicarbonates	
Trona	$Na_3(CO_3)(HCO_3) \cdot 2H_2O$
Nahcolite	$NaHCO_3$
Natron	$Na_2CO_3 \cdot 10H_2O$
Gaylussite	$Na_2Ca(CO_3)_2 \cdot 5H_2O$
Borates	
Borax	$Na_2B_4O_7$
Nitrates	
Soda niter	$NaNO_3$

a major sediment component. Anaerobic bacteria, especially *Desulfovibrio*, thrive in anoxic bottom brines of saline lakes and marine basins, metabolizing dissolved or precipitated sulfate and carbonate.

Chemistry of Evaporites

Evaporate a column of sea water—what precipitates, and in what order? This is the crux of Usiglio's experiments of 1849 (Fig. 10-1). The experiments tell us that a column of sea water, if left to evaporate completely (at a temperature of 40°C into dry air), without any renewal of water or dissolved substances from the outside, precipitates first $CaCO_3$, then $CaSO_4 \cdot 2H_2O$, then NaCl, $MgSO_4$, $MgCl_2$, and NaBr, and finally KCl. Evaporation of a meter depth of normal sea water to dryness produces about 15 millimeters of precipitated salts, mostly halite. A natural body of sea water, if evaporated under similar conditions, presumably makes a deposit having a thickness roughly 1 percent

flats inundated only during unusual high tides or storms precipitate evaporites during intervening periods of desiccation. Probably the greater thicknesses and volumes of ancient evaporites were deposited in deep bodies of water partially isolated from the ocean, which serves as a virtually unlimited source of dissolved ions. Arid conditions promote evaporation from the surface of such a body of water, and salinity increases. Sea water of ordinary salinity enters through some narrow passage or over some submerged sill or shelf, floating over the dense brines and replacing water volumes evaporated from the surface of the partially enclosed basin. Sea water may enter also through an underground pore system. Certain lateral and vertical salinity gradients develop within the basin, which drive density currents, and the resulting dynamical and compositional structure controls the kinds of evaporite minerals precipitated, their geographic distribution in the basin, and the rates of their precipitation.

High salinity does not necessarily exclude organic production. Phleger (1969) described modern salina deposits in Baja California, Mexico, supporting a large standing crop of green algae, grasses (*Salicornia*), and calcareous foraminifers on the salt flats. Kirkland and Evans (1981, p. 182) lucidly described salt lakes in India and Kenya, their verdant waters colored by algae that support enormous flocks of flamingos. In Great Salt Lake, Utah, the fecal pellets of brine shrimp are

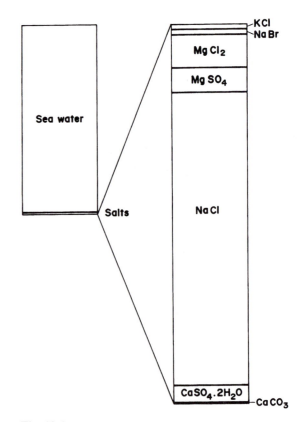

Fig. 10-1. Usiglio's (1849) results.

of the original depth of the water and comprising a layer of calcium carbonate at the base, overlain by a layer of gypsum, then halite, etc. As precipitation proceeds, the composition of the remaining solution changes, and its density gradually increases. The most soluble compounds precipitate last, of course. Moreover, the lower layers are deposited in the deeper water, and the higher layers in the shallower water that remains in the late stages of evaporation.

The constraints that Usiglio placed on his experimental system make his results largely inapplicable to natural systems, however. It is difficult to conceive of some deep body of sea water, fully detached from the ocean and undergoing evaporation to dryness, though an approximation to this is thought to have occurred in the Mediterranean basin in Miocene time, as Hsu and his co-workers have described (see below). The Zechstein beds at Stassfurt, Germany contain fairly well-organized successions of evaporite minerals, but evaporite deposits with ''Usiglio successions,'' or even partial Usiglio successions, are rare.

van't Hoff experimented over many years with simple chemical models of seawater, in an effort to understand the evaporite deposits at Stassfurt. He dealt with calcium-free experimental systems; composition of seawater was adjusted by removing all bicarbonate as calcite, and all remaining calcium ion as gypsum (see Harvie, et al., 1980). This seemed a valid approach, since late-stage brines contain very little calcium.

Order of crystallization depends on relative solubilities and relative concentrations. Complete evaporation of seawater at 25°C, if equilibrium is preserved (that is, no fractionations and no supersaturations), leads theoretically, according to van't Hoff and others, to crystallization of halite, small amounts of calcium salts (precipitated early as calcite, dolomite, gypsum, or anhydrite), the magnesium sulfates kieserite and bischofite, and the magnesium-potassium sulfate carnallite (Krauskopf, 1979, p. 278). If equilibrium is not maintained or fractionations occur, then other minerals are present as well, including the magnesium sulfate epsomite and the magnesium-sodium sulfate bloedite. A brine that is sufficiently concentrated to precipitate the magnesium salts is also saturated with NaCl, and halite precipitates continuously until the very end. The temperature of the system has important consequences on the stability fields of all mineral phases, causing some fields to shrink or disappear, others to expand.

Sea water is complex, and physical chemistry does not explain satisfactorily the associations, proportions,

and stratigraphic sequence or order of crystallization of minerals in natural evaporite deposits. Sylvite (KCl), for example, though a major component of some evaporite deposits (and the most important commercial source of potassium), should never precipitate from sea water, according to the physical chemist, and the ratio of anhydrite to halite is much higher in natural deposits than experiments predict. Part of the problem, certainly, is that the system is so complex that simplifications must be made in experimental or theoretical analyses. The influence of organic compounds surely is important but has been neglected in most theoretical studies. Certain organic compounds in solution are known to promote the precipitation of anhydrite and of sylvite, for example. Evaporite-precipitating environments are not closed systems, and they generally are not equilibrium systems. If the early-formed minerals remain in contact with the brine, then there are certain back-reactions that cause ions already precipitated to re-enter the system; this is the circumstance that the physical chemist ordinarily strives to maintain. But if early-formed minerals are fractionated—that is, sequestered from the brine—then the composition of the active part of the system keeps changing; this is what typically occurs in natural systems. Sustained or intermittent additions of fluids from outside the basin are disruptive; so is outflow of fluids already partially evaporated. As sea water enters a marine-marginal basin and begins to evaporate, it precipitates the least soluble salts first (gypsum or anhydrite). If any of the brine then happens to flow back to the sea, it carries much of its load of the more soluble salts (such as halite) with it; the result is an evaporite basin that, with respect to volume of sulfate minerals, is deficient in chloride minerals. Finally, there is the possibility of evaporites undergoing syndepositional recycling (Hardie, 1984, p. 199), and selective intrastratal dissolution and later diagenetic alterations after burial.

Relative humidity is an important influence on evaporite deposition; some of the differences between marine-marginal and inland-basin evaporites are due to differences in this. Near the sea, relative humidity is typically 70 or 80 percent, but intracontinental desert basins may have atmospheric relative humidities close to zero. At the higher humidities, water stops evaporating once its ionic concentration reaches a certain level; only at low humidities can water evaporate to the point of precipitation of halite, sylvite, and other late-stage evaporite minerals. Evaporation from large, deep basins, such as the Mediterranean Sea or Red Sea, is greatest in winter, when cold air passes over warmer water. Warm air that comes in contact with cold water, in contrast, commonly cools to the dew point, and a fog forms in the water-saturated air. No evaporation can take place under such conditions.

As halite precipitates out of an evaporating brine, the concentrations of dissolved ions other than Na^+ and Cl^- gradually increase. Br^- is an important example; it has a concentration of 0.065‰ in normal seawater, but at the point of NaCl saturation, Br^- concentration is 0.54‰. As halite is precipitating, it incorporates some Br^- as substituents for Cl^-, and the extent of Br^- substitution depends mainly on the salinity of the brine. We can measure a *partition coefficient*, c, defined as

$$c = (\text{conc of } Br^- \text{ in halite})/(\text{conc of } Br^- \text{ in brine})$$
$$= 0.14$$

approximately. In a closed system, the first halite to precipitate from a seawater-derived brine should, therefore, contain $0.14 \times 0.54 = 0.075$‰ Br^-; last-formed halite should be much enriched in Br^-. By measuring the quantity of bromine in a particular sample of halite, we obtain an estimate of the salinity of the brine from which that halite precipitated. According to Hardie (1984, p. 228), the predicted concentration of bromine in marine halite is 0.075 to 0.5‰ (75 to 500 ppm), a wide range. Syndepositional recycling or diagenetic recrystallization of halite redistributes bromine (generally halite looses bromine under these circumstances).

Evaporation concentrates not only the inorganic salts but the nutrients as well. Kirkland and Evans (1981, p. 186) indicated that nutrients in evaporated water become concentrated in direct proportion to the increase in salinity. Wengerd and Strickland (1954, pp. 2169, 2171, 2173) pointed to isolated bands of black (organics-rich) shale in the thick halite body of the Paradox Formation (Permian) of the Paradox basin in Utah and Colorado. (These can be correlated in the subsurface over a broad region by means of gamma-ray logs.) Under anaerobic conditions, sulfate-reducing bacteria combine evaporitic calcium sulfate with organic matter; this process releases hydrogen sulfide:

$$CaSO_4 + 2CH_2O \rightarrow$$
$$CaCO_3 + H_2O + CO_2 + H_2S \qquad (10\text{-}1)$$

Calcium ion released by the bacterial reduction precipitates as micritic calcite or aragonite, which accumulates in the bottom sediment. The hydrogen sulfide

gas bubbles off or combines with dissolved iron to form pyrite. Organic matter is destroyed in the process, but H_2S retained in sediment pores helps to preserve some of it. Moreover, according to Kirkland and Evans, sulfate-reducing bacteria can metabolize only a few organic compounds, mainly those of low molecular weight. Gornitz and Schreiber (1981, p. 790) found that, in the Dead Sea, H_2S rises to the surface and reoxidizes to sulfate; thus, sulfate in solution in the deeper waters is recycled upward.

Nodular and Varved Anhydrite and Gypsum

Gypsum ($CaSO_4 \cdot 2H_2O$) and anhydrite ($CaSO_4$) are the most abundant evaporite minerals in many evaporitic deposits. Both can precipitate directly from an evaporating brine, and each can alter to the other as physical conditions change. Gypsum is by far the more common precipitate, but at salinities upwards of 260‰ and high temperatures, above 22°C, anhydrite precipitates instead, and existing gypsum tends to recrystallize to a mush of fine lath-shaped crystals of anhydrite. Certain organic compounds in solution, such as amino acids and sugars, retard gypsum precipitation (Sonnenfeld, 1991, p. 162), and precipitated gypsum dehydrates to anhydrite only in the presence of such compounds. Park (1977, p. 493) and Warren and Kendall (1985, p. 1014) noted that anhydrite occurs only in supratidal sediments in the marine sabkha at Abu Dhabi City on the Persian Gulf, formed either by direct precipitation, or by alteration of gypsum. Anhydrite generally dominates above the water table, and gypsum always below. Apparently, anhydrite formation requires a hot and dry (or drying) surface; should anhydrite be re-immersed, as a result of either rising water table, or extraordinary high tide, or rainfall onto the surface, anhydrite reverts back to gypsum.

Gypsum (selenite) is precipitated either displacively within desiccating muds, or as pore-filling cement. Both kinds precipitate in the zone of capillary water above the water table, which ordinarily lies very near to the sediment surface (typically within 1 meter). Kinsman (1966, 1969) and Shearman (1966), having examined sulfates forming today on the Trucial coast of the Persian Gulf in Saudi Arabia, argued that gypsum nodules grow by displacement of the carbonate mud or sandy sediment that hosts them. Shearman and Fuller (1969, p. 500) described the nodules as soft, like putty, though some lying close to the surface are cemented with halite. Nodules expand as new crystals grow in the interstices between older laths. The force of crystal growth shoulders earlier crystals aside, which push against the host sediment and align themselves thereby parallel to the margins of the nodule. The nodules may come to be closely fitted, having squeezed aside the carbonate or clay muds that host them. Much anhydrite of ancient evaporite deposits shows this so-called "chickenwire" structure of closely fitted nodules. But large (even tens of centimeters) poikilotopic crystals of gypsum may form within porous carbonate sediment without much disruption of sediment textures. Park (1977, p. 493) observed that some gypsum of Persian Gulf sabkhas appears as large disks within stromatolitic and peloid muds in the lower intertidal zone, enclosing rather than displacing the carbonate sediment. Gypsum that precipitates on the floors of salinas or playas commonly forms thick beds consisting of large, vertically oriented crystals. These crystals grew above the sediment-water interface rather than within the sediment, and they contain very little contaminating carbonate or clay.

Marine coastal salinas (shallow brine ponds) of South Australia (Warren and Kendal, 1985) have made shoaling-upward successions of gypsiferous and aragonitic carbonate mud 5 to 10 meters thick. Coarse gypsum at the base passes upward to vertically elongate but horizontally laminated selenite crystals. The laminae result from periodic accumulation of aragonite peloids, which come to be enclosed poikilotopically within the selenite crystals as they continue their upward growth. Columnar gypsum is overlain by laminated or ripple-laminated gypsum sands. Algal mats with tepees and domal stromatolites are present in some of the upper layers. The Australian salinas, areally up to 20 × 12 kilometers, have bull's-eye sediment patterns, with gypsum dominant in the center, carbonate on the margins. This pattern persists throughout the filling of a salina, and the lateral contact between gypsum- and carbonate-dominated sediments is nearly vertical.

Halite and gypsum (or anhydrite) commonly are interlayered in basinal deposits. Halite, which is more soluble in cold water than in warm, is probably a summer deposit, but gypsum, more soluble in warm water than in cold, precipitates in winter. Thus, a halite-gypsum couplet is a varve. Varves in evaporite deposits may also result from seasonal change in rate of influx of water to a basin or rate of evaporation. Kushnir (1981, p. 1194) described varved evaporites in a Re-cent shallow hypersaline pool in the Sinai, comprising summer layers of gypsum mush and winter layers of carbonate.

Under burial to depths of about 1000 meters, gypsum dehydrates and recrystallizes to anhydrite (see Chap. 11). In spite of the recrystallization, the original shapes of gypsum crystals or crystalline aggregates may be retained, at least approximately. Recrystallization at depth produces anhydrite nodules that are vertically elongate (Elliott and Warren, 1989, p. 1319) and that might retain vestiges of fishtail or swallowtail twinning. Anhydrite of this origin contains very little (carbonate) matrix.

Maiklem et al. (1969) offered a classification of anhydrite textures and structures based on form of crystalline masses, anhydrite-to-matrix relations, bedding character, and features related to distortion (Fig. 10-2). (Consult also Comte des Techniciens, 1981, for numerous photographs, both macro- and micro-, of evaporite textures and structures.) There are crystal-shaped (crystallotopic) anhydrite masses, typically pseudomorphous after gypsum, and nodular and bedded or laminated masses, commonly brecciated or contorted in some way. Some anhydrite consists of blocky, lath-shaped, or anhedral crystals, more or less distinct from one another, embedded in limestone or dolomite. In nodules and beds, anhydrite has a microcrystalline texture similar to that of micrite or is an aggregate of small felted or aligned laths, typically 0.1 millimeter in length, flattened parallel to {100}. Anhydrite also commonly *replaces* calcite, including lime mud, skeletals, and peloids. This replacement can happen syngenetically, but diagenesis under moderate burial may be the more important mechanism.

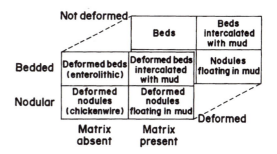

Fig. 10-2. Classification of anhydrite structures. (Greatly modified after Maiklem et al., 1969.)

Ancient Marine and Marginal-Marine Evaporite Deposits

Many ancient basinal evaporite deposits show a concentric zonation (Fig. 10-3), with the most soluble minerals at the center; this is due either to gradual shrinkage of a brine pond, which becomes more and more saline, or to a density stratification such that the heaviest, most concentrated brines lie in the deeper, more central parts of an evaporite basin. Some late salts (such as sylvite) precipitate from brines trapped in shallow depressions or pans at the end of basin filling. Basins maintaining restricted connection with the open sea can accumulate greater thicknesses of evaporites. Sea water continually enters the basin, replacing water lost to evaporation. Being of lower density than basin brines, an inward-flowing plume of sea water keeps to the surface, gradually evaporating and becoming more and more saline as it flows across the basin. As its salinity increases, it precipitates first carbonate minerals near to the basin entrance, then gypsum, and finally halite in the landward part of the basin. Thus, mineral zonation occurs in these basins also (Fig. 10-3). There may be sapropels in the basal part of a thick evaporite succession, as in the Permian of the Zechstein basin in Germany, reflecting an early, pre-evaporite time of stagnation and anoxia; H_2S produced under these conditions may bring down base metals, such as copper. The Kupferschiefer at the base of the Zechstein evaporite succession has always been considered as one of the best examples; but a recent study by Jowett et al. (1987) has cast some doubt.

Evaporite deposits of the arid tidal flat (sabkha) commonly are cyclical, reflecting periods of sediment progradation punctuated by times of flooding. Even a minor sea-level rise or subsidence causes rapid transgression of the flat surface of the sabkha. Bosellini and

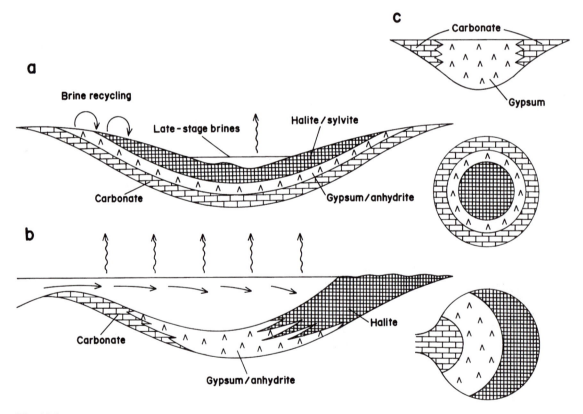

Fig. 10-3. Evaporite facies patterns (**a**) in enclosed basins and (**b**) in restricted marginal-marine basins; (**c**) distribution of carbonate and sulfate in a modern Australian salina. (According to Warren and Kendall, 1985, fig. 4.)

Hardie (1973) described evaporitic cycles in the Bellerophon Formation (Upper Permian) of the southern Alps. Individual cycles range in thickness up to 3 meters, and comprise burrowed micritic dolomite, dasycladacean dolomite with gypsum nodules, nodular or ptygmatic (enterolithic) gypsum, and nodular (chickenwire) gypsum at the top. Cycles were interpreted as regressive bodies produced in arid marginal-marine environments. Each cycle is a result of progradation of a tidal flat over a shallow lagoon; a sabkha environment then advances over the tidal flat. Bosellini and Hardie suggested an autocyclic mechanism operating under continuous subsidence.

Evaporites have precipitated also in deeper water. Dense, evaporated sea water forms at the margin of a restricted basin and sinks to the bottom (Schmalz, 1969, p. 802), where it stagnates, becomes anoxic, and deposits organic matter and sulfides. Evaporation in the open waters causes crystals of gypsum to form at the surface, which sink toward the bottom and redissolve. The basin eventually becomes saturated with respect to gypsum, then halite, and these minerals, in turn, begin to accumulate on the bottom. Evaporites in the Muskeg (Prairie) evaporite succession (Middle Devonian) of the Elk Point basin of western Canada were deposited in deep water, according to Davies and Ludlum (1973). Laminite consisting of couplets of calcite or dolomite or anhydrite and a dark film of organic matter occurs at the base of the succession in the deeper parts of the basin, passing upward into thick anhydrite and halite deposits. The couplets resemble varves and can be traced laterally for distances up to 25 kilometers. They were deposited from a density-stratified water column, anoxic at the bottom, and seasonally precipitating carbonate and sulfate. Interspersed among these couplets are graded beds of clastic limestone, these beds being thicker and more numerous near to basin margins. The graded beds are ordinary carbonate turbidites; from time-to-time skeletal and other carbonate material originating on marginal shoals descended rapidly down slopes as great as 25° and resedimented in water approximately 50 meters deep.

In many basins, evaporite forms most abundantly on the shoals. If steep slopes lie beyond the evaporite-precipitating shoals and flats, then the evaporite deposits may slip away from their place of origin and resediment downslope. Schlager and Bolz (1977) described graded turbidites, mass-flow breccias, and slump breccias, all composed largely of anhydrite, in the Zechstein, and proposed that these clastic evaporites were deposited on the steep flanks of *sulfate platforms* that are analogous to the carbonate platforms of open tropical seas. Hardie and Eugster (1971) described and illustrated the remarkable textural and structural diversity of the Miocene evaporitic beds of Sicily, including resedimented deposits. There are laminated gypsum sandstones and mudstones whose laminae are delineated by carbonate partings or by grading or reverse grading; flat-laminated and cross-laminated gypsum-calcite sandstones containing foraminiferal tests; and gypsum-crystal conglomerates with scattered intraclasts. Schreiber et al. (1976), who also examined the downslope resedimented facies of the Sicilian evaporite beds, described graded gypsum turbidite beds and various coarse, chaotic, gypsum debris-flow beds deposited in deeper waters.

Hsu (1972) and Hsu et al. (1977) described an almost incredible example of shallow-water evaporite deposition in a deep-oceanic basin setting. It is a case of a deep marginal sea disconnecting from the ocean, gradually drying up, and becoming ultimately a shallow desert playa 1000 meters below sea level, all the while depositing salt. A widely distributed Miocene evaporite succession exists under the floor of the Mediterranean Sea, discovered by the *Glomar Challenger* oceanographic drilling ship in 1970. Some 2000 meters thick, the so-called Mediterranean Evaporite is underlain and overlain by deep pelagic sediments. A homogeneous ''Main Halite'' unit, over 1000 meters thick, comprises halite and potash salts; it seems to have been deposited in deep water at a time when the Mediterranean was saturated with salt but maintained connection with the Atlantic Ocean. But then the connection failed, and continued rapid evaporation caused the water level of the Mediterranean Sea to fall. European and African rivers draining to the Mediterranean basin incised the flanks of the Main Salt, and became salt creeks that drained to the very deepest basins as the Mediterranean dried up completely. Salts recycled from the margins of the Main Salt accumulated disconformably on Main Salt of central regions, to thicknesses of several hundred meters. Halite in these upper deposits has a much lower bromine content than the Main Salt, apparently reflecting the recycling. The uppermost sediments bear the marks of a desiccated environment, like a desert salina—there are displacive nodular anhydrite, cross-laminated aeolian silt, stromatolites, and desiccation cracks.

Numerous submarine canyons on the Mediterranean

continental slopes, mappable to depths of 2000 meters or more, appear to be extensions of river valleys on the surrounding lands and mark the upper reaches of the salt creeks that redistributed Main Salt into the shallow salinas and salt pans. Desiccation of the Mediterranean basin led to deep incision of the Nile River, a fact not realized until engineers began their probes for bedrock at the Aswan High Dam site, some 800 kilometers upstream of the delta. It was at a far greater depth, and there was a correspondingly greater thickness of alluvial sediments, than anyone could have imagined; only years later was the explanation found—a base level that had fallen 1000 meters or more. On the European side of the basin, gradual falling of sea level promoted active groundwater movement and creation of the famous Yugoslavian karst (it is this region that gave karst its name).

Lacustrine Evaporite Deposits

Evaporite deposits of intracontinental desert regions differ in many respects—stratigraphically, mineralogically, and biologically—from the marine evaporites. In the arid Basin and Range region of the western United States, ephemeral lakes in intermontane basins are nourished mainly by spring snowmelt in adjacent mountains. Throughout the remainder of the year, the basin may experience a cloudburst or two or even no rainfall at all. Many lakes in the arid and semiarid West of the United States were much larger, deeper, and fresher in the recent past. During the Pleistocene Epoch, when much of North America was covered with glacial ice, there was much more rainfall in the region south of the ice margin, and this, augmented by glacial meltwaters, caused many large, internally drained basins to fill with water, even to the point of spilling over elevated mountain passes. As climate changed and glaciers receded, these so-called *pluvial lakes* began to dry up; fresh lakes became saline as they became smaller. Great Salt Lake in northern Utah is one that has not yet quite disappeared, but dry Lake Sevier is one that has.

Large, shallow desert lakes (playas) may completely evaporate and be seasonally (or less frequently) renewed. Crystals of gypsum or halite formed in such environments may attain spectacular sizes. Gypsum beds may comprise large, multiple "swallowtail" twins that, standing upright, impart a coarse columnar fabric to the bed; halite also commonly forms "corner-

oriented" columns (see Chap. 2) that grow upward parallel to (111). The large crystals grow intermittently over a period of many years and bear internal zones demarcated by fluid inclusions or other impurities or by dissolution surfaces.

It is often observed on desert playas that crystals of halite nucleate and grow at the water surface, forming small hoppers. Individual hoppers congregate into thin, loose "rafts." Surface tension holds these crystals to the water surface, but when they have grown too heavy to be supported in that way or are agitated by the wind, they fall to the bottom. There they act as seeds for larger crystals precipitating out of the bottom waters.

Hardie (1968) studied a small intermontane playa in Saline Valley, southern California, noting a concentric zonation of evaporite minerals. Water entering the basin from surrounding highlands precipitates calcite in alluvial-fan sediments outside the playa. Gypsum precipitates at the playa margin; toward basin center, gypsum gives way to glauberite, thence to a halite-glauberite mixture at basin center. Bristol Dry Lake, also in southern California, is underlain by at least 300 meters of salt-bearing mud, according to Handford (1982, p. 241). A concentric pattern of sediment types occurs here too. Alluvial-fan sands and gravels at the playa margin grade inward to gypsum- and anhydrite-bearing muds, which, in turn, grade to halite-bearing red or green clay interbedded with halite. Gypsum crystals up to 10 centimeters long are present throughout the playa, and isolated halite hoppers attain dimensions of 20 centimeters. Handford considered that the salt hoppers grew displacively in the mud, and that the beds of salt were bottom-nucleated precipitates out of standing water. Some saline lakes have an outlet to another saline lake. The least soluble minerals precipitate in the higher lake, the more soluble minerals in lakes further along the chain.

Hardie et al. (1978) envisioned a process for evaporite precipitation from deep lakes similar to that suggested by Schmalz (1969) for marine basins. Evaporation creates a dense, concentrated brine at the surface; this brine advects to the bottom, displacing less dense bottom-water upward. Eventually the entire body becomes saturated with halite or other evaporite mineral, at which stage continued evaporation precipitates that mineral. The heaviest brines probably form in the shoals, then flow downslope as bottom-hugging density currents; this process could give rise to a density stratification, such as the Dead Sea maintains

(Gornitz and Schreiber, 1981, p. 788). In the Dead Sea, which is a large, deep, inland lake with no outlet, there is (or has been) a marked permanent *pycnocline*—that is, a zone of rapid density increase with depth, separating undersaturated waters above from saturated waters below. Apparently the upper layer is kept undersaturated by continual dilution by Jordan River inflow. On that part of the floor of the lake below 40 meters there is a thick layer of halite precipitated from the underlayer. Gypsum is precipitated almost continuously from the surface waters.

Differences between Marine and Lacustrine Evaporites

Lacustrine evaporite systems display a wide range of composition, according to Hardie (1984, p. 215), but they seem to divide into two major groups—alkaline systems and neutral systems. *Alkaline systems* consist of Na-K-CO_3-SO_4-Cl, and are poor in Ca and Mg. Sodium carbonate minerals such as trona, nahcolite, natron, and gaylussite are characteristic precipitates. The Green River Formation (Eocene) of Utah and Colorado locally contains such evaporite deposits. They are not likely to be confused with marine evaporites. *Neutral systems* are characterized as Na-K-Ca-Mg-SO_4-Cl brines poor in bicarbonate and carbonate ion. They precipitate gypsum, halite, and various sodium and magnesium sulfate minerals. Some neutral systems, such as the Great Salt Lake basin in Utah and Death Valley in California, are calcium-poor; other neutral systems, such as the Dead Sea in Israel, are sulfate-poor.

Criteria for distinguishing between ancient marine and non-marine evaporite deposits (Hardie, 1984, p. 204) include kinds of fossils contained in the deposit, nature of associated facies, the mineral assemblage and vertical (stratigraphic) succession of mineral assemblages, and various trace element, isotope, and fluid inclusion characteristics. Many of the distinctive characteristics of ordinary lake deposits (see Chap. 8) are shared by saline lake deposits as well: evidence of terrestrial (that is, non-marine) environments all around, lack of a tidal influence, presence of varves. Bromine content in halite may provide a useful distinction between marine and non-marine evaporites. Evaporites with very much lower or very much higher bromine contents than predicted for sea water are, according to Hardie, probably non-marine.

Bedded Cherts

Silica is a dichotomous sort of substance that moves through the sedimentary cycle as mechanically durable, chemically inert, seemingly indestructible grains of quartz and as recondite solutions or pseudo-solutions that mysteriously materialize as cell walls of tiny plants and animals. All biogenic chert begins as opal, an amorphous or structurally disordered material that contains a certain amount of water; eventually, this opal crystallizes to chert. Some silica enters sediments diagenetically; it is an important chemical cement in many quartz sandstones. Chert replaces certain carbonate skeletal materials and occurs as nodules, sometimes prolifically, in carbonate rocks. As opal or chalcedony, silica replaces fossil wood and bone.

Extensive bedded cherts seem mostly to be products of sedimentation of planktonic diatoms and radiolarians; some siliceous sedimentary rocks contain large numbers of siliceous sponge spicules (Fig. 10-4). Much silica has been deposited in marine upwelling locales where organic productivity is high; this chert typically occurs with phosphorites and organics-rich shales. Some oceanic cherts seem to be linked to ocean-floor volcanism; radiolarian cherts commonly are associated with mafic volcanic flows and tuffs. Chert is also importantly associated with Precambrian banded iron formation. Bedded cherts are forming inorganically in certain shallow lakes. Silica chemically replaces marine carbonate sediments, forming chert nodules and stringers or silicified skeletal grains, and

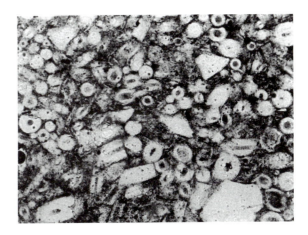

Fig. 10-4. Spicular chert of the Arkansas Novaculite.

there is abundant evidence that this can occur quite soon after deposition, even within depositional environments.

Oceanic siliceous sediments (and phosphatic and organics-rich sediments as well) accumulate most importantly in places where deep marine waters well up to the surface. Coastal upwelling occurs where winds blow parallel to a coast that lies to the left of the wind in the Northern Hemisphere or to the right of the wind in the Southern Hemisphere. Open-ocean upwelling occurs at marine divergences associated with atmospheric low-pressure systems, also at convergences of easterly and westerly winds in a single hemisphere (zonal convergence), and finally at the equatorial convergence. Deposition of siliceous (and phosphatic and organic) sediments is most rapid in coastal upwelling systems, but Parrish (1982, p. 757) noted that open-ocean divergence also leads to elevated biogenic silica in underlying deep-marine sediment.

Chemistry of Silica

At pH less than 9, dissolved silica is present only as monosilicic acid, H_4SiO_4 [alternatively represented as $Si(OH)_4$]. At higher pH, however, H_4SiO_4 dissociates to $H_3SiO_4^-$ and $H_2SiO_4^{2-}$, and silica solubility increases rapidly with pH. At ordinary temperatures and at pH below 9, the solubility of H_4SiO_4 is around 120 to 140 ppm (Fig. 10-5) and is not appreciably influenced by presence of other ions in solution (including the hydrogen ion). A solution at such concentrations is greatly supersaturated with respect to crystalline quartz, but it appears that direct inorganic precipitation of quartz from surface waters occurs only under rather unusual circumstances.

Monosilicic acid polymerizes slowly in water to form a *sol*. Bond formation that brings about this polymerization releases water molecules:

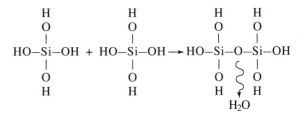

Continued polymerization results in cross-linking, gradual exclusion of water, and formation of a *gel*, which eventually hardens into opaline silica. Mizutani (1970, 1977) showed how diagenesis transforms amorphous silica into quartz, via the various forms of opal (see Chap. 11), to chalcedony, thence to chert or macroquartz. The process is irreversible and exceedingly slow under earth-surface conditions.

Dissolved silica in (fresh) groundwaters and rivers is derived largely from the weathering of feldspars and other silicate minerals and dissolution of siliceous skeletals in shales and limestones. Concentrations typically are between 10 and 30 ppm but locally may be much higher, particularly in volcanic terrains and around hot springs. In the modern ocean, concentration of dissolved silica is quite low at the surface, where silica is consumed by the planktonic siliceous skeleton-builders—the diatoms, silicoflagellates, and radiolarians that live in the open sea. But concentration increases with depth, being approximately 2 ppm in the deep waters of the Atlantic and about 10 ppm in deep Pacific waters. (The two oceans differ because the Atlantic loses deep water to other oceans in exchange for silica-poor surface water; the Pacific gains deep water from other oceans and loses silica-poor surface water; see Berger, 1970, p. 1386.)

Sources of dissolved silica in the ocean are river influx, submarine volcanism and hydrothermal activity, and submarine weathering (halmyrolysis) of ocean-floor sediments. Silica leaves the ocean as siliceous oozes, which are accumulations of siliceous skeletons that formed in surface waters and have sunk to the ocean floor or, in the case of the siliceous

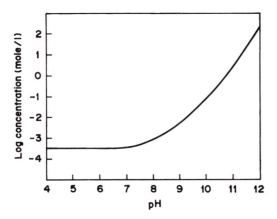

Fig. 10-5. Solubility of H_4SiO_4 as a function of pH.

sponges, that have formed on the ocean floor. Also, fine-grained terrigenous detritus (notably, clay particles) adsorbs some dissolved silica.

Siliceous Sedimentary Deposits

Pelagic siliceous oozes that ultimately become bedded cherts accumulate mainly beneath biologically productive surface waters, as in regions of upwelling associated with eastern boundary currents. In the Phosphoria Formation and associated units (Permian) in Idaho, Wyoming, and Montana (Cressman and Swanson, 1964), bedded chert appears to be composed largely of siliceous sponge spicules (diatoms had not yet evolved). These cherts are interbedded with black shales and phosphorites, a clear sign of high organic productivity. Another well-known bedded chert is the Arkansas Novaculite (Devonian and Mississippian), a thick-bedded, very light gray, spicular chert deposited on the submarine slopes of the Ouachita geosyncline. McBride and Thomson (1970) made detailed description of the Caballos Novaculite and associated sedimentary rocks of the Marathon region in Texas, this unit being partially equivalent to the Arkansas Novaculite. They concluded that the silica was derived from the accumulation of opaline skeletal grains, especially radiolarians and sponges, and that these particles were altered during burial into the present compact cherty texture. Deposition occurred in deep water, probably below the carbonate compensation depth, and sedimentation was very slow, less than 0.5 cm/Ka. Sponge spicules that survived the diagenesis are current-oriented. There are colored cherts associated with the (white) novaculite; these cherts contain terrigenous clay and other land-derived impurities.

Ribbon bedding is characteristic of many deep-water cherts. Jenkyns and Winterer (1982) suggested that ribbon-bedded cherts, which commonly rest directly on ophiolite and contain traces of hydrothermal sulfate minerals, were deposited in the axial regions of active oceanic rift zones. Tada (1991) described rhythmic light-dark banding in middle Miocene bedded cherts of northern Japan and suggested that complex cyclicities reflect an interplay between biogenic productivity cycles and Milankovitch-controlled cycles of dilution by terrigenous detritus (see Chap. 12). Barrett (1982, pp. 361, 364) interpreted certain Jurassic bedded cherts (interbedded with siliceous mudstones) in Italy as turbidites. Iijima et al. (1985, p. 229) also called on turbidity currents to explain chert-shale interbedding

in upper Paleozoic and lower Mesozoic formations in Japan. No modern ribbon-bedded siliceous sediments have ever been recovered from the ocean floor, and there are some unanswered questions about their origins.

The Monterey Formation (Miocene) in California is notable for its thick and extensive diatomaceous deposits. It contains opaline cherts rhythmically interbedded with thin graded sandstones, dolomites, and rhyolitic tuffs (Isaacs et al., 1983, p. 252). Diatoms are the most abundant skeletals, but radiolarians, silicoflagellates, and siliceous sponge spicules are locally abundant; also, there are some foraminifers, fish scales, and marine mammal bones. Pisciotto and Garrison (1981) described varves in some of these beds, reflecting seasonal cycles of marine upwelling and riverine runoff. Larger-scale cycles record climatic fluctuations with periods up to about 2100 years.

The Monterey strata were deposited in small, fault-bounded basins along a transform plate margin; basins nearest to land received hemipelagic muds rich in terrigenous detritus, much of it deposited as large turbidite fans. Distal basins accumulated mostly the biogenous pelagic oozes. Transition from deep-marine foraminifer-coccolith chalks beneath the siliceous beds to shallow-marine sandstones above indicates gradual basin filling and shoaling. White et al. (1992, p. 431) suggested that a change from non-upwelling to upwelling regime during expansion of the Antarctic ice cap in middle Miocene time might have contributed to the transition from early calcareous-phosphatic-organic sedimentation to later siliceous sedimentation.

A close association between pelagic bedded cherts and deep-sea volcanics has been noted, as, for example, in the Franciscan Formation, a Jurassic-Cretaceous melange in California. Bailey et al. (1964, p. 55) pointed out that bedded chert comprises less than 1 percent of the Franciscan, but that it is most abundant where volcanic rocks are common. The close association of chert with volcanic rocks suggested to Bailey et al. that the cherts owe their origin to volcanic eruptions, having been precipitated out of the dissolved silica of superheated waters issuing from deep submarine vents. The rinds of basalt pillows are depleted in silica relative to the cores of the pillows, and it appears that sea water in contact with hot lava dissolved some of the silica, then precipitated it inorganically as chert (presence of scattered radiolarians indicates that some precipitation was mediated by organisms). Under the pressures obtaining at a depth of 4000 meters, water does not boil and can superheat to 350°C or more; at this temperature, the water is capable of holding as much as 2000 ppm of silica in solution (Bailey et al., 1964, p. 67). The hot water convects rapidly to the surface and cools quickly. Cooling causes precipitation of the dissolved silica. Silica drops out and sinks to the ocean floor, at least in part as plankton skeletals, forming a chert bed. Individual chert beds might represent separate eruptions.

Bedded siliceous deposits occur also in shallow, non-marine environments. Peterson and von der Borch (1965) described silica precipitating as gelatinous opal-cristobalite (opal-C) in lakes associated with the Coorong Lagoon, South Australia. pH varies both above and below the critical value of 9. Ordinarily, vigorous photosynthesis keeps the pH high, around 10, and detrital quartz sand (and perhaps some feldspar and clay) in the lake sediments dissolves. But rotting vegetation just below the sediment surface maintains lower pH, and silica precipitates in the zone between high and low pH. Chert precipitation occurs also during seasonal decline in vegetal productivity and lake drying. Eugster (1967) described hydrous sodium silicates precipitated out of alkaline Lake Magadi in Kenya; he considered these inorganic chemical deposits to be precursors of bedded chert. Eugster (1969, p. 23) suggested that some Precambrian banded iron formations have formed in this way.

Siever (1983) outlined the relationships between sites of important siliceous sedimentation and plate-tectonic settings: In continental rift valleys, distinctive lacustrine siliceous deposits (such as Eugster described) are characteristic; typically they are associated with volcaniclastics, which are a major source of dissolved silica. In arid climates, the silica is a chemical precipitate (as magadiite or other alkali silicate, which eventually changes to bedded chert). But in humid climates, silica might be deposited as lacustrine diatomite.

With further rifting and establishment of oceanic basins, marine diatomaceous and radiolarian oozes come to dominance. Again, volcaniclastics might be closely associated. Thin chert beds and nodular cherts of marine-shelf carbonates are deposited on passive continental margins, influenced by marine upwelling and nutrient-rich riverine waters. On active margins, forearc and backarc basins accumulate thick successions of diatomite and radiolarite. Here, too, volcaniclastics may be interbedded.

Phosphatic Sediments

Phosphate-rich sediments are almost exclusively a chemically precipitated product of marine environments, though there are some lacustrine phosphatic sediments, as in the Green River Formation (Eocene) in Wyoming, and terrestrial bat- and bird-guano deposits of caves and oceanic islands. Certain organisms make phosphatic skeletons, including some brachiopods and molluscs and the vertebrates; certain worms precipitate phosphatic tubes in which they live, and some crustaceans line their burrows with phosphatic material.

Chemistry of Sedimentary Phosphate

Sedimentary apatite is mostly non-stoichiometric hydroxy-fluorine carbonate apatite with a chemical formula $Ca_5(PO_4,CO_3,OH)_3(F,OH)$. Up to about 6 percent of PO_4^{3-} is replaced by CO_3^{2-} in marine sedimentary apatite. The phosphate group has the shape of a tetrahedron, and when it is replaced by the flat triangle of the carbonate group, the "empty corner" takes up an OH^- or F^- ion that maintains charge balance. Carbonate fluorapatite is called *francolite*. An x-ray-amorphous or cryptocrystalline mixture of apatite minerals is called **collophane**, the typical inorganic phosphate of modern sediments.

In virtually all depositional environments, phosphorus is but a trace element. Dissolved phosphorus in the modern ocean, occurring mostly as HPO_4^{2-}, is most concentrated in the oxygen-minimum layer, typically between 500 and 1500 meters depth; even here its concentration is only 50 to 100 ppb. Phosphatic skeletal debris on the ocean floor typically is corroded, indicating that bottom sea water is undersaturated in phosphate. It is difficult to imagine primary precipitation from such waters. Phosphate is, however, considerably more concentrated in organic matter, and it appears that Recent phosphorites form just beneath the sediment-water interface in sea water–saturated pores. Calcite in the sediment supplies the Ca^{2+} necessary for apatite formation, and calcitic materials in phosphorite commonly appear to have been replaced to some extent by apatite or collophane. Martens and Harriss (1970) showed experimentally that Mg-ion inhibits the formation of apatite, Mg^{2+} entering some Ca^{2+} lattice sites and distorting the embryonic apatite lattice to such a degree as to inhibit further crystal growth. (Mg^{2+} also complexes with phosphate, forming MgHPO$_4$, which reduces concentration of apatite-making HPO$_4^{2-}$ in solution.) In ordinary sea water, which certainly *does* contain Mg^{2+}, only an amorphous calcium phosphate (collophane) forms, and it can change to crystalline apatite only if Mg ion is removed from the waters in contact. Interstitial water of marine sediment commonly is depleted in dissolved Mg owing to the early diagenetic formation of Mg-bearing minerals such as dolomite or Mg-rich calcite cement or by adsorption of Mg on certain clay minerals; these processes might promote crystallization of apatite. According to Jarvis (1992, p. 79), crystallization to francolite occurs within a few tens of years. With more time, francolite loses carbonate and become fluorapatite. McArthur et al. (1980) presented carbon- and oxygen-isotope evidence suggesting that some francolite is authigenic and some is diagenetic.

Phosphatic Sedimentary Deposits

Phosphorite is essentially a phosphate-rich sedimentary rock of marine origin, containing phosphate mainly in the mineral apatite. Phosphorite nodules and crusts form today in marine environments of high organic productivity, such as those associated with upwelling of cold, nutrient-rich marine waters on continental slopes and outer shelves, in water depths less than 400 meters (Fig. 10-6). Phosphorite is characteristically associated with spicular or radiolarian chert and pyritic, organics-rich sediment. The silica and the organic carbon clearly are accumulated and concentrated by biological processes, and it seems to follow

that the phosphate is too. Early degradation of organic matter in sediments releases phosphate to the interstitial water, presumably building up such high concentrations that phosphorite can form. An anoxic chemical environment and a pH somewhat below that of normal sea water seem to be important; anaerobic organic decay creates the required low Eh and pH. At upwelling sites, an intense oxygen-minimum layer develops on the continental margin as a result of oxidation of settling organic matter (Jarvis, 1992, p. 80). This layer of oxygen-depleted water intersects the sea floor at outer shelf and upper slope. Reducing conditions create sulfide on the bottom, and sulfide-oxidizing microbial mats may thrive. O'Brien et al. (1981) have shown that such mats can be important in concentrating and precipitating phosphorite in modern continental-margin environments.

Modern phosphorite accumulates also in low-productivity systems. Sediments are characteristically poor in organic carbon, but glauconite may be abundant (see Jarvis, 1980, p. 718). Bottom waters are well oxygenated, in contrast to high-productivity systems, and oxidation of organic matter in surficial sediments releases phosphate, which is scavenged by ferric oxyhydroxides. Burial to a few centimeters depth brings the phosphate-laden ferric substances into an anoxic environment, where these substances are reduced and dissolved. Phosphate released back to the pore waters in this way combines with sedimentary carbonate and makes phosphorite nodules or peloids. The reduced (ferrous) iron then diffuses upward toward the oxygenated surface layer of sediment, where it reoxidizes

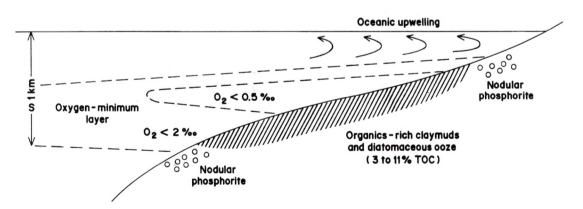

Fig. 10-6. The gross structure of a coastal upwelling oceanic system, modeled after that of the present Peruvian shelf and slope. (After Demaison and Moore, 1980, fig. 12.)

and precipitates once again as ferric oxyhydroxides, ready to scavenge more phosphate. Loughman (1984, p. 1153) suggested that phosphate authigenesis occurs under moderate or fluctuating oxicity, such as maintained at *top* or *base* of the oxygen-minimum layer (OML). Presumably pH must occasionally rise, permitting incorporation of carbonate in the sediment, which appears to be necessary in the fabrication of phosphorite. Organic carbon content of the sediments at these boundaries is low relative to that within the OML, and glauconite is present but pyrite is not.

As Sheldon (1981, p. 265) has pointed out, apatite ooids mixed with and nucleated on quartz sand in the Phosphoria Formation (see below) indicate that some apatite *is* chemically precipitated in depositional environments and is apparently *not* a product of early pore-water diagenesis in muddy sediments. But the chemistry of direct precipitation of apatite from sea water has not been determined.

Of the ancient phosphorites, the Phosphoria Formation (Permian of Idaho, Montana, and Wyoming) is the most famous and certainly the most voluminous. In the Phosphoria Formation phosphorite is interbedded with spicular chert and carbonaceous shale (Fig. 10-7). Much of the phosphate is in the form of ooids and irregular peloids or nodules; phosphate is contained also in the black, organics-rich shales. Cressman and Swanson (1964, p. 381) concluded that the ooid-peloid phosphorites were deposited in shallow marine water as phosphorus-rich deep water welled up onto a shallow shelf and supersaturated. The carbonaceous phosphatic shales were deposited in waters considerably deeper. Upwelling of cold, subarctic waters nourished a large plankton crop, which supplied the large quantities of organic carbon. Riggs (1980, p. 746) suggested a similar setting for the origin of the Miocene phosphate rocks in Florida.

Guano is more or less modern phosphate-rich sediment composed of animal excrement, especially that of bats and of maritime birds. Bat guano occurs almost exclusively in caves, where bats dwell and where the excrement is protected from the rain. The major bird guano deposits occur in marine coastal places, including islands, especially where upwelling of phosphate-rich sea water occurs (which ensures a plentiful food supply for the birds). Rainfall leaches the phosphate of guano deposits and washes some of it into the pores of underlying rocks, where it reacts with carbonate or

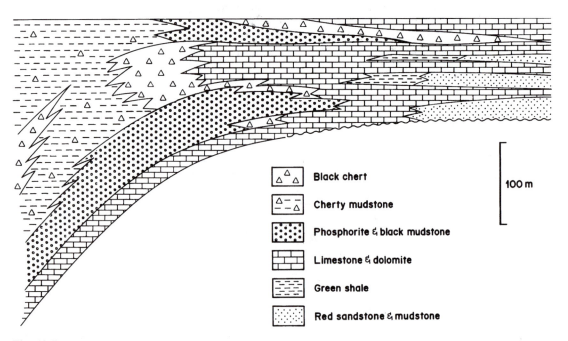

△ △	Black chert
△ - - △	Cherty mudstone
::::	Phosphorite & black mudstone
🧱	Limestone & dolomite
---	Green shale
::::	Red sandstone & mudstone

100 m

Fig. 10-7. Stratigraphic profile of the Permian phosphorite-bearing strata of Idaho and Wyoming. (After McKelvey et al., 1959, fig. 5.)

silicate to form various calcium, aluminum, or iron phosphate minerals such as apatite, variscite, vivianite, and wavellite.

Minor local phosphate accumulations occur in continental environments. There are various local vertebrate bone beds, including graveyards, mires, ravinement lags, and river-current-gathered concentrates. In western North America, there are recent bone beds at the bases of *pishkins* (buffalo jumps), cliffs over which the Indians drove herds of buffalo so as to harvest the hides and meat. At least one of these deposits (in Montana) was quarried for phosphate for use as fertilizer.

Iron-Rich Sedimentary rocks

Iron in sediments exists as oxides, hydroxides, sulfides, and certain silicates and carbonates. Iron-rich sediments are most abundant in the Precambrian, possibly a response to the accumulation of oxygen in earth's atmosphere brought about by organic photosynthesis. It is precipitated today, slowly, along with manganese, over large tracts of the deep ocean floor. Iron accumulates in soils in humid climates, the ferric oxides and hydroxides being among the least soluble of weathering products. Iron occurs in most sedimentary rocks in small quantities and is responsible for much of the color that sedimentary rocks have.

Hematite and magnetite are the principle iron minerals of the Precambrian *iron formations*. Magnetite is present also in the Jurassic iron ores of Lorraine, Luxembourg, and Germany. Goethite and limonite, in contrast, are the principle constituents of the Phanerozoic iron-rich sedimentary rocks, commonly occurring as ooids, and closely associated with chamosite, siderite, or calcite. Hematite occurs abundantly in the shallow marine Clinton Formation (Silurian) in the Appalachian region of the United States, the ''oolitic iron ore'' that was the basis of a steel industry in Birmingham, Alabama. Its occurrence as ooid cortices suggests that it can precipitate directly from sea water, possibly through a colloidal intermediary. But it also replaces calcite skeletals in the same rocks or occurs as peloids that seem to have been chamosite originally. Goethite is more common in Mesozoic and Cenozoic sedimentary rocks and seems to share with hematite many of its occurrences. Siderite, the iron carbonate of the calcite group, occurs in some Precambrian iron formations in the Lake Superior region and also as concretions (clay ironstones) in certain Phanerozoic

shales, notably shales of coal cyclothems. Chamosite and berthierine are iron-rich clay minerals; they commonly form ooids, these dark green objects set in a matrix of siderite or calcite in Phanerozoic ironstones. Glauconite, another iron-rich clay mineral, occurs in many Phanerozoic marine sedimentary rocks, including sandstones and limestones. Greenalite, also a clay mineral, is the only one of the iron silicate minerals that occurs abundantly in Precambrian iron formations. Pyrite, the most important iron sulfide, occurs in small amounts in many sedimentary rocks, generally as scattered, isolated crystals of diagenetic origin, or as tiny (3 to 30 micrometers) globular aggregates (framboids), authigenic, in mudstones. Marcasite forms flat concretions in shale layers in Pennsylvanian coal cyclothems of the Appalachian region.

Chemistry of Sedimentary Iron

Stability of various iron minerals is governed by pH and Eh (Fig. 10-8). The ferric oxide hematite is stable at positive Eh. The ferrous minerals pyrite, siderite, and magnetite are stable at negative Eh, and their presence depends on the activities of carbonate and sulfide ions. Activities of the anions HCO_3^- and HS^- or SO_4^{2-} (Fig. 10-9) also are important.

At approximately neutral pH, pyrite forms if sulfide activity is high, typically a result of bacterial reduction of sulfate. Oxidation of organic matter helps to maintain the negative Eh. Pyrite, or its dimorph marcasite, occurs in some rocks that, though deposited under well-aerated conditions, contained local reducing micro-environments. In the Fort Hays Limestone (Cretaceous of western Kansas and eastern Colorado), for example, the white, foraminiferal chalks contain vertical burrows filled with pyrite; in outcrops, this sulfide has weathered to limonite. The burrow apparently was a place of low Eh, brought on by bacterial decay of the burrow maker or its waste products.

Siderite forms where sulfide activity is zero and Eh is low, as in the fresh waters of many lacustrine and deltaic environments. According to Maynard (1983, p. 43), siderite might form where decomposition of organic matter has led to depletion of oxygen in the sediment but has not progressed to the point of reduction of sulfate. Siderite is an important cement in some rocks, occurring as coarse spar, aggregates of extremely fine crystalline rhombs, or spherulites of radiating fibrous crystals. Magnetite, an important sed-

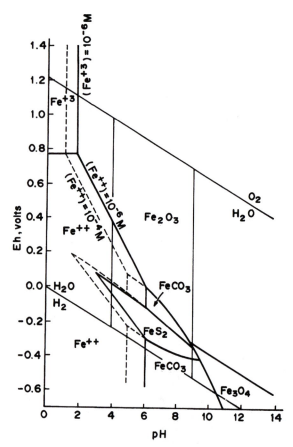

Fig. 10-8. Stability of sedimentary iron minerals with respect to Eh and pH. (After Garrels and Christ, 1965, fig. 7.23; Krauskopf, 1979, fig. 9-4.)

The Precambrian Banded Iron Formations

Most iron-rich sedimentary rocks are Precambrian. Mineralogically, they consist variously of oxides, silicates, carbonates, and sulfides, including hematite, goethite, magnetite, chert, greenalite, chamosite, siderite, and pyrite, and metamorphic derivatives such as stilpnomelane and minnesotaite.

Banded iron formation (BIF) is a sedimentary deposit consisting of iron-rich bands alternating with bands of chert; it seems to be restricted to the Precambrian. Three major types have been recognized: In the *Algoma-type BIF*, shaly sulfides, black cherts, and bituminous shales with siderite concretions are typical; the oxide minerals are generally lacking. Cherts commonly show centimeter banding with millimeter lam-

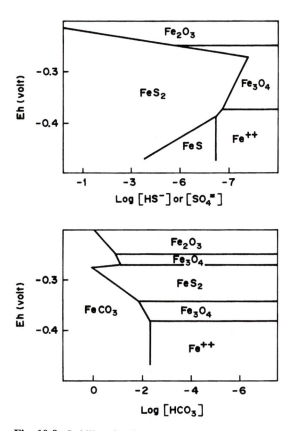

Fig. 10-9. Stability of sedimentary iron minerals with respect to Eh, $[HCO_3^-]$, $[HS^-]$, and $[SO_4^{2-}]$. (According to Curtis and Spears, 1968, figs. 1, 2.)

imentary mineral in some Precambrian rocks, forms under reducing conditions where carbonate and sulfide ions are not available.

Weathering of crystalline rocks containing abundant iron silicate minerals seems to be the major source of iron in the sedimentary environments. It is transported as colloids stabilized by organic matter, chelated in organic molecules, or adsorbed on clay particles. Much of it is precipitated as hematite or limonite cement in sandstones or is adsorbed on the surfaces of clay particles in mudstones. Some dissolved iron derives from submarine volcanic exhalations.

ination, and some seem to have turbidite characteristics. Deposits are thin and not very extensive; most are late Archean and associated with submarine volcanics in greenstone belts. The Michipicoten district in Ontario contains them. *Superior-type BIF* are thick, laterally continuous deposits typically bearing textures and sedimentary structures indicative of shallow-water environments, such as ooids, ripple lamination, and stromatolites; some appear to be associated with evaporitic sulfate. All mineral facies are represented: oxide, carbonate, silicate, sulfide. Their ages are mostly in the range of 2500 to 1900 Ma (early Proterozoic). They are associated with basic volcanics, or with shallow-marine carbonate or terrigenous rocks. Examples are the large deposits in the Lake Superior district in the United States, the Transvaal basin in South Africa, and (arguably) the Hamersley district in Australia. *Rapitan-type BIF* are mainly oxide facies of shallow marine basins at non-volcanic (or weakly volcanic) rift

zones (James, 1992, p. 574); typically they consist of bands of hematite alternating with chert bands. All appear to be late Proterozoic. They are associated with glacimarine deposits (Maynard, 1991, p. 143) and other terrigenous rocks and with minor basic volcanics. One example occurs in the McKenzie Mountains in northwest Canada, and there are others in Namibia, Australia, and Brazil.

Drever (1974) suggested an origin of Superior-type iron formations that is reminiscent of the suggested origin of marine phosphorites and bedded cherts (Fig. 10-10). Sedimentary structures and textures in Superior-type BIF are similar in some respects to those of carbonate rocks and phosphate rocks deposited under the physical process of upwelling of deep water onto a marine shelf in other times. The Proterozoic was a time of stabilization of cratonic areas and widespread development of continental shelves, and this appears to have been an essential development, as it provided

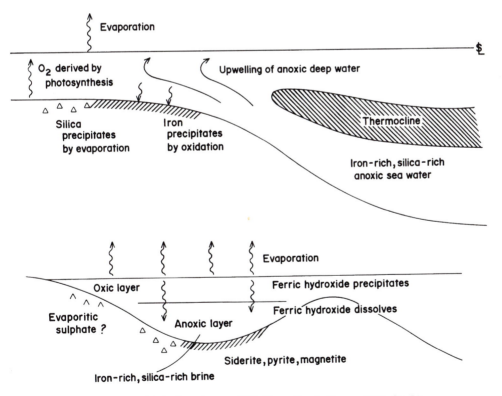

Fig. 10-10. Depositional model for Superior-type BIF. (According to Drever, 1974, fig. 5.)

the shallow marine depositional sites in which up-welling deep marine waters could discharge their dissolved loads.

If atmospheric partial pressure of oxygen were a tenth of its present value, the oceans below the thermocline would be anaerobic, but oxidizing in the shallower depths. The deep water would contain high concentrations of dissolved iron derived presumably from rock weathering on land surfaces, from submarine volcanism, or released hydrothermally from ocean rifts (James, 1992, p. 571). In the absence of silica-secreting organisms, dissolved silica would also have been abundant in the deep water. Upwelling of deep marine water onto shallow shelves would lead to precipitation of the iron by oxidation and of silica by evaporation. Oxidation or iron releases H^+, thereby inhibiting precipitation of carbonate. Organic carbon, according to Drever, would control mineral facies—at low levels of organic carbon hematite or magnetite would precipitate, at intermediate levels ferrous silicates would form, and at high levels, siderite (Fig. 10-11). Shallow, restricted, marginal marine basins not influenced by upwelling might have developed stratified water columns with anoxic bottom layers. Ferric hydroxide raining down from an upper layer would redissolve as ferrous iron in the anoxic lower layer and then precipitate as siderite, pyrite, or magnetite. Evaporation from

such basins may have hastened the precipitation of silica, perhaps as magadiite, which, under diagenesis or metamorphism, becomes chert (sodium released during the transformation would be available for fabrication of riebeckite, a soda amphibole present in Transvaal and Hamersley BIF).

Organisms appear to have been important, not as agents of direct precipitation of iron or silica but as founders of a chemical environment conducive to inorganic precipitation of these substances. Algoma-type iron formations, in which oxide minerals are rare, appear to have been deposited in the virtual absence of oxygen. During late Archean and early Proterozoic time, photosynthesizing organisms caused a slow, gradual accumulation of oxygen in the atmosphere and oceanic surface waters (see Chap. 12). Superior-type BIF formed during a part of earth's history when atmospheric oxygen partial pressure had attained a certain level; it could not have been deposited at other times. Rapitan-type iron formation seems to reflect higher oxygen levels.

The Phanerozoic Ironstones

The iron-rich sediments of the Phanerozoic do not bear the distinctive chert bands of the (dominantly Precambrian) banded iron formations and are different in other ways as well. In recognition of these differences, we apply to them a different name—*ironstone*. Besides detrital contaminants and precursor materials such as quartz grains, organic matter, and carbonate allochems, they are composed variously of hematite or goethite and berthierine or chamosite; some are altered diagenetically to siderite and pyrite. Some ironstones are of shallow marine origin; others formed in lakes or bogs. In contrast to the Precambrian banded iron formations, the immediate source of iron for the Phanerozoic ironstones probably was not the sea but the land, where iron is released during the weathering of rocks.

Many of the shallow marine ironstones are stratigraphically and petrographically similar to carbonate deposits of shallow-marine barrier-island and lagoon complexes; many are oolitic or contain marine skeletal grains. Carbonate stratigraphers and petrologists have, of course, examined many ancient and modern lagoons and sand banks of shallow marine shelves and know them well. In the ironstones, the ooids, peloids, skeletals, intraclasts, muds, and cements, even if not of carbonate, formed in the same way as their carbon-

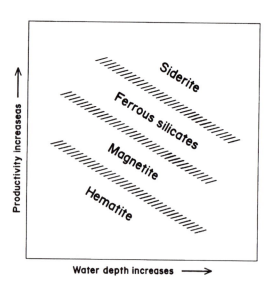

Fig. 10-11. Relationship of iron mineralogy to water depth and organic productivity in BIF. (After Drever, 1974, fig. 6.)

ate counterparts or, in many cases, it is clear, by early diagenetic replacement of their carbonate forebears. But the precipitation of ferric oxides, hydroxides, or silicates in marine (or marine-marginal) environments is not yet understood. We might call on systems that are dynamically and morphologically similar to those that have been called upon to explain early dolomitization (see Chap. 9)—lagoonal or estuarine "reaction vessels," where chemical solutions are prepared under special conditions, isolated somewhat from surrounding "normal" environments, and leaking their productions into the encompassing sediments or neighboring depositional environments.

Ironstones are formed also in bogs and coal swamps. Those of the Jurassic Coal Measures of Yorkshire are siderite-berthierine ooid grainstones and packstones grading to siderite mudstone (Spears, 1989, p. 19). The *blackband ironstones* are laminated siderite interbedded with or grading to cannel coal (brown spore-rich coal). They are considered to have formed in acidic, chemically reducing lakes or bogs.

Redbeds

Some sandstones and mudstones are colored by hematite, and there are many examples in the rock record of extensive, thick detrital deposits with deep red or maroon color. These distinctive deposits are referred to as *redbeds*. They are not particularly iron-rich; hematite concentrations even as low as 0.1 percent are sufficient to impart to a sediment a deep red color. Certainly some deep red sedimentary material is formed in the humid-climate soils called laterite and terra rosa. But T.R. Walker (1967) concluded that the hematite in most redbeds derives from dissolution and oxidation of iron-rich detrital grains (magnetite, hornblende, biotite, certain rock fragments) in *arid*-climate weathering and early diagenetic environments. Walker pointed to the close association of evaporites with redbeds and observed that modern alluvial, aeolian, and intertidal evaporitic sands and muds in Baja California, Mexico, gradually redden in the arid climate. The process begins in the weathering environment and persists into the diagenetic realm. Recent alluvium is not red but reflects the gray color of the granitic parent rocks. In somewhat older alluvium, mild chemical alteration has resulted in some cementation and a reddish color imparted by pigments that are x-ray amorphous. Pliocene alluvial sediments, in contrast, are deep red, and well-crystallized hematite is present. Petrography reveals that Recent alluvium contains several percent of detrital iron silicate grains, especially hornblende; but in older alluvium, many types of detrital iron-bearing mineral grains have undergone intrastratal dissolution. Hornblende, for example, is heavily altered to smectite, and relict grains have haloes of hematite. Walker estimated that more than 60 percent of the original hornblende in the red Pliocene sediments has been destroyed by intrastratal alteration.

Walker and Honea (1969, p. 542) concluded that all sediments, regardless of parent material or conditions at the source, contain enough iron to become bright red and will do so if the interstitial environment during or after deposition favors the decomposition of iron-bearing detrital grains and development and preservation of iron oxide. A depositional environment that is generally oxidizing may locally or momentarily become reducing and produce a redbed succession with non-red interbeds.

Marine Ferromanganese Nodules

Iron is precipitated slowly today on the ocean floor, along with manganese and other heavy metals—such as copper, cobalt, nickel, and cadmium—as porous, spongy nodules (concretions) and crusts. Many of the concretions reportedly have nucleated around whale earbones and shark teeth. Certain bacteria or other organisms apparently contribute to their growth. Rates of accretion vary between about 1 millimeter per year and 1 millimeter per million years. The nodules seem to be most abundant where other marine sedimentation is slowest, and in regions of active submarine volcanism. There are extensive fields of them on the abyssal plains of the Pacific and on the Blake Plateau in the Atlantic. Though they are abundant on the sediment surface, the sediments just beneath the surface typically contain few of them.

Nodules apparently begin to form by direct precipitation of manganese and iron from sea water. Iron oxide precipitates out of seawater at an appreciable rate at pH greater than 6, but manganese oxide only at pH above 8.5. Thus in the deep ocean, where pH is intermediate between these values, iron oxide tends to precipitate and manganese tends to remain in solution. But solid iron oxide catalyzes manganese precipitation (see Yamamoto, 1992, pp. 712, 716), and solid manganese oxide stimulates continued manganese precipitation.

Manganese and other heavy metals are brought to

pelagic and hemipelagic settings also in clay and carbonate particles and organic detritus. Once these particles have been deposited, the manganese and iron begin to diffuse out of them. Gradients in redox potential exist in the upper few centimeters of a deep-marine deposit. The surficial sediment, in close contact with oxygenated sea water, gives way to a "suboxic" condition a few centimeters below. Within this suboxic layer, iron and manganese contained in sediment particles is reduced and goes into solution. The dissolved metals diffuse upward and precipitate as nodules and crusts within the more oxic layer. Oxidation of organic matter in the surface layer may contribute other heavy metals, such as copper and nickel.

The nodules and crusts themselves are subject to dissolution as ongoing slow sedimentation eventually buries them and brings them into the suboxic zone. The manganese and iron go back into solution there, diffuse upward to the sediment surface, and precipitate again. In effect, ferromanganese nodules are continually recycled to the top of the sediment layer. Over the long term, the sediment surface gathers more and more of them, but few of them survive in the sediment layers beneath.

It appears that the heavy-metal content of modern manganese nodules—notably the Mn/Fe ratio and quantity of copper, cobalt, and nickel—reflects depositional environment (Cronin et al., 1991, p. 664). Nodules formed by diffusion from suboxic to oxic sediment layers, called *type i nodules*, are manganese-rich (Mn/Fe ratios up to 30) but poor in all other elements. They occur in areas of "rapid" organics-rich sedimentation. Those that have received metals remobilized from organic matter in oxic sediment layers, *type ii nodules*, have Mn/Fe ratios of 1 to 5; also, they are copper-rich and cobalt-poor. These typically form near continental margins in equatorial high-productivity zones. Those that form by direct precipitation from sea water, *type iii nodules*, have Mn/Fe ratios less than 1; they are poor in nickel and copper, and richer in cobalt than other nodules. They occur in areas where sedimentation rate is very low and organic matter in short supply, as on the Blake Plateau. Some nodules are of intermediate or mixed origins (Dymond et al., 1984). Certain relationships between composition and depth apparently are controlled by depth and degree of development of the oceanic oxygen-minimum layer (see Klinkhammer and Bender, 1980; Landing and Bruland, 1980).

Ferromanganese nodules and crusts are very rare in exposed ancient rocks on land, but Jenkyns (1967) found them in Jurassic marine pelagic rocks of Sicily. Cronin et al. (1991) described ferromanganese nodules in Jurassic limestones in Hungary.

Organic-Carbon-Rich Sedimentary Rocks

Inorganic carbon enters the biosphere by way of photosynthesis, a process that takes place on land and at sea. Most of it gets recycled within the biosphere or degrades into atmospheric or hydrospheric components, but some becomes entrapped in sediments and survives, in altered form, for hundreds of millions of years under deep burial.

Coal

Coal is a sedimentary accumulation of organically fixed carbon, a product of diagenesis of plant matter. Schopf (1956, p. 527) defined coal as "a readily combustible rock containing more than 50 percent by weight and more than 70 percent by volume of carbonaceous material, formed from compaction or induration of variously altered plant remains similar to those of peaty deposits." Most coals are composed of the altered remnants of the higher terrestrial plants and are of late Paleozoic age or younger, reflecting progress in the evolution of the plant kingdom. There are some coals as old as Devonian and even a Precambrian algal "coal" deposit in Siberia. Individual seams of Tertiary-age lignites range up to more than 200 meters in thickness, but most seams of hard coal are less than 2 meters thick. Most of the late Paleozoic coals were formed in tropical or subtropical paleolatitudes and most of the Mesozoic and Cenozoic coals in temperate and polar paleolatitudes.

Coal forms from **peat**, an essentially in situ accumulation of plant matter preserved under conditions that have prevented or retarded its oxidation and bacterial transformation. Submergence in stagnant water preserves plant matter as peat. Low Eh and low pH both are necessary to the preservation of organic matter in the depositional environment—and, to a large degree, fallen plant matter creates the chemical environment conducive to its own preservation. A certain amount of bacterial decay of submerged plant matter creates anoxia in the water; pH declines to low values as tannic acids and other organic acids are leached into the water and as biogenic H_2S and CO_2 accumulate.

Peat forms today in tropical swamps and coastal marshes and in bogs of temperate and polar regions. Many coals are intimately associated with fluvial and deltaic depositional environments. The composition of the plant community and its productivity are controlled by whether the water is fresh, brackish, or marine and whether the climate is humid or dry, warm or cold. These factors, in turn, control the rate of accumulation of the peat and the characteristics of the coals that ultimately result. Whereas warm climates are conducive to high plant productivity, they also promote rapid decay and destruction of organic matter. In cool climates, productivity is lower, but so is rate of destruction. Typically, only about 10 percent of plant matter production in peat-forming environments is actually preserved as peat.

High sulfur content (mainly as pyrite or marcasite) is associated with marine peat-forming environments. The lower pH of freshwater peat-forming environments, on the other hand, leads to low sulfur and low ash content. The chemical environment of peat formation may fluctuate and so create a succession of alternately high- and low-sulfur beds. Sulfur content may also change laterally, reflecting geographic variation in the coal-forming environment.

Like limestones, coals are in large measure a product of diagenesis; in fact, they are classified (partly) on the basis of the extent to which diagenesis has progressed, a property called *rank*. Coalification, or diagenetic conversion of peat to coal, includes early biological decomposition of plant matter, compactive expulsion of moisture, and thermal alteration under burial, which causes loss of volatile components. Increase in diagenetic rank from *peat* to *lignite* to *bituminous coal* to *anthracite* involves progressive darkening and increasing opacity and reflectance (luster), decreasing porosity and increasing density, decreasing solubility, decreasing volatile and moisture content, increasing fixed carbon concentration, and increasing calorific value. Within the bituminous stage there is an abrupt evolution of methane.

Analogous to the mineral grains of sandstones, or the allochems of limestones, plant *macerals* are the constituent particles of coal. Macerals are classified into three broad *maceral groups*: *Vitrinite* is a glossy maceral derived from wood and bark, leaves and roots; most shows well-preserved plant-cellular structure. *Exinite* or *liptinite* is derived from the waxy and resinous parts of plants, the spore cases (exines), cuticles (waxy protective layers on leaves and stems), and res-

inous excretions (sap). A certain kind of exinite, called *alginite*, is derived from algae. *Inertinite* is oxidized vitrinite and, like vitrinite, commonly preserves cellular structure, but it is hard and brittle. Some of it is created by fires in the peat-forming environment. Macerals are mainly distinguished and identified on the basis of their shape and structure, and their reflectance as observed microscopically; chemical etching methods, x-ray diffraction, and fluorescence also are occasionally resorted to. Exinites or liptinites have the lowest reflectance, inertinites the highest. Vitrinite and exinite reflectances increase more or less linearly with coal rank (as defined by volatile content). In blue and near ultraviolet light, exinites fluoresce, green in lower-rank coals changing to yellow and thence to red with increasing rank. As coals become more mature, their constituent macerals come to look more alike.

Coals typically are finely banded, each lenticular lamina composed of one of the four lithotypes *vitrain*, *clarain*, *durain*, or *fusain*. The bright layers in a coal are composed of glossy vitrain and silky clarain. The dull, ashy or sooty layers are durain and fusain; they are the components of coal that soil the hands.

Many coals contain early-formed carbonate concretions, up to 40 centimeters in diameter, called *coal balls*. They are composed generally of dolomite (Stach et al., 1982, p. 164) impregnating plant matter, which commonly is very well preserved within these concretions.

Other Organics-Rich Sediments

Of course there is lots of plant-derived organic matter that is substantially diluted by the mineral component in sediments. The solid organic material buried in sediments called *kerogen*, from which nature makes oil and gas, is composed of the remnants of the lower marine plants and bacteria and of the higher terrestrial plants. Kerogen is composed predominantly of five elements, carbon, hydrogen, oxygen, nitrogen, and sulfur; different kerogens are distinguished from one another in the relative proportions of these elements (particularly the first three). A distinction is made between kerogen and *bitumen*, which is the organic matter of sediments that is extractable by organic solvents; kerogen is the insoluble stuff. Different kerogens have different chemistries, of course, and their molecular compositions are quite complex. But there are simple chemical variations that can distinguish these components, namely atomic H/C ratio and O/C ratio, which together embrace the three most abundant chemical elements in kerogen. Most kerogens that have not been substantially affected by diagenesis have H/C ratios between 0.5 and 1.7, and O/C ratios between 0.05 and 0.3.

Four types of kerogen are recognized on the basis of relative proportions of carbon, hydrogen, and oxygen. *Type I* is kerogen with high H/C, about 1.5, and low O/C, generally less than 0.1. It contains a high proportion of lipids (waxes and fats or fatty acids). Type I kerogen is a product of algal remains or the severe biodegradation of non-lipid components of organic matter during its deposition. It corresponds to the coal maceral alginite and is the predominant organic matter of the so-called oil shales. Far more abundant is *type II* kerogen, with H/C between 1.0 and 1.5 and O/C about the same as in type I. It is a product of marine phyto- and zooplankton deposited in reducing bottom environments and is characteristic of marine sediments, particularly the dark shales and limestones. It is comparable to the coal maceral exinite. As a precursor of petroleum, it is by far the most important kerogen type. *Type III* kerogen has relatively low H/C between 0.5 and 1.0 and O/C as high as 0.2 or 0.3. It is derived from terrestrial plants, mainly the oxygen-bearing lignin and cellulose components; its coal-maceral equivalent is vitrinite. Not much oil is generated from type III kerogen, but it *is* a precursor of natural gas. *Type IV* kerogen, with the lowest H/C ratios, is essentially oxidized type III and is similar to inertinite macerals of coals.

These H/C and O/C ratios are conveniently displayed on the *van Krevelen diagram*. Not only are the various kerogen types and maceral groups clearly distinguished (Fig. 10-12) but also the trends in major-element composition with increasing maturation or coal rank. As the different kerogens undergo diagenesis, their O/C ratios first decline rapidly, then H/C values decline. Unaltered wood has high O/C, but unaltered cuticles and exines have high H/C and low O/C. Alteration of wood to vitrinite and inertinite and cuticles and spore cases to exinite is accompanied by gradual decline in both H/C and O/C. As coal rank advances, these ratios tend toward low values. Kerogens that initially are quite different gradually become quite similar.

Some organics-rich rocks are referred to as *oil shales*, because they can be made to produce "oil" in a retort. Some of these rocks actually *are* shales, but many are closer to limestone in composition; most of

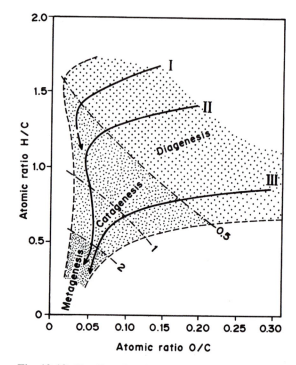

Fig. 10-12. Van Krevelen diagram, showing pathways of thermal alteration for types I, II, and III organic matter; dashed lines show approximate vitrinite reflectances (percent). (After Tissot and Welte, 1984, fig. II.5.1.)

them are of lacustrine origin. There may be recognizable remains of small freshwater or brackish-water green algae (*Chlorophyta*), but much of the kerogen is amorphous, probably owing to microbial degradation of the organic remains in the depositional environment. An oil shale must not have been buried so deeply that its organic matter has already been converted to petroleum naturally. If an oil shale is to be of economic interest, its product must have a higher calorific value than the energy required to accomplish the pyrolysis. According to Tissot and Welte (1984, p. 255), pyrolysis requires 250 cal/g (of rock), and the calorific value of kerogen is 10 kcal/g (of kerogen). Therefore, the thermal "break-even" oil-shale richness is 2.5 percent kerogen by weight. There are other costs, of course, and a more realistic threshold is 5 percent kerogen, corresponding to 25 liters "shale oil" per metric ton (moreover, the economics is closely circumscribed by prevailing or projected price of ordinary petroleum). In some parts of the world, as in China, Brazil, and Indonesia, ancient lacustrine oil shales seem to be the major organic source beds of petroleum.

Sedimentary organic carbon occurs in substantial concentrations in "black shales," such as the Hershey Formation and equivalents (Middle Ordovician) in the Appalachian basin, Chattanooga Shale and equivalents (Upper Devonian-Lower Mississippian) in the eastern interior United States, and Bakken Formation (Upper Devonian-Lower Mississippian) in the Williston basin. Broadhead et al. (1982) examined the Ohio Shale (Devonian) of northern Ohio, part of an extensive complex of black shales that underlie a thick wedge of mudstones, sandstones, and conglomerates shed from the Acadian orogenic highland in Devonian and Mississippian times (Fig. 10-13). They considered that the black shales were deposited in marine waters at least 200 meters deep, mainly on the basis of their occurrence at the downdip, distal end of a thick body of turbidite beds (Chagrin and Brallier Formations). The black shale contains 15 to 37 percent of organic matter, and there are numerous fossils of a planktonic alga *Foerstia* and the acritarch *Tasmanites*.

Demaison and Moore (1980) enumerated the factors influencing the accumulation of organic matter with mineral detritus in aquatic environments. Major factors are primary biological productivity of the upper water layers and of adjacent land surfaces, modes of transport of organic matter to depositional sites, sedi-

ment particle sizes and sedimentation rates, and biochemical degradation of sedimented dead organic matter in contact with bottom waters. *Concentration* of organic matter in sediments is determined largely by rate of deposition of inorganic materials relative to rate of organic-matter deposition.

The phytoplankton crop is a major aquatic source of sedimentary organic matter of marine black shales; another source is terrestrial plant matter transported to the sea by rivers. Bacteria play a major role in decomposing this organic matter in the water column, in the digestive tracts of higher animals, and in the interstitial waters of sediments. Where oxygen is available, organic matter is oxidized by aerobic bacteria; CO_2 is evolved. But in anoxic environments, anaerobic bacteria oxidize organic matter by reducing nitrate; this releases molecular gaseous nitrogen. When the nitrate supply has been exhausted, then anaerobic bacteria use sulfate as the oxidant, releasing H_2S. And when sulfate is gone, then fermentation is resorted to; methane is a product of some kinds of fermentation. Preservation of organic matter in sediments (Fig. 10-14) implies only partial or interrupted bacterial degradation.

The length of time that organic matter remains in oxygenated transporting or depositional environments influences the degree of oxidation that it undergoes, of course. Thus, short transport and rapid deposition promote the preservation and accumulation of organic matter. Deep marine sediments contain less organic

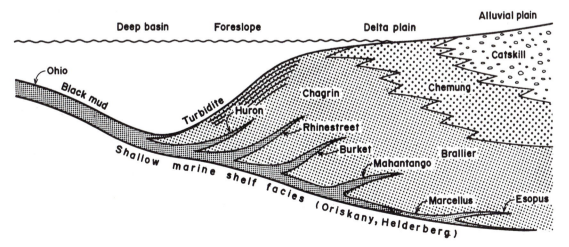

Fig. 10-13. Stratigraphic relationships of black shales in Devonian terrigenous succession in northern Appalachians. (After Broadhead et al., 1982, fig. 8, with permission of AAPG.)

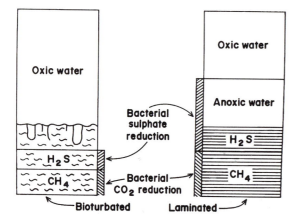

Fig. 10-14. Patterns of black shale deposition under oxic and anoxic conditions. (After Demaison and Moore, 1980, figs. 3, 4.)

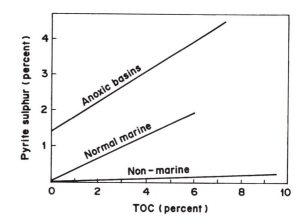

Fig. 10-15. Relationship between pyrite sulfur and organic carbon content in modern black shales. (After Holser et al., 1988, fig. 4-16.)

matter than shallower marine sediments, in part because of the long times required for organic detritus to settle through a long column of water—so much of it oxidizes or is consumed by animals during the descent. Once deposited, sediments are reworked by benthonic organisms as long as bottom waters are reasonably oxic (at least 0.3 ml/l dissolved oxygen). The deposit-feeding bottom fauna largely vanishes at lower oxygen concentrations, and the suspension-feeding fauna disappears at oxygen levels below 0.1 ml/l. Diffusion or mixing of oxidants into the sediment ceases when bioturbation ceases, and bacterial reduction of nitrate and sulfate also is curtailed.

Sulfides, especially pyrite, in a sediment are a result of bacterial reduction of sulfate (mostly in solution in pore waters). Rate and extent of sulfide production depends on the presence of sulfate, of course, and on the presence of sedimented organic-carbon nutrients for the bacteria. Plots of total organic carbon (TOC) concentration in modern sediments against concentration of sulfide sulfur appear to be useful in distinguishing between freshwater, normal marine, and dysaerobic marine environments of deposition. In freshwater environments, sulfate concentrations are low, and pyrite thus appears only in minute quantities, no matter what the organic matter content of the sediment (Fig. 10-15). In marine environments, concentration of organic matter, rather than sulfate, is rate-limiting, and pyrite-sulfur content correlates with TOC. Sediments of nor-

mal (oxic) marine environments give straight-line correlations passing through the origin. But where anoxic bottom waters exist, sulfide forms both within the sediment and in the anoxic water column that overlies. Thus, even if little or no organic matter actually reaches the basin floor, the sediment can still contain pyrite that formed in the water column. Pyrite content relative to TOC in the sediment is generally higher than in oxic environments, so plots have non-zero (positive) intercepts, but the two quantities are less well correlated.

Degens and Stoffers (1980) described organics-rich sediments accumulating today (or in the recent past) in the deep waters of the Black Sea. A catch basin for river discharge from half of Europe and part of Asia, it is the only modern example of a major silled basin turned anoxic at depth. A salinity stratification inhibits circulation and mixing, so the deep waters remain anoxic. On the bottom is a layer of dark-colored, fine-grained sediment, a few centimeters in thickness, containing in excess of 10 percent organic matter. Microlaminae, up to 200 per centimeter, seem to be a manifestation of seasonal plankton blooms and their subsequent mass mortality and sedimentation. As with coal-forming environments, it is organic matter itself that creates the anoxic conditions that ensure its preservation. Examples of much smaller silled basins with organics-rich anoxic bottom layers, are many of the fjords of Norway and British Columbia.

11

Diagenesis and Porosity of Rocks; Pore Fluids

From the time that a body of sediment is deposited until the time of its palingenesis (metamorphism or anatexis), or until it is weathered and eroded, that deposit undergoes many changes in texture, structure, and composition. These changes, called *diagenesis*, are responses mainly to the elevated pressures and temperatures that come with burial and to the long-term effects of fluids within or circulating through the deposit. Among the important diagenetic processes are *compaction*, *cementation*, *dissolution*, *recrystallization*, and *mineral transformation* or *replacement*. Sometimes diagenesis almost completely destroys original textures and even original mineralogy; more often, fortunately, it only gently modifies original features and even enhances them in some ways. By and large diagenesis is a process that we cannot directly observe; for a long time, it had been ignored by sedimentologists; only in recent years has it become the object of detailed study.

Diagenesis takes place over an extended period, generally under physical and chemical conditions that are gradually changing. Many diagenetic results appear to come about through a multistage process; later effects are overprinted on earlier effects. Different diagenetic pathways may come to similar results. Some chemical diagenesis may be effected by a "shot" of acid or other reactant that performs its office and then vanishes, so that the present chemistry of the pore fluid cannot be called upon to explain the present diagenetic state of the rock.

Pore Fluids

Chemical diagenesis is almost wholly a complex of reactions between rocks and the pore fluids that they contain. Pore fluids create the chemical cements of sedimentary rocks and also can dissolve or alter grains or cements. Some of these fluids are altered meteoric waters with high carbon dioxide contents; some are sea water or altered sea water with high salinity. Shales release much water during compaction; this water may be charged with sodium, calcium, magnesium, iron, carbonate, sulfate, and silicate, and other ions that had been only loosely bound to the clay particles in shales, or that were released during clay-mineral transformations, or that were derived from carbonate or siliceous skeletals or from authigenic minerals such as pyrite. Shales also typically contain organic matter, which, under diagenesis, releases water, carbon dioxide, and a host of reactive organic fluids that move out of the shales and into associated sandstones and limestones. Apparently, these fluids cause much of the cementation, alteration, and dissolution that takes place under deep burial in sandstones and limestones. Oil and gas, also originating largely in shales, may ultimately displace the reactive pore fluids in a porous rock body and effectively terminate chemical diagenesis, much as oiling a hammer keeps it from rusting.

Because the concentrations and solubilities of various ions in pore waters commonly are quite low (even if *total* salinity is *high*) and because a slightly oversaturated fluid quickly becomes merely saturated as it precipitates part of its solution load, we must invoke large-scale movements of subsurface fluids in order to explain their large volumetric effects on rocks. Galloway and Hobday (1983, pp. 230–232; see also Coustau et al., 1975, p. 106) made a hydrologic zonation of sedimentary basins, distinguishing three stages or zones called meteoric regime, compactional regime, and thermobaric regime (Fig 11-1). In the *meteoric regime*, pore spaces are filled early on with water diverted from the surface and continually circulating through the sedimentary fill, recharging at outcrops,

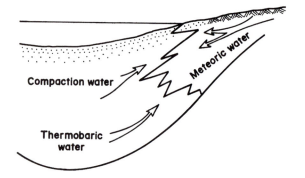

Fig. 11-1. Pore-water regimes. (According to Galloway and Hobday, 1983, fig. 11-3.)

discharging at hydrologic base level (commonly at or near a coastline). The ***compactional regime*** embraces that part of a basin fill that is expelling connate pore waters upward and outward as a result of compaction. Such flow tends to be focused through the more permeable layers. In the ***thermobaric regime*** of the deeper regions of a basin fill, waters and other fluids bound up in clay minerals and in organic matter are being released to the pores. Pore water also undergoes thermal expansion; this may cause thermal convection under some circumstances, but in general convection seems not to be an important process (Bjorlykke et al., 1989, p. 247). The volume of compactional and thermobaric waters, though large, is finite; meteoric water, in contrast, continually renews itself. In mature basins, after basin filling has ceased, the compactional and thermobaric regimes begin to contract as the meteoric regime expands; meteoric waters may ultimately reach strata buried to depths of thousands of meters. Temporal and spatial distribution of pore-water regimes in a basin, and volumes and rates of flow, are strongly controlled by sedimentary basin tectonics, including subsidence and uplift, extension and compression. Probably many diagenetic events, such as a particular cementation or dissolution, can be directly related to some tectonic event.

Temperature plays an important role. Solubility of silica varies with temperature, being greater in hot water; solubility of calcite, on the other hand, is temperature-retrograde, being greater in cold water. Thus, rising pore waters transport dissolved silica upward, depositing it in cooler zones; descending pore waters carry dissolved carbonate downward and precipitate it in warmer zones. Vugs formed by dissolution of car-

bonate at shallow depth may later become cemented by carbonate under deep burial. The structure and composition of clay minerals change gradually as they are buried and subjected to increasing temperatures. Some of these changes cause changes in composition and volume of the pore waters. Gypsum dewaters and recrystallizes to anhydrite above a certain temperature.

We can use certain fossils, such as spores, pollens, and conodonts, as indexes of the degree of thermal alteration of the sediments that host them. As these objects are heated under diagenetic conditions, they change color progressively and irreversibly (like a slice of bread in the toaster), so that dark conodonts, for example, come from rocks more intensively heated than rocks containing light-colored conodonts. Thermal annealing of fission tracks in detrital apatite grains seems to be of some value as an index of diagenetic temperatures. Fission tracks have a characteristic length, but the elevated temperatures of diagenesis gradually shorten them. Some have claimed that the statistical distribution of track lengths gives a thermal *history* (see Naeser et al., 1989).

The value of neutral pH varies with temperature. pH is defined as the negative common logarithm (that is, base 10) of hydrogen ion activity. An aqueous solution is neutral if its hydrogen ion activity equals its hydroxyl ion activity, that is, if pH = pOH. The solution is acidic if pH > pOH and basic if pH < pOH. Water molecules dissociate, forming hydronium ions H_3O^+ and hydroxyl ions OH^-, that is

$$2H_2O \rightleftharpoons H_3O^+ + OH^-$$

The extent of this dissociation is indicated by the equilibrium constant for this reaction:

$$K_w = [H_3O^+] [OH^-]/[H_2O]^2$$

At 25°C, $K_w = 10^{-14}$, very nearly, and so $pK_w = 14$. But this equilibrium constant varies with temperature, $pK_w = 14.94$ at 0°C and 12.32 at 100. Corresponding neutral pH at these temperatures are 7.47 and 6.16 (neutral pH is 1/2 pK_w). Thus some deeply buried hot pore fluid having pH of 7 is not neutral but basic. Solubilities of minerals in contact with pore fluid change as a result of temperature-driven pH change.

Stages of Diagenesis

One of the goals of the sedimentologist is to determine the history of diagenesis of a deposit, placing each event into its temporal sequence. A ***paragenesis*** is a narrative that describes all diagenetic stages and processes that have affected a given sediment body. Studies of this kind are of particular importance to the petroleum industry, especially insofar as diagenesis controls or modifies the porosity and permeability of rocks. The times and depths at which significant porosity-enhancing or porosity-reducing processes occur with respect to time of generation of oil and gas (itself a diagenetic process) to a large extent determine whether, where, and how much oil and gas accumulates as commercially viable pools in the subsurface. Paragenesis of the Frio Formation (Oligocene) of the Texas Gulf Coast has been carefully worked out (Galloway et al., 1982), mainly because of the importance of its sandstones as petroleum reservoirs. Lindsay and Roth (1982) enumerated no fewer than 22 diagenetic events in the Mission Canyon Limestone (Mississippian), an important petroleum reservoir in the Williston basin of the United States and Canada. Moore and Druckman (1981) described a long chain of diagenetic events in another economically important carbonate rock, the oolitic Smackover Limestone (Upper Jurassic) in Arkansas and Louisiana.

An important aid in the working out of diagenetic parageneses, including (or especially) the generation of petroleum, is the burial history diagram (described below). Pressure increases with depth of burial, at rates dependent on density or weight of overlying sediment. Temperature also increases with depth, being a function of thermal conductivities of sediments and rate of heat production in the earth's interior (among other things). If past and present geothermal gradients are known, then a burial history yields a thermal history. We know approximately the temperatures and pressures at which various diagenetic reactions occur, such as release of interlayer water from clays or conversion of kerogen to oil. A burial history diagram discloses the *timing* of these diagenetic processes.

Diagenesis has been divided into three stages (Choquette and Pray, 1970, p. 215): **eogenesis** is a geologically brief but very important early diagenesis that occurs near the surface; **mesogenesis** is long-term diagenesis under deeper burial; and **telogenesis**, a late-stage process, occurs again near the surface, after unroofing or exhumation of old, long-buried rocks that have already undergone prolonged and vigorous mesogenesis. The results of shallow diagenesis might be spatially more heterogeneous than the results of deep diagenesis, which more uniformly affects large volumes of rock. Moreover, conditions fluctuate more slowly at depth, and sediments generally spend very

long periods under deep burial. One suspects that sediments at depth and their pore fluids are, by and large, in chemical equilibrium and closely adjusted to prevailing temperatures and pressures, even as these conditions change over the long term.

The distinctions between diagenesis and metamorphism on the one hand and between diagenesis and weathering on the other are poorly defined. Metamorphism is rather arbitrarily taken to embrace processes occurring at temperatures above 200°C, which is about the temperature at which certain silicate mineral grains begin to change crystallographically and chemically. The clearest signal that metamorphism has begun is that different mineral grains begin to react chemically with one another in the solid state; glaucophane (metamorphic) facies is an early product of such reactions in certain sedimentary rocks. But many sedimentary rocks contain carbonate or sulfate minerals which undergo essentially metamorphic reactions at considerably lower temperature. Carbonate minerals react with certain clay minerals at such low temperatures and pressures. At conditions normally obtaining at depths of only 1000 meters or so, gypsum, monoclinic $CaSO_4 \cdot 2H_2O$, dehydrates and inverts to anhydrite, orthorhombic $CaSO_4$.

Weathering embraces a host of chemical and physical processes that destroy rocks, including sedimentary rocks, causing rocks to enter or re-enter the sedimentary cycle. It takes place under conditions similar to those that promote telogenesis, but telogenesis brings about textural and compositional changes in sedimentary rocks without necessarily breaking them down. Under telogenesis, certain components of a sedimentary rock recrystallize or undergo mineralogical replacement, or new cements are added; anhydrite that formed mesogenetically reverts back to gypsum. But near-surface selective dissolution of certain rock components could be viewed either as a telogenetic process or as a weathering process.

Pores in Rocks

Pores are spaces in rocks; they are always filled with fluids, such as air, fresh water or brine, oil or gas; they are of many different origins; they are subject to diagenetic alteration, just as the solid (mineral) components of a rock are. They may become filled with minerals (cements) and cease to be pores, though even these extinct pores are recognizable and certainly contribute to our general knowledge and understanding of pore-forming and pore-modifying processes. In the following, we will not make distinction between pores that are empty (open) and those that have been filled with chemical cements.

Pores are essentially geometrical objects; important aspects of their geometry are their quantities, sizes, shapes, and interconnectedness. Some pores are large enough for a person to walk through (limestone caverns), but most are so small that proper study of them requires a microscope or even an electron microscope. The study of very small or narrow pores in thin section or polished surface is enhanced if the sample has been impregnated under high pressure with a fluorescent dye (Soeder, 1990).

Kinds of Pores and Their Origins

The classification of pores in sedimentary rocks is based fundamentally on the spatial relationships between the pores and the solid material that hosts them. These characteristics bear upon their origin. Choquette and Pray (1970) made a classification of porosity in carbonate rocks based on whether or not they are controlled by texture or fabric of their host and on the sizes and shapes of the pores (Fig. 11-2). This classi-

fication can be applied equally well to terrigenous rocks, though in these rocks, porosity is not quite so varied or so complex as in carbonate rocks.

Intergranular pores (Fig. 11-3) are spaces between grains; they are there because grains cannot, short of drastic deformation, be brought together into an aggregate that completely fills all of space and forces out all fluids. They are by far the most important pore type in terrigenous rocks and in many carbonate rocks composed mostly of allochems. In granular materials unmolested by pore-destructive compaction or cementation, the number of pores equals or exceeds the number of grains; that is, for each grain in a granular aggregate there is at least one pore. The sizes and shapes of intergranular pores are controlled by the sizes and shapes of the grains themselves and by the ways in which the grains are arranged in the aggregate; thus, intergranular pores are said to be texturally controlled or fabric-controlled. Most pores are smaller than the grains that bound them. Probably most are roughly polyhedral, and surrounded by four, five, or six grains. Intergranular pores are well connected, that is, not isolated from one another but forming a continuous network of empty space in the aggregate. They could be viewed collectively as a single, pervasive

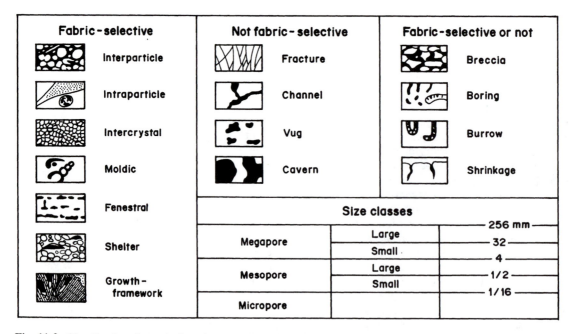

Fig. 11-2. Classification of porosity in carbonate rocks. (After Choquette and Pray, 1970, fig. 2, with permission of AAPG.)

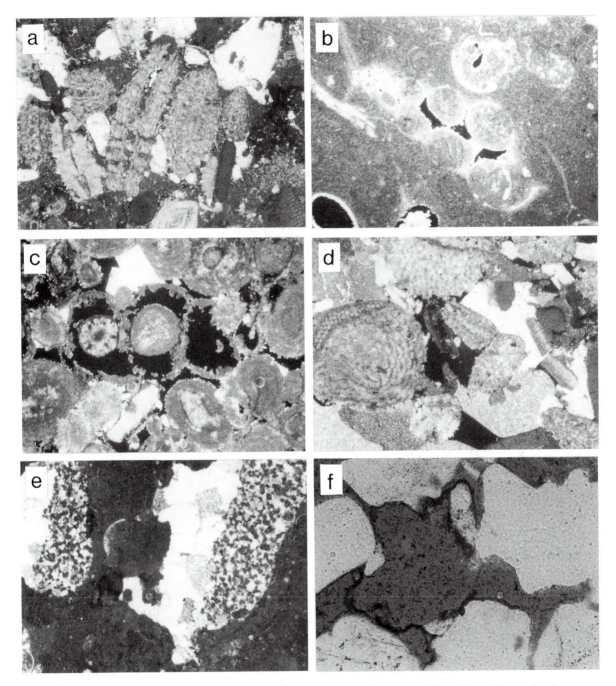

Fig. 11-3. Various types of pores in rocks: (**a**) intergranular pores in a skeletal grainstone; (**b**) skeletal-moldic pore in a limestone—high-spired snail mold partially reclosed before cement was added, suggesting that the lime mud matrix had not yet lithified at time of dissolution; (**c**) ooids partly dissolved—note compaction effect (up is to left); (**d**) primary intergranular pores enlarged under mesogenetic dissolution, then partially cemented; (**e**) large fenestrae or stromatactis with geopetal peloids and calcite cement (up is to left); (**f**) ''missing grain''—a grain has dissolved completely but leaves a thin wall of alteration product.

pore of very complex shape, characterized by poly-hedral nodes or swellings bridged by narrow constrictions called *pore throats*. A rock containing intergranular porosity is electrically conductive if the pore space is filled with a conducting fluid, such as salt water. This property of electrical conductivity depends on both quantity of pore space and its interconnectedness and provides an important means of measuring porosity of rocks in the subsurface.

Shelter pores and "*keystone pores*" result from unusual grain shapes or packings; they are larger than the typical intergranular pore in the rock. A shelter pore is a space roofed over by some large, flat, more or less horizontally disposed grain, such as an intraclast or a brachiopod valve, that, acting like an umbrella, has prevented sedimentation into the space beneath it. A keystone pore could be considered as an outsize intergranular pore resulting from accidental arching over, by a collection of grains, of some space comparable in size to or even larger than the grains themselves. Probably most such pores result from entrapment of small gas bubbles in loose sand, or from crystallization of halite or ice between grains (which holds or pushes grains apart) and subsequent dissolution of the halite or melting of the ice (such pores might more properly be called moldic pores in some cases, or fenestrae; see below).

Intercrystalline pores are spaces between mineral crystals. In some crystalline sedimentary rocks, the crystals are packed similarly to grains and the pore space is analogous to intergranular pore space. Reflecting this similarity, intergranular and intercrystalline porosity are brought together under the term *interparticle porosity*. In other crystalline rocks, the crystals are grown together so that perhaps only a few isolated polyhedral openings occur, bounded by the free surfaces of four or (rarely) more separate crystals. The boundaries between adjacent crystals are essentially planar, that is, extensive in two dimensions; no porosity should exist there, but if the crystals should contract (perhaps as fluids diffuse out of them) or the crystalline rock as a whole should expand (as by release of burial pressure), these crystal boundaries can open up into a network of intercrystalline *sheet pores*. Pittman and Duschatko (1970), and Wardlaw (1976), among many others, have pictured sheet pores, and other interparticle pores, by means of plastic or metallic pore casts.

In very fine-grained rocks such as mudstones interparticle pores are very abundant, but so small that only the scanning electron microscope (SEM) can resolve them. Such *micropores* in a clay mudstone typically are smaller than the clay particles, and their sizes and shapes are strongly dependent on the fabric of the clay particle aggregate. Micropores occur abundantly in many fine-crystalline dolomites and in chalks composed of very small skeletals, such as coccolithophores. These pores are inherited primary pores, though typically modified by micro-scale recrystallization and dissolution along crystal boundaries (Mosier, 1989). In thin sections of micrites and chalks impregnated with blue-dyed epoxy, a pervasive system of micropores is manifested as a blue haze. The presence of a micropore system is also indicated by low density of a rock.

Porosity might also occur within the grains or crystals themselves; this is called *intragranular* or *intraparticle porosity*. It is especially important in carbonate rocks containing fossils or skeletal grains. A brachiopod or ostracod comprises two valves that enclose space between them; a gastropod partly encloses space that sedimentary processes may fail to fill completely; a cephalopod shell or foraminifer test contains chambers almost completely isolated from one another and from the exterior. Bryozoan zoaria and the extensive skeletons built by colonial corals contain countless pores of very specific size and complex shape that were each the residence of an individual organism. The skeletal plates of echinoderms each contain a microscopic pore reticulum that comprises some 50 percent of the bulk volume of the skeletal grain. All of these are examples of intragranular pores.

In both terrigenous rocks and carbonate rocks, intragranular porosity can result from partial dissolution of grains under diagenesis. This is especially evident in detrital feldspars and rock fragments of some sandstones and in certain carbonate skeletal grainstones and packstones.

Moldic pores result from complete dissolution of some grains in the rock, either detrital grains or allochems, especially skeletals and ooids. Some result from decay of the soft tissue of an organism buried in sediment that is firm enough not to collapse into the void (or that is propped up by the gaseous products of the decay), or from dissolution of concretions or nodules. Moldic pores retain the shapes of the objects that they have replaced.

Development of moldic porosity requires, or indicates, a diagenetic chemical environment that causes certain types of grains in the rock to dissolve while leaving other types of grains or cement or matrix unaffected. Thus it may happen that in an oo-biomicrite (ooid-skeletal packstone) the ooids dissolve, leaving small, spheroidal moldic pores, but the skeletals and matrix remain intact. Occasionally, an ooid with displaced nucleus is observed. It appears that the cortex has dissolved, and the nucleus, resistant to dissolution, has fallen to the floor of the moldic pore; then the pore was cemented. In another limestone containing both molluscan skeletals and brachiopod skeletals, all molluscans might have dissolved, leaving empty or calcite cement-filled moldic pores, but no brachiopods have dissolved. This condition likely indicates diagenesis of the limestone in a vadose or fresh phreatic zone, where the calcite skeletons of brachiopods are quite stable but the aragonite skeletons of molluscs dissolve readily. Sorby, in 1879 (cited by Sandberg, 1975, p. 520), probably was the first to notice this, commenting that ''When we come to study some limestones in which casts [molds] are preserved we clearly see that the aragonite shells alone have disappeared. . . . '' In the Newman Limestone (Upper Mississippian) of northeastern Kentucky, shells of the gastropod *Straparolus* are common. The shells have two layers, an outer layer that almost perfectly retains its original ultrastructure, and an inner layer that is replaced by coarse calcite cement. Here it appears that diagenetic processes have made a clear distinction between shell material that was originally calcite and material originally composed of aragonite. The latter was dissolved and cement-replaced, but the former was unaffected.

Walter (1985) determined experimentally the relative dissolution rates of various modern carbonate skeletals in carbonate-undersaturated sea water; she found that, besides mineralogy, skeletal microstructure plays an important role. Thus, the aragonite of the green alga *Halimeda* is more soluble than the aragonite of bivalves and gastropods, and the aragonite of corals is less soluble than the high-magnesium calcite of red algae, and the high-magnesium calcite of echinoids is less soluble than the low-magnesium calcite of barnacles. Although some high-magnesium calcite is as soluble or more soluble than aragonite in fresh water, it nearly always transforms to low-magnesium calcite instead of dissolving and making moldic pores.

A special kind of moldic pore occurs in some carbonate grainstones, a pore enclosed within a thin and fragile *micrite envelope*. Skeletal grains commonly are subjected to surface micritization in the depositional environment (see Chap. 9). Micrite envelopes are associated with dissolution of aragonitic skeletals in fresh water diagenetic environments, strongly suggesting that the carbonate rock was emergent and exposed to meteoric water. The micritic surface layer seems to resist dissolution, even while the unmicritized material within is dissolving readily. Perhaps the individual particles of micrite are coated and bound together with algal slimes that protect them from corrosive fluids; if the skeletal grain is composed of aragonite, it is naturally quite soluble in fresh water, while the calcitic micrite layer is not. Thus, dissolution of a surface-micritized skeletal grain can leave behind this empty micrite envelope, which either remains empty or later fills with some chemical cement (Bathurst, 1975, p. 327).

Skeletal-moldic pores commonly occur in dolomitized rocks and in some sandstones, where skeletal calcite has been scavenged for fabrication of dolomite in nearby matrix or intergranular pores.

In a feldspathic sandstone, grain-dissolution processes may cause the nearly complete destruction of detrital feldspar grains but leave the quartz grains untouched. In a lithic sandstone containing grains of macroquartz and of chert, the chert grains might dissolve because they are more finely crystalline, while the quartz grains, again, are unaffected. There are quartz-cemented carbonate breccias in which all of the clasts have been dissolved, leaving only the cement; these ''boxworks'' are natural pore casts.

How, we ask, can we know what substance it was, or what grain type it was, now vanished, that the moldic pore represents? Commonly we cannot know, but we can guess it, based on the shape and size of the pore. This often works in limestones, where many grains have distinctive shapes. A dissolved detrital feldspar grain in a sandstone commonly is only partly dissolved or leaves a tenuous skeleton of undissolved material; or a distinctive alteration product coats the pore that evidences its precursor.

Fenestrae are pores in carbonate rocks created by entrapment of gas or other fluids or by desiccation, dissolution, or burrowing. Fenestrae are larger than intergranular pores (by definition) and generally larger than framework grains if such are present. Pores that occur in muddy limestones lacking grains cannot, of course, be intergranular. Typically, fenestrae are lenticular, about a millimeter thick, and horizontally oriented; where fenestrae are present, they are abundant. Fenestrae occur in intertidal or supratidal peloid packstones and grainstones and in lime mudstones and algalaminites. Fenestrae commonly are closely associated with subaerial exposure surfaces. Organic burrows and borings occasionally escape sedimentary filling, cementation, or compaction and survive as pores; some fenestrae apparently originate in this way. Some fenestrae appear to be due to dissolution and might be considered as vugs. In certain Mississippian mud-mound limestones of the Williston basin, in Montana and Saskatchewan, there are numerous fenestrae that appear, in thin section, to be separate and distinct from one another, but they are all cemented with anhydrite spar having the same optical orientation, indicating that all these fenestrae actually are connected. Occurring in thromboid packstones that may have contained substantial quantities of algal slimes and other labile organic matter, these fenestrae might have originated through inflation of the soft, sticky carbonate mass by biogenic gas evolved internally.

Grover and Read (1978) apparently were intrigued by the numerous fenestrae that they observed in the New Market Limestone (Middle Ordovician) in Virginia and tried to distinguish them according to their shapes—whether tubular, laminoid, or irregular—suggesting that their shapes are related to their origins. Read's (1973b, figs. 15–19) photographs of fenestrae in Devonian carbonate rocks of Western Australia are representative of the type. Some geologists refer to fenestra-bearing rocks as "birds-eye" limestones, or as *loferites* (Fischer, 1964).

The air bubbles that form in sand or mud under certain conditions (see Chap. 5) perhaps merits separate classification. They might be compared to the far more common gas vesicles of volcanic flow rocks.

Stromatactis (Lowenstam, 1950; Bathurst, 1959), peculiar pores contained in certain muddy carbonate mounds (Lees, 1964; Cotter, 1965; Ross et al., 1975), is generally described as horizontal or inclined lenticular openings, several millimeters to a few centimeters thick, having flat floors and digitate or irregular roofs, and in most cases cemented with calcite, of a peculiar type called radiaxial cement. Their origin has been much debated (see Chap. 9).

Vugs and *caverns* are large pores formed by dissolution, the latter being sufficiently large that a person may occupy them. Caverns are almost always associated with limestone karst, ancient or modern, though most of them eventually fill with red mud (terra rosa) or collapse. Vugs typically are equant pores, a centimeter or two in diameter, in fine-grained carbonate rocks. Probably most vugs result from dissolution of nodules (such as nodules of evaporitic gypsum); many vugs are subsequently lined or filled with chalcedony or with crystals of quartz, zeolites, calcite, celestite, pyrite, or other silicate, carbonate, sulfate, or sulfide minerals—these vug fillings are called *geodes*.

Fractures in rocks—other than those "early fractures" formed by desiccation, syneresis, or syndepositional slumping—result from mechanical stress of sufficient magnitude to cause brittle failure of small rock components or large rock volumes. Stress may be imposed by burial (overburden or burial stress), by tectonism (crustal deformation), or by forcible motion of internal fluids under high pressure. Some fractures are a result of loss of support under deep burial owing to intrastratal dissolution or of major expansions and contractions that accompany mineral alterations or phase changes.

Early formed fractures typically are crooked, having developed before lithification or in its early stages, when sharp contrasts in competence between the various sediment components still existed (lithification reduces these contrasts). Some fractures form in the brittle components of an otherwise soft, unlithified sediment. Space-enclosing ostracod or gastropod shells, for example, may collapse under gentle overburden pressure; ooids may spall or firm peloids or intraclasts may crack if pressed into close mutual contact. The fractures in these grains do not extend into the surrounding cement or matrix—cement had not yet been emplaced and matrix not yet lithified at time of fracture of the grains. Indeed, cementation or lithification of matrix likely would have prevented early grain failure.

The *microfracture* concept was introduced by Snarsky (1962), who indicated that tectonic forces and temperature increase can cause increase in pore pressure, which leads to fracturing of the rock and the opening of cracks. This is fracturing due to forces within. Tissot and Pelet (1971, p. 41) suggested that increased pore pressure is in many cases a direct result of conversion of kerogen to oil or gas, and that if this pressure comes to exceed the mechanical strength of the rock, then microfractures are produced. In other words, microfracturing is associated with *overpressure*. They performed experiments with blocks of organics-rich clays that seem to bear this out. According to Tissot and Welte (1984, p. 321), microfracturing due to evolution of oil or gas is restricted to "relatively deeply buried, compacted, low-permeability rocks such as shales or tight carbonates.'' Microfractures also are subject to enlargement by dissolution. Surdam et al. (1984) showed that kerogen maturation (that is, diagenesis of buried organic matter) can generate carboxylic acids that are very reactive. These and other organic acids released by maturing kerogen tend to widen microfractures by dissolution, even as these fractures are being made.

Fractures created in hard (brittle) rock by regional crustal stress are far more extensive, commonly traceable for many kilometers, generally are planar, are vertical or steeply inclined, and are multiple, constituting sets or families of large numbers of subparallel individuals. Generally there are two or more intersecting sets of these fractures, or joints, one set perhaps slightly offsetting the other, and revealing some pervasive regional stress field. The so-called butt- and cleat-joint sets of the Pennsylvanian coals of the Allegheny basin of the eastern United States maintain nearly constant orientations over the entire extent of the coal fields. The surfaces of large fractures typically display certain patterns of low relief, picturesquely referred to as *hackle plumes*, which trace out the patterns of propagation of the fractures through the rocks.

Fractures commonly come to be enlarged by dissolution as deep intrastratal fluids move through them, or they get cemented. Some fractures, having been cemented, reopen under renewed stress and are then cemented again, perhaps this time by a different material.

Primary and Secondary Porosity

It should be clear by now that some porosity exists from the time of deposition of sedimentary materials, and some types of porosity are introduced later, under diagenesis. The former is called *primary porosity* and the latter *secondary porosity*. Some secondary pores result from enlargement of primary pores and are not particularly fabric-controlled. Other secondary pores result from dissolution of individual grains or matrix or cement and may be very selective, affecting only materials of a certain composition. Much secondary porosity is a result of recrystallization of rock materials, such that primary pore space simply is redistributed and not necessarily accompanied by a change in total pore volume. This happens during dolomitization, for example, which transforms a system of micropores in a lime mudstone into a new system of larger but fewer intercrystalline pores of a coarse dolomite. (In the case of dolomitization, some additional

secondary porosity appears because the molar volume of dolomite is less than that of the calcite that it has replaced.) In a given rock body, secondary pores are typically (though by no means necessarily) larger than the primary pores. Mazzullo and Harris (1992, p. 616) admitted that it is often difficult to distinguish between primary and secondary pores in carbonate rocks.

Early and late pores commonly have different cements; primary pores may contain an early cement that secondary pores lack. The secondary pores did not yet exist at the time of emplacement of the early cement phase. When secondary pores do finally appear, they receive different cements, or no cements at all, reflecting a different diagenetic environment. Early pores, already cemented by this time, do not receive the late cements, unless, of course, early cementation was not complete, in which case these pores might contain early cements overlain by late cements.

Compaction of Sediments

Compaction is an irreversible diagenetic process that reduces original thickness and porosity of sediments under pressure. Some compaction results from early adjustments of grains into mechanically more stable positions than they had initially, and from early dewatering, that is, escape of some of the water initially deposited with (entrapped between) the grains, and tending to hold grains apart. Other compaction comes later, as sediment is buried, and the weight of overburden forces particles into closer contact, or even causes brittle or plastic deformation of the grains, or reshapes them by very local dissolution-precipitation processes. Compaction is less if grains are composed of strong materials or if grains are strong because they are large. Compaction always causes compression or expression of interstitial fluids from the compacting material; if for some reason these fluids cannot be compressed or expelled, then compaction will not occur.

Shales seem to be the major petroleum source beds, and petroleum, or some petroleum-like substance, moves out of shales and toward reservoir rocks during shale compaction. Thus, to know how and when shales undergo compaction is of considerable economic significance. Similarly for sandstones and limestones, which serve generally as the reservoir rocks; compaction reduces their porosity and permeability, rendering them less valuable as containers of oil and gas.

Compaction of Shales

Shales compact quickly during early burial and continue to compact over the long term, though at ever declining rates. Initial porosity of sedimented clay is about 80 percent (Perrier and Quiblier, 1974, p. 507), decreasing to near-zero values only at depths exceeding about 8 kilometers. Most of the compaction occurs very early, during the first few hundred meters of burial. Meade (1966, p. 1094) described experiments suggesting that early compaction (burial depths of less than 1/2 kilometer) involves rearrangement of clay particles. Initial edge-to-face fabrics (see Chap. 6) give rise to horizontal preferred orientation of the particles; initial face-to-face fabrics give rise to *oriented domains*—that is, distinct clusters of oriented clay particles, their orientation varying from cluster to cluster. According to Heling (1970, pp. 245, 258), only at shallow depths (<1000 meters) is shale porosity reduced as a result of clay-particle reorientation. Initial water content seems to be important, water apparently acting as a necessary medium for particle rearrangement. If muds dry early, there is not much chance of clay particle reorientation under early compaction.

Compaction in shales is controlled by three factors: permeability, which declines as compaction increases, total grain-to-grain contact area, which increases with increasing compaction; and viscosity of the pore water, which effectively increases as compaction increases. In the early stages of compaction, shale has high permeability and pore fluid escapes readily; but later, after substantial compaction, permeability has been reduced to low levels and pore fluid is less readily expressed. Total grain-to-grain contact area per unit volume increases as porosity decreases, so that compactive stress is distributed over a greater load-bearing surface. Thus, ever-increasing overburden pressures are required to effect the next incremental compaction. The viscosity of water increases toward a boundary; water molecules near a solid surface tend to bind to the surface and to one another, owing to electrical forces. Water near a clay particle is nearly eight times as viscous as normal, and the pore water in a compacted clay, where clay particles are very close together, has higher average viscosity than the pore water in a less compacted clay. This high-viscosity fluid is not readily displaced from the pores of a material that has already undergone substantial compaction, though the higher temperatures of deeply buried shales counteract this effect to some extent. (Non-wetting pore fluids, such as oil, do not behave in this manner.) Thermal expansion of pore water with increasing burial depth also retards compaction of low-permeability rocks.

Compaction of Sandstones and Limestones

In sandstones, other processes must be at work. Magara (1980) pointed out that, although shale porosity reduction decreases with increasing burial depth, sandstone porosity reduction is quite linear (Fig. 11-4); Sherer (1987, p. 488) observed a linear relationship for depths between 500 and 5000 meters. Compaction of sands is, of course, affected by cementation, and much porosity reduction in sands is achieved by this rather than by compaction. Gradual cementation, by occluding pore space and by gradually strengthening the aggregate, renders it less compactible and gives to it a compaction history that differs fundamentally from

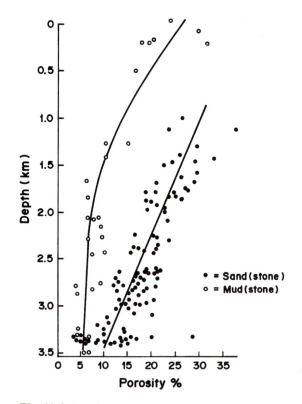

Fig. 11-4. Porosity (a measure of degree of compaction) as a function of depth for sandstones and mudstones. (After Magara, 1980, fig. 8, with permission of Scientific Press, Ltd.)

that of the mudstones. If early cementation does not occur or is weak, then pressure dissolution (see below) might substantially reduce original porosity, in part because grains change shape and come closer together and in part because the dissolved silica precipitates in the pores.

Fine sands compact more rapidly than coarse sands. Well-sorted sands and angular sands have higher initial porosities than poorly sorted sands and round sands and are thus potentially more compactible (Meade, 1966, p. 1096). Early compaction in sandstones, just as in clay muds, involves shifting of particles into denser packing arrangements, and this process is influenced by sorting and roundness of the grains. Sandstones that contain soft or malleable grains, such as shale fragments or carbonate peloids, can compact and loose much porosity as these grains deform under burial pressures (Wilson and McBride, 1988; Pittman and Larese, 1991). In quartz sandstones that contain carbonate skeletal grains, the carbonate grains may dissolve under pressure (Fig. 11-5f) and contribute to carbonate cementation between the quartz grains. The rock compacts because space originally occupied by carbonate grains closes up.

Limestones may compact early or late or show very little evidence of compaction. Many carbonate sediments contain soft peloids that apparently deform very early, before any cementation or other lithification. The peloids flatten and come to fill all intergranular space. Ooids also might deform if they are not cemented early. Resulting textures have been referred to as *overpacked* (Wilson, 1975, p. 62, 66). Many carbonate sediments are cemented early, even before burial, and this greatly inhibits subsequent compaction. Lime muds with high interstitial water content may undergo some early compaction, but these sediments also tend to lithify early. Early dolomitization inhibits compaction, a continuous framework of dolomite crystals serving to ''freeze'' a carbonate sediment (Choquette and Steinen, 1980, p. 187). Ooids or skeletals in carbonate grainstones may undergo pressure dissolution (Fig. 11-5e) under moderate burial depths if early cements are weak; stylolite-forming pressure dissolution at greater burial depths is an important compacting process in many limestones.

Evidence of Compaction

Evidences of early compaction are corrugated or telescoped desiccation cracks; flattened horizontal bur-row tubes or corrugated vertical tubes; space-enclosing skeletals that are fractured or collapsed and filled with undeformed early cement; intraskeletal spaces filled with originally soft material that is less compacted than similar materials outside the skeletal; peloids, intraclasts, and other originally soft materials that are plastically deformed or overpacked; spalled ooids whose ruptured surfaces and shards are coated with early cement; skeletals or other originally hard grains with fractures that do not extend into surrounding matrix or early cement. If a skeletal grain embedded in mud matrix dissolves early, before the matrix has lithified, the moldic pore may partially (or fully) reclose under early compaction.

Evidences of later compaction include grains closely packed (many grain-to-grain contacts, for example), pressure-dissolved grains, fractured or mechanically twinned grains or cements, and prominent stylolites; in some sandstones and limestones grain-moldic pores are partly reclosed under a compaction that follows or accompanies the secondary pore-forming process.

Some features, such as bed-draping around concretions or dip reduction in cross laminae, reveal that compaction has occurred but do not necessarily indicate when. *Undercompaction* of grains ordinarily indicates that early cementation prevented later compaction. Pore-fluid overpressure in deeply buried sediments indicates that compaction *would* occur if the pore fluids could be released. Production of petroleum from shallow overpressured reservoirs has occasionally resulted in damaging subsidence at the surface, owing to compaction of the reservoir rock as it is produced.

Pressure Dissolution and Stylolitization

The solubilities of quartz and calcite and other rock-forming minerals seem to increase under increased pressure. Pressure is not uniformly distributed over the surface of a sand grain buried among other sand grains in an uncemented or lightly cemented deposit. Pressure at a grain-to-grain contact might be close to lithostatic, but the free surfaces of grains (that is, the walls of intergranular pores) experience hydrostatic pressure, which is of lesser magnitude. Thus, very local *pressure dissolution* tends to occur at a grain-to-grain contact, the dissolved material immediately precipitating on adjacent pore walls. Grains change shape and come into closer mutual contact, and the sandstone

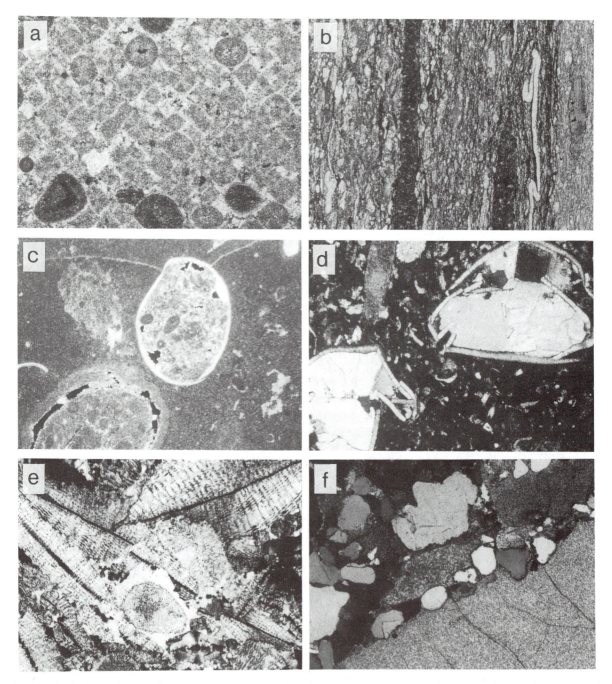

Fig. 11-5. Various evidences of early and late compaction in sedimentary rocks: **(a)** overpacked ooid grainstone; **(b)** overpacked skeletal-peloid packstone—dark bands are organic macerals, thin, contorted object is a flattened spore case (up is at left); **(c)** sediment-fill in small bivalves is less compacted than surrounding matrix; **(d)** ostracods(?) damaged by early compaction, then cemented with calcite and dolomite; **(e)** large foraminifers in stylolitic (sutured) contact; **(f)** quartz grains penetrating a pelmatozoan plate;

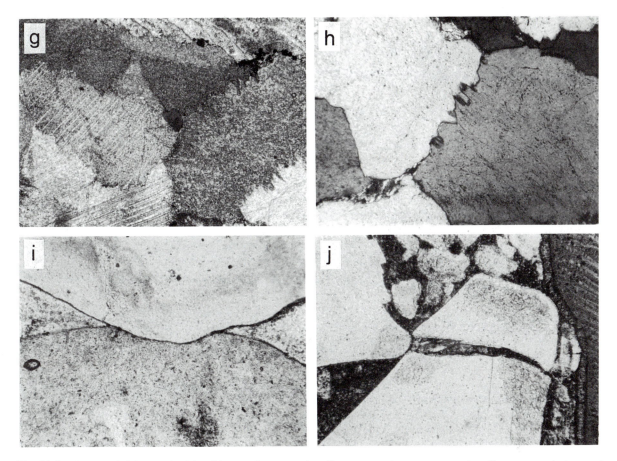

Fig. 11-5. (*Continued*) (**g**) sutured calcite; (**h**) sutured quartz grains; (**i**) a quartz grain penetrates another; (**j**) a quartz grain fractured under compaction.

body compacts. Porosity loss is due partly to grain dissolution and partly to cementation. Both processes increase the surface areas of grain-to-grain contacts, and this gradually reduces the rate of pressure dissolution. In medium-grained quartzarenites, pressure dissolution becomes important at depths greater than about 1.5 kilometers, according to Schmidt and McDonald (1979; compare their figs. 11 and 21), and continues to reduce the bulk volume and porosity of such a rock down to a depth of about 3 kilometers, where porosity declines to a small, irreducible level.

Renton et al. (1969) made a series of informative experiments with pressure dissolution in sandstones, subjecting various natural and artificial sands to gradually increasing hydrostatic loads up to 12,000 psi (over 800 atm) and temperatures up to 550°C for periods up to about 50 days. These experiments indicate

that higher temperatures promote pressure dissolution. Composition of the pore fluid also is important; though pressure dissolution occurred in distilled water, saline and alkaline pore solutions greatly increased the rate of pressure dissolution.

Compaction rates decreased with time. Finer-grained sands compacted more rapidly than coarser sands. Angular sands compacted considerably faster than round sands; after pressure dissolution, the textures of originally angular sands and originally round sands became rather similar. In artificial mixtures of quartz sand and chert, the chert grains are substantially more susceptible to pressure dissolution than is macroquartz.

Feldspars in sandstones are more resistant to pressure dissolution than is quartz. Micas and clay minerals, tourmaline and zircon, and pyrite, also quite resistant to pressure dissolution, accumulate on stylolite seams.

Pressure dissolution in quartz sandstones might be localized and promoted by associated clays. Weyl (1959, p. 2005) postulated that clay films on quartz grains promote diffusion of dissolved silica. Thomson (1959, p. 105) suggested that release of potassium from illite particles near to quartz grains in stressful contact increases the pH of the pore fluid in that neighborhood, thereby increasing the solubility of quartz. Sibley and Blatt (1976, p. 891) concluded that clay content controlled pressure dissolution in the Tuscarora Sandstone (Silurian) in Pennsylvania and West Virginia. In the sandstones that Houseknecht (1988, p. 236) examined, pressure dissolution was most extensive in those having illite-coated quartz grains. One possible effect of clay coats is that it prevents the pressure-dissolved silica from precipitating as cement on the quartz grains (see Wilson, 1992, p. 222), which thus must go elsewhere; as long as the sand remains largely uncemented, pressure dissolution can continue.

Stylolites are extensive surfaces or thin seams resulting from pressure dissolution of mineral matter under directed pressure; typically the surfaces are marked by piston-like projections, or teeth. In most cases there is a thin, dark film or coating on the stylolite surface. Such films appear to be residues of insoluble material (clays, sulfides, and asphaltenes). The general orientation of stylolite surfaces and their teeth indicates the direction of the stress that caused them. In some rocks, different stylolites have contrasting tooth orientation, or teeth are curved. This suggests that structural reorientation of rock masses occurred during stylolite growth; stylolite teeth not normal to bedding likely are younger than those that *are* normal.

Teeth are the result of small-scale spatial variations in rock solubility. The material at the crown of a tooth is presumably less soluble than the opposing material. As soon as dissolution occurs at a grain contact under stress, the load that had been borne by this contact is transferred to adjacent grain contacts, which then undergo pressure dissolution in turn; thus the stylolite propagates laterally, approximately in the plane normal to the principal stress. Tooth margins are inscribed with little slickensides created by movement toward one another of the rock masses in stylolitic contact (Fig. 11-6). Differential movement of small rock volumes along a stylolite causes numerous thin fractures to develop roughly perpendicular to the seam; larger fractures result from major volume changes. The ma-

Fig. 11-6. Stylolites in limestone.

terial that dissolves at the stylolite surface moves to adjacent pore space (including the fractures associated with stylolites), where it precipitates as cement, reducing rock porosity and permeability and increasing rock strength, and eventually terminating stylolite growth thereby. Stylolitization surely cannot proceed in a rock fully cemented and lacking porosity. Metamorphic conditions apparently never produce stylolites.

Stylolitization results in reduction of bulk volume of the sediment, manifested as a thickness loss; tooth relief summed over all stylolites gives minimum value of loss of thickness, which is as high as 25 or 30 percent in some limestones. One might estimate thickness loss by reconstructing large skeletal grains abridged by stylolites, by measuring the apparent lateral displacement of inclined veins or fractures transected by the stylolites, or by measuring the quantity of insoluble material that has accumulated along the seams compared to concentrations in unaffected parts of the rock.

Stylolitization is sensitive to lithic type. Stylolites are most common and best developed in limestones, gypsiferous rocks, and glacial ice; they are less frequently noted in quartzose sandstones, and probably never occur in shales (though the *cone-in-cone* structures of shales might be related). Heald (1955, p. 101) noted that stylolites in "clean" sandstones are more widely spaced and have greater amplitudes than those of argillaceous or calcareous sandstones. Skeletal limestones with clay matrix commonly display pressure-dissolution fabrics, while adjacent non-argillaceous limestones do not; Oldershaw and Scoffin (1967, p. 315) described such a relationship in the Wenlock and Halkin Limestones (Silurian and Carboniferous) in England and Wales. A limestone containing many stylolites of high tooth relief may be flanked by a different limestone containing no stylolites whatsoever. This can create stratigraphic relief (lateral variation in thickness of a deposit) that might be falsely attributed to depositional topography; in a seismic section it might look like a reef. Dolomite seems to be quite resistant to pressure dissolution; prominent stylolites contained in dolomites were probably inherited from the precursor limestone.

Diagenesis of Mudstones and of Clay Minerals

Clay minerals are formed, by and large, in surficial environments and are, therefore, stable in these environments. But, more than most other detrital minerals, the clays are sensitive to the changing physical and chemical conditions that accompany burial, and their chemical compositions and crystal structures change accordingly. Under increasing diagenetic stress, there is a gradual tetracoordination of aluminum (Dunoyer de Segonzac, 1970, p. 334); that is, aluminum in octahedral coordination (as in kaolinite) becomes less important and tetrahedrally coordinated aluminum becomes more important. Cations from interlayers or from the pore waters move to octahedral sites, and dioctahedral clays (see Chap. 6) change toward trioctahedral clays. Interlayers lose water and preferentially adsorb magnesium (at the shallower depths) or potassium (at greater depths). The water released occupies less volume than it had as structured water layers within the clays. At the highest diagenetic grades, crystal growth becomes important.

The illites are a heterogeneous group in depositional environments, where they are compositionally and structurally "degraded" by weathering processes; but diagenesis tends to reconstitute them and make them more uniform. The "crystallinity" of the illites increases with increasing diagenesis, as manifested by a gradual narrowing of the 10 AU x-ray diffraction peak. Several polymorphs of illite are recognized: The *1M illite* has a "one-layer" monoclinic unit cell; there are two forms of illite with "two-layer" monoclinic unit cells, designated *2M$_1$* and *2M$_2$*; and there is a rather rare form, *3T*, with a three-layer trigonal cell. A common disordered illite produced in the weathering environment is designated *1M$_d$*. Velde and Hower (1963) and Maxwell and Hower (1967) have described simple x-ray methods for distinguishing among these polymorphs. With increasing depth of burial, illites change from 1M$_d$ to 1M to 2M forms. As diagenesis advances into the realm of metamorphism, illite particles become larger and change into flakes of sodium-bearing mica generally referred to as sericite.

Smectite and illite-smectite mixed-layer clays are important components of muds and young mudstones; but in older and deeply buried rocks, smectite is not abundant. Under the elevated temperatures of deep burial, smectite gives up its interlayer water to the interstitial pore space and exchanges cations with cations in the pore water. Gradual changes in structure and composition lead generally to formation of illite or chlorite. The smectites are not abundant in older or deeply buried sediments, presumably because they have altered diagenetically.

In Quaternary and Tertiary muddy sediments of the U.S. Gulf Coast Perry and Hower (1970) found an almost linear decline with increasing temperature in the percentage of smectite layers in illite/smectite mixed-layer clays. Numerous other examples have been described since then (for example, Ramseyer and Boles, 1986; Glasmann et al., 1989). Random mixed layers exist at temperatures less than 80°C, but above this temperature, interlayering becomes regular (and the linear relationship disappears). It appears also that random interlayering changes to ordered interlayering as the proportion of illite increases beyond about 60 percent. As smectite converts to illite, it takes up potassium, which might be provided by decomposition of detrital feldspars or biotite. The smectite-to-illite conversion also requires that aluminum be substituted for silicon in the tetrahedral layers; silicon, it appears, is released to the pore water. Collapse and dehydration of smectite evolves much water. According to Powers (1967, p. 1251) and Perry and Hower (1972, figs. 6, 7), dehydration of illite/smectite mixed-layer clays is a two-stage process, some water being released at temperatures of 90 to 100°C that are characteristic of burial depths of 6 or 7 kilometers and additional water being released at about 120°C at a burial depth of 10 kilometers. Pytte and Reynolds (1989, p. 137) suggested that the smectite-to-illite transformation is useful as a geothermometer, recording maximum temperature attained during a diagenesis.

In the deeply buried sandstones and shales of the

Wilcox Formation (Eocene) in southern Texas, Boles and Franks (1979, p. 68) found that detrital kaolinite becomes gradually less abundant with depth, breaking down apparently at temperatures between 170 and 210°C. Chlorite, in contrast, increases markedly at the greater depths and higher temperatures.

Cements and Cementation—Methods of Study

Cement is material that has chemically precipitated in the pores of sediments. Cementation may occur at almost any stage of diagenesis. Even the depositional environment may have such chemistry as to promote cementation. Most deeply buried sediments are cemented, and if a deeply buried sediment is *not* cemented, one must enquire *why* it is not. Some deep cements are derived from the grains themselves, by pressure dissolution or alteration. Otherwise, cements are derived from the outside, and cementation requires substantial movement of pore water. Many rocks contain more than one kind of chemical cement, commonly precipitated under greatly contrasting conditions. Diagenesis under deep burial (mesogenesis) may also cause *decementation*, or removal of some cements, followed perhaps by a new episode of cementation.

Much cementation of calcareous and terrigenous sediments occurs very shortly after deposition and before significant compaction. Pore waters generally are meteoric, or mixtures of fresh water and sea water. Changes in concentration of CO_2 in pore waters control the precipitation and dissolution of early cements.

But not all cementation is eogenetic, and not all CO_2 is atmospheric. Mesogenetic pore fluids are, in general, not the same pore fluids that effected early diagenesis; pore water in a deeply buried sandstone or limestone is connate water, typically altered by long contact with the surrounding mineral grains. Or the pore water might have been derived from associated shales; some of it was shale pore water, some was clay-mineral interlayer water. Other ''new'' waters may appear as a result of the dehydration of gypsum and its recrystallization to anhydrite.

Waters squeezed out of shales may have high concentrations of Na^+, Ca^{2+}, Mg^{2+}, and Fe^{2+}. Carbonate in a buried mudstone reacts with organic matter or with clay minerals to form CO_2, which is released to pore waters also during the smectite-to-illite reaction and during the diagenesis (maturation) of organic mat-

ter buried with shales. These dissolved substances leave the mudstones during compactive expression of the pore fluids and enter surrounding limestone and sandstone formations, where they perform much of the work of mesogenetic chemical diagenesis in these rocks, including dolomitization, cementation and decementation, and grain dissolution and alteration. During cementation of a porous body of sediment by through-flowing waters, more water flows out than flowed in, as cement gradually displaces water that originally occupied the pores.

The means of discriminating among different cements, time of their emplacement, and clues to their origins, are contained in cement crystal habits, relationships between cements and their substrates, types and abundances of inclusions in cement crystals, and minor-element content. Let us summarize the textural evidence of time of cementation and examine also some other lines of evidence for time and environment of cementation.

Petrography and Stratigraphy

Ordinary petrography is, of course, the standard method of study of cement fabrics and cement-to-substrate relationships. Much of the evidence of sequence of cementation and of time of cementation relative to other diagenetic events also is obtained petrographically. The familiar rules of relative dating, such as superposition and various cause-and-effect relationships, apply just as validly here as they do to stratigraphy.

Petrographic evidence of *early* cementation includes rocks not fully compacted or otherwise unaffected by mesogenesis because early cement has gotten in the way; cements affected by early dissolution or buried by internal sediment; cements that have protected soft grains from deformation or that have plastically deformed or penetrated adjacent uncemented grains due to post-cement compaction; and cements deformed by compaction. Limestones may contain undeformed fragile grains, such as micrite envelopes or gracile ostracods, that are filled with load-bearing cement. Had these envelopes not been cemented early on, compaction would have broken and collapsed them. Early cement inhibits compaction and occasionally inhibits emplacement of later cements; if early cement completely occludes pore space, then there is no room for later cements.

Cements binding grains tightly compacted, cements spatially associated exclusively with stylolites, or cements filling tectonic fractures are *late* cements. Of course, late cements can never underlie cements that have formed earlier. Some cements may fill (or partially fill) those secondary pores that for various reasons are thought to have formed under mesogenetic conditions; such cements necessarily are late. Cements containing inclusions of petroleum are generally considered to be late cements.

A lower bound on age of a cement can sometimes be fixed by age of an ensuing erosional event: Suppose that a rock unit is overlain by an unconformity; cementation occurred before erosion if the unconformity truncates the cement; or cement under the unconformity is partially dissolved or otherwise weathered; or the formation has broken up in a way that implies that it was hard and brittle, and thus cemented and lithified, at time of erosion; or distinctive cements in the overlying formation do not appear below the unconformity, because the underlying formation was already cemented. The surface of a bed may be bored or micritized, suggesting that the bed was cemented while still in or near to its marine depositional environment; hardgrounds of carbonate successions typically show these features.

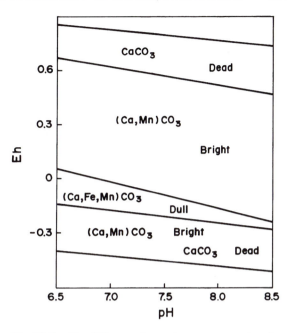

Fig. 11-7. pH and Eh control on carbonate cement mineralogy and cathodoluminescence. (According to Frank et al., 1982, fig. 2.)

Cathodoluminescence and Chemical Staining

Additional evidence, essentially chemical, for time of calcite cementation is provided by cathodoluminescence (CLM), the emission of visible light by a substance that is being bombarded by a beam of electrons (Smith and Stenstrom, 1965). Calcite containing small amounts of manganese as substitution ions in its crystal lattice emits orange luminescence when excited by low-energy electrons (cathode rays). Presence of iron in the lattice quenches or poisons this luminescence. Commonly, cements show complex luminescence zoning.

A geochemical model for zoned luminescence in calcite cements was devised by Carpenter and Oglesby (1976). Manganese and iron take on different oxidation states depending on the ***redox potential*** (Fig. 11-7) of their chemical environment, manganese existing as Mn^{2+} or Mn^{3+} or Mn^{4+}, iron as Fe^{2+} or Fe^{3+}. The Mn^{2+} and Fe^{2+} ions can readily substitute for Ca^{2+} in the calcite crystal lattice. The higher oxidation states of Mn and Fe, however, are not readily accommodated, because they have the wrong charge. Thus, the manganese and iron content of a calcite cement is controlled by the redox potential of the environment in

which this mineral crystallized. Calcite cements that form in well aerated, oxidizing environments, as in a vadose zone, take up very little manganese and tend to be non-luminescent. Brightly luminescent calcite cement forms under lower redox potentials, where Mn^{2+} is an abundant or dominant oxidation state for manganese. At still lower redox potential, Fe^{2+} exists and enters the calcite crystal lattice alongside Mn^{2+}, diminishing and altering the luminescence to dull orange or brown. Low redox potential is generally associated with the phreatic zone, that is, the region below water table, where all rock pores are completely filled with water, but reducing conditions can be created at or near the surface also, in the vadose zone, by decay of organic matter. If sulfur is present, it can cause iron to precipitate as pyrite or marcasite, making iron unavailable for substitution in the calcite lattice.

Frank et al. (1982) made application of this model to the Taum Sauk Limestone (Cambrian) of southeast Missouri. Grover and Read (1983) worked out a complex cementation history in Ordovician limestones in Virginia. They made a plot of Mn versus Fe content in different calcite cement zones, as determined by electron microprobe; this scatter plot separated cements of different luminescence properties into distinct fields (Fig. 11-8).

The presence of Fe^{2+} in calcite cement can also be made manifest by staining with potassium ferricyanide, which imparts a blue color to Fe-rich calcite or dolomite cements. This works better than CLM techniques if iron content is very high.

It is common for the very earliest calcite cements (crystals or layers in direct contact with grains) to be non-luminescent, having been precipitated either in shallow, well-aerated marine waters or in an actively circulating freshwater vadose zone. Syntaxially superposed on this non-luminescent early cement is brightly luminescent cement, precipitated under moderately reducing conditions of a stagnant phreatic zone. Dull orange or brown calcite cement forms under local or ephemeral reducing conditions near the surface or under deep burial, precipitating from oxygen-poor, iron-rich connate waters or deeply circulating meteorically derived waters, also oxygen poor. In some rocks close interlayering of bright and dull cement occurs, suggesting some rapidly fluctuating condition or environmental instability of some kind.

Skeletal and other carbonate grains ordinarily do not luminesce unless they have been replaced by or impregnated with luminescent cement, in part because these grains ordinarily form in oxidizing environments, and because organisms actively exclude heavy metals, such as manganese, from participating in skeleton-forming processes.

Cathodoluminescence microscopy is useful also in the study of the silica cements of terrigenous rocks (Sippel, 1968). It is often difficult to distinguish by ordinary optical methods between a detrital quartz grain and its syntaxial cement overgrowth; CLM can often make this distinction, with quite startling effect, and disconcerting to the confident and complacent petrographer who assumes that he or she can distinguish between detrital grain and syntaxial cement by ordinary petrography. Quartz luminescence is generally quite dull, either red or blue (see Chap. 6). Overgrowths on detrital quartz grains ordinarily luminesce not at all, or their luminescence contrasts with that of the host grain. Moreover, cathodoluminescence reveals that the ordinary petrographic criteria for pressure dissolution are misleading, and that healed fractures in quartz grains often are missed.

Stable Isotopes

The isotope chemistry of cements also can aid in determining the chemical and physical environments under which they precipitated (Fig. 11-9). Variations in the ratios of stable isotopes in natural waters are a

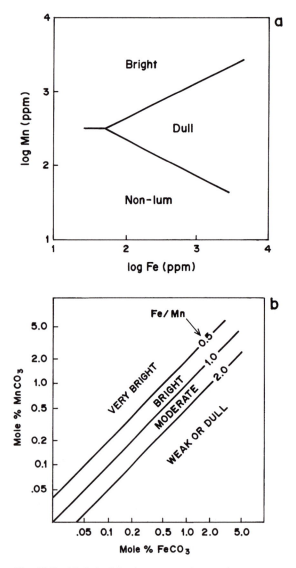

Fig. 11-8. (a) Cathodoluminescence and trace-element content in cements of Middle Ordovician carbonate rocks. (According to Grover and Read, 1983, fig. 2, with permission of AAPG.) (b) General cathodoluminescence properties of carbonate cements, as related to trace-element content. (After Frank et al., 1982, fig. 3.)

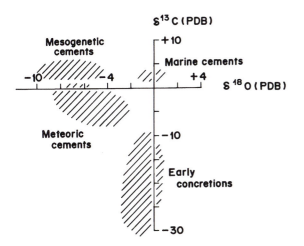

Fig. 11-9. Carbon and oxygen isotope signatures for various carbonate cements.

$$\delta^{18}O = [(\text{sample } {}^{18}O/{}^{16}O)/(\text{standard } {}^{18}O/{}^{16}O) - 1]$$
$$\times 1000‰$$

Positive delta values indicate enrichment in the heavier isotope with respect to the standard; negative values indicate depletion of the heavier isotope. One must always indicate which standard was used ($\delta^{18}O$ SMOW = 1.304 $\delta^{18}O$ PDB + 30.86).

In natural surficial environments, phase change (freezing, melting, evaporation, or condensation) is the dominant fractionating process involving water molecules. When water evaporates, the heavier water molecules enter the gaseous phase more slowly than do the lighter ones. Thus, the water vapor is deficient in ^{18}O relative to the water from which it formed, and the water not yet evaporated is enriched in ^{18}O. For this reason, meteoric water typically has lower ^{18}O concentrations than does sea water (that is, negative delta values with respect to SMOW), and hypersaline brines created by evaporation in restricted marginal marine environments have higher ^{18}O concentrations than does normal sea water (positive delta values). When water freezes, the heavier isotopes enter the solid phase disproportionately, and the remaining unfrozen liquid is isotopically light. When a dissolved substance precipitates, its isotopic content reflects that of the solution from which it crystallizes, though again there is a disproportionation that concentrates heavier isotopes in the crystalline substance relative to the solution. Calcite that precipitates from normal sea water has a $\delta^{18}O$ SMOW value that is slightly positive. Water that is more saline than normal sea water presumably has been evaporated; one expects calcite precipitated from hypersaline waters to have positive delta values, the higher values reflecting higher salinities. Calcite that precipitates from fresh water has a strongly negative value, down to about −60‰.

The oxygen isotope ratio in a carbonate or silicate cement depends not only on the source of the carbonate or silica but also on the temperature (depth) at which the cement was formed. The isotopic fractionation that occurs during precipitation of calcite, including skeletal calcite, is temperature dependent, being more pronounced at low temperatures. Thus, isotopic proportions can serve as a geological thermometer. For example, $\delta^{18}O$ values in the calcite tests of marine planktonic foraminifers now resting in marine sediments have shown that Pleistocene ocean surface temperatures were lower during glacial stages than during interglacial stages, as expected. Carbonate

result of processes that can partially sort out water molecules (or other molecular or ionic species) according to their masses. Most water molecules are composed of atoms of ordinary "light" hydrogen, ^1H, and the common oxygen isotope, ^{16}O; but some water molecules contain an atom of "heavy" hydrogen, ^2H (also known as deuterium) instead, or an atom of ^{18}O. Similarly, most CO_2 molecules are built of ^{12}C atoms and ^{16}O atoms, but a few contain the heavier ^{13}C or ^{18}O atoms. Thus, some water molecules are a bit heavier than most; some CO_2 molecules are heavier than most. *Fractionation* is a process that achieves some partial unmixing of the heavier molecules from the lighter ones. In the laboratory, isotopic fractionation is achieved by a diffusion process (it is one means by which fissionable ^{235}U is separated from ^{238}U in nuclear-fuel enrichment plants); lighter atoms or molecules diffuse through a porous medium, such as a chromatograph column, more quickly than do heavier ones. The fractionation effect is smaller in heavier elements, where differences in masses of the different isotopes are proportionately less.

The isotopic proportion in a substance is generally given in terms of its departure from the proportion in some standard substance. For the oxygen isotopes, two different standards are in common use among geologists, the Standard Mean Ocean Water (SMOW), and the calcite shell of a certain Cretaceous belemnite from the Peedee Formation of South Carolina (PDB). Departure of a sample from a standard is given by

cements that form at the elevated temperatures associated with deep burial have increasingly negative $\delta^{18}O$ values with increasing depth.

Carbon has two naturally occurring stable isotopes, ^{12}C and ^{13}C. Biological processes achieve significant fractionation of these carbon isotopes. The standard most widely used for carbon isotopes is again the PDB, in which $^{13}C/^{12}C$ is close to that of atmospheric CO_2 dissolved in normal seawater. In organic tissues, $\delta^{13}C$ ranges between about -15 and -30 permil. Petroleum, natural gas, and coal, all derived from such tissues, show similar or even greater depletions of ^{13}C. The large fractionation that occurs during photosynthesis is reflected even in the organic matter of Precambrian sediments over 3 billion years old. A carbonate cement that is isotopically light might have been derived from skeletal material. As with oxygen isotopes, higher temperatures cause more negative delta values.

Partition of the sulfur isotopes ^{32}S and ^{34}S between sulfate and sulfide reflects biological influences. Sea water, and gypsum or anhydrite that forms by evaporation of sea water, average about $+20$ permil (with respect to cosmic ratio, as contained in meteorites; source of crustal sulfur is the mantle, which is presumed to be isotopically similar to the cosmos as a whole). Sulfur-reducing bacteria preferentially reduce the lighter isotope, so that sedimentary pyrite or marcasite contains sulfur that is lighter than that contained in sulfates. (Igneous and hydrothermal sulfides of deep-crustal provenance, in contrast, have delta values close to zero.)

Fluid Inclusions

One other line of chemical evidence of time and environment of cementation is provided by fluid inclusions. Many fluid inclusions in a cement are tiny entrapped samples of the fluid from which the cement precipitated. It is possible to obtain chemical analyses of the fluids contained in inclusions, but useful information is also contained in freezing (or melting) points (controlled by colligative properties of the solution), and in temperature of homogenization of two-phase inclusions. Some inclusions consist of liquid and gas; one assumes that at the time of entrapment, there was but a single phase. The experimental temperature at which the inclusion is homogenized is an estimate of the minimum temperature at time of crystallization (a correction for lithostatic pressure might be necessary).

Roedder (1972) has presented many excellent photomicrographs of fluid inclusions in rock materials.

Cement Stratigraphy

Crystals of simple form may harbor a complex history of crystal growth. Chemical staining techniques, CLM, and study of solid and fluid inclusions help to reveal these complexities.

Luminescence zones commonly can be correlated from one cement crystal to another, from one cement-filled pore to another in a given thin section, from sample to sample, and even across localities. Correlation and dating of cement zones is called *cement stratigraphy*. Regional patterns of cementation and relationships between cement stratigraphy and ''depositional'' stratigraphy help us to understand how cementation of a rock body progresses, as controlled, for example, by marine transgression or regression. Cement stratigraphy is also applicable, in a somewhat different way, to cave dripstones. These are composed of carbonate cement arranged in chronological concentric layers, just like the rings of a tree. The entrapped pollen grains or dust particles that they contain, and the isotopic contents of individual layers or rings, are organized records of changes of climate and atmospheric composition. Sandberg (1975, p. 503) suggested that inter-ooid correlation of the concentric growth layers of ooids in marine and lacustrine limestones might be possible, based on growth layers and disconformities in the ooid cortices (revealed by SEM).

In the Mississippian carbonate rocks in the Sacramento Mountains, New Mexico, that Meyers (1974 and 1978) described, petrography (including CLM) revealed several stages of cementation. Cement zones are regionally persistent, and bodies of rock can be delineated on the basis of presence or absence of a particular cement. Occasionally, cementation was interrupted momentarily by a dissolution event; this creates a *cement disconformity* (Fig. 11-10i). It is an etched surface within a cement; some zones may be truncated, or missing altogether. Meyers was able to correlate cementation events, including cement disconformities, over distances of 30 meters vertically and 16 kilometers horizontally. Goldstein (1988) was able to correlate cement zones in the Holder Formation (Pennsylvanian) in New Mexico for tens of meters vertically. The formation contains as many as 15 distinct exposure surfaces, each with its own genetically related cement in the underlying few meters of rock. Cementation patterns were useful in correlating the exposure surfaces between stratigraphic sections typically a kilometer apart. Grover and Read (1983) described regional cementation patterns, based in large part on CLM, in the Middle Ordovician limestones of Virginia, determining that these patterns are closely linked to depositional environment, climate, regional tectonics, and burial history, especially insofar as these factors controlled distribution, composition, and movement of pore waters.

Cementation of a single vug is sometimes a microcosm of cementation of the rock as a whole, showing more clearly than anywhere else a complete sequence of precipitation of chemical cements. In the vug there is room for the full succession of different cements, in contrast to smaller pores, which the earliest cements might completely occlude.

Suppose that gravity has controlled the disposition of a vug filling, that only the lower part of the large pore is filled, or that a pore is floored with internal sediment, which is overlain by chemical cement or empty space. These special pore fillings are *geopetals*, because they ''point'' toward the center of the earth, by virtue of the arrangement of the fill (Fig. 11-3e). Geopetals of this kind are probably made in a vadose zone (that is, above water table). They can be useful in establishing up-direction in a structurally complex

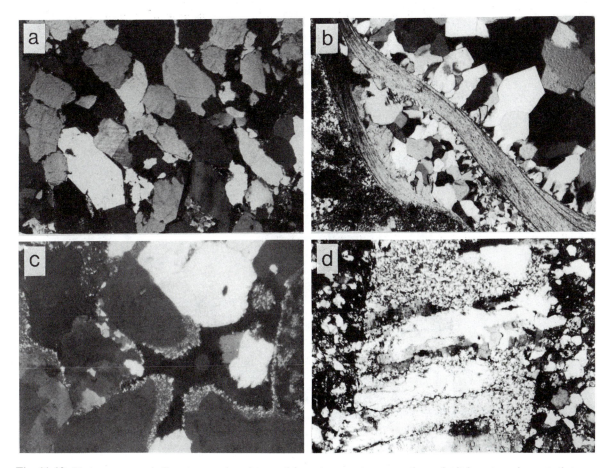

Fig. 11-10. Various cements in limestones and sandstones: (**a**) quartz cement overgrowths on detrital quartz grains; note that some cement has crystal faces; (**b**) quartz cement in a limestone, occupying shelter pores between bivalve fragments; (**c**) kaolinite cement on quartz grains; (**d**) quartz and kaolinite cements intergrown in a fracture;

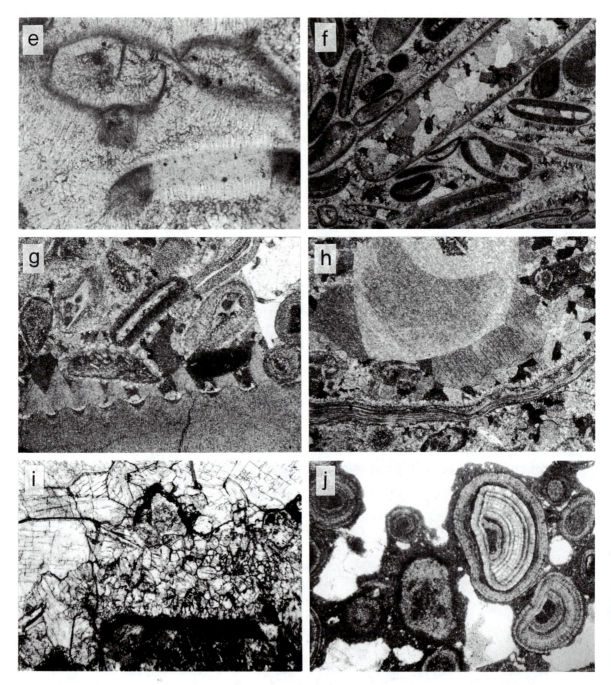

Fig. 11-10. (*Continued*). (**e**) marine acicular calcite; (**f**) micrite envelopes cemented with calcite spar; (**g**) articulation surface of pelmatozoan columnal plate contains thin coat of lime mud in the furrows, but the ribs received scalenohedral calcite, which is overlain by ferroan calcite; (**h**) pelmatozoan plate with (syntaxial) rim cement; (**i**) a cement disconformity; older cement was degraded before second-generation cement was emplaced; (**j**) carbonate grains with microcrystalline calcite meniscus cement, this overlain by pore-occluding calcite spar (up is to left).

terrain, and they have been used in conjunction with bedding planes to demonstrate the existence of steep depositional slopes.

Cementation of Sandstones

Many quartzose sandstones are cemented by quartz. Ordinarily, quartz cement grows syntaxially, that is, in optical continuity with the grain that it lies upon. In this way quartz grains grow and change shape, and in many cases it is not easy to see the boundary between grain and cement, and to know the original size and shape of the grain (and of the pore). Rings of ''dust'' or tiny liquid inclusions may delineate the boundary. Again, CLM is useful here.

Origins of Silica Cement

Much quartz cementation of sandstones has been attributed to pressure dissolution of detrital quartz grains under deep burial. But Siever (1959), having examined certain Pennsylvanian sandstones, raised some doubts. He concluded that quartz cements were the first diagenetic precipitates, predating every other authigenic material. Though pressure dissolution may have been the source of this cement, the lack of correlation between quantity of quartz cement and degree of grain interpenetration, according to Siever, suggests instead that pore waters imported dissolved silica from an external source. Houseknecht (1987) also noted mismatch between grain suturing and quartz cement, and argued (Houseknecht, 1988) that silica moves from bed to bed. This contrasts with ionic transport of only a few grain diameters (or less than a grain diameter) thought to be characteristic of pressure dissolution. Simple geometric arguments, according to Manus and Coogan (1974), indicate that unrealistically large bulk volume reductions are required if significant quantities of quartz cement are to be produced by pressure dissolution. Mitra and Beard (1980, p. 1358) pointed out that if all pressure-dissolved silica is precipitated locally as pore-filling cement in the system, then total destruction of porosity implies thickness reduction equal to original porosity, which typically is in the range of 20 to 30 percent; evidence of such large volume reductions in sandstones is generally lacking. Cathodoluminescence commonly reveals that the grains in quartz-cemented sandstones are quite unabridged and that little compaction at all has occurred.

Quartz cement may come from amorphous aluminosilicates (abundant in soils), from alteration or dissolution of feldspars and other detrital grains (Land, 1984, p. 56), from dissolution of diatoms and radiolarians buried in shales or limestones, or from shale transformation and dewatering. The smectite-to-illite transformation may be the source of much of the silica involved in mesogenetic cementation. In sandstone-shale successions studied by Fuchtbauer (1978), sandstone beds are more heavily silica-cemented near to their contacts with shales, suggesting that the shales were the source of the cement. Moreover, early formed concretions in the shales contain up to 100 percent more silica than adjacent shale, again suggesting that the shale exported silica.

Early cementation of sand by silica in arid regions has been documented. Over large expanses of the Sahara Desert, quartz sand at the surface is cemented by silica. Kikoine and Radier (1949 and 1950) described, in the Sudan, a surficial ''case-hardening'' effect that creates a ''flagstone'' crust a few centimeters in thickness. Shifting of the unconsolidated sands beneath has caused the crust to break up into slabs, rather reminiscent of beachrock.

Heald and Renton (1966) cemented quartz sands experimentally by passing concentrated silica solutions through samples of natural and artificial sands. They found that rate of cementation was initially greater in fine sand than in coarse, but that later on cementation proceeded more rapidly in the coarse sand than in the fine. This is because the coarse sand is able to maintain the requisite permeability for longer periods than the fine sand. Angular sands had greater initial porosity, but lesser final porosity, than round sand (of course, original grain shape becomes less and less meaningful and influential as grain surfaces come to be covered over with cement). Polycrystalline grains, such as fragments of quartzite or chert, cemented more slowly than unit quartz grains, and rate of cementation decreased with decreasing crystal size in the sand grains. Cementation of mixtures of quartz and feldspar was slower than cementation of quartz sands.

Other Cements

Sandstones may be cemented also by other silicate minerals (feldspar, zeolites, chlorite), or by carbonate minerals (calcite, dolomite, siderite), sulfate minerals (gypsum, anhydrite, barite), or iron oxide minerals (limonite, hematite). Much early cementation of sand is a by-product of weathering nearby. Weathering puts various substances into solution, and the solutions then flow into the porous and permeable sands and precipitate part of their dissolved load. Some carbonate is carried in from surrounding limestone terrains, and accomplishes an early cementation of sand in much the same way as many carbonate deposits are cemented early. Some sandstones acquire calcite cements by dissolution and redistribution of carbonate skeletal material contained within the sand itself; this may happen early or under deep burial. Many volcanic sands are cemented early by various carbonate minerals, zeolites, or various forms of silica derived from weathering and devitrification of the unstable volcanic minerals and rock fragments. Some carbonate cements—including calcite, dolomite, ankerite, and siderite—are precipitated under deep diagenesis from pore waters expressed from closely associated compacting shales. Calcium carbonate might originate from dissolution of coccoliths and foraminifers in shales.

In quartz sandstones containing carbonate or other cements in addition to quartz cement, petrography typically indicates that the quartz cement was emplaced first. This may be due to chemical factors. But, on the other hand, it is likely that quartz cement precipitates only on a quartz substrate (such as a reasonably clean quartz grain), and if other cements have already buried the surfaces of quartz grains, then subsequent quartz cementation might be impossible. In arkosic or lithic sandstones, quartz grains commonly are overlaid with syntaxial quartz cement, but other grain surfaces are cemented with calcite or not cemented at all.

The clay minerals are important cements in many sandstones and in some carbonate rocks. All major types of clay minerals are known to form authigenically in sandstones (Wilson and Pittman, 1977, p. 7). Kaolinite cement can form in sandstones under shallow or deep burial, according to Weaver, 1989, p. 446). Typical appearance (SEM) is "books" of well-formed hexagonal sheets; in thin sections, kaolinite, with its low birefringence, looks something like chert (Fig. 11-10d). Chlorite cements commonly appear early, apparently precipitating directly out of connate pore waters, but some chlorite is an alteration of kaolinite. The chlorite typically forms a honeycomb or cellular pattern of plates standing on edge on a detrital grain surface. Authigenic illite can form from kaolinite at temperatures above 120°; it commonly overlies other clay (or non-clay) cements, as distinctive tangled filaments or whiskers. Authigenic smectite is most common in sandstones with volcanic components. It

makes crinkly coatings on detrital grain surfaces, or else resembles chlorite. (See Welton's 1984 SEM atlas for ample illustrations.) Wilson and Pittman (1977) and Wilson (1992) suggest numerous criteria by which (authigenic) clay cements can be distinguished from detrital clay coatings.

Boles and Franks (1979) described mesogenetic diagenesis in deeply buried Wilcox sandstones (Eocene) in southern Texas, which are closely associated with shales. Cements include quartz, calcite, ankerite, kaolinite, and chlorite. Quartz cement, as overgrowths on detrital quartz grains, is common at all burial depths. A major source of the silica, according to Boles and Franks was the smectite-to-illite conversion. Transforming clays also released calcium ions, which combined with CO_2 in the pore waters to make calcite cement; iron and magnesium released to the pore waters by transforming clays reacted with calcite and kaolinite to form ankerite and chlorite.

Land (1984) determined that most quartz cementation in the Frio Sandstone (Oligocene) of the Texas Gulf Coast preceded most calcite cementation, and that much of the silica and carbonate was derived from underlying basinal shales as they were heated under deep burial to temperatures above 80 or 100°C. Calcite cement was emplaced closer to the source shales, and hence at higher temperatures, than the quartz cement.

Cementation of Carbonate Rocks

Cementation of carbonate rocks is in many respects more complicated than cementation of terrigenous rocks. The overwhelming majority of cements in carbonate rocks are calcium carbonate or calcium-magnesium carbonate, but silica cements and evaporite minerals such as halite and anhydrite are locally important. Multiphase cementations seem to be the rule, and it is not unusual for a carbonate rock to display four or five different cement types, each precipitated under a different diagenetic environment, perhaps distantly separated in time. Much attention has been paid to cementation and decementation of carbonate rocks in recent years; some of the results of these researches can be applied to terrigenous rocks as well.

Eogenetic Cementation

Probably most cementation occurs early in the long course of diagenesis (eogenesis). This is especially true of carbonate rocks, but carbonate sediments of the deep ocean floor (abyssal oozes) can remain uncemented for millions or tens of millions of years. Cementation of many or most carbonate sediments begins almost as soon as the sediment forms. Many very young limestones, only decades or centuries old, are already extensively cemented. Many ancient limestones show almost no evidence of compaction or pressure-dissolution, apparently because they were fully cemented early. Carbonate cements may be emplaced in the marine environment, in a freshwater vadose zone or a freshwater phreatic zone. Some cement is emplaced much later under deep burial, from connate or altered connate waters. The facts are not all in, and the origins of some cements are still in debate.

A critical problem in carbonate rocks is the source of large volumes of calcium carbonate needed to achieve the high degree of early cementation of limestones. Limestones may contain as much as 40 or 50 percent of carbonate cement, and the origin of the large quantities of dissolved carbonate needed to accomplish this cementation must be addressed. Many carbonate petrologists conclude that large-scale early carbonate cementation requires destruction of large volumes of roughly contemporaneous carbonate sediment (*donor limestone* of Bathurst, 1975 p. 451), that the transfer of all this carbonate to a *receptor limestone* requires large-scale pore-water circulation (10,000 to 100,000 pore volumes), and that the process takes place within a few meters of the surface and within a few hundreds or thousands of years of deposition of the sediment. If carbonate sediment has initial porosity of 50 percent, then half of its bulk volume must dissolve in order to cement completely the other half. Dissolution of aragonite cement might be the source of much early calcite cement. Various skeletal grains also are composed of aragonite, and these too undergo dissolution by fresh water. James and Bone (1989), suggested that some limestones, such as those deposited in cool waters of temperate zones, are only lightly cemented because they initially contained very little aragonite. Since aragonite was not as important a skeletal material in ancient times as now, cool-water carbonate sediments, as opposed to modern tropical carbonates, might be the better model for early diagenesis of ancient limestones. Today it is becoming apparent that cementation of some initially very porous limestones, especially reefal and reef-talus limestones, is achieved in the marine environment, where the source of the cement is the vast oceanic reservoir of dissolved ions, rather than some coeval or contiguous mass of carbonate rock in the weathering environment.

Folk (1974) and Folk and Land (1975) considered that composition and morphology of carbonate cement are controlled by rate of crystallization, salinity, and Mg/Ca ratio (Fig. 11-11). The higher rates of crystallization lead to smaller crystals, higher Mg content, and greater disorder. Salinity is important, apparently, insofar as it determines ionic activities, which control rate of crystallization. Mg/Ca ratio varies widely in natural waters, being highest in seawater and in the hypersaline brines derived from seawater by evaporation, and much lower in meteoric waters. The higher values favor the formation of acicular aragonite and high-Mg calcite, and dolomite. Calcite, it appears, can form only if Mg/Ca is less than about 2/1, and if the ratio is near to 2/1, then high-Mg calcite is produced. If Mg/Ca exceeds 2/1, aragonite forms instead of calcite; some high-Mg calcite may form. Dolomite forms at low salinity, or at high salinity if Mg/Ca is as high as 5/1 or 10/1. Such high Mg/Ca ratios are achieved when normal seawater is evaporated; total salinity gradually rises, and when it reaches approximately five times normal salinity, gypsum ($CaSO_4.2H_2O$), or anhydrite ($CaSO_4$), is precipitated, using up much of the Ca present in the solution and thus increasing Mg/Ca ratio. Very fine crystalline, non-stoichiometric, disordered dolomite may precipitate from such brines, or calcite in contact with such brines may alter to (disordered) dolomite.

In meteoric waters, including percolating near-surface ground water, and the waters of streams and fresh lakes, salinity and Mg concentration are low; it is the chemical environment of low-Mg calcite unit rhombohedra, and of small platy crystals with important basal pinacoids, and of micrite, and, if Mg/Ca is sufficiently high, *limpid dolomite rhombs*, small euhedra, which are quite striking, for their water-clarity, in thin section. Pore waters evolved away from original marine waters are generally of high salinity but commonly have low Mg/Ca values. These waters crystallize low-Mg calcite as prismatic or scalenohedral palisades and as coarse equant (blocky) euhedral or anhedral spar.

Given and Wilkinson (1985, p. 109) maintained that kinetics is important in controlling crystal habit of carbonate cements. They noted a strong correlation between the rate at which a crystal grows and its morphology, growth rate determined largely by local saturation. In vadose pores where evaporation or rapid degassing of CO_2 from the pore water leads to high degrees of supersaturation, only acicular crystals are precipitated. Similarly, the acicular crystals of arago-

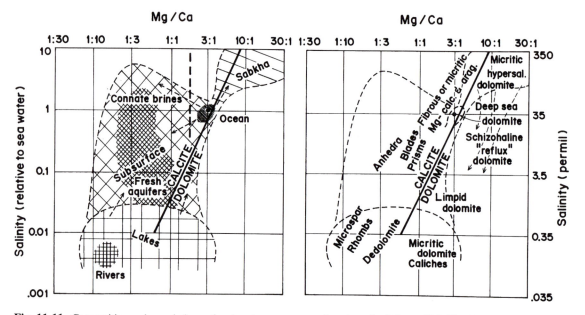

Fig. 11-11. Composition and morphology of carbonate cements as a function of salinity and Mg/Ca. (According to Folk and Land, 1975, figs. 1, 2, with permission of AAPG.)

nite and high-Mg calcite made in agitated, shallow marine environments derive from sustained precipitation from waters that are continuously supersaturated. In cold, deep marine waters, just barely saturated, equant low-Mg calcite is precipitated.

Highest precipitation rates seem to favor aragonite. In certain limestones examined by Mazzullo (1980), pore spaces are cemented with botryoidal acicular aragonite and high-Mg calcite, the latter overlying the former, without exception. Both cements, isotope studies suggest, were precipitated from normal sea water. It seems that the aragonite was precipitated while pores were still large and well connected, and pore waters were continually refreshed. As pores closed up, dissolved ions entered the pores more slowly, and the lower crystallization rate favored calcite. That crystallization rate controls cement type seems to be the reason that Mg content in calcite does not exceed about 17 mole percent; crystallization rate that might lead to higher Mg content in calcite causes aragonite to precipitate instead. Other studies have shown that high CO_2 content or presence of Mg^{2+} in cement-forming solutions seems to favor aragonite over calcite.

Calcite cements contain Mg, Sr, Fe, Mn, Na, and other elements in proportions that are related to the environment under which the cement was precipitated, but much more study is needed. Moore (1989, p. 58) decided that use of trace-element content of carbonate cements as a source of information on the composition of the cement-precipitating fluids is an ''exercise in futility'' at the present state of knowledge.

Four major near-surface diagenetic environments of carbonate cementation can be defined, being (1) marine phreatic, (2) mixing zone, (3) fresh (meteoric) phreatic, and (4) fresh vadose (see Longman, 1980, p. 463). Many carbonate sediments experience each of these environments in turn.

In the ***marine phreatic zone*** sediment pore space is filled with sea water, and the sediment is not far removed from its depositional environment. Normal sea water is saturated with respect to calcium carbonate, and if circulation through the sediment pores can be maintained, then this water might be induced to precipitate some of it in the pores. Two kinds of cement can emerge from this process, high-Mg calcite and aragonite. The ooid is an example of a marine precipitate, being a small particle (nucleus) surrounded by thin layers or coats (comprising the cortex) of aragonite or other mineral, the grain acquiring its cement even while still in transport in the depositional environment. Other examples of cement precipitation from normal seawater are cement-hardened peloids (fecal pellets), grapestones (small cemented aggregates of carbonate grains), beachrock of wave-splash zones, some of the cements in reef rocks and carbonate hardgrounds. Aragonite and high-Mg calcite are generally unstable outside the marine environment, where they quickly dissolve or recrystallize to low-Mg calcite.

Isopachous layers of acicular or fibrous calcite on the walls of pores in grainstones (Fig. 11-10e) are generally regarded as marine phreatic (high-Mg) calcite cement or a neomorphism of marine aragonite cement. Cements of this kind are occasionally noted in ancient grainstones and reefal boundstones. They are always the earliest of cements and commonly are overlain by equant calcite spar or other cement precipitated in other diagenetic environments. Reefs undergo extensive (early) marine cementation, even as they are growing. This requires that large volumes of seawater pass through the pores; probably wave and tidal pumping is the mechanism. With this in mind, Marshall (1986, p. 23), explained that the outer margins of individual reefal bodies in the Great Barrier Reef of Australia are well cemented, but the interior parts, more removed from the pumping action of the surf, contain relatively little marine cement. Marshall (1983, p. 1145) similarly noted that cementation in the southern part of the Great Barrier Reef is more advanced on the windward margin than on the leeward margin. Aissaoui, Buigues, et al. (1986, p. 34), working in the Mururoa Atoll, French Polynesia, observed that the peripheral islands and outer slope are highly cemented, but the carbonate sediments beneath the lagoon, even 100 meters down, are almost totally uncemented. Other factors may be involved—sea water tends to degass CO_2 under agitation, and precipitate carbonate cement as a result; sediments intermittently moistened by sea water, as on a beachface, might be incrementally cemented during the drying phases.

In modern carbonate-depositing environments, acicular aragonite or high-Mg calcite fringe, not necessarily isopachous, is a characteristic cement of beachrock, a consolidated grainstone forming on the beach, and typically underlain by unconsolidated carbonate sand. Taylor and Illing (1969, p. 95) described beachrocks on the Qatar peninsula in the Persian Gulf, containing both radial aragonite and high-Mg calcite cements with stalactitic and perched configurations. Donaldson and Ricketts (1979), summarizing the char-

acteristics of some modern beachrocks, pointed out that there are commonly two layers of cement, a micritic rind and an overlying fibrous aragonite.

Another kind of marine cement is *radiaxial* calcite. There are two forms—*radiaxial-fibrous* (not to be confused with radial-fibrous cements) and *radiaxial-mosaic*. These cements are very distinctive aggregates of gently divergent crystals, but with fast vibration directions that *converge* within these aggregates (Bathurst, 1959, p. 511; 1975, p. 426); in mosaic style, crystals are polysynthetically twinned, and the {0112} composition planes are curved convexly. (These are growth twins, not mechanical twins.) Henrich and Zankl (1986, p. 256) presented excellent photomicrographs of both kinds of radiaxial spar from Triassic reefs of the Bavarian Alps. Radiaxial calcite cements seem to be nearly ubiquitous in the stromatactoid cavities of Paleozoic Waulsortian mounds (Lees, 1964; Cotter, 1966; Ross et al., 1975; Miller, 1986). Kendall (1977) described cements similar in appearance to radiaxial spar, but in which fast vibration directions *diverge* away from the substrate and twin lamellae are *concave*.

Many petrologists have considered radiaxial cement to be a neomorphism, that is, a recrystallization, of acicular aragonite or high-Mg calcite cement (Cotter, 1966, p. 143; Kendall and Tucker, 1973, p. 377 ff). But Kendall (1985) later suggested that radiaxial cement is precipitated directly out of seawater as high-Mg calcite. Others, including Sandberg (1985, p. 45) and Saller (1986, p. 758) have concurred. Curvature and the peculiar compositeness of the crystals are presumed to be caused by a complex process of sideward growth and "split crystallization."

The *mixing zone* environment lies at the boundary between marine phreatic and fresh phreatic zones; mixing is a result of ionic diffusion and (more importantly) a seasonal or longer-term fluctuation of the boundary. As we learned above (Chap. 9), dolomitization and dolomite precipitation can occur in mixing zones. The mixing zone can also be a place of limestone dissolution. Dolomite spar cement, called *saddle dolomite* (Radke and Mathis, 1980) or *baroque dolomite* (Folk and Asseretto, 1974), because of its curved cleavage surfaces and sweeping extinction (Fig. 11-12), may be an early cement in some limestones, a product of the mixing zone diagenetic environment.

The *fresh phreatic zone*, where all pore space is filled with meteoric water, typically has an upper layer

Fig. 11-12. Saddle dolomite cement.

of active water circulation, below which water is essentially stagnant. Water in the upper layer is largely water that has recently percolated from the surface through the overlying vadose zone, and it very likely is saturated with respect to calcite.

Probably most of the calcite spar cements of most limestones are precipitated from fresh water in a phreatic zone. Calcite scalenohedral fringe or isopachous palisade of limpid calcite crystals (*dogtooth spar*) oriented approximately perpendicular to pore walls is commonly the earliest fresh phreatic cement. The crystals are sparse or crowded, projecting into empty pore space or overlain by blocky spar mosaic; commonly they are epitaxial on brachiopod shells or bryozoans but never on aragonitic corals or molluscan shells, though a skeletal grain composed of aragonite may have had a thin (perhaps imperceptible) coating of lime mud that has nucleated a scalenohedral fringe. Early scalenohedral fringe typically is overlain by or grades to *equant spar* (also called *blocky spar*) that coarsens away from pore walls.

Coarse *rim cements*, that is, syntaxial overgrowths on echinoderm plates, are an important mode of spar cement in many Paleozoic limestones. Echinoderm skeletal plates, consisting of single crystals of calcite, are a preferred substrate for calcite spar cementation; commonly the cement grows well beyond the adjacent pores, poikilotopically enclosing many other (non-echinoderm) grains. This is probably an early fresh phreatic cement in most instances. The extent of spar cementation in Paleozoic limestones is controlled in part by quantity of echinoderm debris; beds containing fewer echinoderm plates and correspondingly larger quantities of bryozoan and brachiopod fragments, peloids, etc., typically are less well cemented, and more porous, than closely associated beds in which echinoderm plates are abundant.

A transition from marine to freshwater cement may be reflected in a sharp boundary between cloudy (inclusion-ridden; fluid inclusions particularly) cement and overlying clear cement. Meyers and Lohmann (1978) argued that cloudy syntaxial cements on echinoderm plates in the Paradise Formation, Mississippian in New Mexico, were precipitated in a zone of mixed marine and fresh water, and that inclusion-free (clear) cements, which overlie the cloudy cements, were precipitated from fresh phreatic water. James and Klappa (1983, p. 1070) described early prismatic calcite cements comprising an inclusion-rich, ferroan first phase, overlain by an inclusion-free, non-ferroan second phase. The earlier phase apparently was a marine high-magnesium calcite, the later phase apparently precipitated as low-magnesium calcite in a freshwater diagenetic environment.

The phreatic zone passes upward quite abruptly into the *freshwater vadose zone*, where pores are only partly filled with water, or perhaps saturated part of the time and quite dry at other times. The boundary between phreatic and vadose zones is called the *water table* or *phreatic surface*. It may be quite smooth and stable or uneven and fragmented, and bobbing up and down in response to seasonal variations in rainfall. It is clear that calcite dissolution and precipitation can occur rapidly in a freshwater vadose zone—it occurs abundantly in limestone caverns, and in arid climates it forms caliche.

Indications of vadose-zone dissolution and precipitation are meniscus and pendent cements and vadose crystal silt. *Meniscus cements* and *pendent cements* (Dunham, 1969b and 1971) are little masses of calcite configured to the capillary water of moist vadose zones (they are not restricted to carbonate rocks; some sandstones have meniscus cements). Meniscus cements bridge the narrow gaps between adjacent grains; pendent cements (also called *gravity cements* or *microstalactites*) make thin "contact lenses" that cling to the undersides of grains. These cements are generally quite fine crystalline, but Schroeder (1973, fig. 7d) illustrated *sparry* meniscus cements (it seems to indicate that blocky spar can form in the vadose zone). Presence of meniscus cements implies that water menisci are frequently renewed or recharged with calcite-saturated water. A meniscus is probably part of a more-or-less continuous grain-enclosing film in which water is moving slowly downward, at least some of the time. Dunham (1969a; 1969b) described *vadose silt* or *crystal silt* in the Capitan reef complex, an internal sediment deposited in pores in a vadose zone. (Dunham's reports contain many outstanding photomicrographs.) Some internal sediment is perched on the tops of carbonate grains. Some forms flat layers or little conical piles on the floors of some of the larger pores; it may actually show tiny foreset laminae, indicating that water was moving rapidly through the pores, at several millimeters per second, at least intermittently (this is possible in a vadose zone, but probably not in a phreatic zone). Some crystal silt is interlayered with chemical cement. The color of internal sediment commonly differs from the color of the host rock.

Calcite precipitates also around plant roots in a vadose zone to form distinctive *alveolar structures* (Esteban and Klappa, 1983, p. 36). *Clay cutans* on plant roots have been mentioned by Esteban and Klappa (1983, p. 42) and by Meyers (1988, p. 316) as preserved in ancient soil zones.

Mesogenetic Cementation

Cementation that occurs under deep burial is delayed cementation. It completes the occlusion of pores only partially cemented eogenetically, or it occludes pores newly created mesogenetically. Cements in mesogenetic secondary pores are late cements, obviously postdating development of the secondary pores. Ankerite or other iron- and magnesium-rich carbonates are common mesogenetic cements, as is kaolinite (Franks and Forester, 1984, p. 64; Land and Dutton, 1978). Iron and magnesium for the late-stage carbonate cements may derive from transforming clays. Kaolinite is precipitated from solutions free of potassium, such as might be present in andesitic or basaltic volcanic sandstones or expressed from smectite shales.

Pressure dissolution seems to be an important local source of carbonate cement in limestones. Wong and Oldershaw (1981) examined deep-burial carbonate cements in Devonian reefal limestones in Alberta, Canada, and found that there is an increase in degree of cementation, and corresponding decrease in porosity and permeability, in the vicinity of stylolites. Moreover, cement stratigraphy indicates that older cement is confined to zones close to stylolites, while the younger cements predominate at greater distances. It is clear that stylolites were the source of the cement, and that convergence of rock masses at stylolites continually recycles carbonate grains, matrix, and cement into newer cement. Finkel and Wilkinson (1990) determined that stylolitization accounts for a significant amount of the cement in the Salem Limestone (Mississippian; subsurface) of west-central Indiana. Their conclusion was based not only on petrographic textural observations but also on trace-element mass-balance considerations.

Hattin (1981) presented a simple, but elegant example of diagenesis in a chalk formation, the Smoky Hill Member of the Niobrara Formation (Upper Cretaceous) in western Kansas. Compaction was the main diagenetic event in this lightly cemented, friable, foraminifer-peloid packstone and wackestone; the tests of foraminifers show varying degrees of crushing; fecal pellets are, in general, severely flattened. There was only minor cementation, and it occurred sometime after the compaction, apparently under mesogenetic conditions. Calcite spar cement fills or partially fills the chambers of the foraminifer tests, having been emplaced, it is clear, sometime after the compactional crushing. Small amounts of calcite cement occur also as overgrowths on various small skeletals, including coccoliths. Source of the calcite cement, according to Hattin, may have been aragonitic skeletal material, such as the nacreous layers of inoceramid shells.

But mesogenetic cements of some other limestones

are imported. In many globigerinid-bearing shales of Indonesia, foraminifer loculi are cemented with strongly ferroan calcite (or with kaolinite). Eogenetic cementation probably does not occur in the deeper-water (carbonate-undersaturated) marine environments under which these rocks were deposited, and foraminifer tests probably remained uncemented for substantial periods. Eventually, cements were precipitated mesogenetically from shale-compaction waters.

Oldershaw and Scoffin (1967, p. 318) found that mesogenetic calcite cements in certain Paleozoic limestones in England are iron-rich near to stratigraphically associated shales and iron-poor away from shale beds. This suggested that the iron was derived from the shales, and that the calcite cementation was a direct result of mesogenetic shale compaction and dewatering. In Middle Ordovician limestones of Virginia (Grover and Read, 1983), dull orange- or brown-luminescent coarse spar is the last pore-filling calcite cement, overlying various early cements and also filling tectonic fractures. In downslope carbonate buildups (mud mounds), for example, dull-luminescent ferroan dolomite overlies rim cement or bright calcite spar. The late dull cements, both calcite and dolomite, again may have precipitated out of pore fluids released from shales compacting and dewatering under deep burial. Some of it is contaminated by hydrocarbons, indicating that it was emplaced during and after generation and migration of petroleum, a mesogenetic process.

In the subsurface Kaybob reef complex (Devonian) in Alberta, Canada, Wong and Oldershaw (1981) found coarse anhedral mosaics of clear dolomite and calcite spar cements overlying isopachous and microstalactitic cements and other products of early diagenesis. Staining revealed sharp compositional zoning in the coarse spar cements, seven precisely correlatable zones in the dolomite, and eleven in the calcite (CLM did not work well here, the cements containing too much Fe^{2+} and too little Mn^{2+}). The dolomite cement occurs mainly on the outer reef margin and reef slope, the calcite cement mainly in the reef interior and inner margin. Dolomite precipitated first; this cement and the earliest calcite cement contain most of the iron. That the ferroan and non-ferroan zones could be correlated over section thickness of 23 meters and well to well indicated to Wong and Oldershaw that the cements precipitated from pore fluids that were essentially continuous and, at any given time, of similar composition and origin throughout. Clearly, cements everywhere precipitated at the same times and at about the same rates. Fluid-inclusion studies on similar cements in other organic buildups of the same age and similar burial history indicated that the cements formed at 150°C and 4 to 5 kilometers depth, derived at least in part from closely associated basinal shales. Close compositional zoning was a result of Eh fluctuations brought about by somewhat unsteady introduction of basin-derived organic matter carried in with carbonate-bearing fluids.

Neomorphism

Aggregates of very small crystals or very elongate crystals are thermodynamically unstable and under diagenesis tend to change to aggregates of larger, more equant crystals. We observe the results of this process in old snow; even yesterday's snow exposed to today's warm sunshine is texturally very much different from the downy flakes that made the original snow blanket. The recrystallized snow is grainy, the grains are larger and more equant than the original snowflakes.

Neomorphism of Carbonate

A similar, but slower, textural change occurs in carbonate minerals. Certain minor compositional changes or even structural inversion may accompany the textural alteration. Aragonite and high-magnesium calcite, as marine cement or as skeletal material, are unstable outside the marine environment, and if they do not just dissolve, they recrystallize to low-magnesium calcite. Fine carbonate mud, whether calcite or aragonite, also recrystallizes to coarser calcite. Solid-state change of crystal size and shape, including those changes that accompany polymorphic inversion, is called *neomorphism*.

It appears that neomorphism takes place at a thin liquid film through which ions diffuse. Original carbonate (aragonite, let us say) on one side of this film dissolves; the ions in solution diffuse to the other side of the film and precipitate as calcite. Thus, the liquid film migrates across original material and leaves new material in its wake. Conversion of aragonite to calcite results in an 8% molar volume increase. It is not clear how this affects the process; probably the excess ions leak out of the liquid film into adjacent pore space. Or the volume increase might create a crystallization pressure that physically displaces materials involved in the conversion process. Certainly the expansion would help to keep the diffusion film thin; if there were a contraction instead, then the film would get gradually wider and more dilute and the neomorphism process would quickly self-terminate.

Evidently, liquid films are not necessarily closed systems, and ions diffuse laterally into and out of them. Neomorphism of aragonite releases Sr^{2+}, which apparently leaks into the pore space (aragonite structure readily accommodates the large Sr^{2+} ion, but the calcite structure does not). There are examples of fabric-retentive alteration of calcite (fibrous cement or skeletal) to dolomite, which presumably involves diffusion across thin liquid films. This certainly cannot be a closed system; Ca^{2+} must leave the system and Mg^{2+} must enter. In this case there *is* a contraction (about 10% on a mole-for-mole CO_3^{2-} basis); perhaps this is the reason that neomorphism of calcite to dolomite is rather rare. Some alterations of calcite to chalcedony are fabric-retentive, and open-system liquid films again are implied.

High-magnesium calcite is unstable in fresh water, and it recrystallizes to low-magnesium calcite. The magnesium may exsolve to tiny dolomite rhomb inclusions, which are partly responsible for the cloudiness of some calcite cements. Microdolomite inclusions that Lohmann and Meyers (1977, p. 1081) described are subhedral or euhedral rhombs in crystallographic (optical) continuity with the host calcite. Echinoderm plates, which are initially composed of high-magnesium calcite, also may develop dolomite inclusions; such inclusions are lacking in the rim cements.

Folk (1974, p. 48) indicated that, during neomorphism of lime mud, the magnesium ions exsolve to the surfaces of the tiny micrite crystals, where they tend to inhibit crystal growth beyond a certain small size (in the micrite range), and this, according to Folk, has ensured the worldwide uniformity of crystal size of micrite. Any process that removes the magnesium "cage" that surrounds micrite grains promotes, or at least permits, further crystal growth. The coarser aggregate that results, called ***microspar***, consists of grains typically 5 to 10 micrometers across and less uniform in size than the micrite, and grain boundaries are wavy or interlocked (Fig. 11-13a). Meteoric waters may flush the magnesium from outcropping micritic rocks. Deep connate waters commonly are low in dissolved magnesium, and also may strip Mg^{2+} ions from micrite grain surfaces. Folk (1965, p. 41) noticed a

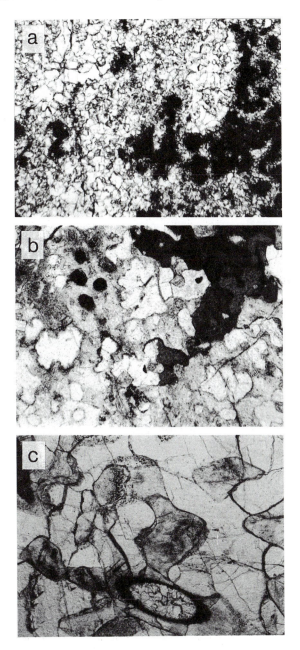

Fig. 11-13. (a) Microspar replacing peloids and mud; (b) neomorphosed coral skeleton. This whole view contains (parts of) only three crystals, interlocked. (c) Coral skeleton partly neomorphosed to coarse pseudospar, partly cemented with calcite spar.

strong association of microspar fabrics with carbonate rocks that are interbedded with shale, the shale apparently acting as a magnesium sink. Alternatively, magnesium exsolved from recrystallizing lime mud might appear as tiny dolomite crystals intimately mixed with the calcite microspar, a common feature in recrystallized limestones.

Neomorphic crystal size aggradation can go even beyond the microspar stage. Folk (1965, p. 42) referred to the coarser recrystallized calcites as *pseudospar*. In pelmatozoan-bearing wackestones, columnal plates occasionally develop euhedral overgrowths, especially, parallel growths of calcite crystals on the articulation surfaces. (Pelmatozoan columnal plates are single crystals, and the crystallographic c-axis is parallel to axis of the column.) Clearly the overgrowth is not an ordinary pore-filling cement—it has grown into a region occupied by lime mud. The columnal plate seems to have nucleated and catalyzed the recrystallization of adjacent lime mud into large calcite crystals.

Neomorphism also transforms complex organic skeletals, such as those of corals and coralline algae, into aggregates of coarse crystals (Fig. 11-13b,c), these aggregates commonly retaining ghosts of the original skeletal microstructure. Neomorphism of skeletal aragonite to calcite usually results in large crystals; probably this is because the aragonite host provides few seeds for calcite crystals. Schroeder (1973, fig. 10) illustrated a gastropod shell (originally aragonite) neomorphosed to calcite spar that he considered to have nucleated on blocky calcite *cement* already contained in the whorl of the shell; the neomorphism shows relict shell structure, which clearly distinguishes it from the cement.

Some neomorphism surely occurs very early, even *before* early cementation. Extensive neomorphism has occurred in certain Miocene reef rocks of Java, Indonesia. Coral skeletons are locally replaced by aggregates of coarse calcite crystals that partly preserve the original fabric. In some rocks, coarsely recrystallized skeletons are overlaid with eogenetic blocky spar cement. One occasionally sees two widely separated blocks of this cement syntaxial on a single large pseudospar crystal; clearly the pseudospar nucleated the cement, and not vice-versa, as in Schroeder's example; the neomorphism predates the early cement. In one sample examined, a large, calcite-cemented dissolution vug transects a boundary between recrystallized and unrecrystallized rock (Fig. 11-14). Where vug-filling calcite cement is in contact with unrecrystallized

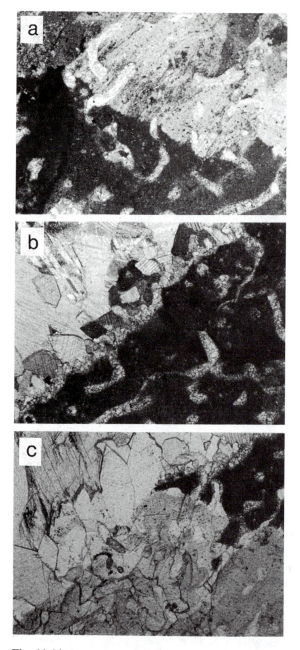

Fig. 11-14. Neomorphism, dissolution, cementation in an Indonesian rock. **(a)** Coral fragment and dark mud fill; in the lighter region the wall material and mud fill are neomorphosed to coarse pseudospar. In the darker region the wall material dissolved and the moldic pores were then cemented with calcite spar. **(b)** The same coral fragment, here cut by large vug; note inward-coarsening of vug-filling calcite cement. **(c)** The same vug also transects the neomorphosed part of the coral; vug-filling cement has grown syntaxially on the coarse neomorphic spar. Cementation of skeletal-moldic pores and of the large vug is post-neomorphism.

material it is fine-crystalline (and coarsens away from the pore wall). But where the calcite cement is in contact with neomorphosed material it is syntaxial with the neomorphic crystals, in effect propagating the coarse neomorphic crystal fabric into the pore; there is no fine-crystalline fringe at the pore wall. This indicates, again, that the neomorphism occurred before the cementation (and also shows the important influence of substrate on cement-crystal size and shape).

Davies (1977) described fibrous calcite cement in Pennsylvanian-Permian bryozoan limestones of Ellsmere Island, Canada; James and Klappa (1983, p. 1070) observed spherulites and fans of calcite needles, and rinds of fibrous calcite in Cambrian reefal limestones in Labrador, Canada; and Mazzullo (1980) found similar cements in Permian reefal limestones in New Mexico and Texas. These fibrous cements are botryoidal (not isopachous); high concentration of strontium in the cement suggests an aragonite precursor. Recrystallization of the aragonite has preserved the chemistry (a high Sr content), and the microfabric in detail. Tucker (1983) described fibrous cements in Precambrian dolomite rocks in Death Valley, California. These cements appear to have been high-magnesium calcites that have dolomitized. Tucker pointed out that early dolomitization of high-magnesium calcite usually is fabric-retentive.

Distinction between Neomorphisms and Cements

Cements conform to the shapes of pores; cements nearly always show a fabric progression from pore wall to pore center. Crystals in contact with the pore wall typically are very small, and show various orientations; they are buried in coarser crystals that commonly are prisms or columns oriented normal to the pore wall. In tabular pores there may be a clearly defined medial seam, where crystals that grew from opposite sides of the pore meet. There are, according to Bathurst (1975, p. 490), many enfacial junctions (see Chap. 2). Cement crystals can show compositional zoning that reflects the growth of crystal faces and fluctuations in pore fluid properties during crystallization.

In contrast to cement crystals, size of neomorphic crystals is not constrained by pore size. Moreover, neomorphic crystals grow from materials already on site; they do not rely on import of raw materials from the outside. Thus, these crystals can be much larger than cement crystals. Neomorphic crystals occur in places

where cements would not be; for example, an overgrowth on an echinoderm plate facing matrix cannot be a cement. In a rock composed of crystalline calcite with floating grains, the spar is necessarily a product of neomorphism. There is generally no inward-coarsening in a neomorphic polycrystalline mass. Intercrystalline contacts in aggregates of neomorphic crystals are compromise boundaries; only rarely is a crystal face expressed; enfacial junctions reportedly are few. Neomorphic crystals commonly contain relict or "ghost" fabrics of the materials that they have replaced; these are inclusions of organic matter or mineral particles that remained unaffected by the replacement process. Whereas neomorphic crystals might be quite cloudy, associated cement crystals might be very clear. Neomorphic spar might have a trace-element composition (including isotopic ratios) that differs from that of coeval cement. In fine polycrystalline aggregates (microspar or fine pseudospar resulting from recrystallization of lime mud or micrite), there are random areal changes in crystal size, and perhaps small stellate aggregates; boundaries between regions of recrystallized and unrecrystallized micrite are diffuse and irregular.

Neomorphism of Silica

Opal is an amorphous silica or extremely fine-crystalline aggregate of the metastable polymorphs of silica, α-tridymite (orthorhombic) and α-cristobalite (tetragonal). The strictly amorphous opal (x-ray amorphous), called *opal-A* (classification of Jones and Segnit, 1971), occurs in the siliceous skeletal materials of diatoms and certain sponges. *Opal-C*, which reveals the cristobalite structure under x-ray diffraction analysis, is associated mainly with volcanic rocks. *Opal-CT*, a disordered stacking of cristobalite layers and tridymite layers, occurs in sedimentary rocks as cement. To distinguish among these types of opal, x-ray diffraction methods are required. Biogenic opal of the siliceous sponges, radiolarians, and diatoms undergoes neomorphism, opal-A crystallizing gradually to opal-CT, thence to fully crystalline chert (Mizutani, 1977). Stein (1982, p. 1281) found that opal of silicified wood crystallizes in much the same manner. According to Thiry and Millot (1987), opal-A and opal-CT in opaline regoliths (silcretes) in the Paris basin, France, transform to chalcedony, lutecite, and chert, then to limpid quartz crystals. Experiments of Kastner et al. (1977, pp. 1054, 1056) indicated that conversion of opal-A to opal-CT requires presence of Mg^{2+} and OH^-, and that the rate of conversion is much higher in carbonate sediments than in clay-rich sediments. Moreover, opal-CT of carbonate rocks is less disordered than that of argillaceous rocks, and it converts more readily to quartz.

Carbonate Concretions

Carbonate concretions are present in some rocks, especially shales. In many cases they are concentrated on a few distinct, laterally persistent stratigraphic horizons in the shale and rare or absent at all other levels. They are objects made by local cementations, taking place under conditions quite unlike those of the more ordinary, pervasive kind of cementation that changes large bodies of sediment into sedimentary rock. Many carbonate concretions in shales are syndepositional or early diagenetic and form close to the sediment-water interface, while the sediment is still soft. Some concretions are perforated with borings or encrusted with barnacles or serpulid worm tubes, indicating that these concretions formed in the depositional environment, or that they formed a little way below the sediment-water interface, but shortly afterward were exhumed to the depositional surface by erosion of the soft muds overlying them. Concretions occasionally have been transported and disoriented in the depositional environment, either by some storm-generated current or by foraging animals. Oxygen isotope signatures commonly suggest that concretions have formed at the same temperatures as skeletal material (mollusc shells and such) in the host sediment, rather than at the higher temperatures attendant to deep burial.

Laminae preserved in concretions are continuous with laminae in the enclosing shale, but are deformed in ways that indicate that the concretions grew during early compaction of the host mud. Sass and Kolodny (1972, p. 269) examined some concretions that are thicker than their laterally equivalent sediment layers by a factor of more than 4. Volume-percent carbonate in a concretion corresponds approximately to the porosity of the host sediment at time of concretion growth (but some concretions contain very little detrital material). McBride (1988, p. 1809) used this kind of evidence to show that some concretions grow over a longer term, perhaps several million years. Many concretions are oblate, not spherical; this is *not* because they have been compacted under burial pressure but because their growth depended on chemical dif-

fusion through an anisotropic medium. Concretions formed in uncompacted muds are more spherical than those formed in compacted muds because compaction renders muds less isotropic. Raiswell (1971, p. 170) noted that concretions in isotropic sandstones *are* nearly spherical.

Raiswell (1976, p. 228) suggested that concretion growth is driven by microenvironments of localized carbonate supersaturation. Dickson and Barber (1976) determined that appropriate chemical microenvironments are created by decay of organisms, and they distinguished between concretions associated with individual large carcasses and those associated with concentrations of many tiny organisms or microscopic organic detritus. Hecht (1933) showed that decomposition of a fish carcass releases ammonia or amines, which raises pH and leads to the precipitation of $CaCO_3$. Experiments by Berner (1968) seem to confirm that calcium carbonate concretions form by decomposition of organic matter. Early in the process of organic decay, and as long as anaerobic conditions can be maintained, ammonia is released, which keeps pH at sufficiently high levels to allow dissolved calcium to precipitate as Ca-fatty-acid salts (that is, calcium soaps), which later becomes $CaCO_3$. It is common to find strongly negative $\delta^{13}C$ in the calcite of concretions (see, for example, Sass and Kolodny, 1972, pp. 272, 273); this strongly suggests that the carbon of calcite is derived from organic matter. Just outside of the concretion-forming microenvironment, where there is less organic matter, conditions may well be aerobic. Decay of organic matter under aerobic conditions produces CO_2, which, in contrast to conditions inside the concretions, tends to *dissolve* carbonate or *prevent* its precipitation.

Fossils commonly are very well preserved in concretions (Weeks, 1953; Blome and Albert, 1985), even if very poorly preserved outside, in the host rock. Gutschick and Wuellner (1983) and Sandberg and Gutschick (1984, p. 147) advanced the concept of concretion-based biostratigraphy. Sandberg and Gutschick observed that some concretions in the Delle Phosphatic Member of the Brazer Dolomite (Mississippian) in Utah contain the well-preserved remains of planktonic and nektonic pelagic organisms, land-plant detritus, and benthonic organisms that are less abundant and less well preserved in surrounding sediment; some contain animal burrows and other sedimentary structures. Concretions encase fossil material in an impermeable, incompressible tomb, protecting this material from predation, chemical attack, and deformation, in some cases so soon post-mortem that the flesh has not even yet fully decayed. A concretion is a "snapshot," a record of life (or death) at a virtual point in time and space, a record probably more accurate and complete than that preserved outside the mass. Comparison of fossil biota within a concretion with that just outside suggests the degree to which (post-concretion) diagenesis has diminished or altered the fossil record. The preservational value of **coal balls**, which are early formed dolomite concretions in coals, is well known to the coal petrologist.

Aside from their ability to preserve delicate fossils, concretions also provide clues to the progress of diagenesis of the sediments that contain them. A concretion tends to protect the sediment that it encloses against later diagenesis. Thus, degree of compaction inside an early formed concretion is less than outside. The concretion excludes pore waters, which nonetheless may actively circulate in the host sediment, bringing about various alterations, cementations, and dissolutions there. The composition of detrital clay minerals inside a concretion commonly differs from that outside. The quantity and variety of detrital heavy minerals commonly is greater inside concretions than outside. Any contrasts between the concretion interior and the sediments outside are likely a result of post-concretion diagenesis.

There is a widely distributed horizon (or two) of carbonate concretions in the Blue Hill Shale (Upper Cretaceous) of western Kansas and eastern Colorado, deposited in marine water about 20 meters deep. The concretions seem to be related to thin bentonite beds. Volcanic ashfall onto the surface of the sea probably altered the turbidity and chemistry of the seawater sufficient to cause mortality of nekton and benthon over a wide region, the dead bodies nucleating many carbonate concretions simultaneously. A short-term mass mortality might be caused also by algal blooms, similar to the so-called "red tides" that occasionally stain and poison the modern neritic waters. Maybe some concretion horizons of ancient rocks record mass mortalities brought on by a burst of cold water or anoxic water. Kraus (1988) described concretions in laterally persistent zones in Tertiary alluvial deposits in the Bighorn basin, Wyoming, that encase standing tree trunks. Here, a forest appears to have been drowned by a river flood.

Large (2- or 3-meter), nearly spherical concretions have weathered out of cross-bedded Cretaceous sand-stone at "Rock City" in central Kansas (Fig. 11-15). Though a few have rolled or toppled, many maintain the positions and orientations that they had originally in the poorly cemented sandstone that hosted them. Cementation of the sand clearly proceeded centrifugally from centers spaced laterally a few meters apart, and the absence of compound objects suggests that there was some pattern of chemical diffusion or of pore-water flow that inhibited mergers of adjacent concretions.

Chert Nodules and Silicifications

Much quartz precipitation and cementation is an early diagenetic process, predating other kinds of chemical diagenesis. Rain water that has flowed or percolated only a short distance through quartz sand dissolves quartz as monosilicic acid, H_4SiO_4. The resulting mildly supersaturated solution with few impurities can precipitate macroquartz on a quartz nucleus. Water that has percolated farther develops higher concentrations of dissolved silica and might take various metal cations into solution, such as Na^+, Ca^{2+}, Mg^{2+} and Al^{3+}. Silica goes into aqueous solution during the weathering of silicate minerals such as the feldspars, and groundwaters commonly are greatly supersaturated with respect to quartz. These solutions can give rise to cryptocrystalline varieties of quartz, either by precipitation or by replacement of carbonate minerals. If foreign ions are abundant in the solution, then opal precipitates, with its more disordered tridymite-cris-

Fig. 11-15. Large concretions at Rock City in Kansas.

tobalite structures. Silica is precipitated typically as opal where clay is abundant.

The replacement of wood by opal seems to involve some kind of organic chemical process (Evans, 1964, p. 269; Stein, 1982, p. 1277), and it occurs very early, as the wood is decaying. Much bedded chert forms from biologically precipitated opal, deposited as skeletal parts; chert crystallizes from this material diagenetically. Many workers have suggested that chert nodules form in depositional or very early diagenetic environments from a colloidal form of silica, that is, a silica gel. Some chert nodules have systems of radial and concentric cracks, apparently owing to syngenetic dehydration (syneresis) and shrinkage of blobs of this gel precipitated in the depositional environment. Some chert nodules are concentrically banded, and this banding has been attributed to diffusion of impurity ions through a gel.

Elouard and Millot (1959; see Millot, 1970, p. 294) described silicifications in the western Sahara (in Mauritania), where arenaceous limestones are transformed into chert and chalcedony. Chalcedony replaces calcareous material, whether calcite or dolomite, but quartz sand that lacks calcareous cement or matrix acquires syntaxial rims of macroquartz. Some (macro)quartz-cemented sandstone contains nodules of chalcedony that apparently were calcareous originally. In the Fazzan, western Libya, sandy limestones and calcareous sandstones commonly are surficially silicified in ways that seem to be topographically controlled (Muller-Feuga, 1952). On isolated plateaus (hammadas), quartz sand grains are cemented with syntaxial macroquartz. But on the lower slopes of these plateaus and in the lowlands near to water table, there are opaline or chalcedonic cements that appear to have replaced calcite. Quartz grains commonly are severely corroded, and enveloped in opal that is in various stages of crystallization to chalcedony or macroquartz.

Folk and Pittman (1971) enumerated several types of chemically precipitated quartz, including three distinct kinds of chalcedony (originally identified in the 1890s by Michel-Levy and Munier-Chalmas). Chalcedony is fibrous quartz; one type, *chalcedonite*, is length-fast, where c-crystallographic axis is perpendicular to length of the fiber. Other chalcedony is length-slow, where c-axis is parallel to the fibers; this is called *quartzine*. Note that ordinary (macro)quartz is length-slow. In a third type of chalcedony, called *lutecite*, c-axis is at a 30° angle to the fibers. Length-fast chalcedony is, in virtually all cases, pore-filling—

that is, a chemical cement; typically, it forms neat, regular spherulites of delicate, radially arranged fibers. In length-slow chalcedony, the fibers are thicker, cruder, and not so neatly radial. Folk and Pittman discovered that length-slow chalcedony is an almost unequivocal guide to the former presence of evaporites in a carbonate succession, and that nodules of this chalcedony, so common in some carbonate rocks, are replacements of evaporitic gypsum nodules, though some exceptions have been noted by others (see Keene, 1983). Lutecite, which displays a crossed-lamellar structure, is chiefly a replacement of calcite skeletals or calcite cement, but also replaces evaporite minerals.

Richter (1972) noted that authigenic macroquartz can contain gypsum or anhydrite inclusions where no gypsum or anhydrite exists outside. Clearly evaporite minerals originally present outside the authigenic quartz have dissolved, and quartz authigenesis *preceded* the dissolution. Richter also described calcite inclusions in quartz crystals within dolomite rocks, indicating that quartz authigenesis preceded the dolomitization also. A quartz crystal replacing skeletal carbonate may contain inclusions of high-magnesium calcite, even though the shell material outside the crystal is low-magnesium calcite. This indicates that the shell material was originally high-magnesium calcite and that the quartz crystal formed before other diagenetic processes transformed the shell to low-magnesium calcite.

Silicification of Carbonate Grains

Carbonate rocks, or certain components of carbonate rocks, commonly are partially replaced by silica (Fig. 11-16). In many cases it is an early diagenetic process. But telogenetic superficial silicification of limestone or dolomite, or of certain skeletals only, also is common and is familiar to most geologists who have examined outcrops of carbonate rocks. Silica replacement of carbonate grains in limestones appears to be driven mainly by the insolubility of crystalline silica (quartz, cristobalite) relative to amorphous silica, and by the decreased solubility of silica at low pH or low temperature, under which conditions calcite solubility is high.

Newell et al. (1953, p. 171 ff) described silicification of skeletal grains in the Capitan reef carbonates (Permian in West Texas). They were impressed by its selectivity, a peculiarity apparent to any carbonate petrographer or stratigrapher. Not only are fossils silic-

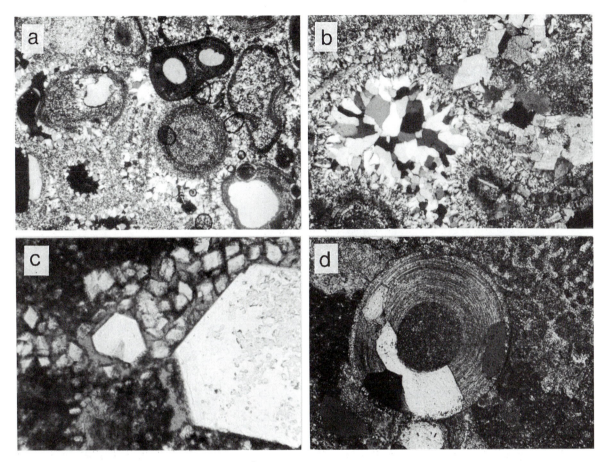

Fig. 11-16. Examples of silicification: **(a)** cherty replacement of grains and matrix in a packstone; **(b)** coarse quartz is probably a pore-filling cement—note unreplaced dolomite rhombs; **(c)** euhedral authigenic quartz in a sandstone; **(d)** section of a brachiopod spine partly replaced by quartz.

ified in preference to carbonate matrix, but certain types of fossils are favored over others: Bryozoans, tetracorals, tabulate corals, and punctate brachiopods are the most readily silicified, according to these authors, followed in turn by impunctate brachiopods, molluscs, echinoderms, foraminiferans, and last, calcareous sponges and dasycladaceans.

Just as calcite commonly is replaced by silica, so is silica occasionally replaced by calcite. This is most frequently expressed as embayment or etching of detrital quartz grains in calcareous sandstones; in fact, quartz grains may be so reduced in size chemically that they come to float in the calcite matrix (Fig 11-17). This is probably a mesogenetic process, brought about by alkaline pore waters or due to movement of

pore waters from regions of lower temperature, where silica is less soluble and calcite more soluble than at the higher temperatures. Walker (1960, p. 148) suggested that it can happen at any time in the history of a sedimentary rock, and Walker (1962) described several examples of reversible chert-carbonate replacement. Hesse (1987) described reversible replacement in certain carbonate-bearing turbidites.

Nodular Cherts

Chert nodules or lenticular stringers of chert are common components of many limestone formations. Some nodules are very hard and dense, with sharply defined

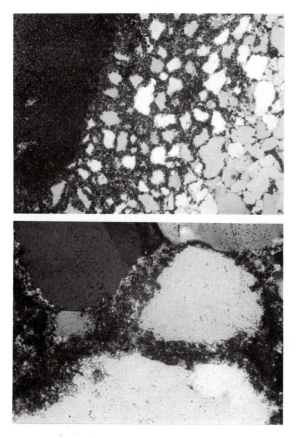

Fig. 11-17. Corroded quartz grains; fine calcite fills spaces between the grains.

edge; other nodules show a spongy character, enclosing much carbonate material that escaped silicification. The nodules commonly retain vestiges or ''ghosts'' of sedimentary structure, or of fossils or other grains that without doubt were originally composed of calcite or other carbonate mineral. Commonly, bedding or lamination in the host rock partially drapes the nodule, in a manner that indicates that the object formed early, before much compaction had taken place. Relict grains inside an early nodule are substantially less compacted than the grains on the exterior, and thus less deformed and less condensed (that is, farther apart). Some chert nodules show concentric light-dark color banding. In the ***wood-grained cherts*** of DeCelles and Gutschick (1983), banding was attributed to a chemical diffusion-precipitation phenomenon such as described by Liesegang in 1898.

Dietrich et al. (1963) described silicification in the Knox Dolomite (Cambrian and Ordovician) near Blacksburg, Virginia, that interrupted the progress of dolomitization. The chert, occurring typically as lenticular beds and as nodules, contains euhedral dolomite crystals. Moreover, the chert masses commonly are broken up, brecciated apparently as a result of desiccation or syneresis, the angular fragments then scattered about in some cases, and even imbricated, by waves or currents. Overlying carbonate layers show draping or supratenuous folds that resulted from differential compaction around the chert nodules. The authors interpreted these features to mean that precipitation of silica, and its lithification into hard, rigid nodules, occurred early on; in fact, it occurred on the depositional surface. Dolomitization had already begun; silica locally replaced calcium carbonate sediment but did not replace any dolomite crystals. Dolomitization continued, eventually affecting all of the carbonate sediment except that already replaced by chert.

Carozzi and Gerber (1978) also saw evidence of very early chert in the form of synsedimentary chert breccia in the Burlington Limestone (Mississippian) in Missouri. They suggested that a layer of early chert was ripped up and the angular fragments strewn about by storm waves. Gao and Land (1991) described early chert nodule formation in ooid grainstones in the Arbuckle (Ordovician) in Oklahoma, in which silicification preceded dolomitization.

It would have been a source of some wonder, for geologists wandering about the beach in Paleozoic time, to encounter these hard siliceous nodules embedded in the soft carbonate muds and sands. Would the source of the silica have been evident to them? Are such objects forming today?

Knauth (1979) proposed that many nodular cherts in limestones formed in sea-water–freshwater mixing zones; it is much like the dorag model for dolomitization (see Chap. 9). Opal-A of sponge spicules or other siliceous skeletals dissolves in fresh water or sea water, but mixtures containing between 5% and 50% seawater can be highly supersaturated with respect to opal-CT and quartz (see Knauth's fig. 2). Thus, where silica-bearing fresh water mixes with sea water, silica is precipitated, usually along bedding planes or other permeability channels. Knauth supported his proposal with oxygen-isotope data. This model seems to account for the common association of early chert and early dolomite.

Quartz Geodes

Another form of silica in carbonate rocks is the **geode**, a spheroidal vug-filling or lining. That quartz geodes occurring in considerable abundance in certain sedimentary rocks are replacements of evaporitic gypsum or anhydrite nodules (Folk and Pittman, 1971) seems to be well established. Chowns and Elkins (1974) described "cauliflower" cherts and quartz geodes in Mississippian rocks in Tennessee that were formed by silicification of anhydrite nodules. Milliken (1979) also examined these nodules. Her nice photomicrographs show how distinctive these special nodules can be petrographically. In a typical geode there is an outer zone of interlocked spherules of quartzine (length-slow chalcedony) and a mosaic of macroquartz inside that has rhombohedral crystal terminations and shows undulose, radial extinction patterns. Anhydrite inclusions are abundant in the macroquartz. A peculiar fibrous quartz with zebra-striped extinction (lutecite?) forms crusts that line the internal cavities of some of the geodes. Oxygen isotope studies suggest that the quartzine precipitated from a mix of sea water and meteoric water and that the macroquartz and "zebra"-quartz formed in meteoric water. Commonly the vugs or other large pores in a rock are only partly cemented, being lined with crystals with well-formed terminations that face empty space on the interior. Perhaps such a pore never completely filled because the initial cement lining sealed the pore against further entry or throughflow of cementing fluid.

Maliva (1987) examined the multitudinous quartz geodes of the Harrodsburg Limestone (Mississippian) in southern Indiana (Fig. 11-18) , again concluding that they are silicified anhydrite nodules. Maliva recognized an outer zone of length-slow chalcedony that gives way inward to a zone of xenotopic quartz, then idiotopic macroquartz at the center. Anhydrite inclusions are most abundant in the zone of xenotopic quartz.

Concentrated silica solutions commonly occur in the vugs and vesicles of volcanic rocks where glassy components are devitrifying. These solutions may precipitate banded chalcedony nodules (agates) in large numbers. Some agate nodules are concentrically banded on the margin and have horizontally banded (geopetal) chalcedony inside, perhaps overlain by ordinary macroquartz. Amorphous silica (that is, opal) precipitates from very concentrated solutions as tiny spheres (lepispheres) in suspension (see Rau and

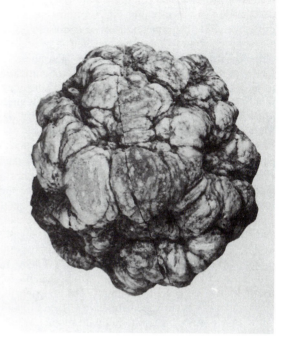

Fig. 11-18. Exterior of a cauliflower geode. (Courtesy Lee J. Suttner.)

Amaral, 1969); these spheres then fall slowly to the floor of the vug. Chalcedony and more coarsely crystalline quartz precipitate from weaker solutions. One suspects that the concentric bands in agate nodules were precipitated directly as chalcedony and the horizontally banded material was deposited as opaline lepispheres (transformed now to chalcedony).

Late Dolomite

Above (Chap. 9), we examined the mechanisms thought to be responsible for early dolomitization of carbonate sediment, mechanisms that create fluids having high Mg/Ca ratio and that cause these fluids to move through the pores of carbonate sediment. Now we look to dolomitization that takes place in lithified rocks under deep burial.

Origin and Characteristics

Late dolomite or **epigenetic dolomite** results from alteration of lithified carbonate rock by deep connate waters or hydrothermal solutions. Pore fluids at depth can develop sufficiently high Mg/Ca ratios by cation exchange with clay minerals or transformation of smectite clays into illite. The elevated temperatures promote the dolomitization reaction, and there is plenty of time for large volumes of dolomitizing fluid to move through the pores. Deep dolomitization commonly is rather patchy; there may be irregular boundaries between dolomitized and undolomitized rock; Mg/Ca ratio in the dolomite might vary widely over short distances; mean size of crystals may vary markedly from place to place. The rocks are coarser than syndepositional dolomites and show non-uniform grain size; there may be a porphyrotopic texture, the rock being a mixture of large crystals and small ones. The dolomites tend to be vuggy or cavernous; skeletals and other original grains commonly are preserved only as molds. Zenger (1983) described deep-burial dolomites of the Lost Burrow Formation (Devonian) in California and presented various textural criteria for recognition of epigenetic dolomites. The coarse crystalline texture certainly is distinctive. Also, a "net" fabric may develop, wherein dolomite rhombs are arranged in seams that enclose millimeter- to decimeter-size cells of dolomicrite. There may be truncated or relict stylolites transected by dolomite rhombs, and do-

lomite-filled fractures. On a larger scale, sharp dolomite "fronts" and tongues of coarse dolomite discordant to limestone beds also indicate epigenetic dolomitization.

Many dolomites are conspicuously mottled, among them several lower Paleozoic dolomite formations of Montana and Alberta. Beales (1953) described one of these, the Palliser Formation (Devonian) of Banff and Jasper National Parks in Alberta (Fig. 11-19), and determined that mottling reflects differential dolomitization favoring burrow fills and "algal colonies." This is probably in part a permeability effect; bioturbation causes a certain amount of dilation of an unconsolidated sediment, increasing its porosity and permeability, and alga-precipitated carbonate probably is quite porous at first.

Some workers, for example Land (1986, p. 37), have recently suggested that late dolomitization is insubstantial, that compacting shales or transforming clay minerals (or other deeply buried sources) do not provide sufficient magnesium for bulk dolomitizations. Nonetheless, numerous petrographic and stratigraphic and geochemical results indicate that late dolomitization does occur, even on a large scale, and that shales are a major source of the dolomitizing fluids. Probably some shales are magnesium exporters, others

Fig. 11-19. Mottled dolomite.

not. It is possible that much epigenetic dolomite is only a recrystallization (neomorphism) of syngenetic dolomite; in such cases externally derived magnesium is not required, of course.

Mattes and Mountjoy (1980) described dolomitization under deep burial, exemplified by a stromatoporoid buildup (reef) in the Upper Devonian Cairn Formation in Jasper National Park, Alberta, Canada. They suggested that the major dolomitization was a long-lasting event that began after deep burial, that degree of dolomitization was controlled by permeability of precursor limestone, and that the magnesium was supplied mainly by stratigraphically associated compacting shales and to a minor extent by the precursor limestone itself. Machel (1986, p. 348) similarly called upon expression of dolomitizing fluids from basinal shales to explain the dolomites in Nisku (Devonian in Alberta) buildups. Dolomites of seaward slope facies have higher boron contents than reefal facies to landward, boron being characteristically associated with marine shales. Machel and Anderson (1989) suggested that Nisku dolomitization occurred at burial depths of 300 to 1000 meters and at temperatures of 40 to 50°C. Woronick and Land (1985) described coarse baroque dolomite in the Glen Rose Formation (Cretaceous) in southern Texas; it contains up to 15 mole percent of $FeCO_3$, which was considered to have come from surrounding shales. Certain Pennsylvanian limestones in Iowa (subsurface) contain increasing proportions of dolomite near shale formations or within argillaceous lenses or interbeds (McHargue and Price, 1982, p. 879). In Dinantian (Carboniferous) strata in England, Gawthorp (1987, p. 552) noted that dolomite is most abundant near contacts with shales and that dolomitized rocks have higher iron and magnesium contents than associated undolomitized rocks. Again, this suggests that the shales were the source of the magnesium. Addison et al. (1985) provided some interesting evidence for timing of dolomitization. In the same limestone-dolomite succession that Gawthorp examined, they found that the limestones have a *reversed* remanent magnetic polarity acquired during or shortly after sedimentation; but dolomitized parts of the succession have a *normal* polarity overprint acquired during dolomitization that must have come much later.

Gregg and Shelton (1990) studied Cambrian dolomites of southeastern Missouri and saw evidence of both early and late dolomitization (as did Mattes and Mountjoy in their Devonian rocks of Alberta). They

suggested that much early dolomite was neomorphosed under mesogenetic conditions (thus, no late dolomitization, but merely a recrystallization of early dolomite). As neomorphic crystals grew, the distribution of crystal size became coarse-skewed and a net fabric developed, such that coarse dolomite formed three-dimensional networks or "skins" around cores of fine-crystalline (unrecrystallized or only lightly recrystallized) precursor dolomite, just as Zenger had found in the Lost Burrow Formation.

Dolomite Textures and Porosity

As a dolomite crystal grows in a porous host, the surface of the crystal advances not only across calcitic rock material, which transforms and becomes part of the dolomite crystal, but also across pore space, which in effect retreats before the advancing front. Thus, original (primary) pore space accumulates between the growing dolomite rhombs as secondary pore space. This also happens, of course, when existing porous dolomites recrystallize to coarser dolomites. These secondary pores are larger than the pores that existed before the dolomitization. Some additional pore space appears because there is a molar volume decrease of about 13 percent when calcite changes to dolomite.

Depending on supply of magnesium and of carbonate ions and on original rock fabric, dolomitization may yield a porous aggregate of small (typically <0.1 mm) idiomorphic crystals, or growth of dolomite crystals continues until all space is filled with a mosaic of irregular, tightly-fitted crystals ranging up to several millimeters in size. Importation of carbonate ions from some distant source must surely be required in the making of coarse xenotopic dolomites of low porosity.

Many dolomites contain the large secondary pores (larger than the usual intercrystalline pore) called vugs. Some vugs have distinctive shapes that suggest that they resulted from dissolution of skeletal grains or ooids, or of replacive minerals, such as anhydrite, or dissolution of gypsum or anhydrite nodules. Some fully dolomitized rocks contain skeletal molds (of echinoderm plates especially); in nearby rocks only partially dolomitized these molds are not present. It appears that echinoderm plates, being more resistant to dolomitization than other calcitic components in the original limestone, remain until all else has been dolomitized (and dolomites with calcite echinoderm plates are not at all uncommon). In the end, though, they might succumb to dissolution, perhaps contributing to the formation of the very last dolomite in intercrystalline pores nearby.

It is not unusual to find crystalline dolomites with abundant intercrystalline porosity partly occluded by calcite spar cement (Fig. 11-20). It is clearly post-dolomitization cement, *not* old calcite that escaped dolomitization or subsequent dissolution. Commonly this cement forms large, irregular single-crystal units, each poikilotopically enclosing many dolomite rhombs. A deep dolomitizing pore fluid, rich in Mg, gradually gives up some of its Mg^{2+} ions, exchanging them for Ca^{2+} ions, and probably also picks up CO_3^{2-} in the bargain. It seems logical to assume that the late calcite cement was precipitated by ''spent'' dolomitizing fluids. In many epigenetic dolomites another post-dolomitization calcium-rich cement, anhydrite, is precipitated instead of, or in addition to, calcite.

Dedolomite and Dedolomitization

Dedolomite is calcite that has resulted from the calcitization of dolomite. It is achieved by a reaction similar to

$$CaMg(CO_3)_2 + 2H^+ \rightarrow$$
$$CaCO_3 + Mg^{2+} + CO_2 + H_2O \qquad (11\text{-}1)$$

von Morolot, who seems to have been the first to report on dedolomite (in 1847), suggested that it is a product of chemical reaction between dolomite and anhydrite (or gypsum):

Fig. 11-20. Epigenetic dolomite with late calcite spar cement.

$$CaMg(CO_3)_2 + CaSO_4 \rightarrow$$
$$2CaCO_3 + Mg^{2+} + SO_4^{2-} \qquad (11\text{-}2)$$

Sulfate to drive such a reaction could be derived alternatively from oxidation of pyrite.

Shearman et al. (1961) and Evamy (1967, p. 1205) described a kind of dedolomite consisting of rhomb-shaped mosaics of several small calcite crystals; clearly, the rhomb shape was inherited from precursor dolomite, and the calcite is a cement pseudomorph after dolomite. Another kind of dedolomite is coarse crystalline calcite mosaic bearing irregular inclusions of dolomite, or showing relict dolomite texture (such as rhomb outlines). Texture of this kind of dedolomite is coarser than that of the original dolomite rock. *Calcite cores* in rhombs of dolomite may be a partial dedolomitization.

Fabric indicators of dedolomite are calcite rhombs, presumably pseudomorphic after dolomite; polycrystalline calcite rhombs, which are, apparently, dolomoldic pores cemented by calcite; dolomite rhombs with corroded edges facing calcite, or partial dolomite rhombs floating in calcite, or dolomite rhombs with interior patches or cores of calcite; and palimpsest textures in which rhombic zones of ferric oxides or grain boundaries of precursor dolomite crystals survive as ghosts within calcite spar. In short, dedolomite seems to be taken as any rhomb, or part of a rhomb, or aggregate of rhombs, that is composed of calcite. Some dolomite-replacive calcite is, according to Budai et al. (1984, p. 277), optically continuous with unreplaced dolomite. (Of course this cannot always be so, certainly not in a polycrystalline rhomb; here, a calcite crystalline substrate, if it exists, is *outside* the rhomb.) Evamy (1967, p. 1210), addressing rhombohedral pores in dolomite rocks, indicated that these are dissolved *dedolomite* rhombs, not dissolved dolomite; leaching selectively dissolved dedolomite, leaving dolomite rhombs and matrix untouched.

Cotter (1966, p. 145) found large dolomite rhombs with zones of dedolomite, and other rhombs extensively altered to calcite, in Waulsortian mound carbonate rocks in the Lodgepole Formation (Mississippian) in Montana. Goldberg (1967) and Katz (1971) described them from Jurassic dolomites in Israel, and Koch and Schorr (1986, p. 235) found them in Jurassic reefs in Germany. Frank (1981) reported on zoned dedolomite in the Taum Sauk Limestone (Cambrian) in southeast Missouri, made especially evident by CLM. SEM reveals a complex microstructure of intergrown calcite, dolomite, and iron oxide, which argues against any possibility that the calcite zones simply precipitated syntaxially on dolomite rhombs during some interruption in the growth of the dolomite. The calcite zones are in fact calcitized dolomite, apparently selectively replacing only heavily ferroan zones in the precursor dolomite.

Many workers have argued that dedolomite forms under (near-surface) telogenetic diagenesis (Evamy, 1967; Katz, 1968; Chafetz, 1972, p. 325). Wolfe (1970, p. 46) described a short (less than 1-meter) vertical profile of dedolomitized coccolith chalk below an unconformity, which documents a gradation from dolomoldic pores and fully calcitized rhombs at the unconformity, to dolomite rhombs with calcitized cores, to unaltered dolomite rhombs at the base of the profile. The cores of dolomite rhombs were calcitized first, and calcitization has partially *regenerated the original texture* of the coccolith micrite! Apparently the orientation of minute calcite inclusions in the dolomite acts as a kind of memory that allows dedolomite crystals to restore, at least approximately, the original pre-dolomitization texture.

Dedolomites formed at or near weathering surfaces commonly are pervaded by yellowish brown or dark brown limonitic or goethite crusts. Evamy (1963, p. 169) proposed that the iron is ejected from ferroan dolomite during calcitization. Al-Hashimi and Hemingway (1973) established a genetic relationship between dedolomite and rusty weathering crusts on Carboniferous rocks exposed in Northumberland, England; as dolomite calcitizes it releases ferrous iron, which then oxidizes to insoluble yellowish or brownish ferric oxide or hydroxide. Warrak (1974, p. 239) likewise described a close association between dedolomite and ferric oxide in Triassic carbonate rocks of the French Alps.

But Budai et al. (1984) presented evidence (association with stylolites and late fractures; oxygen and carbon isotope content) that some dedolomite forms under deep burial. Dedolomite has been recovered from deep wells, from formations that dolomitized at depth, and there can be no question that the dedolomitization *also* occurred at depth. For example, dolomite with rhombohedral zones of dedolomite occurs at considerable depth in the Arbuckle Group carbonate rocks in the Arkoma basin of Arkansas and Oklahoma (Fig. 11-21). Woronick and Land (1985, p. 268) found deep dedolomite in the Cretaceous Glen Rose Formation in Texas, at depths of 600 to 2800 meters.

Fig. 11-21. Zoned dedolomite from a deep well in Arkansas.

Diagenesis of Gypsum; Anhydrite and Halite Cements

At the elevated pressures and temperatures that obtain under burial depths of a few hundred meters, sedimentary gypsum becomes unstable and converts rapidly to anhydrite. Presence of chloride-saturated pore water hastens the process. The inversion results in a molar volume reduction of solid phase by 38 percent and release of large quantities of water. Any porosity that was present in the original gypsiferous sediment is lost in the process. Some have estimated that gypsum beds loose two-thirds of their bulk as a result of compaction and conversion to anhydrite. Reduction of solid volume in the subsurface, and generation of large quantities of water can have far-reaching consequences, including fracturing and intrastratal dissolution of surrounding rocks, with attendant porosity and permeability enhancement. In high-permeability systems there can be substantial flow of the evolved water into associated carbonate rocks. The water apparently can redistribute calcium sulfate, precipitating much anhydrite spar in newly formed pore space, and promoting the replacement of calcite grains and cement by anhydrite. Replacive anhydrite occurs in carbonate rocks as isolated subhedral or euhedral porphyroblasts up to several millimeters in length (Fig. 11-22), or as very irregular poikilotopic crystals, commonly forming interlocked aggregates, or "snowflakes." In some limestones, anhydrite makes entry as pore-occluding cement, but then "overfills" its pore by replacing surrounding calcite matrix. In its replacive mode it may

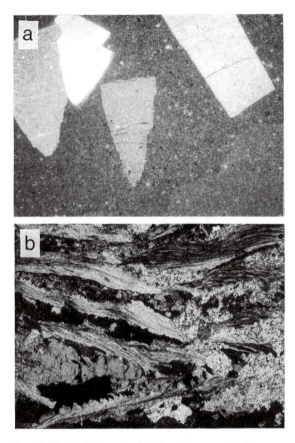

Fig. 11-22. Replacive anhydrite in a limestone: **(a)** large anhydrite crystals replacing lime mud; **(b)** anhydrite non-selectively replaces mollusc fragments and lime mud matrix.

develop crystal faces. Should the anhydrite dissolve later, it leaves pores that preserve the (typically blocky) shape of anhydrite euhedra.

In contrast to the inclusion-free pore-filling anhydrite cement, replacive anhydrite spar is rife with inclusions of incompletely replaced grains, matrix, and cement. Commonly there are tiny rhombs of dolomite inside the replacive anhydrite. These may be unreplaced crystals that were present from the start. Or, dolomite rhomb inclusions might form during the replacement process. As anhydrite replaces magnesium-bearing calcite, magnesium ions and carbonate ions presumably are released. These ions might exsolve as dolomite rhombs inside the anhydrite host.

Clark (1980b, p. 196) described replacement and cementation of Zechstein (Permian) carbonate sedi-

ments in the Netherlands by halite. Halite appears to have been passively emplaced in secondary pores under deep burial. Some limestones in the upper part of the Charles Formation (Mississippian) in northeastern Montana (subsurface) are extensively replaced by halite. Ragged remnants of calcitic skeletals and cement are floating in coarse-crystalline halite, which constitutes 40 to 60 percent of the rock. The details of such a replacement seem not to have been addressed by geochemists; it seems to occur eogenetically, within the halite-precipitating depositional environment.

Hovorka (1992) described pseudomorphic replacement of gypsum by halite, occurring very early, within the evaporite depositing environment. Closely associated anhydrite, however, remains unaltered.

Pyrite

Much of the pyrite contained in sedimentary rocks, especially the very small, disseminated material, is authigenic (formed in the depositional environment) or early diagenetic. It seems to require presence of organic matter in the sediment, sulfate in solution in the pore water, and a locally anaerobic chemical environment. The anaerobic environment is created by decay of organic matter. Bacterial reduction of sulfate under these conditions produces H_2S, according to a reaction (see Berner, 1984, p. 606)

$$2CH_2O + SO_4^{2-} \rightarrow H_2S + 2HCO_3^- \quad (11\text{-}3)$$

The H_2S reacts with iron in the sediments, ultimately forming FeS_2. According to Berner, reactivity of the sedimented organic matter exerts the major rate control on pyrite production in marine environments. Decay of organic matter occurs most rapidly near the sediment-water interface; by the time sediment has been buried a few centimeters, most of its reactive organic matter has already been consumed, and pyrite formation ceases. In freshwater environments, in contrast, low dissolved sulfate concentration limits pyrite formation. Iron content appears to be a limiting factor only in carbonate-depositing environments far from sources of iron-bearing terrigenous detritus, and in euxinic marine environments, where sulfide is produced rapidly and abundantly.

Pyrite in marine sediments commonly occurs as *framboids*, tiny spheroidal bodies 10 to 20 micrometers in diameter, consisting of many very small intergrown pyrite crystals. Some pyrite forms so early that it is transported, and concentrated as placers in the sediment (Love, 1971). Framboids may occur as geopetal cumulates in the loculi of globigerinid tests; this pyrite is of course earlier than any cements that the loculi contain.

Sweeney and Kaplan (1978) made pyrite framboids in the laboratory and compared them with natural framboids. Pyrite, it appears, begins as FeS in sediments, made by reaction of H_2S with dissolved iron. This changes gradually in the presence of elemental sulfur to pyrrhotite (a hexagonal, non-stoichiometric iron sulfide), then to greigite (Fe_3S_4), and finally to pyrite, with the lowest Fe/S ratio. Apparently the spherical form of the pyrite framboid is inherited from greigite; pyrite that does not form from greigite does not make framboids. Some framboids seem to be made inside bacteria, and Sweeney and Kaplan noted that some framboids collected from modern sediments of the San Diego trough are enclosed in a transparent sac or membrane. Nonetheless, these workers were able to synthesize abiotically in the laboratory all the kinds of natural framboids that they observed in the sediments.

Other pyrite is mesogenetic, precipitated under a reducing sulfide-rich environment created by the entry of petroleum or hydrothermal fluids into the rock system at depth. Such pyrite typically forms larger and more coarsely crystalline masses, occupying secondary pores or overlying other cements in primary pores.

Feldspar

Authigenic potassium feldspar is not unusual in sedimentary rocks. Swett (1968) described a dolomitic siltstone containing as much as 10 percent of authigenic orthoclase and suggested that the potassium was displaced out of illite during dolomitization of an overlying carbonate formation. Baskin (1956, p. 136) noted that all the carbonate rocks he observed that contain authigenic feldspars are recrystallized or dolomitized. Some authigenic orthoclase derives from volcanic ash in the sediment, and some is closely associated with potassium-rich evaporites. In sandstones, authigenic feldspar occurs as overgrowths on detrital feldspar grains. Smith (1974, p. 262 ff) summarized some of the numerous reports of albite in sedimentary rocks. It appears that some authigenic albite is tabular on {010} (a so-called albite-type habit), and some on {001} (pericline-type habit). The albite law,

in combination with a certain penetration twin called *X-Carlsbad*, gives some crystals a ''four-ling'' appearance (Fig. 11-23). Authigenic feldspars are euhedral with pinacoidal faces; the crystals cut across original textures. They always bear carbonate inclusions, but aside from this they are extraordinarily pure and crystallographically highly ordered (Kastner, 1971, p. 1425). Authigenic feldspars are never perthitic. Whereas detrital alkali feldspars derived from igneous rocks are bright blue or red under cathode-ray excitation, authigenic feldspars are not cathodoluminescent.

Detrital potassium feldspars and calcic plagioclases may alter diagenetically to albite. Boles (1982) described the *albitization* of calcic plagioclase in the Frio and Wilcox Sandstones (Tertiary) of the Gulf of Mexico coast, taking place at burial temperatures of 100 to 120°C. Sodium is provided by pore fluids; quartz is consumed in the process, and kaolinite and calcite are by-products.

Alteration and Dissolution; Secondary Porosity in Sandstones

As sediment undergoes burial, it moves into regions of higher pressure and higher temperature and a more aggressive chemical environment, where rock materials that resisted destruction or degradation at the surface now experience rapid alteration to other minerals or even complete dissolution. In a complex rock, cer-

Fig. 11-23. Authigenic albite, one or two with cross-Carlsbad twin.

tain components undergo alteration or dissolution, while other components remain largely untouched. We have all seen quartz sandstones that contained shale ''rip-up'' clasts or scattered shells, now dissolved; or boxworks created when limestones with cement-filled fractures dissolved, leaving the fracture fillings intact. The sand grains and the fracture fillings that remain show little or no evidence of dissolution. Secondary porosity in some sandstones is largely a result of dissolution of their cements, and in other sandstones a result of dissolution of their grains.

The plagioclase grains and glass and fine-grained lithic fragments in volcanic sandstones are so unstable that they undergo extensive and rapid alteration eogenetically to zeolites, clay minerals, and calcite (Surdam and Boles, 1979, p. 231). Many ''limestones'' in Indonesia appear to have had a volcanic ash matrix that is now almost completely altered to ferroan calcite or ferroan dolomite. The peculiar product is a rock containing volcanic quartz and plagioclase grains and skeletal grains (such as large foraminifers) floating in ''spar.''

In most sandstones, dissolution and alteration of detrital grains is mesogenetic. Feldspars are an important detrital grain type susceptible to deep diagenetic alteration. Pore-water CO_2 causes alteration of grains of potassium feldspar, according to the reaction

$$4KAlSi_3O_8 + 4CO_2 + 14H_2O \rightarrow 4K^+ + Al_4Si_4O_{10}(OH)_8 + 8H_4SiO_4 + 4HCO_3^- \quad (11\text{-}4)$$

Note that this reaction releases silica; the K^+ could go toward reconstitution of illite. Calcic plagioclase might alter to kaolinite by a similar reaction:

$$2CaAl_2Si_2O_8 + 4CO_2 + 6H_2O \rightarrow Al_4Si_4O_{10}(OH)_8 + 2Ca^{2+} + 4HCO_3^- \quad (11\text{-}5)$$

The calcium released by destruction of calcic plagioclase probably precipitates as calcite. [Compare the reaction in Eq. (11-4) with the weathering reaction in Eq. (7) of Chap. 6.] Instead of dissolving and altering to kaolinite, calcic plagioclases may instead become more sodic, a process called albitization:

$$CaAl_2Si_2O_8 + Na^+ + H_4SiO_4 \rightarrow NaAlSi_3O_8 + Ca^{2+} + Al(OH)_4^- \quad (11\text{-}6)$$

The sodium might be supplied from the connate water, which commonly is entrapped sea water, or it might come from the smectite-illite transformation.

Organic diagenetic reactions seem to be important in promoting secondary porosity, but we have only begun to understand them. Perhaps most of the CO_2 that is required for alteration and dissolution of silicate and carbonate components of deeply buried rocks derives from diagenesis of organic matter and generation of petroleum (a process ordinarily referred to as organic maturation) buried with the sands and shales (Tissot and Welte, 1984, p. 204, 246; Hunt, 1979, p. 166; Siebert et al., 1984, p. 168). As carbonic acid, this CO_2 advances into the rocks surrounding the petroleum source beds, clearing pathways of porosity and permeability though which the oil and gas can migrate on its way to reservoirs. A pressure drive is created by conversion of solid organic matter into liquid and gaseous products that occupy more volume.

Surdam et al. (1984) suggested that carboxylic acids (acetates and oxylates) formed during the maturation of kerogen may be important in the mobilization of aluminum, which is generally regarded as highly *immobile* in weathering and diagenetic processes. These organic acids are very soluble in water, and they solubilize aluminum by complexing with it. An example of a possible reaction (Surdem et al., 1984, p. 129), causing the complete dissolution of anorthite, is

$$CaAl_2Si_2O_8 + 2H_2C_2O_4 + 8H_2O + 4H^+ \rightarrow 2H_4SiO_4 + 2(AlC_2O_4 \cdot 4H_2O)^+ + Ca^{2+} \quad (11\text{-}7)$$

wherein oxalic acid in water-solution puts aluminum into solution as oxalate and promotes the dissolution of silica. Stoessell and Pittman's (1990) experiments, on the other hand, suggest that carboxylic acids are insignificant as agents of dissolution under deep burial.

Much secondary porosity is caused by deep-circulating meteoric waters. Bjorlykke et al. (1989, p. 247) described its extensive effects on Jurassic and Cretaceous rocks (mostly sandstones) of the North Sea basin. Mazzullo and Harris (1992, p. 609) considered it to be a cause of mesogenetic dissolution in carbonate rocks.

Among the evidences of deep dissolution of grains and cements are the following (Schmidt and McDonald, 1979, p. 222; Schenk and Richardson, 1985, pp. 1066, 1074): porous sandstones only lightly compacted (undercompacted), or inhomogeneously compacted, because early cements, since dissolved, supported the grains against compaction during burial; hackly or rounded cement crystal surfaces; alteration of cement surfaces resulting in some loss of birefringence (as in anhydrite); disjoint but optically continuous cement (see below); etched or embayed grains; oversize elongate pores or channelized pores; missing grains or grain molds or grains that have been reduced to tenuous skeletons; fractured grains and cleaved cements resulting from the loss of mechanical strength that comes with appearance of secondary pore space; fractures that are too wide, or whose opposite walls do not match because the fracture has been dissolution-enlarged. The large-scale stratigraphic distribution of porosity may suggest secondary origin. A plot of porosity versus depth may reveal an abrupt increase at some considerable depth, interrupting a trend of gradual porosity decline with depth in the overlying interval. Secondary pores commonly are larger than most primary pores, and certain modern well-log combinations can distinguish between the larger and smaller pores in a borehole.

Dissolution and alteration of a detrital grain begins at the margin of the grain, commonly resulting in a rind of alteration product. Later the material inside this rind may dissolve and leave an empty envelope that outlines the space originally occupied by the grain. Sandstones composed mostly of chemically unstable grains may undergo such extensive grain alteration and dissolution that a tenuous network of alteration rinds is all that remains; some volcanic sandstones have progressed to this state. In some cases a heap of alteration product accumulates on the floor of the moldic pore left by the dissolution and alteration process. Selective dissolution occasionally occurs within a single rock fragment; in a fragment of silty shale, for example, the clays may dissolve, and leave the surviving quartz silt on the floor of the moldic pore.

Sandstones of the Wilcox Group (Eocene) and Frio Formation (Oligocene), important petroleum reservoirs of the Texas Gulf Coast, are porous mainly because much of the detrital feldspar and rock fragments that they contain has been dissolved under deep burial. Louks et al. (1984) observed that alteration of feldspars began in the detrital source area and continued in the depositional basin even as burial depths exceeded 3400 meters. At the shallower depths, detrital feldspars were replaced by calcite or dissolved, leaving grain-moldic pores, some of which filled with calcite cement. At greater depths some of this calcite went back into solution, restoring moldic pores and creating additional porosity.

Schenk and Richardson (1985) described deep intrastratal dissolution of anhydrite spar cement from both dolomite rock (San Andres Formation, Permian,

New Mexico) and sandstone (Minnelusa Formation, Permian, Wyoming). Partially dissolved anhydrite crystals show a comb-like fringe or complex rectangular (stairstep) reentrants (Fig. 11-24c). Many of the cement crystals were large poikilotopic porphyroblasts; dissolution reduces such a crystal to a few widely separated remnants that all bear the same crystallographic orientation. There are dolomite rhombs on the deep reentrants, crystals that apparently formed after the dissolution. Markert and Al-Shaieb (1984, p. 382) also observed tiny crystals of ferroan dolomite in secondary pores in the Minnelusa.

In the San Andres, Schenk and Richardson noted pores shaped like replacive anhydrite crystals, that is, rectangular pores in thin section. Here, creation of secondary porosity was a two-stage process: First, some carbonate was replaced by coarse crystals of anhydrite; this resulted in some reduction of porosity, as cementation of pore space typically accompanies replacement of the grains. Then, under different diagenetic conditions, the anhydrite crystals dissolved; the sizes, shapes, quantity, and distribution of the resulting pores were determined not by original textures but by the replacement fabric. Lindsay and Kendall (1985, p. 185), working in subsurface Mission Canyon Formation (Mississippian) in western North Dakota, noted skeletal grains replaced by anhydrite, which was leached later, leaving moldic pores.

Diagenesis of Organic Matter

Organic matter in sedimentary rocks is derived mostly from the terrestrial higher plants, and from marine and lacustrine plants and animals of generally low phylogenetic levels. Some of it is called *kerogen* (see Chap. 10), being the raw material from which petroleum is made; it is subject to much alteration or degradation under diagenesis. Some of this alteration, especially in the early stages of diagenesis, is biological, owing to the actions of anaerobic bacteria, but thermal alteration is by far the more important diagenetic process in deeply buried organic matter. Thermal alteration of buried kerogen in so-called *petroleum source rocks* gives rise to mobile fluids rich in hydrocarbons and other organic molecules. Some of the products of this alteration are, we have learned, important agents of chemical diagenesis. The oil and gas produced during the thermal alteration, or *organic maturation*, of buried organic matter may migrate though the pore sys-

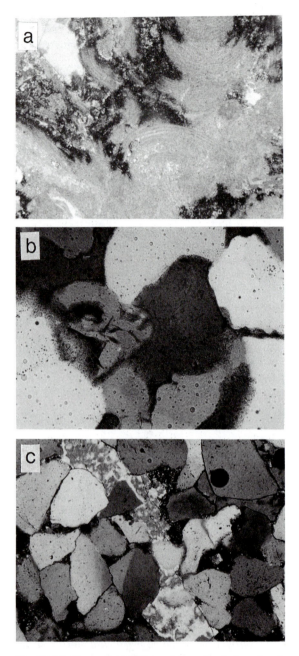

Fig. 11-24. Examples of effects of deep dissolution of grains or cement: **(a)** Coralline alga masses partly dissolved; **(b)** "missing grains" and truncated grains in a sandstone; **(c)** partially decemented sandstone—disjoint masses of anhydrite cement are all syntaxial and are parts of a single crystal now mostly dissolved.

tems, and accumulate in a trap, displacing the heavier fluid (saline connate water, usually) already present in those pores. Mature oil and gas are chemically unreactive, and presence of these fluids severely retards the further diagenesis of the host (reservoir) rock. By occupying the pore space it excludes other fluids that bring chemical cements to the pores, or that cause alteration or dissolution of the mineral components of the rock.

Organic maturation is a slow process, so degree of maturation depends on length of time that organic matter has been exposed to elevated burial temperatures. A Russian geochemist, N. V. Lopatin, modeled the maturation process on a first-order kinetics law (Arrhenius equation) and introduced a certain time-temperature index (TTI). This index, based on geothermal gradient and burial history, expresses the accumulated effects of a time-varying temperature over the entire history of the organic matter. Lopatin calibrated the TTI index against vitrinite reflectance (R_o), a widely-used measure of organic maturity, using data from the Munsterland-1 borehole in Germany. Waples (1980) subsequently corrected and adjusted Lopatin's calibration, drawing on reflectance data from 31 wells worldwide. He found that a TTI value of about 10 or 15 corresponds to $R_o = 0.65$, generally regarded as signaling the onset of oil generation, that TTI of 75 corresponds to $R_o = 1.00$, indicating peak oil generation, and that TTI of 160 corresponds to $R_o = 1.30$, indicating end of oil generation. Higher values mark generation of gas and destruction of oil. A depth-versus-time plot of appropriate TTI values (Fig. 11-25) establishes oil- and gas-generation windows, giving estimates of times and depths at which these important diagenetic events occur. The simple concept became quite popular among petroleum explorationists (see Guidish et al., 1985), and has undergone various refinements over the years since 1980.

Telogenesis

Telogenesis, that is, alteration of rocks under exhumation to the surface after deep burial, is in large measure a catena of late-stage freshwater effects, such as those giving rise to caliches. It is easy to confuse telogenesis with weathering, and the two types of processes could be viewed as intergradational. Commonly it is difficult also to distinguish between mesogenetic results and telogenetic or weathering results, espe-

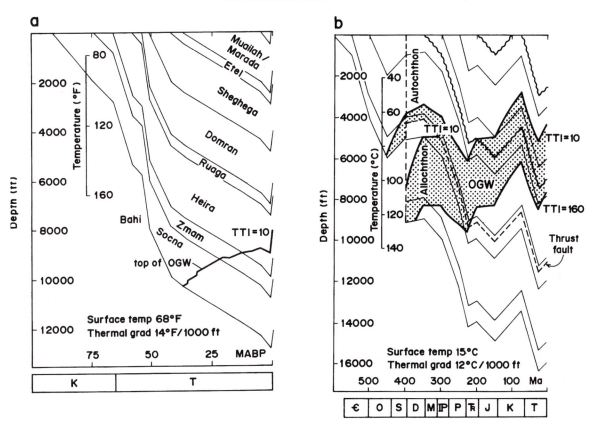

Fig. 11-25. Burial-history diagrams showing oil-generation windows. **(a)** Simple burial history from a famous oil-producing region; Socna, Zman, and Heira Formations entered the OGW (oil-generation window) in late Tertiary time. **(b)** A more complicated burial history, with unconformities and thrust plate; rapid (instantaneous) tectonic burial has hastened maturation in lower-plate (allochthonous) rocks. OGWs obtained by automatic computation and plotting of TTI integral.

cially in sandstones. Did dissolution of labile grains in an outcropping lithic sandstone, for example, occur at depth or after exhumation? Sometimes this question can be answered by comparing samples taken from the weathered surface with samples taken a few centimeters in from the rock face. The effects of weathering or telogenesis diminish inward from the weathered rock face; mesogenesis would not produce such a gradient.

Telogenesis of a carbonate rock can cause aggradational neomorphism of calcite and dedolomitization. Surface silicification may take place, being a replace-

ment, by chert, of the weathered surface of a limestone, or of some calcitic component of exposed rock, typically to a depth of only a few millimeters. Limestones weathered for long periods in arid (or semiarid) climates can develop a "punky," coarse-crystalline rind that apparently is a result of recrystallization or neomorphism. Dedolomitization is stimulated by contact of dolomite with fresh surface waters, which have low Mg/Ca ratio. Chafetz (1972) considered dedolomite, coarse-crystalline rinds, and surficial chertifications as marks of buried unconformities in carbonate successions.

12

The Big Picture

> All our knowledge brings us nearer to our ignorance.
> —''The Rock''; T. S. Eliot, 1934

Although this final chapter deals with ''the big picture,'' we do not neglect the small-scale elements of sedimentary rocks; rather, we put them into the context of large-scale features and the world-embracing dynamical systems that have molded and in various ways unified, the body of sediment over billions of years. Processes of astronomical scale can have microscopic effects. Rock properties that change imperceptibly from bed to overlying bed can, over great stratigraphic distances and times, add up to striking trends and contrasts.

Origin of Atmosphere, Hydrosphere, and Biosphere

In the beginning, the earth accreted out of a diffuse cloud of extrastellar material, the debris of bygone stellar explosions called supernovae, flowing in turbulent eddies around the sun. When small particles collided, they sometimes bounced away from one another, sometimes shattered one another, and sometimes stuck together, creating larger objects. Bodies that had managed to grow larger than most, and to increase their gravitational fields accordingly, were able to sweep up and retain ever larger quantities of the smaller debris. In this way the earth and the other planets and their moons came into being. Toward the end of the major accretion, the planets and moons consumed most of the ''runners-up'' in the competition for growth, in a gravity-driven crescendo of big stone against little stone that left the ''rocky'' planets and moons with their characteristic cratered surfaces.

The earth melted as it grew and continuously differentiated into core, mantle, and chaotic crust. The heat-producing elements K, U, and Th collected in the crust; Fe-rich liquid sank to earth's center. The moon, probably accreting alongside the earth in the same way and at the same time, also made a core and mantle.

The earth's surface was cool throughout most of the period of accretion, though a worldwide magma ocean might have existed at some stage. The early crust was a mixture of ultrabasic, basic, calc-alkaline, and possibly even some K-rich granitic rocks; continental-type crust consisted of small patches of scum concentrated over convergences of small polygonal thermal convection cells. At first, all crust was continually recycled into the upper mantle, but as the mantle cooled and style and scale of convection changed, continents began to grow by accretion of oceanic crust, augmented by basaltic magmatism directly out of the mantle below. Volcanism was intense.

Probably the primordial earth had an atmosphere, perhaps chemically similar to the present atmosphere of Jupiter, composed largely of methane (CH_4) and ammonia (NH_3). But, compared to cosmic abundances, earth's present atmosphere is grossly depleted in the noble gases, and this suggests that any proto-atmosphere that had accreted to earth's rocky sphere during the formation of the solar system leaked off, back into interplanetary space (taking the noble gases with it). Perhaps the proto-atmosphere puffed away during an early catastrophic impact. A new and different kind of atmosphere subsequently appeared, outgassed from the rocky interior. If the composition of modern volcanic exhalations is any guide, then steam and CO_2 would have been the dominant gases, mixed with small quantities of SO_2, SO_3, N_2, H_2S, CO, H_2, CH_4, NH_3, and HCl. Many of the outgassed substances were quickly removed from the atmosphere—water vapor condensed to form the oceans; carbon dioxide, sulfur dioxide and trioxide, hydrogen sulfide, and hydrogen chloride dissolved in the water, making acids; these acids quickly reacted with crustal rocks, mostly silicates, and the oceans thus became salty and mildly alkaline. The atmosphere remained a dense mixture of water vapor, CO_2, and some N_2. Molecular nitrogen, rather unreactive and relatively insoluble in water, continues to accumulate in the atmosphere as a result of volcanic outgassing.

Special conditions near the top of the atmosphere—namely, high temperature, low pressure, and high influx of solar ultraviolet radiation—stimulated certain chemical reactions. One product, considered by some workers as important in the development of the complex organic molecules of life, was carbon monoxide; another was molecular hydrogen. When laboratory mixtures of N_2, H_2, and CO are subjected to the conditions presumably obtaining in the earth's early upper atmosphere, a mixture of simple organic compounds is produced; among the most interesting is hydrogen cyanide (HCN). It has been detected spectroscopically in interstellar gas clouds, and Voyager I found it in the atmosphere of Titan, a moon of Saturn. At ordinary temperatures and pressures of earth's surface, in solutions similar to sea water, HCN reacts with water when irradiated with ultraviolet light. A typical reaction is

$$5HCN + 7H_2O \rightarrow C_3H_7O_3N + 4NH_3 + 2CO_2$$

one of the products being serine, an amino acid. Similar reactions produced other amino acids. The formation of simple proteins from amino acids by nonliving systems also has been demonstrated in the laboratory, and a great variety of other organic molecules essential to living systems have likewise been synthesized by bringing together very simple compounds that might have existed in earth's early atmosphere or hydrosphere. Besides various amino acids, there are fatty acids, nucleic acids, sugars, and porphyrins. Many of these compounds are hydrophobic or surfactant, meaning that they do not dilute in water but coagulate into oily drops or slicks on a water surface or on the surfaces of mineral grains. In concentrated form, these compounds polymerize into large, very complex molecules, perhaps most efficiently in the presence of clay mineral substrates behaving as catalysts. Many complex organic compounds are preserved in the oldest of rocks.

Self-assembly of polyamino acids into various structures that in some respects resemble biological cells also is readily achieved in the laboratory. There are the *proteinoid microspheres*, with two-layer outer walls, and the *coacervate droplets*, formed when positively charged and negatively charged colloids are brought together (see Fox and Dose, 1972, p. 196 ff). These little objects are not inert; they clump together or interact in other ways and undergo certain changes such as budding and fission. Similar structures have been discovered in the very oldest rocks.

The first self-replicating molecules may have evolved in some particularly favorable but rare environment. There may have been other molecules in the environment that could catalyze the replication. Once a molecule had developed the ability to replicate itself, it naturally became more abundant. It was important that replication be not too accurate, so that some degree of evolution could take place. Self-replicating molecules might have evolved quickly away from their progenitors, finding opportunities in less rare environments. Sugars, or rather their polymers, turned out to be useful structural materials; most important, they became the backbone for the large complex molecules DNA and RNA, upon which replication and genetics are based.

Almost no molecular oxygen was present in the early atmosphere. The first living organisms apparently did not use oxygen, that is, they were anaerobic and probably depended on fermentation for their livelihood. They were heterotrophic, assimilating ready-made organic substances available in their surroundings. Some of these anaerobic *heterotrophs* were more successful than others and, through natural selection, proliferated and eventually (over 3000 million years ago) gave rise to some *autotrophs*, organisms capable of manufacturing their own food from inorganic substances. Thus, photosynthesizers emerged, organisms that use light energy to convert water and carbon dioxide into living tissue. Fixed nitrogen is required in the synthesis of chlorophylls, and so anaerobes had to find ways of extracting molecular nitrogen from the atmosphere and fixing it as NH_3 (solar ultraviolet radiation dissociates atmospheric ammonia; in the absence of an ozone shield there could not have been much ammonia in the early atmosphere). Only the anaerobes can do this today, fixing nitrogen as ammonia or as nitrate.

Photosynthesis involves the splitting of water molecules:

$$CO_2 + 2H_2O^* \rightarrow CH_2O + H_2O + O_2^*$$

Oxygen is released by this reaction, derived from water molecules (as the asterisks are intended to show). This waste product of photosynthesis was undoubtedly toxic to the early photosynthesizers, which might have made use of dissolved Fe^{2+} in their environment to dispose of it, precipitating the iron as ferric oxide or hydroxide. Such a mechanism has been proposed (Cloud, 1972, p. 544) as the origin of the (Superior-type) banded iron formations (see Chap. 10), which occur in 2200 ± 300 million year Precambrian terrains but not in younger or older successions. Eventually, the photosynthesizers evolved enzymes that could mediate the oxygen directly to the atmosphere, and they also developed certain protections against this poison. By the beginning of Phanerozoic time, partial pressure of oxygen may have been about 1 percent total atmospheric pressure, compared to the present value of about 20 percent; atmospheric oxygen content might have reached half its present value by Devonian time. Some of the O_2 in the upper atmosphere recombined into ozone, O_3, which thereafter shielded earth's surface from the energetic ultraviolet rays that are destructive of nucleic acids. Cyanobacteria, which came into existence before an ozone shield developed, today have special DNA-repair mechanisms, and ultraviolet-absorbent sheath pigments. Water absorbs ultraviolet light efficiently; an organism immersed in a few meters of overlying water is completely shielded from the harmful rays.

The earliest organisms lacked a cell nucleus, that is, a membrane-enclosed genetic package. They are called *prokaryotes*. Prokaryotes survive (and flourish) today, of course, as the bacteria (the methanogens and the cyanobacteria included, comprising the kingdom *Monera*). The fossil record is not clear about when the *eukaryotes* evolved, but these organisms with cell nuclei and organelles, such as mitochondria, certainly were present well before the beginning of the Cambrian Period.

Heterotrophs eventually learned to use oxygen as a source of energy. All the eukaryotes are aerobic. The metabolic pathways of the eukaryotes consist essentially of anaerobic pathways with various steps appended. For example, glycolysis (carried out in the absence of oxygen) in the eukaryotes is remarkably similar to fermentation accomplished anaerobically by the prokaryotes. This suggests that as soon as free oxygen became available, living systems experimented with it, modifying and building upon the existing an-

aerobic pathways and processes, rather than trying to invent new aerobic processes and syntheses out of whole cloth.

The organelles, which are the "vital organs" of eukaryotic cells, may have evolved independently of their hosts and entered the prokaryotic cell as parasites or endosymbionts. Eukaryotic cells may have gained flagella and chloroplasts in this manner. (Mitochondria and chloroplasts contain a small fragment of DNA that is similar to the DNA of prokaryotes.) It took at least 2000 million years for cells with nuclei to evolve, and the all-important development of genetics by sexual recombination and meiosis. The single-celled eukaryotes, which comprise the kingdom *Protista*, had, by the beginning of the Paleozoic Era, given rise to the multicellular organisms, comprising three kingdoms (following Whittaker and Margulis, 1978)—*Plantae*, *Fungi*, and *Animalia*—the first embracing the autotrophs and the other two the heterotrophs. Multicellularity seems to have evolved more than once in each of the three multicellular kingdoms, either by the joining together of protistans into a colony (the concept favored by most biologists today) or by division of a single protistan into many cells that failed to separate. In either case, various members of the aggregate developed specializations, eventually became mutually dependent, and were no longer able to survive as separate organisms.

Was the life-forming process fortuitous or was life inevitable? The speed at which the first living systems developed (relative to age of the planet) suggests that, given a suitable physical environment and the necessary raw materials and energy sources, the process is inevitable. But if it is inevitable, why has not the spontaneous generation of life out of non-living materials happened again and again throughout earth's history? Evidently, once life evolved, it changed the earth in such a way as to preclude the process from happening again. The similarities among existing life forms (for example, all organisms use about the same set of 23 amino acids, all left-handed; they all use ATP for energy storage) suggests that all life on earth has a single ancestor.

Brief Lives—Extinction Episodes of the Phanerozoic

During the 600 million years of Phanerozoic time, the biosphere has undergone a series of apparent mass extinctions (Raup and Sepkoski, 1982; 1984; 1986), unusually large numbers of species vanishing over periods of a few million years or much less. Among the major extinction episodes are the following (Fig. 12-1): *Late Ordovician*, with loss of 21 trilobite families, 13 cephalopod families, 12 articulate brachiopod families, and 10 crinoid families; *Late Devonian*, with loss of 25 rugose and tabulate coral families, 17 articulate brachiopod, and 14 cephalopod families; *Late Permian*, a most devastating loss of half of marine families, including 24 coral, 19 cephalopod, 18 bryozoan, and 42 crinoid families; *Late Triassic*, with loss of 31 cephalopod families and disappearance of the conodont animals; and *end of Cretaceous*, with decline of many marine families and extinction of the dinosaurs.

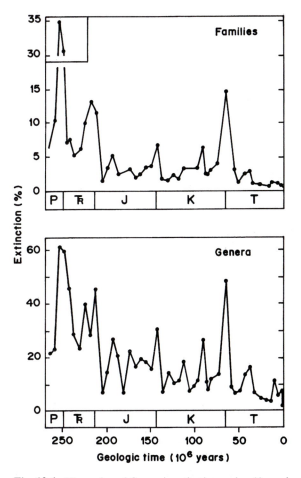

Fig. 12-1. Mesozoic and Cenozoic extinction cycles. (According to Raup and Sepkoski, 1986, fig. 1, with permission of AAAS.)

There were numerous less profound extinction episodes. It appeared to these workers, moreover, that, at least as far back as Late Permian (about 250 Ma), major extinction episodes were cyclical, with a period of 26 million years. Some scientists (Rampino and Stothers, 1984) have reanalyzed Raup and Sepkoski's data and found that a period of 32 ± 1 million years fits better.

Newell (1982) cautioned of the danger of illusory extinction episodes (his proviso applies also to global correlation of stratigraphic sequence packages). Part of the problem, according to Newell, is the way that biostratigraphers gather their data. Commonly they record species ranges to the nearest formally recognized biostratigraphic boundary; this creates the false impression that many species appear or vanish together. Moreover, stratigraphic successions are considered in general to embrace numerous small gaps—obscure stratigraphic discontinuities, called paraconformities—that are recognized solely by paleontology. "Paraconformities," according to Newell (p. 261), "can produce a false appearance of mass extinction by bringing last occurrences together."

Nonetheless, there can be little doubt that extinctions of the scale documented by Raup and Sepkoski are real. That there have been profound and rather sudden changes in the taxonomic content of the biosphere, especially at the end of the Permian and end of the Cretaceous Periods, is reflected in our division of geological time, established long ago by the pioneers of our field, into Paleozoic (meaning ancient life), Mesozoic (middle life), and Cenozoic (recent life) Eras.

One of these extinctions had always been of particular interest, namely, the one at the end of the Cretaceous Period, when the dinosaurs vanished. Alvarez et al. (1980) discovered fortuitously a peculiar geochemical anomaly at the K-T boundary at Gubbio, Italy. Here the boundary is marked by a 1-centimeter clay layer in a succession of pelagic foraminiferal limestones. Could the clay layer have been a result of a short-term contraction in foraminiferal productivity, or was there a momentary increase in rate of clay sedimentation? Alvarez's group favored the former hypothesis, but to test their suspicion they obtained a set of iridium analyses of the clay layer and of the limestones above and below. Variations in sedimentation rate would be reflected in variations in the concentration of iridium, which is considered to enter sediments as a steady infall of cosmic dust. If the ratio of iridium to clay is constant, then the clay layer is the result of reduced plankton productivity. Now, instead of the expected 10-fold increase in iridium concentration across the clay layer, there was a 160-fold increase. This was a big surprise. The only explanation, finally, after other possible explanations had been dismissed, seemed to be that there had been an "iridium event," and that this event was accompanied by a momentary lapse in planktonic carbonate production. Meteorites contain iridium in important quantities. It appeared that a large asteroid (or maybe a comet) had transported a large quantity of iridium to earth all at once.

Alvarez et al. (1980) proposed that, 65 million years ago, the cosmic shooting gallery visited upon earth's surface a 10-kilometer asteroid, which smacked the earth at a speed of several tens of kilometers per second. The event was so catastrophic to living things that it terminated the age of the dinosaurs and brought to a close the Mesozoic Era. A similar event, of even greater magnitude, might have ended the Paleozoic Era 160 million years earlier, with the extinction of 70 percent of all species living at the time.

A disaster (the word means literally "bad star") such as described by Alvarez and his associates, releases the energy of a magnitude 12 (or greater) earthquake (see McLaren and Goodfellow, 1990, p. 127), and if the impact occurs in the ocean can create tsunami of unimaginable proportions, 4 or 5 kilometers high, and moving at 500 meters per second, with the potential to collapse shelf-marginal sediments. The projectile excavates a crater 100 or 150 kilometers in diameter, displacing some 10 trillion metric tons of crustal and mantle rock; the projectile itself vaporizes explosively on impact. A mass of air, instantaneously heated to 20,000 K, expands explosively away from the impact site, burning up much of the plant and animal life of the planet. The explosion bursts through the top of the atmosphere and carries tens of billions of tons of rock dust and smoke into orbital and suborbital trajectories, dispersing it to all parts of the earth. The atmosphere becomes opaque, and total darkness and bitter cold on earth's surface lasts for weeks or months.

A large projectile landing in the ocean can vaporize a substantial portion of the ocean's water. The water vapor carried aloft would remain in the atmosphere much longer than the smoke and dust. In collaboration with other gases possibly generated by the impact, such as CO_2, it would cause "greenhouse" global warming to follow the "impact winter." Explosive heating of the atmosphere at time of impact might also cause atmospheric nitrogen and oxygen to combine

into nitrogen oxides; these compounds would then be swept out of the atmosphere over the ensuing months or years as a very aggressive "acid rain," which would dissolve the calcareous shells of marine plants and animals.

Such an impact has a devastating effect on the biosphere. Photosynthesis ceases, probably for several months, and food chains collapse. Many creatures are killed outright by the blast, by fires, by suffocation or poisoning; others die of starvation. Probably the larger animals, such as the dinosaurs, with their large food requirements, suffered the more and perished the more quickly. The smaller reptiles and the mammals might have found shelter from the storm and sufficient resources to maintain themselves through the hard times; many of them hibernated, perhaps. Some dinosaur species lingered on, finally succumbing to their reduced populations and changing environments many generations after the disaster.

Evidence supporting catastrophic impact was soon brought to light from the K-T boundary at localities worldwide, not only an iridium spike, but glass spherules (created by sudden melting of target material), and shock-metamorphosed quartz grains (also created by hypervelocity impact). The impact, it seems, was so violent (estimated at 100 million megatons TNT) that it dispersed these materials over the entire surface of the earth. The composition of the glass spherules indicates that the impact site was oceanic; the presence of shocked quartz, on the other hand, suggests a continental impact. Multiple impacts are not out of the question; there are many examples of multiple impact craters on earth, on the moon, and on other planetary surfaces. They may result from two asteroids in mutual orbit, or an asteroid or comet that had begun to break up before it struck the target.

Hildebrand and Boynton (1990) have obtained evidence that the impact site (or one of them) lay in the Colombian basin of the Gulf of Mexico, near the Yucatan peninsula. The K-T boundary deposit in this vicinity comprises two layers. An upper layer, about 3 millimeters thick, apparently of global distribution, containing shocked minerals, spherules, soot, and isotopic anomalies, is called the "fireball layer" (Hildebrand and Boynton, 1990, p. 843). A lower layer, about 2 centimeters thick, occurs only on and near North America; called the "ejecta layer," it contains size-graded peloids up to about a centimeter in diameter, made of poorly ordered smectite, and interpreted as altered tektites of mafic and ultramafic glass made during the impact. A candidate (arguably) for a wholly continental companion impact site is a circular structure near Manson, Iowa, 35 kilometers in diameter and dated at 65 Ma, that is, the end of the Cretaceous.

Some researchers have suggested alternatively that a protracted volcanism, punctuated by explosive eruptions, might have caused the K-T extinctions and the stratigraphic and geochemical anomalies. At about the time of the end of the Cretaceous, there were enormous outpourings of lava in western India, creating the so-called Deccan Traps. More than 10,000 cubic kilometers of basaltic rocks accumulated to a thickness of more than 2 kilometers in a period of only a million years. It is true that catastrophic volcanisms and asteroid impacts can have similar consequences, and some extinctions in earth's past may have been caused or aggravated by volcanism. But there are many problems with the hypothesis of volcanic catastrophe. Proponents contend, for example, that major volcanic explosions can generate shocked quartz, but cathodoluminescence reveals that shocked quartz at the K-T boundary is not volcanic. There is a certain upper limit on the magnitude of volcanic explosion, imposed by strength of rocks and the critical pressure of steam, and sporadic "small-bore" shots fired over a period of a million years or so cannot match the effects of one or two quick rounds from a single "big gun." It does seem possible, on the other hand, that localized massive basaltic outpourings, whether on land or on the ocean floors, and the mantle plumes to which they commonly are attributed, are themselves caused by bolide impact (suggested at least as long ago as the late '70s; see Alt et al., 1990).

If the end-of-Cretaceous extinction was caused by cosmic impact, then perhaps others were as well; and if extinction episodes are periodic, as Raup and Sepkoski suggested, then perhaps catastrophic impacts (or at least, a heightened chance of impact) are too. What, then, could cause periodic as opposed to random bombardment? More specifically, what could cause a 26-million-year (or 32-million-year) periodicity? It seems that only the epicyclic motion of the sun in its orbit around the galaxy has a frequency anywhere near what is required (see Fig. 12-2). Every 38 million years, approximately, the solar system passes through the galactic plane, where cosmic debris is most abundant and where there is presumably a greater risk of bombardment. This cycle is not exactly tuned to the extinction period, nor is it in phase—the last passage of the sun through the galactic plane occurred a million years

ago, in contrast to Raup and Sepkoski's most recent extinction event 13 Ma. Sepkoski (1982b, p. 288) himself had some reservations about the cyclicity. Though statistical tests that he performed indicated clumping of extinction events in time, they did not offer a great deal of support for the hypothesis of constant time interval between successive extinction crises. Weissman (1990, p. 269) wrote that "there does not seem to be any good way to trigger periodic bombardment in the solar system, and there is little or no evidence that such a phenomenon exists anyway," and he suggested that the claimed extinction cyclicity is spurious. (Keep in mind that periodicity is not essential to the cosmic impact hypothesis.)

Stratigraphic Cycles

Every geological process has a characteristic rate or duration; many processes are cyclic, with a regular period; other events recur irregularly. Cycles may be superimposed so as to yield complex patterns; thus, the magnitude of the astronomical tide at a particular place and time represents the influence of position of the moon and of the sun, which vary monthly and daily, and the angular relationship between the orbital planes of moon and sun, which varies with the seasons, and other complex relationships as well, with longer periods (Fig. 12-2). In contrast, the pattern of geomagnetic reversals is decidedly aperiodic, but there are certain fairly well established upper and lower limits on duration of polarity episodes. Glacial advance and retreat requires a certain span of time; so does development of a soil; likewise the geomorphic cycle, or time required for peneplanation of a mountainous terrain.

The Shorter Cycles

Brown et al. (1990) described neap-tide/spring-tide cyclicity in the Salem Limestone (Mississippian) in southern Indiana; successions of laminae exhibit systematic thickening or thinning trends, with periodicities of 19 to 30 laminae, that seem to reflect the monthly cycles of the tides. Kvale et al. (1989) observed rather similar cycles in the Mansfield Formation (Pennsylvanian) in Indiana. Williams (1989) described complex cyclicities in laminae in several rock units of late Precambrian age in South Australia; he concluded (after some vacillation), that these patterns

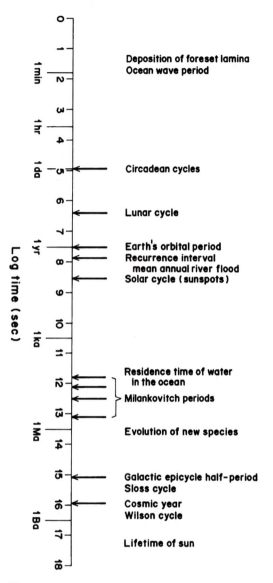

Fig. 12-2. Time scale of various geologic processes.

in Recent travertines of central Italy, each 2- to 4-centimeter band embracing 100 to 200 microscopic laminae; they concluded that the laminae represent daily deposits and that the wider composite bands are annual (the shortfall of daily layers owing to interruption of travertine deposition in the wintertime). Hunter and Richmond (1988) saw daily cycles in foreset laminae of aeolian dunes of the Oregon coast. Gebelein (1969, p. 59) found daily growth layers on algal onkoids of Bermuda. Some banded iron formations have chert beds that are varved; the varves themselves are grouped into composite bands (Trendall, 1968, p. 1536). Commonly the deposits of lakes are varved, having a layered structure that reflects the cycle of the seasons. The 11-year solar sunspot cycle seems to have been recorded in certain varved sediments. Gustavson et al. (1975, pp. 278) and Ashley (1975, p. 310, 311) illustrated varves in Recent and Pleistocene glacial lakes that show very pronounced multi-annual rhythms that might reflect the solar cycle.

The growth rings on coral thecae (Wells, 1963) reflect, we think, not only the annual rhythm but also monthly and daily rhythms. Wells suggested that very narrow and closely spaced ridges on the epitheca of certain Devonian corals are daily growth bands, and that broader bands represent annual growth. But there are about 400 daily bands on each yearly cycle in these and similar corals of Devonian age. Now, there are lots of good reasons for presuming that the length of the year has not changed, so the day had to be shorter if 400 days are to fit into a year; in fact, the day had to have had a length of about 22 hours.

Scrutton (1965) found an intermediate cycle on some of these corals, each comprising about 30.6 daily bands; he interpreted these cycles as related to the lunar month. Pannella et al. (1968) made similar observations and counts on growth layers of various bivalves and algal stromatolites of many different ages, which, together with data from corals, indicated a non-uniformly changing number of days in the month from about 31.5 in early Paleozoic time, to about 30 in Mesozoic time, then to its present value of about 29.5. Apparently, the length of the day and of the month were changing; the rate of the earth's rotation and the moon's revolution have changed, it appears, owing to the tidal friction.

The cycles that these various bands, beds, and incremental layers reflect are cycles with which we are all familiar. They are sufficiently short-term that we experience them again and again in our lifetimes, and

reflect the tidal cycle. Roedder (1979, p. 90) found celestite crystals in vugs in Paleozoic dolomite that show a microscopic, very regular banding due to cyclic variation in barium content; he interpreted these bands as "varves" reflecting annual variations in salinity during growth of the crystals (might these bands actually be daily?). Folk et al. (1985) described bands

we successfully predict their recurrence. These cycles rule our lives at every turn; our clocks and calendars keep track of them. But we find that some of the familiar cycles have changed over the long term, and that much longer cycles exist.

Longer Cycles

There are slower ticks—longer cycles. Milankovitch (1941) suggested that periodic variations of earth's rotational and orbital motions are responsible for climate changes. There is variation in the longitude of perihelion due to the axial precession; this currently has a period of about 26,000 years, but the variation interacts with advance of perihelion to give effective periods of 19,000 and 23,000 years. Axial obliquity varies with a period of 41,000 years. Orbital eccentricity varies over a period of 100,000 years, and there is a variation of lesser amplitude at around 400,000 years. Based on varve counts, van Houten (1964, pp. 509, 529) concluded that cycles in the lacustrine Lockatong Formation (Triassic) in New Jersey represent about 22,000 years; these are grouped into cycles of about 100,000 years duration. Similar cycles are suspected in certain Cretaceous pelagic deposits (see below). It is now widely accepted that Milankovitch cycles stimulated the several glaciations of the Pleistocene. Astronomical motions combined to change the magnitude and distribution of solar energy flux on the earth, causing or promoting glacial advance or retreat, which, in turn, caused falling or rising sea level, which in its turn affected the rate of marine sedimentation and the composition and texture of the sediments deposited. Thus, glaciation mediates between celestial motions and sedimentary processes, and the effects are synchronous worldwide. (Continental glaciation implies the existence of continents in the higher-latitude regions; if the requisite land masses are not there, a glaciation probably does not occur.) Of course there are other, less drastic climatic variations that influence sedimentation.

Hancock and Kauffman (1979) contended that transgressive-regressive pulses can be detected in Upper Cretaceous marine strata, and that these pulses are synchronous on a global scale and hence eustatic. Barron et al. (1985) described bedding cyclicity in the Greenhorn Limestone (a Cretaceous pelagic limestone) near Pueblo, Colorado. There are oscillations in illite/smectite ratios and in the content of fine-grained detrital quartz, reflecting, it seems, cyclic changes in runoff discharge from surrounding lands, in effectiveness of aeolian transport, and in organic productivity of both organic carbon and skeletal calcium carbonate. Glaciation almost certainly was not involved, and the authors stated that the nature of the link between orbital periodicities and cyclic variations in the stratigraphic record during non-glacial times has not yet been defined. Various climate studies suggest that low-latitude insolation (as opposed to high-latitude glaciation) is an important factor. For example, perihelion comes in winter in the Northern Hemisphere, but because of the precession, perihelion came in *summer* a few thousands of years ago, and there was therefore a greater difference between mean summer and mean winter insolation. Various atmospheric and oceanic processes, especially the marine circulation patterns, both shallow and deep, are affected by small climatic changes such as this and tend to amplify the effects of atmospheric climatic change in certain parts of the globe.

Barlow and Kauffman (1985) suspected that the bed succession in the Fort Hays Limestone (Cretaceous), lying a few tens of meters above the Greenhorn, near Pueblo, Colorado, reflects a Milankovitch-driven cyclicity. This rock unit consists essentially of hard chalk beds interlayered with much thinner, soft argillaceous beds. At Pueblo there are 34 chalk/shale couplets arranged in seven bundles. If individual couplets are a response to the precession cycle (roughly 20,000 years), then each of the seven bundles reflects the 100,000-year eccentricity cycle, and deposition of the Fort Hays required, therefore, approximately 700,000 years.

Creer (1975) suggested that secular variations in earth's magnetism are tied to fluctuations in earth's rotational frequency. In late Paleozoic time, according to Creer, there was a strong reversed *polarity bias*; Mesozoic time saw a strong normal bias. The Cenozoic Era has been a time of mixed polarity, with a higher frequency of polarity reversals than in previous times.

Creer proposed that the common link is temperature distribution and topography at the core-mantle interface and phase changes in the mantle or changes in oblateness of the geoid. These internal changes probably are linked also to rates of sea-floor spreading and continental drift, and to volcanic activity, and ultimately affect sedimentation in various ways. Magnetic reversals may affect organic evolution also. Crain

(1971) documented a strong correlation between reversals and a certain "extinction index." Hays (1971) pointed to extinctions of certain radiolarians in the last 2 or 3 million years that seem to coincide with reversals.

Whyte (1977) and Muller and Morris (1986) tied the earth's magnetism to climate, suggesting that glaciation upsets the earth's rotational moments, and that this leads to geomagnetic polarity reversals. Sedimentary rocks can record both the glaciations and the magnetic reversals.

There are cycles of still longer period. The solar trip around the galaxy takes 274 million years (Steiner and Grillmair, 1973, p. 1008), the so-called *cosmic year* (the length of the cosmic year has decreased with time). As the sun makes its great revolution, it bobs from one side of the galactic plane to the other (rather like a carousel horse), the so-called epicyclic motion, with a period of 77 million years (Steiner and Grillmair, 1973, p. 1008). Moreover, the solar orbit is elliptical; at the end of a cosmic year the sun is farther from the galactic nucleus than at mid-year; the next perigalacticum, marking the end of the current cosmic year, comes soon, in about 4 million years. All of these cosmic motions are known to influence, or are suspected of influencing, earth processes. Steiner (1967; 1973) and Steiner and Grillmair (1973) have suggested that cosmic motions have been responsible for the Sloss cycles and for episodic glaciations.

Cycles and Stratigraphy

Because cycles of different periods are superimposed, the resulting pattern of cyclicity is very complicated and rather random-appearing. Fischer et al. (1985, p. 7) suggested that there is a threshold level, and that a cyclic process triggers a depositional event only when it crosses this threshold. In stratigraphic successions, this gives rise to regularly spaced bed couplets, to irregularly spaced couplets, or to bundled couplets (Fig. 12-3), depending on how near the oscillator is to the threshold. Sensitivity of the depositional system to an external process may vary geographically or temporally, and long stratigraphic sections commonly drift from one kind of cyclicity to another. One problem is how to distinguish between the "forced" cyclicities generated remotely and cyclicity generated internally (an autocyclicity), as by some instability in a feedback system. Another problem is discriminating between real and apparent periodicities.

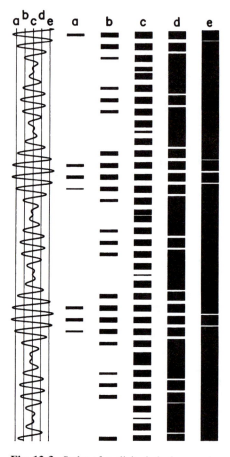

Fig. 12-3. Styles of cyclicity in bed successions. Waveform is a linear combination of three sinusoids having different periods; it represents some sedimentation-controlling set of environmental conditions. Stratigraphic response varies according to proximity of the depositional system to a conceptual shale/limestone threshold. If, for example, the threshold is at **b** on the waveform, such that shale (black in stratigraphic profile) is deposited when waveform is to the left of **b**, and limestone (white) when waveform is to the right of **b**, then stratigraphic profile **b** is expected. Profile **c** shows highest sensitivity, recording nearly every fluctuation.

Donovan and Jones (1979) enumerated the probable causes of worldwide sea-level change: some sea-level variation is due to changes in the total volume of ocean water (resulting from phase change); some is due to changes in volume of the ocean basins. The most obvious mechanism for change in ocean water volume is, of course, continental glaciation, which can effect a total sea-level change of about 150 meters at a rate of about 1 centimeter per year. Desiccation of marginal basins (such as the Mediterranean Sea) that have become detached from the world ocean could cause sea level to rise 15 meters, again at a rate of 1 centimeter per year.

Major changes in volume of the container, that is, the ocean basins, are brought about by changes in the length and activity of mid-ocean ridges. It seems that sea-level fluctuations of 300 meters are possible by this means, but rate is much less, on the order of 1 centimeter per 1000 years. Valentine and Moores (1970, p. 659) considered that major long-term transgressions occur during continental fragmentation and that regressions occur during times of continental plate assembly. This is referred to as *first-order* eustatic sea-level change. The beginning of the Phanerozoic was a time of low-standing sea level (Fig. 12-4). There was a long-term highstand in Cambrian through Mississippian time. A long-term emergence accompanied the assembly of Pangaea, culminating in worldwide lowstand in Permian and Triassic time. A major inundation attended the subsequent breakup of this supercontinent in Cretaceous and early Tertiary time, followed by gradual emergence that has continued to the present.

Fischer (1981) focused on the two Phanerozoic first-order "supercycles" and suggested that oceanic and atmospheric temperature is a balance between the rate at which CO_2 is added to ocean and atmosphere by volcanism and the rate at which it is withdrawn due to weathering of rocks (and storage of CO_2 as carbonate rock components). According to Fischer, CO_2 level is controlled ultimately by rate of mantle convection. During times of diminished convection, continental plates merge and the mid-ocean ridge system is shorter and less active; volcanism diminishes; sea level falls and the surface area over which weathering occurs expands. Thus, atmospheric partial pressure of CO_2 declines and temperatures fall; Fischer referred to this as an *icehouse state* (Fig. 12-5). Conversely, increased mantle convection causes breakup of continental masses, longer and broader oceanic ridges, and higher sea level. Much of the continental area is under water, and out of reach of weathering processes. Increased volcanism elevates CO_2 partial pressures; temperatures rise. This is a *greenhouse state*. Fischer was not able to explain the Ordovician-Silurian glaciation, which occurred during the first Phanerozoic greenhouse state. In an analysis of first-order cycles extended back to 1100 Ma, Veevers (1990) suggested that a time of glaciation occurred within the latter half of *each* (of three) greenhouse states; Veevers included the Neogene glaciation among these.

Some interesting petrographic evidence of alternation between greenhouse and icehouse states is provided by the ooids. Virtually all modern carbonate ooids are composed of aragonite, but most ancient ooids are of calcite. Some of these ancient ooids are composed of equant, anhedral spar, but others have a radial fabric. Some petrographers consider that the equant fabric results from the calcitization of ooids that were aragonite originally; ooids that retain a radial fabric were calcite from the beginning and never experienced a fabric-altering neomorphism. Mackenzie and Pigott (1981, p. 193) presented some evidence that ooid mineralogy has fluctuated between calcite and aragonite during the Phanerozoic, and suggested that fluctuation in atmospheric CO_2 level was the cause (lower levels of CO_2 favor aragonite over calcite). Sandberg (1983, p. 19) made a census of ancient ooids and concluded that calcite-ooid episodes and aragonite-ooid episodes coincide generally with the first-order eustatic-climatic cyclicity. In times of low sea-floor spreading rates (corresponding to Fischer's icehouse state), there is a smaller flux of CO_2 to the atmosphere, and ooids tend to be aragonitic. In times of high spreading rate (and greenhouse state), CO_2 flux to the atmosphere is greater and ooids are calcitic.

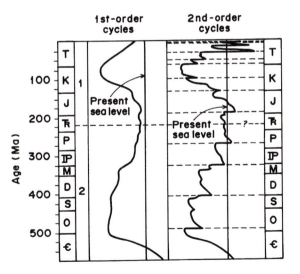

Fig. 12-4. First-order and second-order variation of sea level in Phanerozoic time. (From Vail et al., 1977, fig. 1, with permission of AAPG.)

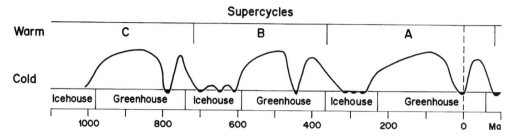

Fig. 12-5. The first-order cycles (supercycles) of long-term global cooling and warming over the past 1000 million years. (After Veevers, 1990, fig.6.)

Wilkinson et al. (1985, p. 181) pointed out that the Mg/Ca ratio in sea water may have enhanced the effect (high Mg/Ca favors aragonite over calcite). Submarine weathering of new oceanic crust consumes large quantities of dissolved Mg ion (Bloch and Hofmann, 1978) and severely reduces the Mg/Ca ratio of sea water. The effect is substantially greater when spreading rates are high; thus, ooids are calcitic. When mid-ocean ridges are less active, sea level is lower, continents are emergent, subaerial weathering increases, and rivers transport Mg ion to the sea in greater quantities; ooids are aragonitic.

A *second-order* cyclicity is superimposed on the first-order sea-level fluctuation. Sloss (1963; 1972) recognized six unconformity-bounded "sequences" on the North American continent (Fig. 12-6) and on the Russian platform, each of these six sequences representing a major continental transgression and subsequent regression. Sloss (1984) has suggested that these cycles are driven by global changes in intensity of tectonism on continents. Transgressions across cratons are caused by cratonic subsidence, because no credible eustatic sea-level rise by itself can accomplish a cratonic transgression of such proportions. Indeed, it might be very difficult to prove that sea level absolute changed at all.

There have been numerous shorter-term transgressions and regressions; they are especially evident in the stratigraphy of continental margins, as revealed by seismic profiles. Vail et al. (1977, p. 86) recognized many *third-order* cycles, recorded stratigraphically as depositional sequences having durations of 1 to 10 million years. As for the case of first and second-order stratigraphic cycles, third-order cycles could also be driven by tectonism, local or otherwise. Cloetingh (1988) appealed to intraplate stresses as a cause of third-order cycles.

Vail and his co-workers have insisted that most of the third-order sequences are global-correlated and therefore controlled by eustatic sea-level change, though they offer no mechanism for this. Others have gone even further, claiming worldwide synchroneity for fourth- and even fifth-order cycles. (It is entirely possible that some high-order cycles are global-synchronous, while some low-order cycles are not.)

Cyclic relative changes in sea level could have been due entirely to uniform or episodic subsidence. It is an old concept. Harbaugh and Bonham-Carter (1970, p. 373 ff) modeled the basin-filling process numerically; more recently, Aigner et al. (1991) made similar nu-

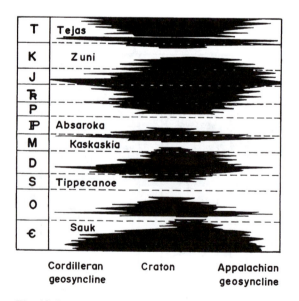

Fig. 12-6. Sloss's (1963, fig. 6) North American stratigraphic sequences.

merical models (Fig. 12-7). It is an autocyclic process; the period of the cycle depends on local thickness and strength of the crust and on rate of sedimentation. Moreover, it is a local cyclicity. A similar process might be occurring in some other corner of the world at the same time and yielding cycles of similar frequency. Though the two systems are totally independent, individual pulses in each system may appear to be synchronous (within the limits of absolute dating), and it is tempting to view these systems as instances of (correlated) global cyclicity.

Why has the sea-level fluctuation been so difficult to quantify? It is, of course, because sea level is our "point of reference" or "origin of coordinates." When we examine an ancient sedimentary deposit, we try to determine where it was deposited with respect to sea level. When sea level itself changes, we have no fixed coordinate system to relate this change to other than a previous sea level. Moreover, unconformities are difficult to date, and the ages and durations of many of them must vary substantially from place-to-place (see Chap. 7).

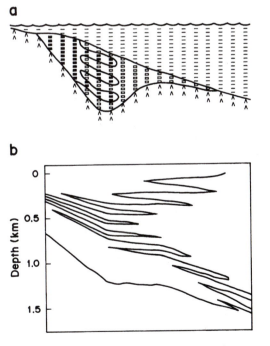

Fig. 12-7. Computer-simulated autocyclicity in a progradational deposit. (After Harbaugh and Bonham-Carter, 1970, fig. 7-10; Aigner et al., 1991, fig. 5.)

Is it conceivable that most of the third-order cyclicities are strictly local? Perhaps some, but not all, of Vail's third-order cycles are global. Many geologists, in their zeal to promote the new sequence stratigraphy, have been too quick to claim global synchroneity and to deny that many depositional sequences are created by strictly local processes (see Reading, 1987). Many sequence stratigraphies seem to be based on pretended chronometric control, and some sequences are correlated globally on the flimsiest of evidence. Vail et al.(1984, p. 139) admit that "If precise age dating cannot be achieved because a significant hiatus is present throughout the study area or because the paleontologic data are poor or lacking, the age of the closest global unconformity is used. . . . " They do not indicate how often this happens, but we should note that their best "correlations" are achieved when they have no age data, and unconformities are globally correlated by default. (Is not the "significant hiatus" in fact the *object* of the dating procedures?)

Still higher-order cycles with periods of a few tens of thousands of years are recognized in parts of the stratigraphic record. Such cycles have generally been attributed to continental glaciations or other climatic rhythms, which, again, seem to be tuned to Milankovitch cycles. But Algeo and Wilkinson (1988) contended that Phanerozoic "cycles" with periods of thousands to hundreds of thousands of years are *randomly* distributed with respect to age, most epochs exhibiting a wide spectrum of cycle periods ranging from 10^4 to 10^7 years. They determined that cycle periods do not coincide preferentially with Milankovitch periods and suggested that non-Milankovitch mechanisms (such as autocyclic processes) are important in the formation of all Phanerozoic cycles. Upper Mississippian and Pennsylvanian cycles are an exception; they have periods clustering between 400,000 and 500,000 years that seem to reflect orbital control through glacio-eustacy.

Global Anoxias

Most stratigraphers have encountered now and again in their stratigraphic field programs the distinctive rock bodies known as black shales (see Chap. 10). Commonly these rocks contrast strongly and sharply with beds above and below. They are of considerable economic interest because many of them supply the organic raw material for the formation of oil and gas.

In North America there are widespread black shales at or near the Devonian-Mississippian boundary, ranging in thickness up to about 100 meters, and known in various regions by various names, including Ohio, Bakken, and Exshaw. Stratigraphers also have been struck by the occurrence of black shale "marker beds" or event strata, widely distributed but relatively thin black shales of a particular age. The Late Cretaceous "black band" in the chalk beds of northeast England, with fish remains and pyrite nodules, reportedly is traceable across the North Sea into Scandinavia and Germany (Jenkyns, 1986, p. 387). It is not, by the way, particularly rich in organic matter, containing only about 1 percent total organic carbon (TOC) as compared with 10 or 15 percent in many other black shales. Such widely distributed black shales suggest to some stratigraphers that widespread or even global episodes of relative anoxia—that is, reduced oxygen concentration levels—have occurred in the marine environment. (Apparently, this concept has not been extended to coal beds.)

Global anoxias have been claimed for parts of Middle Cambrian, Middle Ordovician, Early Silurian, Late Devonian, Early Carboniferous, Early Jurassic, Late Jurassic, and Late Cretaceous times. Anoxia events may relate to times of high sea level and warm oceans. Expansion of the oceanic oxygen-minimum layer could be a factor. According to Schlanger and Jenkins (1976) and Leggett (1980), widespread epicontinental bituminous shales correlate with marine transgressions. Hallam and Bradshaw (1979, p. 158) maintained that black (bituminous) shales occur close to the base of transgressive successions, resting on nonmarine or marginal-marine rocks, or on a "basal conglomerate" or bone-bed, perhaps. Jenkyns (1980, p. 177) suggested that transgressions swamp vast areas of deltas and coastal plains and cause large quantities of terrestrial plant matter to drift out to sea; the world ocean in mid-Cretaceous time (following the greatest marine inundation since Ordovician times) may have been "awash with wood." Bacterial decomposition of organic matter in the fertile epeiric and marginal seas, according to Jenkyns, consumed oxygen; tongues of anoxic water spread from the shelf seas into the ocean basins. In Cretaceous time, solubility of oxygen in sea water was lower because of the higher global temperatures; this also favored bottom-water anoxia. Moreover, without a strong latitudinal zonation of temperature, ocean circulation probably was rather sluggish (Fischer, 1981, p. 108).

Waples (1983) disputed the claim that transgression causes global anoxia, and, moreover, disputed the *evidence* of global or oceanwide anoxia events in general, which, he argued, is based on a flawed sample. Many of the so-called black shales, including some of the Cretaceous pelagic black shales of the Atlantic Ocean floor (considered the best evidence of Cretaceous global anoxia), were deposited under *oxic* conditions and do not indicate anoxia at all. Much of the organic matter that they contain is terrestrial plant matter, which degrades slowly in depositional environments; its presence does not necessarily indicate anoxic conditions, nor does it necessarily cause anoxia in the sediments that contain it. Indeed, many black shales that contain predominantly terrestrial plant matter are extensively bioturbated, suggesting an oxygenated environment. Cretaceous pelagic sediments of the North Atlantic that *do* probably indicate anoxia are of local significance only. In Cretaceous time, the proto-Atlantic was not really an ocean but a small, highly restricted sea. Black sediment layers in North Atlantic bottom samples are most likely the intermittent deposits of stagnant water layers in more or less isolated, oxygen-poor basins surrounded and overlain by normally oxygenated ocean waters. Those anoxic events that did occur are not necessarily synchronous and by no means global.

Pedersen and Calvert (1990) likewise disputed the claim that organics-rich sediments are an indication of global anoxia. High organic-matter content in a marine sediment is a result of high organic productivity in the surface waters. Dysaerobic conditions in a marine water mass or in a marine sediment are not the cause of high organic carbon content in the sediment but a consequence of it.

What might have caused the extensive Devonian-Mississippian black shales in North America? Most likely it was the tectonic milieu at the time—much of the continent was a lowland and detrital sources were limited. There were numerous sediment-starved troughs and isolated sags with restricted circulation but presumably high organic productivity in the surface waters. Organic matter sank to the bottom, where it was not much diluted by terrigenous influx; anoxia in the sediment, a condition created and maintained by decay of some of this organic matter, ensured the preservation of the rest of it.

In Precambrian and early Paleozoic times, there was little molecular oxygen in the atmosphere, and one assumes that the oceans were similarly anoxic. It was an anoxia far more profound than any of those proposed for later times. There was substantial biomass, however, possibly comparable to today's. Are Precambrian and lower Paleozoic shales (and other sedimentary rocks) mainly black and organics-rich, therefore? Why should not Precambrian rocks serve as the best examples of the sedimentological results of global anoxia?

Sedimentation, Orogenesis, and the Global Tectonics

Tectonism creates the topographic relief that drives the sedimentary cycle, and controls the rate of flow of sedimentary materials through the cycle. Tectonism creates the detrital source terrains, determines their elevation and relief, and to a large extent controls their compositions. Tectonism creates the basins where sedimentation occurs and controls their sizes and shapes as well as the rates and durations of their subsidence; it also influences their depositional environments. We have come to a general notion that the sediments of a craton are, in the main, different from the sediments of an orogenic welt, and that the sediments of continental platforms are different from the sediments of ocean basins. There are differences in composition, texture, sedimentary structure, and fossil content; size, shape, and homogeneity of sedimentary bodies; and differences in the grand-scale depositional architecture. Often have we attributed to some tectonic ''hiccup,'' near or far—this thin layer of coarse sand interposed between the samely muds or that sharp bedding break that marks incursion of the sea.

Greenstone Belts

Anhaeusser et al. (1969) tried to imagine what the earliest Precambrian continental crust was like and what tectonics was like in 3400 to 2400 Ma in the oldest sedimentary terrains that we know of and can recognize as such. There are isolated, strongly downfolded bodies of distinctive metavolcanic and metasedimentary rocks embedded in the old granitic cratons. They are called *greenstone belts* (Fig. 12-8). They seem to have formed on elongate downwarps or graben on an unstable, thin, sialic crust. Typically there is an older volcanic succession, composed of ultramafic, mafic, intermediate, and sialic flows and pyroclastics; pillows are common, indicating extrusion of lavas under wa-

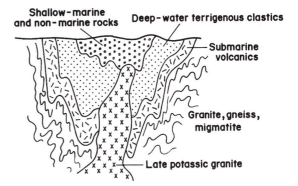

Fig. 12-8. Structure of a greenstone belt. (After Anhaeusser et al., 1969, fig. 2.)

ter. A predominantly sedimentary succession overlies, containing turbidites, banded iron formations, cherts and jaspilites, various conglomerates and breccias, quartzites, and shales. Cross-bedding and mudcracks in the younger sediments indicate that these last layers were deposited in shallow waters, finally filling the deep troughs to the brim. Total greenstone-belt thicknesses of 30,000 meters have been recorded! Some of the granite that surrounds the greenstone belts seems to be younger, having intruded and altered the greenstone fill. Upwelling of granitic magmas may have been coupled with downsagging of the troughs.

These objects occur on all the old cratons. They may have been largely a response to vertical tectonism; perhaps they required a thin crust, rather hotter than today's. But as the crust cooled and differentiated and thickened, greenstone-belt-style tectonism became obsolete, and the (rather more horizontal) plate tectonism took over. On the other hand, Sleep and Windley (1982) suggested that a certain style of subduction tectonics *did* occur in Archean times, involving a hot mantle and an oceanic crust substantially *thicker* than the modern oceanic crust. Engel et al. (1974), who also examined the problem of the greenstone belts, have pointed out that the roots of most of them are oceanic crust and mantle and that the greenstone belts were emplaced *between* fragments of rifted sialic scum.

Geosynclines

A different kind of tectonic furrow, the *geosyncline*, and its filling of stratified rocks, sedimentary and volcanic, seems to have replaced the greenstone belt as crustal (and mantle) evolution progressed. Krynine (1941, 1942) postulated that long-term tectonic development of a geosyncline creates a succession of contrasting rock types. He considered that a geosyncline evolves through three stages—an early quiescent stage during which sedimentation occurs slowly on a stable shelf or craton; a middle stage of subsidence, volcanism, and rapid sedimentation; and a final stage of plutonism, violent deformation, and heterogeneous, episodic sedimentation. Each stage is characterized by a particular association of detrital source terrains, dispersal patterns, depositional topographies, and depositional environments. Each stage, therefore, gives rise to a concinnity of rock types with characteristic compositions and textures, sedimentary structures, thicknesses and shapes, even fossil assemblages. Notably, an evolving tectonics is reflected in the composition of sandstones in a geosyncline (Fig. 12-9). In the first stage—tectonic quiescence—detrital source lands of low relief deliver the compositionally and texturally mature residues of weathered crystalline basement rocks and recycled sedimentary rocks to the cratonic borderlands and shallow marine shelves. Blankets of quartzarenite are deposited, typically interbedded with illitic shales and light-colored limestones. Sedimentary structures indicate slow sedimentation in shallow water or on wind-swept lowlands. Under ensuing rapid geosynclinal subsidence of cratonic margins, sedimentation quickens.

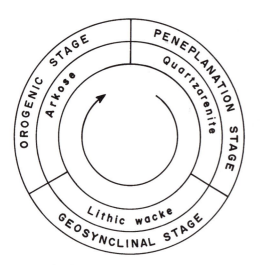

Fig. 12-9. The geosynclinal cycle. (According to Krynine, 1942, fig. 2.)

Rapid subsidence of a geosyncline may at first exceed rate of sedimentation, so that deep marine environments appear. In the middle stage of orogenesis, tectonic welts or volcanic arcs arise beyond the cratonic margin. The new source lands of high relief pour large volumes of immature detritus, perhaps rich in volcanic materials, into the trough and onto the edges of the craton, overwhelming the old cratonic detrital sources. A sedimentary body of great thickness is deposited, configured to the tectonic furrow that received it, and composed of lithic wackes and lithic arenites and dark shales that thin rapidly onto the shelf, and overlap or merge with the very different sediments that mantle the craton. Even as the geosynclinal sediment prism is being deposited, its proximal margin is being metamorphosed, and folded and uplifted, keeping alive the orogenic highland detrital source and reworking detritus craton-ward. Large slabs (olistoliths) may break away from the tectonic welt, and slide into the trough. Soon the trough fills and overflows. Compression and collapse thicken the crust, granitic batholiths intrude, and block faults break up the orogen, exposing its crystalline interior. Detrital sediments of the final stage of the geosynclinal cycle—Krynine's peneplanation stage—are locally derived coarse arkose sands and lithic gravels, spilled into the small fault-bounded intermontane basins and into successor basins of the foreland.

Flysch and *molasse* are terms that originated in studies of geosynclinal sedimentation associated with orogenesis of the European Alps. **Flysch** is a thick marine succession of rhythmically bedded mudstones and graded muddy sandstones, today at least partially synonymous with deep-water turbidite. It is construed as a pre-orogenic or syn-orogenic fill of a tectonically created trough or furrow. It corresponds to Krynine's middle stage. **Molasse** is a thick succession of mostly non-marine (or marginal-marine) conglomerates, cross-bedded sandstones, and various siltstones and shales, limestones, and coals. It is said to be post-orogenic, consisting mainly of detrital materials shed from the newly elevated highlands and deposited on the intermontane and foreland basins (Krynine's final stage). Two well-defined flysch-molasse successions appear in the sedimentary pile of the Appalachian basin, the older one of Ordovician and Silurian age and related to Taconic orogenesis, the younger one of Devonian, Mississippian, and Pennsylvanian age, created by the Acadian orogenic cycle. Delineation of the two great orogenies was an outgrowth largely of studies of the sediment products rather than of structural relationships.

Schwab (1969a; 1969b) and McLane (1972) examined vertical profiles of the sedimentary pile created during the long-term tectonic evolution of the present northern Rocky Mountains region of the western United States. McLane described, in the Phanerozoic sedimentary pile in southwestern Montana, a secular change in composition and texture of the detrital rocks and in the proportions of gross lithotypes (Fig. 12-10): Paleozoic sandstones are generally fine grained quartz-arenites and subarkose arenites; Mesozoic sandstones are dominantly lithic and sublithic arenites and quartz-wackes, generally coarser than Paleozoic sandstones; and Cenozoic sandstones are lithic feldspathic arenites, wackes, and conglomerates of low compositional and textural maturity. Schwab (1969b, p. 1330) pointed to a secular increase in sedimentation rate (or survival rate) and increase in clastic ratio (Schwab, 1969a, p. 152) in a western Wyoming stratigraphic profile. Both workers attributed the secular trends to tectonic evolution of the Cordilleran geosyncline. It is a matter of changing provenance and changing rates of uplift and subsidence.

Plate-Margin Troughs or Basins

With the emergence of the *plate tectonics* paradigm, our concept of the geosyncline and of the nature of

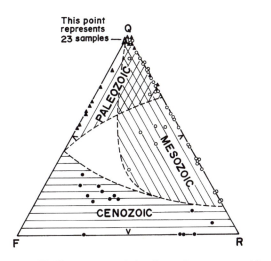

Fig. 12-10. Secular variation in sandstone composition and texture in southwestern Montana. (McLane, 1972, fig. 1, with permission of AAAS.)

orogenesis has changed. ''Geosyncline theory'' has always recognized that an orogen is an evolving set of genetically related, closely spaced or consecutive, more or less elongate and parallel sediment sources and sediment sinks. At various stages, sediment sources include older sedimentary rock, old crystalline basement, newly erupted volcanics, and newly intruded plutons. Plate tectonics theory has not substantially altered this broad view; but it has clarified and explained, and has unified the diverse orogens, past and present, under a single, global mechanism of rifting and subduction of crustal plates. Today, sedimentary basins can be classified on the basis of their plate-tectonic setting (see Ingersol, 1988, for detailed review). Basins of tectonically different origins have different sizes, shapes, and subsidence histories and contrasting kinds of sedimentary fills.

Rifting of continental crust at *divergent plate margins* causes thinning of the crust and development of graben that receive coarse detritus from the margins of the rift, chemical sediments of acidic or alkaline lakes, and acidic and basic volcanics. Continued spreading brings further subsidence, encroachment of the sea, extrusion of tholeitic basalts, hydrothermal sulfide and heavy-metal exhalations, and the beginnings of a new oceanic crust. This new crust (Fig. 12-11) comes to be blanketed by pelagic sediment that thickens and ages away from the spreading axis. On the block-faulted passive continental margin, land-derived detritus is deposited in marine environments as a prograding shelfal succession that gives way seaward to slope deposits, mainly the contourites and turbidites, which merge farther basinward with the pelagic muds and oozes.

It may happen that continental crustal rifting does not last beyond some initial stretching and subsidence. At a *triple junction*, where three spreading axes meet, two of the rifts may succeed in initiating a new tract of oceanic crust, while the third fails. Long-term subsidence in a failed arm, or *aulacogen*, is accompanied by marine sedimentation that is much more rapid than on the flanking platforms of continental crust. The cratonic platforms are, in fact, the source of most of the detritus, which enters the trough from the flanks and moves axially toward the triple junction. Shallow marine carbonate sedimentation dominates the early stages of trough filling, and carbonate shelves build out from the flanks. In the Proterozoic Athapuscow aulacogen of northwest Canada (Hoffman, 1973, p. 560; Hoffman et al., 1974, p. 45), oldest sedimentary

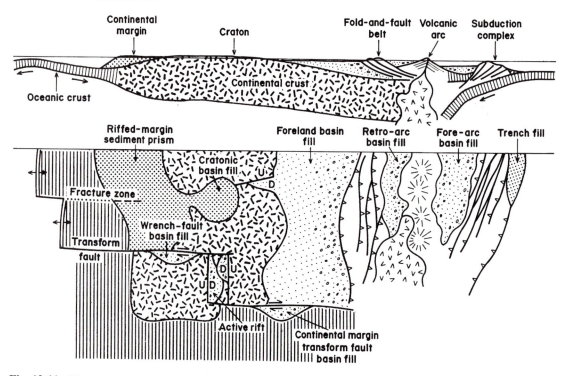

Fig. 12-11. Divergent and convergent plate boundaries. (After Dickinson and Suczek, 1979, figs. 5, 7.)

rocks are shelf quartzites (quartzarenites) and shelf dolomites (Fig. 12-12); these pass upward into thick muddy turbidites (flysch stage), thence to arkose conglomerates and redbeds (molasse stage). Sediments typically are interlayered with basaltic volcanics.

Continental and oceanic crust linked together in a single plate may at some stage decouple, creating a *convergent plate margin*. Convergent (active continental) margins are more complicated and more varied than the divergent margins and passive margins. There is an arc-trench system (Fig. 12-11) created by subduction of oceanic crust beneath the continental crust, with fore-arc basin between arc and trench that receives sediments from the volcanic arc. Spreading from divergent margins transports pelagic sediments (and perhaps micro-continents) to the active margin and creates melange that accretes to the continental margin. Landward of the arc, in a back-arc basin, a thick sediment prism of marine and then non-marine sediments, perhaps accumulated in part during an earlier passive-margin stage, is compressed and broken by thrusts, and tectonically transported toward the

foreland. Edge-loading by tectonic crustal thickening creates a foreland basin, which fills with a molasse. With subsequent relaxation of compressive stress, graben form in the foreland thrust belt and fill up with coarse alluvial and lacustrine deposits.

Obliquity between spreading axis and active continental margin, as in southern California, leads to substantial strike-slip component at the margin, and small-scale plate fragmentation. Commonly, small basins appear (Fig. 12-11), both in the continental plate and in the oceanic plate, the former receiving alluvial gravels and lacustrine deposits, the latter accumulating pelagic muds. Basaltic volcanism may be important.

Dickinson and Suczek (1979, p. 2164) described relationships between sandstone composition and plate tectonics, based on hundreds of modal analyses of many workers. They considered that "the key relations between provenance and basin are governed by plate tectonics, which thus ultimately controls the distribution of different types of sandstones." Plate tectonics processes create three broad provenance types, being (1) continental block, (2) magmatic arc, and (3)

Athapuscow aulacogen   Slave platform

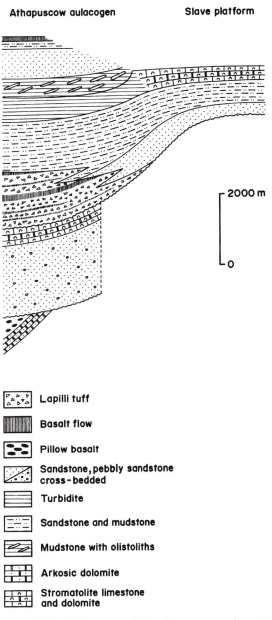

2000 m

0

- Lapilli tuff
- Basalt flow
- Pillow basalt
- Sandstone, pebbly sandstone cross-bedded
- Turbidite
- Sandstone and mudstone
- Mudstone with olistoliths
- Arkosic dolomite
- Stromatolite limestone and dolomite

Fig. 12-12. Sedimentary pile in Athapuscow aulacogen in western Canada. (After Hoffman et al., 1974, fig. 7, with permission of SEPM.)

orogen. Sands derived from continental blocks, including craton interiors and uplifted basement blocks, are quartzarenites and arkoses in which potassium feldspars are dominant over plagioclases. Quartzose sands reflect intense or prolonged weathering on cratons of low relief; they are deposited in cratonic basins or in troughs marginal to the craton. Local relief created by continental rifting or wrench-faulting gives rise to arkosic deposits on cratons. Magmatic arcs create volcanic lithic sands at one extreme and arkoses derived from plutons at the other extreme. The former may give way to the latter with time, as a volcanic arc is gradually dissected, eventually unroofing and revealing great plutons beneath. The undissected arc yields lithic sands rich in plagioclase but generally lacking in quartz, these sands coming to rest in trenches and fore-arc basins, and in shallow seas behind the arc, on the margins of continents. Sands of the dissected arc typically are more complex, containing quartz and potassium feldspars in addition to plagioclases and volcanic lithic fragments; these sands come to rest over the volcanogenic sands in fore-arc and back-arc basins. Orogens consisting of uplifted folded and faulted sedimentary and metasedimentary rocks produce recycled sedimentary detritus that accumulates in foreland basins and in rapidly subsiding troughs associated with subduction of the oceanic plate beneath a continental margin. The orogen may consist partly of sheared ophiolite and pelagic muds and cherts (melange) that, if uplifted and emergent, shed chert-rich lithic sands into trenches and forearc basins. Troughs representing the remnants of closing ocean basins between colliding continental plates can receive turbidite sands. Fold-and-thrust belts tend to shield foreland basins from the magmatic arcs and melange sediment sources; foreland basins receive instead recycled detritus from the deformed belt itself. This detritus typically is quartzose; it may contain abundant chert grains or carbonate rock fragments; but there is little or no feldspar.

On QFR plots (Fig. 12-13), continental-block sands, magmatic-arc sands, and orogenic-deformed-belt sands occupy discrete fields, though the first and last are partly intermingled. As a continental margin evolves, so too do the detrital sources, the sedimentary basins, and the composition of the sands being created and deposited.

Potter (1978), having examined the composition and texture of the sands of 36 modern big rivers worldwide, suggested that sandstone petrology can distin-

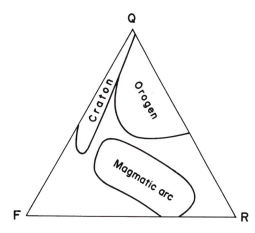

Fig. 12-13. Relationship between tectonic milieu and sandstone composition. (According to Dickinson and Suczek, 1979, figs. 1, 2, with permission of AAPG.)

guish between rivers issuing onto continental-plate trailing edges (passive margins) and leading edges (active margins and collision coasts). Trailing-edge river sands seem to be quartz-rich, contain more potassium feldspar than plagioclase and have relatively small rock-fragment content (mainly sedimentary rock fragments). Sands of leading edges are richer in plagioclase and commonly contain abundant volcanic rock fragments, derived from volcanic island arcs, and the grains are more angular. Rivers draining to marginal seas, such as the Gulf of Mexico, are compositionally and texturally intermediate between the sands of passive and collision margins. Valloni and Maynard (1981) determined that marine sands of different tectonic settings have distinct compositions and found different proportions of quartz, feldspars, and rock fragments in Recent sands of trailing-edge, leading-edge, fore-arc, and back-arc basins.

Secular Variation in the Sedimentary Rock Record

The composition of old sedimentary rocks seems to be different, on average, from the composition of younger sedimentary rocks. The long-term secular trends in composition have been examined in considerable detail by geochemists, but the conclusions reached are tentative. It is difficult to obtain ''average'' compositions of large bodies of sedimentary rock, and there

are large gaps in the rock record. It is difficult or impossible even to know whether the total volume of sediments has changed substantially over the past 3000 or 4000 million years.

There might be many causes of the long-term secular trends in rock composition. Probably earth's crust was thinner, and of a different composition, in the distant past than today, and earth's interior generated heat at a higher rate than today. Oceans and atmosphere were of a different composition. An evolving biosphere gradually altered the composition of atmosphere and ocean and greatly complicated the chemistry of earth-surficial processes. Thus, long-term trends in sediment composition could reflect changes in composition of lithosphere, hydrosphere, atmosphere, and biosphere. Overprinted on these trends are trends or cycles of shorter term (but still embracing tens or hundreds of millions of years) related to tectonic and geomorphic evolution of continents.

Diagenesis creates trends, which might be viewed simply as the effects of aging—older rocks are more severely altered than younger rocks. The trends in clay mineralogy with depth and in K/Na ratio with increasing age that the diageneticists have documented suggest that alteration of clay minerals under deep burial is partly responsible for apparent secular variation. Smectite-type clays change into illite-type clays (taking up potassium from the pore waters in the process, and releasing magnesium). There is a strong secular trend in Mg/Ca ratio in carbonate rocks, declining from about 1/2 in the Precambrian to about 1/50 in the Cenozoic. This too could be largely a diagenetic effect; perhaps much of the magnesium of old carbonate rocks is just the magnesium that has been released from aging shales. But we are not concerned with these diagenetic trends except in the matter of how they might be distinguished from the secular trends that have resulted from an evolving cosmos and an evolving planetary surface.

Evolution of living systems seems to have had profound effects on composition of the ocean and atmosphere, the mechanisms of rock weathering, and the rates of weathering and erosion. Weaver (1967), having documented an increasing illite content and declining smectite content with age in Phanerozoic shales, maintained that the evolution of plant life caused the chemical weathering environment to change over the long term. Land plants would release potassium and sodium from rocks in large quantities but would take up a considerable portion of the potassium in their tissues, recycling it on land rather than through the sea. Thus the younger marine shales contain less potassium and more sodium. (Secular trends in shales are, then, not necessarily diagenetic.) Kramer (1965), Broeker (1974, p. 197) and Wilkinson (1979) have suggested that Mg/Ca ratios were substantially lower in Paleozoic seas than the modern value of about 5 (even though Mg/Ca is higher in Paleozoic than in younger carbonate rocks). Petrographic evidence suggested to Sandberg (1975) that marine skeletal material, ooids, and micrite of Paleozoic rocks were precipitated by-and-large as low-Mg calcite, reflecting relatively low Mg/Ca ratio in Paleozoic seas. Ever since, carbonate-skeleton-building marine organisms have been selectively removing Ca from sea water and causing Mg/Ca ratios to increase. Changing sea-water chemistry has, in its turn, caused a shift from calcite as the dominant raw material in Paleozoic time to high-Mg calcite and aragonite today (Wilkinson, 1979, p. 525), materials that do not precipitate from waters having a low Mg/Ca ratio (see Chap. 9). Maliva et al. (1989) observed that the temporal distribution of sedimentary cherts reflects biological evolution, Proterozoic cherts are chemical or bacterial precipitates, Paleozoic and Mesozoic cherts are mainly (sponge) spicular and radiolarian, and Cenozoic cherts are made mostly by diatoms. Moreover, there is a secular change in the facies distribution and depositional environments of cherts: Lacustrine and shallow marine bedded cherts of the Proterozoic give way to nodular cherts of Paleozoic and Mesozoic open-marine-shelf muddy limestones, thence to the deep-sea diatomaceous bedded cherts of the Cenozoic.

We have noted that certain rock types are more abundant (or less abundant) in the older segments of the stratigraphic record than in the younger. Banded iron formation is the outstanding example of a defunct sedimentary rock type. Pelagic chalks, spiculites, and coals each have limited and fairly distinct time ranges. Organic evolution seems to have been the immediate cause of this. Some workers have noticed a secular decline in stromatolites (Garrett, 1970) and in flat-pebble conglomerates (Sepkoski, 1982a), and have argued that it is due to the evolution and proliferation of alga grazers and sediment burrowers.

McLennan (1982, pp. 336, 344) pointed to a drastic change in detrital rock composition at the Archean-Proterozoic boundary, suggesting that exposed Archean continental crust was more basic than post-Archean crust. It is a pronounced and rather abrupt change 2600 to 2400 million years ago that has been noticed on all the continents (Engel et al., 1974, p. 846) and is reflected in K_2O/Na_2O weight-percent ratios and in the relative abundance of compositionally mature and immature sediments.

Many workers have documented a progressive depletion of Al_2O_3, Na_2O, and MgO and enrichment in SiO_2, K_2O, and CaO in younger detrital rocks of the Phanerozoic. Vinogradov and Ronov (1956) observed a sixfold secular decline in K_2O content in Phanerozoic shales of the Russian platform (see also van Moort, 1972). They considered this trend to be provenance-controlled, noting that the area of exposed granitic basement rocks (a source of potassium) gradually decreased during Phanerozoic time, owing to gradual expansion of the sedimentary mantle. Younger sediments thus received less potassium.

Burke et al. (1982) documented long-term variation in $^{87}Sr/^{86}Sr$. Throughout Phanerozoic time there have been many episodes of increasing and decreasing value. They suggested that these variations reflect changes in the relative contributions of strontium to the sea from two sources: (1) old sialic rocks of continental interiors, with an average isotopic ratio of 0.720, and (2) mafic volcanics and intrusives of ocean basins and active plate margins, with average isotopic ratio of 0.704. Heavier strontium in marine sediments thus indicates greater contribution from continental interiors. During sea-level high stands, continents are submerged, and strontium in contemporaneous marine sediments is lighter.

Holser and Kaplan (1966, p. 120) and Holser (1977) pointed to a worldwide secular variation in $\delta^{34}S$ that they considered to be a result of large-scale transfer of sulfur to and from shale sulfides and evaporitic sulfates. Variations in denudation rates, probably reflecting global tectonic conditions, were considered to be responsible for the wide-ranging excursions in the isotopic ratio. In late Paleozoic time, for example, increase in ^{32}S relative to ^{34}S indicates a monumental 45 percent increase in dissolved sulfate concentration in the ocean. In the Mesozoic there was a net sedimentation of sulfur, preferentially removing light sulfur from the sea.

Garrels and Lerman (1981) also addressed the problem of sulfur-isotope variation, but, in contrast to Holser and Kaplan's views, assumed that the concentration of sulfate in the oceans remains constant. Sulfur moves through the sedimentary cycle either as $CaSO_4$ or as FeS_2, in relative proportions that seem to be linked to the carbon cycle. The sulfur isotope ratio at any given time reflects this ratio of sulfate to sulfide. This, and the ratio between sedimentary sulfur and organic carbon, seem to be linked to sea-level cycles (Berner, 1984, p. 611).

Sedimentary recycling creates secular trends, because more mobile rock components are continually recirculating into newly created sediment, while the less mobile components lag further and further behind, in the older rocks. Perhaps some of the long-term secular trends are due mainly to this. Garrels and Mackenzie (1969) observed that the total mass of sedimentary rocks existing at the present time is not uniformly distributed over geologic time, but that roughly half of the total mass is younger than 600 Ma (Fig. 12-14); in other words, the sedimentary rocks have a *half-age* of 600 million years. They observed further that the present proportions of sedimentary rock types show certain secular trends. Shales are the most abundant sedimentary rock type throughout geological time, but limestones and evaporites make up increasing proportions of younger strata. Garrels and Mackenzie suggested that long-term secular variation in survival rates and lithic proportions are a result of sedimentary recycling and developed a model that predicted in a gross sense the observed distribution of mass of sedimentary rocks.

According to this model, recycling continually remakes older sediments into younger ones; the older rocks gradually disappear as the younger sediments accumulate. Hence, the volume or mass of younger rocks is greater than that of older rocks existing today, even if mean rates of sedimentation have remained constant over the long term. Some rock types, because of their lower resistance to weathering and erosion, recycle more rapidly than others, and secular variation in lithic proportions is due largely to differential recycling of the various lithic types. Carbonate rocks, for example, make up decreasing proportions of older stratigraphic successions, even though their percentage of the total mass or volume of sediments may always have been very nearly what it is today. The trend is due to rapid removal of easily weathered carbonate rock from older parts of the stratigraphic pile and its redeposition as younger sediments. The proportion of more resistant rock types in older strata correspondingly increases. Thus, although the half-age of sediment as a whole is 600 million years, the half-age of carbonate rock seems to be about 300 million years and that of the very soluble and rapidly recycled evaporite rocks only 200 million years. The durable and relatively insoluble cherty rocks recycle more slowly than the shales and sandstones, limestones and evaporite rocks, and reside mainly in the older parts of the rock record.

Veizer (1988, p. 202) agreed that secular variation in rock proportions is a consequence of recycling but suggested that differential recycling is due not so much to the differing durabilities of major rock types as to their differing tectonic associations.

Garrels and Mackenzie assumed that the total volume of earth's sediment has been constant with time. Since crystalline rocks are continually degenerating into new sedimentary material, constancy of sediment volume implies that sediments are being transformed back into crystalline rocks at the same rate. Most of this presumably is a result of subduction and orogenesis. Estimates of the present total volume have been made; most are based on geochemical balance considerations, especially involving sodium. Livingstone (1963) and Gregor (1967, 1968) concluded that the sodium content of the ocean reached a steady state long ago and that the concentration and total mass of sodium in the ocean is fixed by the rate at which crystalline rocks are recycled into sediments and sediments back into crystalline rocks. When crystalline rocks weather, some sodium goes into solution and is lost to the sea, and the sediment derived from the weathering contains less sodium than the unweathered parent rock (only about one-third as much). Sodium returns to the crystalline crust via subduction of marine sediments.

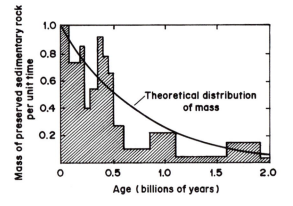

Fig. 12-14. Distribution of mass of sedimentary rocks in geologic time. (According to Garrels and Mackenzie, 1969, fig. 1, with permission of AAAS.)

Under the condition of steady state, the quantity of sodium in the ocean is a measure of the volume of sediment.

But there is much sedimentary material that is recycled not from crystalline rocks but from sediments. Thus, nested within the major cycle is a cycle that operates about six times faster (Gilluly et al., 1970, p. 367). Livingstone (1963, p. 1065) concluded that existing sediments represent a quarter to a sixth of the total volume that has been deposited throughout geologic time, and Garrels and Mackenzie (1969, p. 570) found that sediments have, in the mean, been reworked five times.

Gregor (1968, p. 27) concluded that based on sediment survival rates, denudation rates have changed during Phanerozoic time and that the rate was about five times as great in pre-Carboniferous times as at present, presumably owing to lack of vegetative ground cover in those times. Garrels and Mackenzie (1971b) recognized Gregor's two-cycle mass distribution of Phanerozoic sediments but adjusted his sediment accumulation and destruction figures.

The secular "megatrend" in the global sedimentary pile, in summary, seems to be a result of differential recycling within the sedimentary pile and exchange between the sediments and the crystalline or "primary" crustal rocks, that is, the plutonics, volcanics, and metamorphics that have insinuated or intermingled themselves with the sedimentary pile. Planetary evolution—that is, progressive irreversible changes in earth's atmosphere, hydrosphere, and biosphere, and presumably changes deep within the mantle and core as well—has modified and contributed to the secular megatrend. Superimposed on this, finally, are the local and shorter-term, though not necessarily less pronounced trends reflecting tectonic and geomorphic cycles, and other trends and cycles, long or short, that the earth experiences because it is a part of the cosmos.

REFERENCES

Abbate, E., Bortolotti, V., and Passerini, P., 1970, Olistostromes and olistoliths; Sed. Geol. 4, 521–557

Aboav, D.A., 1972, Foam and polycrystal; Metallography 5, 251–263

Aboav, D.A. and Langdon, T.G., 1969, The shape of grains in a polycrystal; Metallography 2, 171–178

Adams, A.E., MacKenzie, W.S., and Guilford, C., 1984, Atlas of Sedimentary Rocks under the Microscope; Wiley; New York; 104 p.

Adams, J.E. and Rhodes, M.L., 1960, Dolomitization by seepage refluxion; AAPG Bull. 44, 1912–1920

Addison, F.T., Turner, P., and Tarling, D.H., 1985, Magnetic studies of the Pendleside Limestone: evidence for remagnetization and late-diagenetic dolomitization during a post-Albian normal event; Jour. Geol. Soc. London 142, 983–994

Adelseck, C.J., Jr., Geehan, G.W., and Roth, P.H., 1973, Experimental evidence for the selective dissolution and overgrowth of calcareous nannofossils during diagenesis; GSA Bull. 84, 2755–2762

Ahlbrandt, T.S. and Andrews, S., 1978, Distinctive sedimentary features of cold-climate eolian deposits, North Park, Colorado; Paleogeography Paleoclimatology Paleoecology 25, 327–351

Ahlbrandt, T.S., Andrews, S., and Gwynne, D., 1978, Bioturbation in eolian deposits; Jour. Sed. Petr. 48, 839–848

Ahlbrandt, T.S. and Fryberger, S.G., 1982, Eolian deposits; pp. 11–47 in Scholle, P.A. and Spearing, D., eds., Sandstone Depositional Environments; AAPG Mem. 31; 410 p.

Aigner, T., 1982, Calcareous tempestites: storm-dominated stratification in upper Muschelkalk limestones (Middle Trias, SW-Germany); pp. 180–198 in Einsele, G. and Seilacher, A., eds., Cyclic and Event Stratigraphy; Springer-Verlag, Berlin-Heidelberg-New York; 536 p.

Aigner, T., Brandenburg, A., van Vliet, A., Doyle, M., Lawrence, D., and Westrich, J., 1991, Stratigraphic modelling of epicontinental basins: two applications; pp. 167–190 in Dott, R.H., Jr. and Aigner, T., eds., Processes and Patterns in Shelf and Epeiric Basins; Sed. Geol. 69

Aigner, T. and Reineck, H.-E., 1982, Proximity trends in modern storm sands from the Helegoland Bight (North Sea) and its implications for basin analysis; Senkenberg. Marit. 14, 183–215

Aissaoui, D.M., Buigues, D., and Purser, B.H., 1986, Model of reef diagenesis: Mururoa atoll, French Polynesia; pp. 27–52 in Schroeder, J.H. and Purser, B.H., Reef Diagenesis; Springer-Verlag, Berlin-Heidelberg-New York; 455 p.

Aissaoui, D.M., Coniglio, M., James, N.P., and Purser, B.H., 1986, Diagenesis of a Miocene reef-platform: Jebel Abu Shaar, Gulf of Suez, Egypt; pp. 112–131 in Schroeder, J.H. and Purser, B.H., Reef Diagenesis; Springer-Verlag, Berlin-Heidelberg-New York; 455 p.

Aitken, J.D., 1967, Classification and environmental significance of cryptalgal limestones and dolomites, with illustrations from the Cambrian and Ordovician of southwestern Alberta; Jour. Sed. Petr. 37, 1163–1178

Algeo, T.J. and Wilkinson, B.H., 1988, Periodicity of mesoscale Phanerozoic sedimentary cycles and the role of Milankovitch orbital modulation; Jour. Geol. 96, 313–322

Al-Hashimi, W.S. and Hemingway, J.E., 1973, Recent dolomitization and the origin of the rusty crusts of Northumberland; Jour. Sed. Petr. 43, 82–91

Allen, J.R.L., 1963, The classification of cross-stratified units, with notes on their origin; Sedimentology 2, 93–114

Allen, J.R.L., 1964, Primary current lineation in the lower Old Red Sandstone (Devonian), Anglo-Welsh basin; Sedimentology 3, 89–108

Allen, J.R.L., 1965a; Late Quaternary Niger delta, and adjacent areas; sedimentary environments and lithofacies; AAPG Bull 49, 547–600

Allen, J.R.L., 1965c, The sedimentation and paleogeography of the Old Red Sandstone of Anglesey, North Wales; Proc. Yorkshire Geol. Soc. 35, 139–185

Allen, J.R.L., 1965d, A review of the origin and characteristics of recent alluvial sediments; Sedimentology 5, 89–191

Allen, J.R.L., 1966, On bedforms and paleocurrents; Sedimentology 6, 153–190

Allen, J.R.L., 1967, Depth indicators of clastic sequences; Mar. Geol. 5, 429–446

Allen, J.R.L., 1968a, The nature and origin of bed-form hierarchies; Sedimentology 10, 161–182

Allen, J.R.L., 1968b, Flute marks and flow separation; Nature 219, 602–604

Allen, J.R.L., 1968c, Current Ripples; North-Holland, Amsterdam; 433 p.

Allen, J.R.L., 1970, Physical Processes of Sedimentation; an Introduction; Allen & Unwin, London; 248 p.

Allen, J.R.L., 1980, Sand waves: a model of origin and internal structure; Sed. Geol. 26, 281–328

Allen, J.R.L., 1983, Studies in fluviatile sedimentation: bars, bar-complexes, and sandstone sheets (low-sinuosity braided streams) in the Brownstones (L. Devonian), Welsh Borders; Sed. Geol. 33, 237–293

Allen, J.R.L. and Friend, P.F., 1968, Deposition of the Catskill facies, Appalachian region; with notes on some other Old Red Sandstones basins; pp. 21–74 in Klein, G. deV., ed., Late Paleozoic and Mesozoic Continental Sedimentation, Northeastern North America; GSA spec. paper 106; 309 p.

Allen, P.A., 1981, Some guidelines in reconstructing ancient sea conditions from wave ripplemarks; Mar. Geol. 433, M59–M67

Allen, P.A., 1984, Reconstruction of ancient sea conditions with an example from the Swiss Molasse; Mar. Geol. 60, 455–473

Alt, D., Hyndman, D.W., and Sears, J.W., 1990, Impact origin of late Miocene volcanism, Pacific Northwest; GSA 86th ann. mtg. Abstracts with Programs Pacific Section, 2

Alvarez, L.W., Alvarez, W., Asaro, F., and Michel, H.V., 1980, Extraterrestrial cause for the Cretaceous-Tertiary extinction; Sci. 208; 1095–1108

Alvarez, W, Colacicchi, R., and Montanari, A., 1985, Synsedimentary slides and bedding formation in Appenine pelagic limestones; Jour. Sed. Petr. 55, 720–734

Amsbury, D.L., 1962, Detrital dolomite in central Texas; Jour. Sed. Petr. 32, 5–14

Anderson, A.E., Jonas, E.C., and Odum, H.T., 1958, Alteration of clay minerals by digestive processes of marine organisms; Science 127, 190–191

Anderson, R.Y., Dean, W.E., Kirkland, D.W., and Snider, H.I., 1972, Permian Castile varved evaporite sequence; GSA Bull. 83, 59–86

Anderson, R.Y. and Kirkland, D.W., 1966, Intrabasin varve correlation; GSA Bull. 77, 241–255

Andrews, S., 1981, Sedimentology of Great Sand Dunes, Colorado; p. 279–291 in Ethridge, F.G. and Flores, R.M., eds., Recent and Ancient Nonmarine Depositional Environments: Models for Exploration; SEPM spec. publ. 31; 349 p.

Anhaeusser, C.R., Mason, R., Viljoen, M.J., and Viljoen, R.P., 1969, A reappraisal of some aspects of Precambrian shield geology; GSA Bull. 80, 2175–2200

Ashley, G.M., 1975, Rhythmic sedimentation in glacial lake Hitchcock, Massachusetts-Connecticut; pp. 304–320 in Jopling, A.V. and McDonald, B.C., eds., Glaciofluvial and Glaciolacustrine Sedimentation; SEPM spec. publ. 23; 320 p.

Ashley, G.M., 1990, Classification of large-scale subaqueous bedforms: a new look at an old problem; Jour. Sed. Petr. 60, 160–172

Ashley, G.M., Southard, J.B., and Boothroyd, J.C., 1982, Deposition of climbing-ripple beds: a flume simulation; Sed. 29, 67–79

Asquith, D.O., 1974, Sedimentary models, cycles, and deltas, Upper Cretaceous, Wyoming; AAPG Bull. 58, 2274–2283

Awramik, S.M., 1971, Precambrian columnar stromatolite diversity: reflection of metazoan appearance; Sci. 174, 825–827

Baba, J. and Komar, P.D., 1981, Measurements and analysis of settling velocities of natural quartz sand grains; Jour. Sed. Petr. 51, 631–640

Badiozamani, K., 1973, The dorag dolomitization model—application to the Middle Ordovician of Wisconsin; Jour. Sed. Petr. 43, 965–984

Bagnold, R.A., 1937, The sizegrading of sand by wind; Royal Soc. (London) Proc., Ser. A, 163, 252–264

Bagnold, R.A., 1941, The Physics of Blown Sand and Desert Dunes; Methuen, London; 265 p.

Bagnold, R.A., 1954, Experiments on gravity-free dispersion of large solid spheres in a Newtonian fluid under shear; Royal Soc. (London) Proc. Ser. A, 225, 49–63

Bagnold, R.A., 1956, The flow of cohesionless grains in fluids; Royal Soc. (London) Philos. Trans., Ser. A, 249, 235–297

Bagnold, R.A., 1966, An approach to the sediment transport problem from general physics; U.S. Geol. Survey prof. paper 411–I, 1–37

Bagnold, R.A. and Barndorff-Nielsen, O., 1980, The pattern of natural size distributions; Sedimentology 27, 199–207

Bailey, E.H., Irwin, W.P., and Jones, D.L., 1964, Franciscan and related rocks, and their significance in the geology of western California; Calif. Div. Mines and Geol. Bull. 183; 177 p.

Bak, P., Tang, C., and Wiesenfeld, K, 1988, Self-organized criticality; Physical Rev. 38, 364–374

Baker, H.W., Jr., 1976, Environmental sensitivity of submicroscopic surface textures on quartz sand grains—a statistical evaluation; Jour. Sed. Petr. 46, 871–880

Baker, P.A. and Burns, S.J., 1985, Occurrence and formation of dolomite in organic-rich continental margin sediments; AAPG 69, 1917–1930

Baker, P.A. and Kastner, M., 1981, Constraints on the formation of sedimentary dolomite; Sci. 213, 214–216

Baker, V.R., 1973, Paleohydrology and Sedimentology of Lake Missoula Flooding in Eastern Washington; GSA spec. paper 144; 69 p.

Barlow, L.K. and Kauffman, E.G., 1985, Depositional cycles in the Niobrara Formation, Colorado Front Range; p. 199–208 in Pratt, L.M., Kauffman, E.G., and Zett, F.B., eds., Fine-grained deposits and biofacies of the Cretaceous Western Interior seaway: evidences of cyclic processes; 2nd ann. midyear mtg, Golden, Colorado, fld. trip 9; SEPM; 249 p.

Barrett, T.J., 1982, Stratigraphy and sedimentology of Jurassic bedded chertoverlying ophiolites in the North Apennines, Italy; Sedimentology 29, 353–373

Barron, E.J., Arthur, M.A., and Kauffman, E.G., 1985, Cretaceous rhythmic bedding sequences: a plausible link between orbital variations and climate; Earth and Planetary Sci. Letters 72, 327–340

Barth, T.F.W., 1962, Theoretical Petrology, 2nd ed.; Wiley, New York; 416 p.

Barwis, J.M. and Hayes, M.O., 1985, Antidunes on modern and ancient washover fans; Jour. Sed. Petr. 55, 907–916

Bascom, W., 1951, The relationship between sand size and beach face slope; Am. Geophys. Union Trans. 32, 866–874

Baskin, Y., 1956, A study of authigenic feldspars; Jour. Geol. 64, 132–155

Basu, A., Young, S.W., Suttner, L.J., James, W.C., and Mack, G.H., 1975, Re-evaluation of the use of undulatory extinction and polycrystallinity in detrital quartz for provenance interpretation; Jour. Sed. Petr. 45, 873–882

Bathurst, R.G.C., 1959, The cavernous structure of some Mississippian *Stromatactis* reefs in Lancashire, England; Jour. Geol. 67, 506–521

Bathurst, R.G.C., 1975, Carbonate Sediments and their Diagenesis; 2nd ed., Developments in Sedimentology 12; Elsevier, Amsterdam; 658 p.

Bathurst, R.G.C., 1982, Genesis of stromatactis cavities between submarine crusts in Palaeozoic carbonate mud buildups; Jour. Geol. Soc. (London) 139, 165–181

Batschelet, E., 1981, Circular Statistics in Biology; Academic Press, London-New York; 371 p.

Batten, D.J., 1991, Reworking of plant microfossils and sedimentary provenance; pp. 79–90 in Morton, A.C., Todd, S.P., and Haughton, P.D.W., eds., Developments in Sedimentary Provenance Studies; Geol. Soc. (London) spec. publ. 57, 370 p.

Beales, F.W., 1953, Dolomitic mottling in Palliser (Devonian) Limestone, Banff and Jasper National Parks, Alberta; AAPG Bull., v. 37, 2281–2293

Beales, F.W. and Hardy, J.L., 1980, Criteria for the recognition of diverse dolomite types with an emphasis on studies on host rocks for Mississippi valley-type ore deposits; pp. 197–213 in Zenger, D.H., Dunham, J.B., and Ethington, R.L., eds., Concepts and Models of Dolomitization; SEPM spec. publ. 28; 320 p.

Bear, J., 1972, Dynamics of Fluids in Porous Media; Dover, New York; 764 p.

Beard, D.C. and Weyl, P.K., 1973, Influence of texture on porosity and permeability of unconsolidated sand; AAPG Bull. 57, 349–369

Beauchamp, B. and Savard, M., 1992, Cretaceous chemosynthetic carbonate mounds in the Canadian Arctic; Palaios 7, 434–450

Beck, J.W., Edwards, R.L., Ho, E., Taylor, F.W., Recy, J., Rougerie, F., Joannot, P., and Henin, C., 1992, Sea-surface temperature from coral skeletal calcium/strontium ratios; Sci. 257, 644–647

Bein, A. and Weiler, Y., 1976, The Cretaceous Talme Yafe Formation: a contour current shaped sedimentary prism of calcareous detritus at the continental margin of the Arabian craton; Sedimentology 23, 511–532

Bell, 1940, Armored mud balls—their origin, properties and role in sedimentation; Jour. Geol. 48, 1–31

Berger, W.H., 1970, Biogenous deep-sea sediments: fractionation by deep-sea circulation; GSA 81, 1385–1402

Berger, W.H., 1976, Biogenous deep sea sediments: production, preservation, and interpretation; p. 265–388 in Riley and Chester, eds., Chemical Oceanography, v. 5, Academic Press, London-New York-San Francisco; 401 p.

Berner, R.A., 1968, Calcium carbonate concretion formed by the decomposition of organic matter; Sci. 159, 195–197

Berner, R.A., 1984, Sedimentary pyrite formation: an update; Geochim. Cosmochim. Acta 48, 605–615

Beveridge, A.J., 1960, Heavy minerals in lower Tertiary formations in the Santa Cruz Mountains, California; Jour. Sed. Petr. 30, 513–537

Bjorlykke, K., Ramm, M., and Saigal, G.C., 1989, Sandstone diagenesis and porosity modification during basin evolution; Geol. Rundshau 78, 243–268

Blair, T.C., 1987, Sedimentary processes, vertical stratification sequences, and geomorphology of the Roaring River alluvial fan, Rocky Mountain National Park, Colorado; Jour. Sed. Petr. 57, 1–18

Blatt, H., 1967, Original characteristics of clastic quartz grains; Jour. Sed. Petr. 37, 401–424

Blatt, H., 1970, Determination of mean sediment thickness in the crust: a sedimentologic method; GSA Bull. 81, 255–262

Blatt, H., 1985, Provenance studies and mudrocks; Jour. Sed. Petr. 55, 69–75

Blatt, H. and Caprara, J.R., 1985, Feldspar dispersal patterns in shales of the Vanoss Formation (Pennsylvanian), south-central Oklahoma, Jour. Sed. Petr. 55, 548–552

Blatt, H. and Christie, J.M., 1963, Undulatory extinction in quartz of igneous and metamorphic rocks and its significance in provenance studies of sedimentary rocks; Jour. Sed. Petr. 33, 559–579

Blatt, H. and Sutherland, B., 1969, Intrastratal solution and non-opaque heavy minerals in shales; Jour. Sed. Petr. 39, 591–600

Blatt, H. and Totten, M.W., 1981, Detrital quartz as indicator of distance from shore in marine mudrocks; Jour. Sed. Petr. 51, 1259–1266

Bloch, S. and Hofmann, A, 1978, Magnesium metasomatism during hydrothermal alteration of new ocean crust; Geol. 6, 275–277

Blome, C.D. and Albert, N.R., 1985, Carbonate concretions: an ideal sedimentary host for microfossils; Geol. 13, 212–215

Bloss, F.D., 1957, Anisotropy of fracture in quartz; Am. Jour. Sci. 255, 214–225

Bluck, B.J., 1967, Sedimentation of beach gravels: examples from south Wales; Jour. Sed. Petr. 37, 128–156

Bluck, B.J., 1974, Structure and directional properties of some valley sand deposits of southern Iceland; Sedimentology 21, 533–554

Blumberg, P.N. and Curl, R.L., 1974, Experimental and theoretical studies of dissolution roughness; Jour. Fluid Mech. 65, 735–751

Boggs, S., Jr., 1968, Experimental study of rock particles; Jour. Sed. Petr. 38, 1326–1339

Boggs, S., Jr., 1969, Relationship of size and composition in pebble counts; Jour. Sed. Petr. 39, 1243–1246

Bokman, J., 1952, Clastic quartz particles as indices of provenance; Jour. Sed. Petr. 22, 17–24

Bokman, J., 1953, Lithology and petrology of the Stanley and Jackfork Formations; Jour. Geol. 61, 152–170

Bokman, J.W., 1957, Suggested use of bed-thickness measurements in stratigraphic descriptions; Jour. Sed. Petr. 27, 333–335

Boles, J.R., 1982, Active albitization of plagioclase, Gulf coast Tertiary; Am. Jour. Sci. 282, 165–180

Boles, J.R. and Franks, S.G., 1979, Clay diagenesis in Wilcox sandstones of southwestern Texas: implications of smectite diagenesis on sandstone cementation; Jour. Sed. Petr. 49, 55–70

Boon, J.D., Evans, D.A., and Hennigar, H.F., 1982, Spectral information from Fourier analysis of digitized quartz grain profiles; Mathematical Geol. 14, 589–605

Boothroyd, J.C., 1978, Mesotidal inlets and estuaries; 287–360 in Davis, R.A., Jr., ed., Coastal Sedimentary Environments; Springer-Verlag, New York-Heidelberg-Berlin; 420 p.

Bornhold, B.D. and Giresse, P., 1985, Glauconitic sediments on the continental shelf off Vancouver Island, British Columbia, Canada; Jour. Sed. Petr. 55, 653–664

Bosellini and Ginsburg, 1971, Form and internal structure of Recent algal nodules (rhodolites) from Bermuda; Jour. Geol. 79, 669–682

Bosellini, A. and Hardie, L.A., 1973, Depositional theme of a marginal marine evaporite; Sedimentology 20, 5–27

Boulton, G.S., 1990, Sedimentary and sea level changes during glacial cycles and their control on glacimarine facies architecture; p. 15–52 in Dowdeswell, J.A. and Scourse, J.D., eds., Glacimarine Environments: Processes and Sediments; Geol. Soc. (London) spec. publ. 53; 423 p.

Bouma, A.H., 1962, Sedimentology of some Flysch Deposits. A Graphic Approach to Facies Interpretation; Elsevier, Amsterdam; 168 p.

Bouma, A.H., 1972, Fossil contourites in lower Niesenflysch, Switzerland; Jour. Sed. Petr. 42, 917–921

Bourgeois, J., 1980, A transgressive shelf sequence exhibiting hummocky cross-stratification: the Cape Sebastian Sandstone (Upper Cretaceous), southwestern Oregon; Jour. Sed. Petr. 50, 681–702

Bourque, P.A. and Gignac, H., 1983, Sponge-constructed stromatactis mud mounds, Silurian of Gaspe, Quebec; Jour. Sed. Petr. 53, 521–532

Bowen, A.J. and Inman, D.L., 1969, Rip currents 2. laboratory and field observations; Jour. Geophys. Res. 74, 5479–5490

Bowen, A.J. and Inman, D.L., 1971, Edge waves and crescentic bars; Jour. Geophys. Res. 76, 8662–8671

Bowler, J.M., 1973, Clay dunes: their occurrence, formation and environmental significance; Earth Sci. Rev. 9, 315–338

Boyles, J.M. and Scott, A.J., 1982, A model for migrating shelf-bar sandstones in upper Mancos Shale (Campanian), northwestern Colorado; AAPG Bull. 66, 491–508

Bracken, B. and Picard, M.D., 1984, Trace fossils from Cretaceous/Tertiary North Horn Formation in central Utah; Jour. Paleont. 58, 477–487

Braithwaite, C.J.R., 1973, Settling velocity related to sieve analysis of skeletal sands; Sedimentology 20, 251–262

Brayshaw, A.C., 1984, Characteristics and origin of cluster bedforms in coarse-grained alluvial channels; pp. 77–85 in Koster, E.H. and Steel, R.J., eds., Sedimentology of Gravels and Conglomerates; Canadian Soc. Petroleum Geologists mem. 10; 441 p.

Brayshaw, A.C., Frostick, L.E., and Reid, I., 1983, The hydrodynamics of particle clusters and sediment entrainment in coarse alluvial channels; Sedimentology 30, 137–143

Brazier, M.D., 1980, Microfossils; Allen & Unwin; London; 193 p.

Brenner, R.L., 1980, Construction of process-response models for ancient epicontinental seaway depositional systems using partial analogs; AAPG 64, 1223–1244

Brett, C.E. and Baird, G.C., 1986, Symmetrical and upward-shallowing cycles in the Middle Devonian of New York State and their implications for the punctuated aggradational cycle hypothesis; Paleoceanography 1, 431–445

Brice, J.C., 1964, Channel patterns and terraces of the Loup River in Nebraska; U.S. Geol. Surv. prof. paper 422-D, 41 p.

Bridge, J.S., 1978, Origin of horizontal lamination under turbulent boundary layers; Sed. Geol. 20, 1–16

Bridge, J.S., 1985, Paleochannel patterns inferred from alluvial deposits: a critical evaluation; Jour. Sed. Petr. 55, 579–589

Bridge, J.S. and Diemer, J.A., 1983, Quantitative interpretation of an evolving ancient river system; Sedimentology 30, 599–623

Broadhead, R.F., Kepferle, R.C., and Potter, P.E., 1982, Stratigraphic and sedimentologic controls of gas in shale—example from Upper Devonian of northern Ohio; AAPG 66, 10–27

Broeker, W.S., 1974, Chemical Oceanography; Harcourt Brace Jovanovich, New York; 214 p.

Bromley, R.G., 1975, Trace fossils at omission surfaces; p. 399–428 in Frey, R.W., ed., The Study of Trace Fossils; Springer-Verlag, New York-Heidelberg-Berlin; 562 p.

Bromley, R.G. and Ekdale, A.A., 1984, *Chondrites*: a trace fossil indicator of anoxia in sediments; Sci. 224, 872–874

Bromley, R.G. and Ekdale, A.A., 1986, Composite ichnofabrics and tiering in burrows; Geol. Mag. 123, 59–65

Brookfield, M.E., 1977, The origin of bounding surfaces in ancient aeolian sandstones; Sedimentology 24, 303–332

Brown, M.A., Archer, A.W., and Kvale, E.P., 1990, Neap-spring tidal cyclicity in laminated carbonate channel-fill deposits and its implications: Salem Limestone (Mississippian), south-central Indiana, U.S.A.; Jour. Sed. Petr. 60, 152–159

Budai, J.M., Lohmann, K.C., and Owen, R.M., 1984, Burial dedolomite in the Mississippian Madison Limestone, Wyoming and Utah overthrust belt; Jour. Sed. Petr. 54, 276–288

Budai, J.M., Lohmann, K.C., and Wilson, J.L., 1987, Dolomitization of the Madison Group, Wyoming and Utah overthrust belt, AAPG Bull. 71, 909–924

Bull, W.B., 1964, Alluvial fans and near-surface subsidence in western Fresno County, California; U.S. Geol. Surv. prof. paper 437–A; 71 p.

Burke, K., 1972, Longshore drift, submarine canyons, and submarine fans in development of Niger delta; AAPG Bull., v. 56, 1975–1983

Burke, W.H., Denison, R.E., Hetherington, E.A., Koepnick, R.B., Nelson, H.F., and Otto, J.B., 1982, Variation of seawater $^{87}Sr/^{86}Sr$ throughout Phanerozoic time; Geol. 10, 516–519

Burst, J.F., 1958a, "Glauconite" pellets: their mineral nature and applications to stratigraphic interpretations; AAPG Bull. 42, 310–327

Burst, J.F., 1958b, Mineral heterogeneity in "glauconite" pellets; Am. Mineralogist 43, 481–497

Byrne, J.V., LeRoy, D.O., and Riley, C.M., 1959, The chenier plain and its stratigraphy, southwestern Louisiana; Gulf Coast Geol. Soc. Trans. 9, 237–260

Cadee, G.C., 1992, Eolian transport and left/rith sorting of *Mya* shells (Mollusca, Bivalvia); Palaios 7, 198–202

Callender D.L. and Folk, R.L., 1958, Idiomorphic zircon, key to volcanism in the lower Tertiary sands of central Texas; Am. Jour. Sci. 256, 257–269

Cameron, K.L. and Blatt, H., 1971, Durabilities of sand size schist and volcanic rock fragments during fluvial transport, Elk Creek, Black Hills, South Dakota; Jour. Sed. Petr. 41, 565–576

Campbell, C.V., 1971, Depositional model—Upper Cretaceous Gallup beach shoreline, Ship Rock area, northwestern Mexico; Jour. Sed. Petr. 41, 395–409

Campbell, K.A., 1992, Recognition of a Mio-Pliocene cold seep setting from the northeast Pacific convergent margin, Washington, USA; Palaios 7, 422–433

Cant, D.J. and Walker, R.G., 1976, Development of a braided-fluvial facies model for the Devonian Battery Point Sandstone, Quebec; Can. Jour. Earth Sci. 13, 102–119

Carozzi, A.V. and Gerber, M.S., 1978, Synsedimentary chert breccia: a Mississippian tempestite; Jour. Sed. Petr. 48, 705–708

Carey, S.N., 1991, Transport and deposition of tephra by pyroclastic flows and surges; pp. 39–57 in Fisher, R.V. and Smith, G.A. eds., Sedimentation in Volcanic Settings; SEPM spec. publ. 45, 257 p.

Carpenter, A.B. and Oglesby, T.W., 1976, A model for the formation of luminescently zoned calcite cements and its implications; GSA Abs. with Prog. 8, 469–470

Caston, V.N.D., 1972, Linear sand banks in the southern North Sea; Sedimentology 13, 63–78

Chafetz, H.S., 1972, Surface diagenesis of limestone; Jour. Sed. Petr. 42, 325–329

Chafetz, H.S., 1986, Marine peloids: a product of bacterially induced precipitation of calcite; Jour. Sed. Petr. 56, 812–817

Chafetz, H.S., and Buczynski, C., 1992, Bacterially induced lithification of microbial mats; Palaios 7, 277–293

Chamberlain, C.K., 1975, Recent lebensspuren in nonmarine aquatic environments; pp. 431–458 in Frey, R.W., ed., The Study of Trace Fossils; Springer-Verlag, New York-Heidelberg-Berlin; 562 p.

Chandler, R.J., 1972, Periglacial mudslides in Vestspitsbergen and their bearing on the origin of fossil "solifluction" shears in low angled clay slopes; Quart. Jour. Engineering Geol. (London) 5, 223–241

Chao, E.C.T., 1967, Impact metamorphism; pp. 204–233 in Abelson, P.H., ed., Researches in Geochemistry, vol. 2; Wiley, New York; 663 p.

Chepil, W.S., 1965, Function and significance of wind in sedimentology; U.S. Dept. Agr. publ. 970, 89–94

Choquette, P.W. and Pray, L.C., 1970, Geological nomenclature and classification of porosity in sedimentary carbonates; AAPG Bull. 54, 207–250

Choquette, P.W. and Steinen, R.P., 1980, Mississippian non-supratidal dolomite, Ste. Genevieve Limestone, Illinois basin: evidence for mixed-water dolomitization; pp. 163–196 in Zenger, D.H., Dunham, J.B., and Ethington, R.L., eds., Concepts and Models of Dolomitization; SEPM spec. publ. 28; 320 p.

Chowns, T.M. and Elkins, J.E., 1974, The origin of quartz geodes and cauliflower cherts through the silicification of anhydrite nodules; Jour. Sed. Petr. 44, 885–903

Clark, D.N., 1980a, The sedimentology of the Zechstein 2 carbonate formation of eastern Drenthe, The Netherlands; pp. 131–165 in Fuchtbauer, H. and Peryt, T., eds., The Zechstein Basin with Emphasis on Carbonate

Sequences; Contributions to Sedimentology 9; Schweizerbart'sche Verlagsbuchhandlung, Stuttgart; 328 p.

Clark, D.N., 1980b, The diagenesis of Zechstein carbonate sediments; pp. 167–203 in Fuchtbauer, H. and Peryt, T., eds., The Zechstein Basin with Emphasis on Carbonate Sequences; Contributions to Sedimentology 9; Schweizerbart'sche Verlagsbuchhandlung, Stuttgart; 328 p.

Clayton, C.J., 1986, The chemical environment of flint formation in Upper Cretaceous chalks, pp. 43–54 in Sieveking, G. de G. and Hart, M.B., eds., The Scientific Study of Flint and Chert; Cambridge Univ. Press, Cambridge; 290 p.

Cleary, W.J. and Conolly, J.R., 1971, Distribution and genesis of quartz in a piedmont–coastal plain environment; GSA Bull. 82, 2755–2766

Clifton, H.E., 1969, Beach lamination: nature and origin; Mar. Geol. 7, 553–559

Clifton, H.E., 1981, Progradational sequences in Miocene shoreline deposits, southeastern Caliente Range, California; Jour. Sed. Petr. 51, 165–184

Clifton, H.E. and Dingler, J.R., 1984, Wave-formed structures and paleoenvironmental reconstruction; Mar. Geol. 60, 165–198

Clifton, H.E., Hunter, R.E., and Phillips, R.L., 1971, Depositional structures and processes in the non-barred high-energy nearshore; Jour. Sed. Petr. 41, 651–670

Cloetingh, S., 1988, Intraplate stresses: a tectonic cause for third-order cycles in apparent sea level?; p. 19–29 in Wilgus et al., eds., Sea-Level Changes—An Integrated Approach; SEPM spec. publ. 42; 407 p.

Cloud, P.E., 1960, Gas as a sedimentary and diagenetic agent; Am. Jour. Sci. 258-A, 35–45

Cloud, P.E., 1972, A working model of the primitive earth; Am. Jour. Sci. 272, 537–548

Coleman, J.M. and Gagliano, S.M., 1964, Cyclic sedimentation in the Mississippi River deltaic plain; Gulf Coast Assoc. Geol. Soc. Trans. 14, 67–80

Compte des techniciens, 1981, Evaporite Deposits, Illustration and Interpretation of Some Environmental Sequences; Gulf Publ. Co.; 266 p.

Conolly, J.R., 1965, The occurrence of polycrystallinity and undulatory extinction in quartz in sandstones; Jour. Sed. Petr. 35, 116–135

Cook, H.E., 1979, Ancient continental slopes and their value in understanding modern slope development; pp. 287–305 in Doyle, L.J. and Pilkey, O.H., eds., Geology of Continental Slopes; SEPM spec. publ. 27; 374 p.

Correns, C.W., 1949, Growth and dissolution of crystals under linear pressure; Discussions Faraday Soc. 5, 267–271

Cotter, E., 1965, Waulsortian-type carbonate banks in the Mississippian Lodgepole Formation of central Montana; Jour. Geol. 73, 881–888

Cotter, E., 1966, Limestone diagenesis and dolomitization in Mississippian carbonate banks in Montana; Jour. Sed. Petr. 36, 764–774

Cotterill, R., 1985, The Cambridge Guide to the Material World; Cambridge Univ. Press, Cambridge; 352 p.

Coustau, H., Rumeau, J.L., Sourisse, C., Chiarelli, A., and Tison, J., 1975, Classification hydrodynamique des bassins sedimentaires, utilisation combinee avec d'autres methodes pour rationaliser l'exploration dans des bassins non-productifs; Proc. 9th World Petroleum Congr., Tokyo, London; Applied Sci. Publ. vol. II, 105–119

Cox, A., 1969, Geomagnetic reversals; Sci. 163, 237–245

Crain, I.K., 1971, Possible direct causal relation between geomagnetic reversals and biological extinctions; GSA Bull. 82, 2603–2606

Creer, K.M., 1975, On a tentative correlation between changes in the geomagnetic polarity bias and reversal frequency and the earth's rotation through Phanerozoic time; pp. 293–318 in Rosenburg, G.D. and Runcorn, S.K., eds., Growth Rythms and the History of the Earth's Rotation; Wiley, London; 559 p.

Cressman, E.R. and Swanson, R.W., 1964, Stratigraphy and petrology of the Permian rocks of southwestern Montana; U.S. Geol. Surv. prof. paper 313–C, 275–569

Crevello, P.D. and Schlager, W., 1980, Carbonate debris sheets and turbidites, Exuma Sound, Bahamas; Jour. Sed. Petr. 50, 1121–1148

Cronin, D.S., Galacz, A., Midszenty, A., Moorby, S.A., and Polgari, M., 1991, Tethyan ferromanganese oxide deposits from Jurassic rocks in Hungary; Jour. Geol. Soc. (London) 148, 655–668

Crook, K.A.W., 1968, Weathering and roundness of quartz sand grains; Sedimentology 11, 171–181

Crowe, B.M. and Fisher, R.V., 1973, Sedimentary structures in base-surge deposits with special reference to cross-bedding, Ubehebe Craters, Death Valley, California; GSA Bull. 84, 663–682

Curl, R.L., 1966, Scallops and flutes; Trans. Cave Res. Gp. (Great Britain) 7(2), 121–160

Curray, J.R., 1956, Dimensional grain orientation studies of Recent coastal sands; AAPG Bull. 40, 2440–2456

Curtis, C.D. and Spears, D.A., 1968, The formation of sedimentary iron minerals; Econ. Geol. 63, 258–270

Dalrymple, R.W., Zaitlin, B.A., and Boyd, R, 1992, Estuarine facies models: conceptual basis and stratigraphic implications; Jour. Sed. Petr. 62, 1130–1146

Davidson-Arnott, R.G.D. and Greenwood, B., 1974, Bedforms and structures associated with bar topography in the shallow water wave environments, Kouchibouguac Bay, Brunswick, Canada; Jour. Sed. Petr. 44, 698–704

Davies, D.K. and Ethridge, F.G., 1975, Sandstone composition and depositional environment; AAPG Bull. 59, 239–264

Davies, G.R., 1977, Turbidites, debris sheets, and truncation structures in upper Paleozoic deep-water carbonates of the Sverdrup basin, Arctic Archipelago; pp. 221–247 in Cook, H.E. and Enos P., eds., Deep-Water Carbonate Environments; SEPM spec. publ. 25; 336 p.

Davies, G.R. and Ludlum, S.D., 1973, Origin of laminated and graded sediments, Middle Devonian of western Canada; GSA Bull. 84, 3527–3546

Davies, I.C. and Walker, R.G., 1974, Transport and deposition of resedimented conglomerates: the Cap Enrage Formation, Cambro-Ordovician, Gaspe, Quebec; Jour. Sed. Petr. 44, 1200–1216

Davies, P.J., Bubela, B., and Ferguson, J., 1978, The formation of ooids; Sedimentology 25, 703–773

Davis, J.C., 1986, Statistics and Data Analysis in Geology; 2nd ed.; Wiley, New York; 646 p.

Dean, W.E., 1981, Carbonate minerals and organic matter in sediments of modern north temperate hard-water lakes; pp. 213–231 in Ethridge, F.G. and Flores, R.M., eds., Recent and Ancient Nonmarine Depositional Environments: Models for Exploration; SEPM spec. publ. 31; 349 p.

Dean, W.E. and Fouch, T.D., 1983, Lacustrine environment; pp. 98–130 in Scholle, P.A., Bebout, D.G., and Moore, C.H., eds., Carbonate Depositional Environments; AAPG Mem. 33; 708 p.

DeCelles, P.G. and Gutschick, R.C., 1983, Mississippian wood-grained chert and its significance in the western interior United States; Jour. Sed. Petr. 53, 1175–1191

Deffeyes, K.S., Lucia, F.J., and Weyl, P.K., 1965, Dolomitization of Recent and Plio-Pleistocene sediments by marine evaporite waters on Bonaire, Netherlands Antilles; pp. 71–88 in Pray, L.C. and Murray, R.C., eds., Dolomitization and Limestone Diagenesis, a Symposium; SEPM spec. publ. 13; 180 p.

Degens, E.T. and Stoffers, P., 1980, Environmental events recorded in Quaternary sediments of the Black Sea; Jour. Geol. Soc. (London) 137, 131–138

Demaison, G.T. and Moore, G.T., 1980, Anoxic environments and oil source genesis; AAPG Bull. 64, 1179–1209

DeMowbray, T. and Visser, J., 1984, Reactivation surfaces in subtidal channel deposits, Oosterschelde, southwest Netherlands; Jour. Sed. Petr. 54, 811–824

Dennen, W.H., 1964, Impurities in quartz; GSA Bull. 75, 241–246

Dickinson, W.R., 1970, Interpreting detrital modes of graywacke and arkose; Jour. Sed. Petr. 40, 695–707

Dickinson, W.R., 1985, Interpreting provenance relations from detrital modes in sandstones; pp. 333–361 in Zuffa, G.G., ed., Provenance of Arenites; Reidel; 408 p.

Dickinson, W.R., Beard, L.S., Brakenridge, G.R., Erjavek, J.L., Ferguson, R.C., Inman, K.F., Knepp, R.A., Lindberg, F.A., and Ryberg, P.T., 1983, Provenance of North American Phanerozoic sandstones in relation to tectonic setting; GSA Bull. 94, 222–235

Dickinson, W.R. and Suczek, C.A., 1979, Plate tectonics and sandstone composition; AAPG Bull. 63, 2164–2182

Dickson, J.A.D., 1978, Length-slow and length-fast calcite: a tale of two elongations; Geol. 6, 560–561

Dickson, J.A.D., 1993, Crystal growth diagrams as an aid to interpreting the fabrics of calcite aggregates; Jour. Sed. Petr. 63, 1–17

Dickson, J.A.D. and Barber, C., 1976, Petrography, chemistry and origin of early diagenetic concretions in the Lower Carboniferous of the Isle of Man; Sedimentology 23, 189–211

Dietrich, R.V., Hobbs, C.R.B., and Lowry, W.D., 1963, Dolomitization interrupted by silicification; Jour. Sed. Petr. 33, 646–663

Dillon, W.P. and Conover, J.T., 1965, Formation of ice-cemented sand blocks on a beach and lithologic implications; Jour. Sed. Petr. 35, 964–967

Dionne, J.-C., 1969, Tidal flat erosion by ice at La Pocatiere, St. Lawrence estuary; Jour. Sed. Petr. 39, 1174–1181

Dobkins, J.E., Jr. and Folk, R.L., 1970, Shape development on Tahiti-Nui; Jour. Sed. Petr. 40, 1167–1203

Doe, T.W. and Dott, R.H., Jr., 1980, Genetic significance of deformed cross bedding—with examples from the Navajo and Weber Sandstones of Utah; Jour. Sed. Petr. 50, 793–812

Donaldson, J.A., 1976, Paleoecology of *Conophyton* and associated stromatolites in the Precambrian Dismal Lakes and Rae Groups, Canada; pp. 523–541 in Walter, M.R., ed., Stromatolites; Developments in Sedimentology 20; Elsevier, Amsterdam-Oxford-New York; 790 p.

Donaldson, J.A. and Ricketts, B.D., 1979, Beachrock in Proterozoic dolostone of the Belcher Islands, Northwest Territories, Canada; Jour. Sed. Petr. 49, 1287–1294

Donovan, D.T. and Jones, E.J.W., 1979, Causes of worldwide changes in sea level; Jour. Geol. Soc. (London) 136, 187–192

Dorobek, S.L., 1987, Petrography, geochemistry, and origin of burial diagenetic facies, Siluro-Devonian Helderberg Group (carbonate rocks), central Appalachians; AAPG Bull. 71, 492–514

Dorobek, S.L. and Read, J.F., 1986, Sedimentology and basin evolution of the Siluro-Devonian Helderberg Group, central Appalachians; Jour. Sed. Petr. 56, 601–613

Dott, R.H., Jr., 1963, Dynamics of subaqueous gravity depositional processes; AAPG Bull. 47, 104–128

Dott, R.H., Jr., 1964, Wacke, graywacke, and matrix—what approach to immature sandstone classification?; Jour. Sed. Petr. 34, 625–632

Dott, R.H., Jr. and Bourgeois, J., 1982, Hummocky stratification: significance of its variable bedding sequences; GSA Bull. 93, 663–680

Doust, H. and Omatsola, E., 1989, Niger delta; pp. 201–238 in Edwards, J.D. and Santogrossi, P.A., eds., Divergent/Passive Margin Basins; AAPG Mem. 48, 252 p.

Drever, J.I., 1974, Geocheemical model for the origin of Precambrian banded iron formations; GSA Bull. 85, 1099–1106

Driese, S.G. and Dott, R.H., Jr., 1984, Model for sandstone-carbonate "cyclothems" based on upper member of Morgan Formation (middle Pennsylvanian) of northern Utah and Colorado; AAPG Bull. 68, 574–597

Droser, M.L. and Bottjer, D.L., 1986, A semiquantitative field classification of ichnofabric; Jour. Sed. Petr. 56, 558–559

Droser, M.L. and Bottjer, D.L., 1990, Ichnofabric of sandstones deposited in high-energy nearshore environments: measurement and utilization; Palaios 4, 598–604

Droser, M.L. and Bottjer, D.L., 1991, Ichnofabric and basin analysis; Palaios 6, 199–205

Dunham, R.J., 1962, Classification of carbonate rocks according to depositional texture; pp. 108–121 in Ham, W.E., ed., Classification of Carbonate Rocks; AAPG Mem. 1; 279 p.

Dunham, R.J., 1969a, Early vadose silt in Townsend mound (reef), Mexico; pp. 139–181 in Friedman, G.M., ed., Depositional Environments in Carbonate Rocks, a Symposium; SEPM spec. publ. 14; 209 p.

Dunham, R.J., 1969b, Vadose pisolite in the Capitan reef (Permian), Mexico and Texas; pp. 182–191 in Friedman, G.M., ed., Depositional Environments in Carbonate Rocks, a Symposium; SEPM spec. publ. 14; 209 p.

Dunham, R.J., 1971, Meniscus cement; p. 297–300 in Bricker, O.P., ed., Carbonate Cements; The Johns Hopkins Univ. Press, Baltimore-London; 376 p.

Dunoyer de Segonzac, G., 1970, Clay-mineral diagenesis and low-grade metamorphism; Sedimentology 15, 281–346

Dupre, W.R., 1984, Reconstruction of paleo-wave conditions during the late Pleistocene from marine terrace deposits, Monterey Bay, California; Mar. Geol. 60, 435–454

Dymond, J., Lyle, M., Finney, B., Piper, D.Z., Murphy, K., Conard, R., and Pisias, N., 1984, Ferromanganese nodules from MANOP sites H, S, and R—control of mineralogical and chemical composition by multiple accretionary processes; Geochim. Cosmochim. Acta 48, 931–949

Dzulynski, S., 1965, Data on experimental production of sedimentary structures; Jour. Sed. Petr. 35, 196–212

Dzulynski, S., Ksiazkiewicz, M., and Kuenen, Ph. H., 1959, Turbidites in flysch of the Polish Carpathian Mountains; GSA Bull. 70, 1089–1118

Edwards, M., 1986, Glacial environments; pp. 445–470 in Reading, G., ed., Sedimentary Environments and Facies; 2nd ed., Blackwell, Oxford; 615 p.

Edzwald, J.K. and O'Melia, C.R., 1975, Clay distribution in Recent estuarine sediments; Clays and Clay Minerals 23, 39–44

Ehlmann, A.J., Hulings, N.C., and Glover, E.D., 1963, Stages of glauconite formation in modern foraminiferal sediments; Jour. Sed. Petr. 33, 87–96

Ehrlich, R., Brown, P.J., Yarus, J.M., and Przygocki, R.S., 1980, The origin of shape frequency distributions and the relationship between size and shape; Jour. Sed. Petr. 50, 475–484

Ehrlich, R. and Weinberg, B., 1970, An exact method for characterization of grain shape; Jour. Sed. Petr. 40, 205–212

Ekdale, A.A., 1988, Pitfalls of paleobathymetric interpretations based on trace fossil assemblages; Palaios 3, 464–472

Elliott, L.A. and Warren, J.K., 1989, Stratigraphy and depositional environment of lower San Andres Formation in subsurface and equivalent outcrops: Chaves, Lincoln, and Roosevelt Counties, New Mexico; AAPG Bull. 73, 1307–1325

Elouard, P. and Millot, G., 1959, Observations sur les silicifications du Lutetien en Mauritanie et dans la vallee du Senegal; Bull. Serv. Carte Geol. Als. Lor. 12, 15–21

Elrick, M. and Read, J.F., 1991, Cyclic ramp-to-basin carbonate deposits, Lower Mississippian, Wyoming and Montana: a combined field and computer modelling study; Jour. Sed. Petr. 61, 1194–1224

Embry, A.F. and Klovan, J.E., 1971, A Late Devonian reef tract on northeastern Banks Island, Northwest Territories; Canadian Petroleum Geol. 19, 730–781

Emery, K.O., 1968, Positions of empty pelecypod valves on the continental shelf; Jour. Sed. Petr. 38, 1264–1269

Emmel, F.J. and Curray, J.R., 1985, Bengal Fan, Indian Ocean, pp. 107–112 in Bouma, A.H., Normark, W.R., and Barnes, N.E., eds., Submarine Fans and Related Turbidite Systems; Springer-Verlag, New York; 351 p.

Engel, A.E.J. and Engel, C.G., 1960, Progressive metamorphism of the Major paragneiss, northwest Adirondack Mountains, New York, Pt. 2, Mineralogy; GSA Bull. 71, 1–57

Engel, A.E.J., Engel, C.G., and Havens, R.G., 1961, Variations in properties of hornblende formed during progressive metamorphism of amphibolites, northwest Adirondack Mountains, New York; U.S. Geol. Surv. prof. paper 424–C, 313–316

Engel, A.E.J., Itson, S.P., Engel, C.G., Stickney, D.M., and Cray, E.J., Jr., 1974, Crustal evolution and global tectonics: a petrogenic view; GSA Bull. 85, 843–858

Engel, C.G. and Sharp, R.P., 1958, Chemical data on desert varnish; GSA Bull. 69, 487–518

Enos, P. and Sawatsky, L.H., 1981, Pore networks in Holocene carbonate sediments; Jour, Sed. Petr. 51, 961–985

Epstein, J.B. and Epstein, A.G., 1972, The Shawangunk Formation (Upper Ordovician? to Middle Silurian) in eastern Pennsylvania; U.S. Geol. Surv. prof. paper 744; 45 p.

Eriksson, K.A., 1977, Tidal flat and subtidal sedimentation in the 2250 m.y. Malmani Dolomite, Transvaal, South Africa; Sed. Geol. 18, 223–244

Eschner, T.B. and Kocurek, G., 1986, Marine destruction of eolian sand seas: origin of mass flows; Jour. Sed. Petr. 56, 401–411

Esteban, M. and Klappa, C.F., 1983, Subaerial exposure environment; pp. 2–54 in Scholle, P.A., Bebout, D.G., and Moore, C.H., eds., Carbonate Depositional Environments; AAPG Mem. 33; 708 p.

Ethridge, F.G. and Schumm, S.A., 1978, Reconstructing paleochannel morphologic and flow characteristics: methodology, limitations, and assessment; pp. 703–721 in Miall, A.D., ed., Fluvial Sedimentology; Canadian Soc. Petroleum Geologists Mem. 5; 859 p.

Eugster, H.P., 1967, Hydrous sodium silicates from Lake

Magadi, Kenya: precursors of bedded chert; Sci. 157, 1177–1180

Eugster, H.P., 1969, Inorganic bedded cherts from the Magadi area, Kenya; Contrib. Mineral. Petrol. 22, 1–31

Evamy, B.D., 1963, The application of chemical staining technique to a study of dedolomitization; Sedimentology 2, 164–170

Evamy, B.D., 1967, Dedolomitization and the development of rhombohedral pores in limestones; Jour. Sed. Petr. 37, 1204–1215

Evans, G., 1965, Intertidal flat sediments and their environments of deposition in the Wash; Geol. Soc. (London) Quart. Jour. 121, 209–245

Evans, W.D., 1964, The organic solubilization of minerals in sediments; pp. 263–270 in Colombo, U. and Hobson, G.D., eds., Advances in Organic Geochemistry; Macmillan, New York; 488 p.

Everts, A.J., 1991, Interpreting compositional variations of calciturbidites in relation to platform stratigraphy: an example from the Paleogene of SE Spain; Sed. Geol. 71, 231–244

Eynon, G. and Walker, R.G., 1974, Facies relationships in Pleistocene outwash gravels, southern Ontario; a model for bar growth in braided rivers; Sedimentology 21, 43–70

Feder, J., 1988, Fractals; Plenum, New York; 283 p.

Feo-Codecido, G., 1956, Heavy-mineral techniques and their application to Venezuelan stratigraphy; AAPG Bull. 40, 894–1000

Finkel, E.A. and Wilkinson, B.H., 1990, Stylolitization as source of cement in Mississippian Salem Limestone, west-central Indiana; AAPG Bull. 74, 174–186

Fischer, A.G., 1964, The Lofer cyclothems of the Alpine Triassic, pp. 107–141 in Merriam, D.F., ed., Symposium on Cyclic Sedimentation; Kansas Geol. Surv. Bull. 169, 636 p. (2 vols.)

Fischer, A.G., 1981, pp. 103–133 in Nitecki, ed., Biotic Crises in Ecological and Evolutionary Time; Academic Press, New York; 301 p.

Fischer, A.G. and Bottjer, D.J., 1991, Orbital forcing and sedimentary sequences; Jour. Sed. Petr. 61, 1069–1069

Fischer, A.G. and Garrison, R.E., 1967, Carbonate lithification on the sea floor; Jour. Geol. 75, 488–496

Fischer, A.G., Herbert, T., and Silva, I.P., 1985, Carbonate bedding cycles in Cretaceous pelagic and hemipelagic sequences; pp. 1–10 in Pratt, L.M., Kauffman, E.G., and Zelt, F.B., eds., Fine-grained Deposits and Biofacies of the Cretaceous Western Interior Seaway: Evidences of Cyclic Processes; 2nd ann. midyear mtg, Golden, Colorado, fld. trip 9; SEPM; 249 p.

Fisher, R.V., 1977, Erosion by volcanic base-surge density currents: U-shaped channels; GSA Bull. 88, 1287–1297

Fisher, R.V. and Waters, A.C., 1969, Bedforms in base-surge deposits: lunar implications; Sci. 165, 1349–1352

Fisher, W.L. and McGowan, J.H., 1969, Depositional systems in Wilcox Group (Eocene) of Texas and their relation to occurrence of oil and gas; AAPG Bull. 53, 30–54

Fisk, H.N., 1961, Bar-finger sands of Mississippi delta; pp. 29–52 in Peterson, J. A. and Osmond, J.C., eds., Geometry of Sandstone Bodies—a Symposium; AAPG; 240 p.

Fisk, H.N., McFarlan, E., Jr., Kolb, C.R., and Wilbert, R.J., Jr., 1954, Sedimentary framework of the modern Mississippi delta; Jour. Sed. Petr. 24, 76–99

Flemming, B.W., 1988, Zur Classifikation subaquatischer, stromungstrans versaler Transportkorper; Boch. Geol. Geothechn. Arb. 29, 44–47

Flugel, E., 1977, Fossil Algae, Recent Results and Developments; Springer-Verlag; 375 p.

Flugel, E., ed., 1982, Microfacies Analysis of Limestones; Springer-Verlag; 633 p.

Folk, R.L., 1951, Stages of textural maturity, Jour. Sed. Petr. 21, 127–130

Folk, R.L., 1954, The distinction between grain size and mineral composition in sedimentary-rock nomenclature; Jour. Geol. 62, 344–359

Folk, R.L., 1955, Student operator error in determination of roundness, sphericity, and grain size; Jour. Sed. Petr. 25, 297–301

Folk, R.L., 1959, Practical petrographic classification of limestones; AAPG Bull. 43, 1–38

Folk, R.L., 1960, Petrography and origin of Tuscarora, Rose Hill, and Keefer Formations, Lower and Middle Silurian of eastern West Virginia; Jour. Sed. Petr. 30, 1–58

Folk, R.L., 1962, Spectral subdivision of limestone types; pp. 62–84 in Ham, W.E., ed., Classification of Carbonate Rocks; AAPG Mem. 1; 279 p.

Folk, R.L., 1965, Some aspects of recrystallization in ancient limestones; pp. 14–48 in Pray, L.C. and Murray, R.C., eds., Dolomitization and Limestone Diagenesis, a Symposium; SEPM spec. publ. 13; 180 p.

Folk, R.L., 1974, The natural history of crystalline calcium carbonate: effect of magnesium content and salinity; Jour. Sed. Petr. 44, 40–53

Folk, R.L., 1980, Petrology of Sedimentary Rocks; Hemphill's, Austin, TX; 170 p.

Folk, R.L. and Assereto, R., 1974, Giant aragonite rays and baroque white dolomite in tepee-fillings, Triassic of Lombardy, Italy; SEPM ann. mtg. prog., San Antonio, 34–35

Folk, R.L., Chafetz, H.S., and Tiezzi, P.A., 1985, Bizarre forms of depositional and diagenetic calcite in hot-spring travertines; pp. 349–369 in Schneidermann, N. and Harris, P.M., eds., Carbonate Cements; SEPM spec. publ. 36; 379 p.

Folk, R.L. and Land, L.S., 1975, Mg/Ca ratio and salinity: two controls over crystallization of dolomite; AAPG Bull. 59, 60–68

Folk, R.L. and Pittman, J.S., 1971, Length-slow chalcedony: a new testament for vanished evaporites; Jour. Sed. Petr. 41, 1045–1058

Folk, R.L. and Robles, R., 1964, Carbonate sands of Isla Perez, Alacran reef complex, Yucatan; Jour. Geology 72, 255–292

Folk, R.L. and Ward, W.C., 1957, Brazos River bar: a study

in the significance of grain size parameters; Jour. Sed. Petr. 27, 3–26

Force, E.R., 1980, The provenance of rutile; Jour. Sed. Petr. 50, 485–488

Ford, D., 1988, Characteristics of dissolutional cave systems in carbonate rocks; pp. 25–57 in James, N.P. and Choquette, P.W., Paleokarst; Springer-Verlag, New York-Berlin-Heidelberg; 416 p.

Fouch, T.D. and Dean, W.E., 1982, Lacustrine and associated clastic depositional environments; pp. 87–114 in Scholle, P.A. and Spearing, D., eds., Sandstone Depositional Environments; AAPG Mem. 31; 410 p.

Fox, S.W. and Dose, K., 1972, Molecular Evolution and the Origin of Life; Freeman, San Francisco; 359 p.

Francois, R., 1987, The influence of humic substances on the geochemistry of iodine in nearshore and hemipelagic marine sediments; Geochim. et Cosmochim. Acta 51, 2417–2427

Frank, J.R., 1981, Dedolomitization in the Taum Sauk Limestone (Upper Cambrian), southeast Missouri; Jour. Sed. Petr. 51, 7–18

Frank, J.R., Carpenter, A.B., and Oglesby, T.W., 1982, Cathodoluminescence and composition of calcite cement in the Taum Sauk Limestone (Upper Cambrian), southeast Missouri; Jour. Sed. Petr. 52, 631–638

Franks, S.G. and Forester, R.W., 1984, Relationships among secondary porosity, pore-fluid chemistry and carbon dioxide, Texas Gulf coast; pp. 63–79 in McDonald, D.A. and Surdam, R.C., eds., Clastic Diagenesis; AAPG Mem. 37; 434 p.

Franzinelli, E. and Potter, P.E., 1983, Petrology, chemistry, and texture of modern river sands, Amazon River system; Jour. Geol. 91, 23–39

Frazier, D.E., 1974, Depositional episodes: their relationship to the Quaternary stratigraphic framework in the northwestern portion of the Gulf basin; Bur. Econ. Geol., Texas Univ. at Austin; Circ. 74–I; 28 p.

Freeman, T. and Rothbard, D., 1983, Terrigenous dolomite in the Miocene of Menorca (Spain): provenance and diagenesis; Jour. Sed. Petr. 53, 543–548

Frey, R.W., 1972, Paleoecology and Depositional Environment of Fort Hays Limestone Member, Niobrara Chalk (Upper Cretaceous), West-Central Texas; Paleont. Contr. Univ. Kansas 58; 72 p.

Friedman, G.M., 1961, Distinction between dune, beach, and river sands from textural characteristics; Jour. Sed. Petr. 31, 514–529

Friedman, G.M., 1965, Terminology of crystallization textures and fabrics in sedimentary rocks; Jour. Sed. Petr. 35, 643–655

Fritz, W.J. and Harrison, S., 1983, Giant armoured mud boulder from the 1982 Mount St. Helens mudflows; Jour. Sed. Petr. 53, 131–133

Fryberger, S.G., Al-Sari, A.M., and Clisham, T.J., 1983, Eolian dune, interdune, sand sheet, and siliciclastic sabkha environments of an offshore prograding sand sea, Dharam area, Suadi Arabia, AAPG Bull 67, 280–312

Fryberger, S.G., Schenk, C., and Krystinik, L.F., 1988, Stokes surfaces and the effects of near-surface groundwater-table on aeolian deposition; Sedimentology 35, 21–41

Fuchtbauer, H., 1978, Zur Herkunft des Quaerzzements, Abschatzung der Quartzauflosung in Silt- und Sandsteinen; Geol. Rundschau 67, 991–1008

Gaillard, C., Rio, M., Rolin, Y., and Roux, M., 1992, Fossil chemosynthetic communities related to vents or seeps in sedimentary basins: the pseudobioherms of southeastern France compared to other word examples; Palaios 7, 451–465

Gaither, A., 1953, A study of porosity and grain relationships in experimental sands; Jour. Sed. Petr. 23, 180–195

Gale, S.J., 1984, The hydraulics of conduit flow in carbonate aquifers; Jour. Hydrology 70, 309–324

Galloway, W.E., 1975, Process framework for describing the morphologic and stratigraphic evolution of deltaic depositional systems, pp. 87–98 in Broussard, M.L., ed., Deltas; Houston Geol. Soc.; 555 p.

Galloway, W.E., 1989, Genetic stratigraphic sequences in basin analysis 1: architechture and genesis of flooding-surface bounded depositional units; AAPG Bull. 73, 125–142

Galloway, W.E., and Hobday, D.K., 1983, Terrigenous Clastic Depositional Systems; Springer-Verlag, New York-Berlin-Heidelberg; 423 p.

Galloway, W.E., Hobday, D.K., and Magara, K., 1982, Frio Formation of Texas Gulf coastal plain: depositional systems, structural framework, and hydrocarbon distribution; AAPG Bull. 6, 649–688

Gao, G and Land, L.S., 1991, Nodular chert from the Arbuckle Group, Slick Hills, SW Oklahoma: a combined field, petrographic, and isotope study; Sedimentology 38, 857–870

Garrels, R.M., 1965, Silica: role in the buffering of natural waters; Sci. 148, 69

Garrels, R.M. and Christ, C.L., 1965, Solutions, Minerals, and Equilibria; Harper & Row, New York; 450 p.

Garrels, R.M. and Lerman, A., 1981, Phanerozoic cycles of sedimentary carbon and sulfur; Proc. Natl. Acad. Sci. 78, 4652–4656

Garrels, R.M. and Mackenzie, F.T., 1969, Sedimentary rock types: relative proportions as a function of geologic time; Sci. 163, 570–571

Garrels, R.M. and Mackenzie, F.T., 1971a, Evolution of Sedimentary Rocks; Norton, New York; 397 p.

Garrels, R.M. and Mackenzie, F.T., 1971b, Gregor's denudation of the continents; Nature 231, 382–383

Garrett, P., 1970, Phanerozoic stromatolites: non-competitive ecologic restriction by grazing and burrowing animals; Sci. 169, 171–173

Garrison, R.E. and Fischer, A.G., 1969, Deep-water limestones and radiolarites of the Alpine Jurassic; pp. 20–56 in Friedman, G.M., ed., Depositional Environments in Carbonate Rocks—a Symposium; SEPM spec. publ 14; 209 p.

Gaudin, A.M. and Meloy, T.P., 1962, Model of a comminution distribution equation for single fracture; Am. Inst. Mining Metallurg. Petrol. Eng.; Mining Trans. 223, 40–43

Gawthorpe, R.L., 1987, Burial dolomitization and porosity development in a mixed carbonate-clastic sequence: an example from the Bowland basin, northern England; Sedimentology 34, 533–558

Gaynor, G.C. and Swift, D.J.P., 1988, Shannon Sandstone depositional model: sand ridge dynamics on the Campanian western interior shelf; Jour. Sed. Petr. 58, 868–880

Gebelein, C.D., 1969, Distribution, morphology, and accretion rate of Recent subtidal algal stromatolites, Bermuda; Jour. Sed. Petr. 39, 49–69

Gebelein, C.D. and Hoffman, P., 1973, Algal origin of dolomite laminations in stromatolitic limestone; Jour. Sed. Petr. 43, 603–613

Gerhard, L.C., 1985, Porosity development in the Mississippian pisolitic limestone of the Mission Canyon Formation, Glenbern field, Williston basin, North Dakota; pp. 193–205 in Roehl, P.O. and Choquette, P.W., eds., Carbonate Petroleum Reservoirs; Springer-Verlag, New York-Berlin-Heidelberg; 622 p.

Gerhard, L.C., Anderson, S.B., and Berg, J., 1978, Mission Canyon porosity development, Glenburn field, North Dakota, Williston basin; 24th ann. Williston basin symp.; Montana Geol. Soc.; 177–188

Gilbert, R., 1990, Rafting in glacimarine environments; pp. 105–120 in Dowdeswell, J.A. and Scourse, J.D., eds., Glacimarine Environments:Processes and Sediments; Geol. Soc. London spec. publ. 53; 423 p.

Gill, D., 1985, Depositional facies of Middle Silurian (Niagaran) pinnacle reefs, Belle River Mills gas field, Michigan basin, southeastern Michigan; pp. 123–139 in Roehl, P.O. and Choquette, P.W., eds., Carbonate Petroleum Reservoirs; Springer-Verlag, New York-Berlin-Heidelberg-Tokyo; 622 p.

Gilligan, A., 1920, The petrography of the Millstone Grit of Yorkshire; Geol. Soc. (London) Quart. Jour. 75, 251–291

Gilluly, J., Reed, J.C., and Cady, W.M., 1970, Sedimentary volumes and their significance; GSA Bull. 81, 353–376

Given, R.K. and Wilkinson, B.H., 1985, Kinetic control of morphology, composition, and mineralogy of abiotic sedimentary carbonates; Jour. Sed. Petr. 55, p.109–119

Glaister, R.P. and Nelson, H.W., 1974, Grain-size distributions, an aid in facies identification; Bull. Canadian Petroleum Geol. 22, 203–240

Glasmann, J.R., Larter, S., Breidis, N.A., and Lundegard, P.D., 1989, Shale diagenesis in the Bergen High area, North Sea; Clays and Clay Minerals. 37, 97–112

Glennie, K.W., 1970, Desert Sedimentary Environments; Developments in Sedimentology 14; Elsevier; 222 p.

Gloppen, T.G. and Steel, R.J., 1981, The deposits, internal structure and geometry in six alluvial fan-fan delta bodies (Devonian Norway)—a study in the significance of braiding sequence in conglomerates, pp. 49–69 in Ethridge, F.G. and Flores, R.M., eds., Recent and Ancient Nonmarine Depositional Environments: Models for Exploration; SEPM spec. publ. 31; 349 p.

Goldberg, M., 1967, Supratidal dolomitization and dedolomitization in Jurassic rocks of Hamakhtesh Haqatan, Israel; Jour. Sed. Petr. 37, 760–773

Goldhammer, R.K. and Elmore, R.D., 1984, Paleosols capping regressive carbonate cycles in the Pennsylvanian Black Prince Limestone, Arizona; Jour. Sed. Petr. 54, 1124–1137

Goldstein, R.H., 1988, Cement stratigraphy of Pennsylvanian Holder Formation, Sacramento Mountains, Mexico; AAPG Bull. 72, 425–438

Golubic, S., Campbell, S.E., Drobne, K., Cameron, B., Balsam, W.L., Cimeran, F., and Dubois, L., 1984, Microbial endoliths: a benthic overprint in the sedimentary record and a paleobathymetric cross-reference with foraminifera; Jour. Paleont. 58, 351–361

Golubic, S., Perkins, R.D., and Lukas, K.J., 1975, Boring microorganisms and microborings in carbonate substrates; pp. 229–259 in Frey, R.W., ed., The Study of Trace Fossils; Springer-Verlag, New York-Heidelberg-Berlin; 562 p.

Goodchild, M.F. and Ford, D.C., 1971, Analysis of scallop patterns by simulation under controlled conditions; Jour. Geol. 79, 52–62

Goodwin, P.W. and Anderson, E.J., 1985, Punctuated aggradational cycles: a general hypothesis of episodic stratigraphic accumulation; Jour. Geol. 93, 515–533

Gornitz, V.M. and Schreiber, B.C., 1981, Displacive halite hoppers from the Dead Sea: some implications for ancient evaporite deposits; Jour. Sed. Petr. 51, 787–794

Goudie, A.S. and Sperling, C.H.B., 1977, Long distance transport of foraminiferal tests by wind in the Thar Desert, northwest India; Jour. Sed. Petr. 47, 630–633

Gow, A.J., Ueda, H.T., and Garfield, D.E., 1968, Antarctic ice sheet: preliminary results of first core hole to bedrock; Sci. 161, 1011–1013

Gradzinski, R. and Jerzykiewicz, T., 1974, Sedimentation of the Barun Goyot Formation; Palaeont. Polonica 30, 111–146

Graf, W.H., 1971, Hydraulics of Sediment Transport; McGraw-Hill, New York; 513 p.

Graton, L.C. and Fraser, H.J., 1935, Systematic packing of spheres, with particular relation to porosity and permeability; Jour. Geology 43, 785–909

Greenwood, B., 1969, Sediment parameters and environmental discrimination: an application of multivariate statistics; Canadian Jour. Earth Sci. 6, 1347–1358

Greenwood, B. and Mittler, P.R., 1984, Sediment flux and equilibrium slopes in a barred nearshore; Mar. Geol. 60, 79–98

Gregg, J.M. and Shelton, K.L., 1990, Dolomitization and dolomite neomorphism in the back reef facies of the

Bonneterre and Davis Formations (Cambrian), southeastern Missouri; Jour. Sed. Petr. 60, 549–562

Gregor, C.B., 1967, The geochemical behavior of sodium, with special reference to post-Alkonkian sedimentation; Trans. Roy. Netherlands Acad. Sci., 1st ser., 24, no. 2; 67 p.

Gregor, C.B., 1968, The rate of denudation in post-Algonkian time; Proc. Koninklijke Nederlandse Acad. Weterschappen 71, 22–30

Grim, R.E., 1968, Clay Mineralogy; 2nd ed.; McGraw-Hill, New York; 596 p.

Grinell, R.S., Jr., 1974, Vertical orientation of shells on some Florida oyster reefs; Jour. Sed. Petr. 44, 116–122

Grover, G., Jr. and Read, J.F., 1978, Fenestral and associated vadose diagenetic fabrics of tidal flat carbonates, Middle Ordovician Market Limestone, southwestern Virginia; Jour. Sed. Petr. 48, 453–473

Grover, G., Jr. and Read, J.F., 1983, Paleoaquifer and deep burial related cements defined by regional cathodoluminiscent patterns, Middle Ordovician carbonates, Virginia; AAPG Bull. 67, 1275–1303

Guidish, T.M., Kendall, C.G.St.C., Lerche, I., Toth, D.J., and Yarzab, R.F., 1985, Basin evaluation using burial history calculations: an overview; AAPG Bull. 69, 92–105

Gustavson, T.C., Ashley, G.M., and Boothroyd, J.C., 1975, Depositional sequences in glaciolacustrine deltas; pp. 264–280 in Jopling, A.V. and McDonald, B.C., eds., Glaciofluvial and Glaciolacustrine Sedimentation; SEPM spec. publ. 23; 320 p.

Gutschick, R.C. and Sandberg, C.A., 1983, Mississippian continental margins of the conterminous United States; pp. 79–96 in Stanley, D.J. and Moore, G.T., eds., The Shelfbreak: Critical Interface on Continental Margins; SEPM spec. publ. 33; 467 p.

Gutschick, R.C., Suttner, L.J., and Switeck, M.J., 1962, Biostratigraphy of transitional Devonian-Mississippian Sappington Formation of southwest Montana; Billings Geol. Soc. Guidebook, 13th ann. field conf., 79–89

Gutschick, R.C. and Wuellner, D., 1983, An unusual agglutinated foraminiferan from Late Devonian anoxic basinal black shales of Ohio; Jour. Paleont. 57, 308–320

Guy, H.P., Simons, D.B., and Richardson, E.V., 1966, Summary of alluvial channel data from flume experiments, 1956–1961; U.S. Geol. Surv. prof. paper 462–I; 96 p.

Hald, A., 1952, Statistical Theory, with Engineering Applications; Wiley, New York; 783 p.

Hallam, A. and Bradshaw, M.J., 1979, Bituminous shales and oolitic ironstones as indicators of transgressions and regressions; Jour. Geol. Soc. (London) 136, 157–164

Halley, R.B., 1977, Ooid fabric and fracture in the Great Salt Lake and the geologic record; Jour. Sed. Petr. 47, 1099–1120

Hallock, P. and Schlager, W., 1986, Nutrient excess and the demise of coral reefs and carbonate platforms; Palaios 1, 389–398

Hamblin, A.P. and Walker, R.G., 1979, Storm-generated shallow marine deposits: the Fernie-Kootenay (Jurassic) transition, southern Rocky Mountains; Canadian Jour. Earth Sci. 16, 1673–1690

Hampton, M.A., 1972, The role of subaqueous debris flows in generating turbidity currents; Jour. Sed. Petr. 42, 775–793

Hancock, J.M. and Kauffman, E.G., 1979, The great transgressions of the Late Cretaceous; Jour. Geol. Soc. (London) 136, 175–186

Hand, B.M., 1974, Supercritical flow in density currents; Jour. Sed. Petr. 44, 637–648

Hand, B.M., 1975, Supercritical flow in density currents: reply; Jour. Sed. Petr. 45, 750–753

Hand, B.M., Wessel, J.M., and Hayes, M.O., 1969, Antidunes in the Mount Toby Conglomerate (Triassic), Massachusettes; Jour. Sed. Petr. 39, 1310–1316

Handford, C.R., 1982, Sedimentology and evaporite genesis in a Holocene continental-sabkha playa basin—Bristol Lake, California; Sedimentology 29, 239–253

Hanshaw, B.B., Black, W., and Deike, R.G., 1971, A geochemical hypothesis for dolomitization by groundwater; Econ. Geol. 66, 710–724

Hantzchel, W. and Reineck, H.-E., 1968, Faziesuntersuchungen im Hettangium von Helmstedt (Niedersachsen); Mitt. Geol. Staatsinst. (Hamburg) 37, 5–39

Harbaugh, J.W. and Bonham-Carter, G., 1970, Computer Simulation in Geology; Wiley-Interscience, New York; 575 p.

Hardie, L.A., 1968, The origin of the Recent non-marine evaporite deposit of Saline Valley, Inyo County, California; Geochim. Cosmochim. Acta 32, 1279–1301

Hardie, L.A., 1984, Evaporites: marine or non-marine?; Am. Jour. Sci. 284, 193–240

Hardie, L.A. and Eugster, H.P., 1971, The depositional environment of marine evaporites: a case for shallow, clastic accumulation; Sedimentology 16, 187–220

Hardie, L.A., Smoot, J.P., and Eugster, H.P., 1978, Saline lakes and their deposits: a sedimentological approach; in Matter, A. and Tucker, M.E., eds., Modern and Ancient Lake Sediments; Internat. Assoc. Sed. spec. publ. 2; Blackwell, Oxford; 290 p.

Harding, J.P., 1949, The use of probablitiy paper for the graphical analysis of polymodal frequency distributions; Jour. Mar. Biological Assoc. (U.K.) 28, 141–153

Harland, W.B, Cox, A.V., Llewellyn, P.G., Pickton, C.A.G., Smith, A.G., and Walters, R., 1982, A Geologic Time Scale; Cambridge Univ. Press, Cambridge; 131 p.

Harms, J.C., 1969, Hydraulic significance of some sand ripples; GSA Bull. 80, 363–396

Harms, J.C. and Fahnestock, R.K., 1965, Stratification, bedforms, and flow phenomena (with an example from the Rio Grande), pp. 84–115 in Middleton, G.v., ed., Primary Sedimentary Structures and their Hydrodynamic Interpretation—a Symposium; SEPM spec. publ. 12; 265 p.

Harms, J.C., Southard, J.B., Spearing, D.R., and Walker,

R.G., 1975, Depositional Environments as Interpreted from Primary Sedimentary Structures and Stratification Sequences; SEPM short course No. 2; 161 p.

Harms, J.C., Southard, J.B., and Walker, R.G., 1982, Structures and Sequences in Clastic Rocks; SEPM short course no. 9; 249 p.

Harris, P.T., 1988, Large-scale bedforms as indicators of mutually evasive sand transport and the infilling of wide-mouth estuaries; Sed. Geol. 57, 273–298

Harrison, S.C., 1975, Tidal-flat complex, Delmarva Peninsula, Virginia; pp. 31–38 in Ginsburg, R.N., ed., Tidal Deposits; a Casebook of Recent Examples and Fossil Counterparts; Springer-Verlag, New York; 428 p.

Harvie, C.E., Weare, J.H., Hardie, L.A., and Eugster, H.P., 1980, Evaporation of sewater: calculated mineral sequences; Sci. 208, 498–500

Hattin, D.E., 1971, Widespread, synchronously deposited, burrow-mottled limestone beds in Greenhorn Limestone (Upper Cretaceous) of Kansas and southeastern Colorado; AAPG 55, 412–431

Hattin, D.E., 1975, Petrology and origin of fecal pellets in Upper Cretaceous strata of Kansas and Saskatchewan; Jour. Sed. Petr. 45, 686–696

Hattin, D.E., 1981, Petrology of Smoky Hill Member, Niobrara Chalk (Upper Cretaceous), in type area, western Kansas; AAPG Bull. 65, 831–849

Hattin, D.E., 1982, Stratigraphy and depositional environment of Smoky Hill Chalk Member, Niobrara Chalk (Upper Cretaceous) of the type area, western Kansas; Kansas Geol. Soc. Bull. 225; 108 p.

Hayes, M.O., 1980, General morphology and sediment patterns in tidal inlets; Sed. Geol. 26, 139–156

Hays, J.D., 1971, Faunal extinctions and reversals of the earth's magnetic field; GSA Bull. 82, 2433–2447

Heald, M.T., 1955, Stylolites in sandstones; Jour. Geol. 63, 101–114

Heald, M.T. and Larese, R.E., 1973, The significance of the solution of feldspar in porosity development; Jour. Sed. Petr. 43, 458–460

Heald, M.T. and Renton, J.J., 1966, Experimental study of sandstone cementation; Jour. Sed. Petr. 36, 977–991

Hecht, F., 1933, Der Verlieb der organischen Substanz der Tierebei Meerischen Einbettung; Senckenb. Naturforsch. Ges. 15, 165–249

Heckel, P.H., 1974, Carbonate buildups in the geologic record: a review; in Laporte, L.F., ed., Reefs in Time and Space; Selected Examples from the Recent and Ancient; SEPM spec. publ. 18; 256 p.

Hedberg, H.D., ed., 1976, International Stratigraphic Guide; Wiley, New York; 200 p.

Heezen, B.C. and Hollister, C.D., 1971, The Face of the Deep; Oxford Univ. Press, New York; 659 p.

Heezen, B.C., Hollister, C.D., and Ruddiman, W.F., 1966, Shaping of the continental rise by deep geostrophic contour currents; Sci. 152, 502–508

Heim, A., 1924, Uber submarine Denudation und chemische Sedimente; Geol. Rundschau 15, 1–47

Hein, F.J., 1984, Deep-sea and fluvial braided channel conglomerates: a comparison of two case studies, p. 33–49 in Koster, E.H. and Steel, R.J., eds., Sedimentology of Gravels and Conglomerates; Canadian Soc. Petroleum Geologists Mem. 10, Calgary; 441 p.

Hein, F.J. and Walker, R.G., 1982, The Cambro-Ordovician Cap Enrage Formation, Quebec, Canada: conglomeratic deposits of a braided channel with terraces; Sedimentology 29, 309–329

Heling, D., 1970, Micro-fabrics of shales and their rearrangement by compaction; Sedimentology 15, 247–260

Heller, P.L. and Frost, C.D., 1988, Isotopic provenance of clastic deposits: application of geochemistry to sedimentary provenance studies; pp. 27–42 in Kleinspehn, K.L. and Paola, C., eds., New Perspectives in Basin Analysis; Springer-Verlag, New York-Berlin-Heidelberg; 453 p.

Heller, P.L., Komar, P.D., and Pevear, D.R., 1980, Transport processes in ooid genesis; Jour. Sed. Petr. 50, 943–952

Heller, P.L., Peterman, Z.E., O'Neil, J.R., and Shafiquillah, M., 1985, Isotopic provenance of sandstones from the Eocene Tyee Formation, Oregon Coast Range; GSA Bull. 96, 770–780

Henrich, R. and Zankl, H., 1986, Diagenesis of Upper Triassic Wetterstein reefs of the Bavarian Alps; pp. 245–268 in Schroeder, J.H. and Purser, B.H., Reef Diagenesis; Springer-Verlag, Berlin-Heidelberg-New York; 455 p.

Hesse, R., 1986, "Drainage systems" associated with midocean channels and submarine yazoos: alternative to submarine fan depositional systems; Geology 17, 1148–1151

Hesse, R., 1987, Selective and reversible carbonate-silic replacements in Lower Cretaceous carbonate-bearing turbidites of the eastern Alps; Sedimentology 34, 1055–1077

Hesse, R. and Butt, A., 1976, Paleobathymetry of Cretaceous turbidite basins of the east Alps relative to the calcite compensation level; Jour. Geol. 34, 505–533

Hildebrand, A.R. and Boynton, W.V., 1988, Impact wave deposits provide new constraints on the location of the K/T boundary impact; pp. 76–77 in Global Catastrophies in Earth History: An Interdisciplinary Conference on Impacts, Volcanism, and Mass Mortality; Lunar and Planetary Institute contrib. 673; 226 p.

Hildebrand, A.R. and Boynton, W.V., 1990, Proximal Cretaceous-Tertiary boundary impact deposits in the Caribbean; Sci. 248, 843–847

Hill, C.A., 1990, Sulfuric acid speleogenesis of Carlsbad Cavern and its relationship to hydrocarbons, Delaware basin, Mexico and Texas; AAPG Bull. 72, 1685–1694

Hiscott, R.N. and James, N.P., 1985, Carbonate debris flows, Cow Head Group, western Newfoundland; Jour. Sed. Petr. 55, 735–745

Hoffman, P., 1969; Algal stromatolites—use in stratigraphic correlation and paleocurrent determination; Sci. 157, 1043–1045

Hoffman, P., 1973, Evolution on an early Proterozoic continental margin; Phil. Trans. Roy Soc. (London), Ser. A, 273, 547–581

Hoffman, P., 1974, Shallow and deepwater stromatolites in lower Proterozoic platform-to-basin facies change, Great Slave Lake, Canada; AAPG Bull. 58, 856–867

Hoffman, P., 1975, Shoaling-upward shale-to-dolomite cycles in the Rocknest Formation (lower Paleozoic), Northwest Territories, Canada; pp. 257–265 in Ginsburg, R.N., ed., Tidal Deposits; a Casebook of Recent Examples and Fossil Counterparts; Springer-Verlag, New York-Heidelberg-Berlin; 428 p.

Hoffman, P., 1976, Stromatolite morphogenesis in Shark Bay, Western Australia; pp. 261–271 in Walter, M.R., ed., Stromatolites; Elsevier, New York; 790 p.

Hoffman, P., Dewey, J.F., and Burke, K., 1974, Aulacogens and their genetic relation to geosynclines with a Proterozoic example from Great Slave Lake, Canada; pp. 38–55 in Dott, R.H., Jr. and Shaver, R.H., eds., Modern and Ancient Geosynclinal Sedimentation; SEPM spec. publ. 19; 380 p.

Holland, C.H., et al., 1978, A Guide to Stratigraphical Procedure; Geol. Soc. (London) spec. rept. 10; 18 p.

Holser, W.T., 1977, Catastrophic chemical events in the history of the ocean; Nature, 267, 403–408

Holser, W.T. and Kaplan, I.R., 1966, Isotope geochemistry of sedimentary sulphates; Chem. Geol. 1, 93–135

Holser, W.T., Schidlowski, M., Mackenzie, F.T., and Maynard, J.B., 1988, Biogeochemical cycles of carbon and sulfur; pp. 105–174 in Gregor, C.B., Garrels, R.M., Mackenzie, F.T., and Maynard, J.B., eds., Chemical Cycles in the Evolution of the Earth; Wiley, New York; 276 p.

Hooke, R. LeB., 1967, Processes on arid-region alluvial fans; Jour. Geol. 75, 438–460

Horne, J.C., Ferm, J.C., Caruccio, F.T., and Baganz, B.P., 1978, Depositional models in coal exploration and mine planning in the Appalachian region; AAPG Bull. 62, 2379–2411

Horowitz, A.S. and Potter, P.E., 1971, Introductory Petrography of Fossils; Springer-Verlag; 302 p.

Horowitz, D.H., 1982, Geometry and origin of large-scale deformation structures in some ancient wind-blown sand deposits; Sedimentology 29, 155–180

Houbolt, J.J.H.C., 1968, Recent sediments in the southern bight of the North Sea; Geol. Mijnb. 47, 245–273

Houseknecht, D.W., 1987, Assessing the relative importance of compaction processes and cementation to reduction of porosity in sandstones; AAPG Bull. 71, 633–642

Houseknecht, D.W., 1988, Intergranular pressure solution in four quartzose sandstones; Jour. Sed. Petr. 58, 228–246

Hovorka, S., 1992, Halite pseudomorphs after gypsum in bedded anhydrite—clue to gypsum-anhydrite relationships; Jour. Sed. Petr. 62, 1098–1111

Hoyt, J.H. and Henry, V.J., Jr., 1964, Development and geologic significance of soft beach sand; Sedimentology 3, 44–51

Hsu, K.J., 1967, Chemistry of dolomite formation; pp. 169–191 in Chilingar, G.V., Bissell, H.J., and Fairbridge, R.W., eds., Carbonate Rocks—Physical and Chemical Aspects; Developments in Sedimentology 9; Elsevier, Amsterdam-New York; 413 p.

Hsu, K.J., 1972, When the Mediterranean dried up; Scientific Am. 227 (Dec.), 26–36

Hsu, K.J., Kelts, K., and Valentine, J.W., 1980, Resedimented facies in Ventura basin, California, and model of longitudinal transport of turbidity currents; AAPG Bull. 64, 1034–1051

Hsu, K.J., Montadert, L., Bernoulli, D., Cita, M.B., Erickson, A., Garrison, R.E., Kidd, R.B., Melieres, F., Muller, C., and Wright, R., 1977, History of the Mediterranean salinity crisis; Nature 267, 399–403

Hsu, K.J. and Siegenthaler, C., 1969, Preliminary experiments on hydrodynamic movement induced by evaporation and their bearing on the dolomite problem; Sedimentology 12, 11–25

Huang, W.H. and Keller, W.D., 1970, Dissolution of rock-forming silicate minerals in organic acids: simulated first-stage weathering of fresh mineral surfaces; Am. Mineralogist 55, 2076–2094

Hubert, J.F., 1962, A zircon-tourmaline-rutile maturity index and the interdependence of the composition of heavy mineral assemblages with the gross composition and texture of sandstones; Jour. Sed. Petr. 32, 440–450

Huffman, G.G. and Price, W.A., 1949, Clay dune formation near Corpus Christi, Texas; Jour. Sed. Petr. 19, 118–127

Hunt, C.B., 1954, Pleistocene and Recent deposits in Denver area, Colorado; U.S. Geol. Surv. Bull. 966–C, 91–140

Hunt, C.B., 1972, Geology of Soils; Freeman; 344 p.

Hunt, J.M., 1979, Petroleum Geochemistry and Geology; Freeman, San Francisco; 617 p.

Hunter, R.E., 1969, Eolian microridges on modern beaches and a possible ancient example; Jour. Sed. Petr. 39, 1573–1578

Hunter, R.E., 1977a, Basic types of stratification in small eolian dunes; Sedimentology 24, 362–387

Hunter, R.E., 1977b, Terminology of cross-stratified sedimentary layers and climbing-ripple structures; Jour. Sed. Petr. 47, 697–706

Hunter, R.E., 1985, Subaqueous sand-flow cross-strata; Jour. Sed. Petr. 55, 886–894

Hunter, R.E., Clifton, H.E., and Phillips, R.L., 1979, Depositional processes, sedimentary structures, and predicted vertical sequences in barred nearshore systems, southern Oregon coast; Jour. Sed. Petr. 49, 711–726

Hunter, R.E. and Richmond, B.M., 1988, Daily cycles in coastal dunes; Sed. Geol. 55, 43–67

Huntley, D.A. and Bowen, A.J., 1973, Field observations of edge waves; Nature 243, 160–162

Hurst, H.E., Black, R.P., and Simaika, Y.M., 1965, Long-Term Storage: An Experimental Study; Constable, London; 145 p.

Ibbeken, H., 1983, Jointed source rock and fluvial gravels

controlled by Rosin's law: a grain-size study in Calabria, south Italy; Jour. Sed. Petr. 53, 1213–1231

Iijima, A., Matsumoto, R., and Tada, R., 1985, Mechanism of sedimentation of rhythmically bedded chert; Sed. Geol. 41, 221–233

Illich, H.A., Hall, F.W., and Alt, D., 1972, Ice-cemented sand blocks in the Pilcher Quartzite, western Montana; Jour. Sed. Petr. 42, 927–929

Ingersol, R.V., 1988, Tectonics of sedimentary basins; GSA Bull. 100, 1704–1719

Ingram, R.L., 1954, Terminology for the thickness of stratification and parting units in sedimentary rocks; GSA Bull. 65, 937–938

Isaacs, C.M., Pisciotto, K.A., and Garrison, R.E., 1983, Facies and diagenesis of the Miocene Monterey Formation, California: a summary; pp. 247–282 in Iijima, A., Hein, J.R., and Siever, R., eds., Siliceous Deposits in the Pacific Region; Developments in Sedimentology 36; Elsevier, Amsterdam; 479 p.

Jackson, R.G., 1976, Sedimentological and fluid dynamic implications of the turbulent bursting phenomenon in geophysical flows; Jour. Fluid Mech. 77, 531–560

James, H.L., 1992, Precambrian iron formations, nature, origin, and mineralogic evolution from sedimentation to metamorphism; pp. 543–589 in Wolf, K.H. and Chilingarian, G.V., eds., Diagenesis III, Developments in Sedimentology 47; Elsevier, Amsterdam; 674 p.

James, N.P., 1983. Reef environment; pp. 346–440 in Scholle, P.A., Bebout, D.G., and Moore, C.H., eds., Carbonate Depositional Environments; AAPG Mem. 33, 708 p.

James, N.P. and Bone, Y., 1989, Petrogenesis of Cenozoic, temperate water calcarenites, South Australia: a model for meteoric/shallow burial diagenesis of shallow water calcite sediments; Jour. Sed. Petr. 59, 191–203

James, N.P. and Klappa, C.F., 1983, Petrogenesis of Early Cambrian reef limestones, Labrador, Canada; Jour. Sed. Petr. 53, 1051–1096

James, W.C., Mack, G.H., and Suttner, L.J., 1981, Relative alteration of microcline and sodic plagioclase in semiarid and humid climates; Jour. Sed. Petr. 51, 151–164

Jarvis, I., 1980, Geochemistry of phosphatic chalks and hardgrounds from the Santonian to early Campanian (Cretaceous) of northern France; Jour. Geol. Soc. London 137, 705–721

Jarvis, I., 1992, Sedimentology, geochemistry, and origin of phosphatic chalks: the Upper Cretaceous deposits of NW Europe; Sedimentology 39, 55–97

Jenkyns, H.C., 1967, Fossil manganese nodules from Sicily; Nature 216, 673–674

Jenkyns, H.C., 1980, Cretaceous anoxic events: from continents to oceans; Jour. Geol. Soc. (London) 137, 171–188

Jenkyns, H.C., 1986, Pelagic environments; pp. 343–397 in Reading, G., ed., Sedimentary Environments and Facies; 2nd ed., Blackwell, Oxford; 615 p.

Jenkyns, H.C. and Winterer, E.L., 1982, Paleoceanography of Mesozoic ribbon radiolarites; Earth Planet. Sci. Letters 60, 351–375

Johns, D.R., 1978, Mesozoic carbonate rudites, mega-breccias and associated deposits from central Greece; Sedimentology 25, 561–573

Johnson, J.H., 1961, Limestone-Building Algae and Algal Limestones; Colo. School Mines, Boulder, CO; 295 p.

Johnsson, M.J., 1990, Tectonic vs chemical-weathering controls on the composition of fluvial sands in tropical environments; Sedimentology 37, 715–726

Johnsson, M.J., Stallard, R.F., and Meade, R.H., 1988, First-cycle quartz arenites in the Orinoco River basin, Venezuela and Colombia; Jour. Geol. 96, 263–277

Jones, D.J., 1953, Tetrahedroid pebbles; Jour. Sedimentary Petrology 23, 196–201

Jones, J.B. and Segnit, E.R., 1971, The nature of opal. I. Nomenclature and constituent phases; Jour. Geol. Soc. Australia 18, 57–68

Jones, M.L. and Dennison, J.M, 1970, Oriented fossils as paleocurrent indicators in Paleozoic lutites of southern Appalachians; Jour. Sed. Petr. 40, 642–649

Jones, R.L. and Blatt, H., 1984, Mineral dispersal patterns in the Pierre Shale; Jour. Sed. Petr. 54, 17–28

Jopling, A.V., 1965, Laboratory study of the distribution of grain sizes in cross-bedded deposits; pp. 53–65 in Middleton, G.V., ed, Primary Sedimentary Structures and their Hydrodynamic Interpretation—a Symposium; SEPM spec. publ. 12; 265 p.

Jopling, A.V., 1966, Some principles and techniques used in reconstructing the hydraulic parameters of a paleoflow regime; Jour. Sed. Petr. 36, 5–49

Jowett, E.C., Rydzewski, A., and Jowett, R.J., 1987, The Kupferschiefer Cu-Ag ore deposits in Poland: a reappraisal of the evidence of their origin and presentation of a new genetic model; Can. Jour. Earth Sci. 24, 2016–2036

Kah, L.C. and Grotzinger, J.P., 1992, Early Proterozoic (1.9 Ga) thrombolites of the Rocknest Formation, Northwest Territories, Canada; Palaios 7, 305–315

Kahle, C.F., 1974, Ooids from Great Salt Lake, Utah, as an analog for the genesis and diagenesis of ooids in marine limestones; Jour. Sed. Petr. 44, 30–39

Kahn, J.S., 1956, The analysis and distribution of the properties of packing in sand-size sediments; Jour. Geol. 64, 385–395

Kaldi, J., 1980, The origin of nodular structures in the Lower Magnesian Limestone (Permian) of Yorkshire, England; pp. 45–60 in Fuchtbauer, H. and Peryt, T., eds., The Zechstein Basin with Emphasis on Carbonate Sequences; Contributions to Sedimentology 9; Schweizerbart'sche Verlagsbuchhandlung, Stuttgart; 328 p.

Kalinske, A.A., 1942, Criteria for determining sand-transport by surface-creep and saltation; Trans. Am. Geophys. Union 23, Pt. 2, 639–643

Kastens, K.A. and Cita, M.B., 1981, Tsunami-induced sediment transport in the abyssal Mediterranean Sea; GSA Bull. 92, 845–857

Kastner, M., 1971, Authigenic feldspars in carbonate rocks; Am. Mineralogist 56, 1403–1442

Kastner, M., Keene, J.B., and Gieskes, J., 1977, Diagenesis of siliceous oozes. I. Chemical controls on the rate of opal-A to opal-CT transformation—an experimental study; Geochim. Cosmochim. Acta 41, 1041–1059

Katz, A., 1968, Calcian dolomites and dedolomitization; Nature 217, 439–440

Katz, A., 1971, Zoned dolomite crystals; Jour. Geol. 79, 38–51

Kauffman, E.J., 1970, Population systematics, radiometrics and zonation; a new biostratigraphy; pp. 612–666 in North American Paleontological Convention Proc., pt. F

Keene, J.B., 1983, Chalcedonic quartz and occurrence of quartzine (length-slow chalcedony) in pelagic sediments; Sedimentology 30, 449–454

Kellerhals, R., Shaw, J., and Arora, V.K., 1975, On grain size from thin sections; Jour. Geol. 83, 79–96

Kendall, A.C., 1977, Fascicular-optic calcite: a replacement of bundled acicular carbonate cements; Jour. Sed. Petr. 47, 1056–1062

Kendall, A.C., 1985, Radiaxial fibrous calcite: a reappraisal; pp. 59–77 in Schneidermann, N. and Harris, P.M., eds., Carbonate Cements; SEPM spec. publ. 36; 379 p.

Kendall, A.C. and Tucker, M.E., 1973, Radiaxial-fibrous calcite: a replacement after acicular carbonate; Sedimentology 20, 365–389

Kennard, J.M. and James, N.P., 1986, Thrombolites and stromatolites: two distinct types of microbial structures; Palaios 1, 492–503

Kennedy, J.F., 1963, The mechanics of dunes and antidunes in erodible-bed channels; Jour. Fluid Mech. 16, 521–544

Kennedy, W.J. and Garrison, R.E., 1975, Morphology and genesis of nodular chalks and hardgrounds in the Upper Cretaceous of southern England; Sedimentology 22, 311–386

Kenyon, N.H., 1970, Sand ribbons of European tidal seas; Mar. Geol. 9, 25–39

Kikoine, J. and Radier, H., 1949, Quartzites d'alteration au Soudan oriental; Soc. Geol. France, Compte Rendu 14–15, 339–341

Kikoine, J. and Radier, H., 1950, Silicifications au Soudan oriental; le calcaire a silex du Tertiaire Continental (T.C.); Soc. Geol. France, Compte Rendu 9–10, 168–170

King, D.T., Jr., 1986, Waulsortian-type buildups and re-sedimented (carbonate-turbidite) facies, Early Mississippian Burlington shelf, central Missouri; Jour. Sed. Petr. 56, 471–479

Kinsman, D.J.J., 1966, Gypsum and anhydrite of Recent age, Trucial Coast, Persian Gulf; pp. 302–326 in Raup, J.L., ed., Symposium on Salt, 2nd., Northern Ohio Geol. Soc., 530 p.

Kinsman, D.J.J., 1969, Modes of formation, sedimentary associations, and diagnostic features of shallow-water and supratidal evaporites; AAPG Bull., v. 53, 830–840

Kirkland, D.W. and Evans, R., 1981, Source-rock potential of evaporitic environment; AAPG 65, 181–190

Klappa, C.F., 1980, Rhizoliths in terrestrial carbonates: classification, recognition, genesis and significance: Sedimentology 27, 613–629

Klein, G. deV., 1971, A sedimentary model for determining paleotidal range; GSA Bull. 82, 2585–2592

Klein, G. deV., 1975; Tidalites in the Eureka Quartzite (Ordovician), eastern California and Nevada, pp. 145–151 in Ginsburg, R.N., ed., Tidal Deposits; a Casebook of Recent Examples and Fossil Counterparts; Springer-Verlag, New York-Heidelberg-Berlin; 428 p.

Klein, G. deV. and Ryer, T.A., 1978, Tidal circulation patterns in Precambrian, Paleozoic, and Cretaceous epeiric and mioclinal shelf seas; GSA Bull. 89, 1050–1058

Klinkhammer, G.P. and Bender, M.L., 1980, The distribution of manganese in the Pacific Ocean; Earth Planet. Sci. Lett. 46, 361–384

Knauth, L.P., 1979, A model for the origin of chert in limestone; Geology 7, 274–277

Kobluk, D.R. and Risk, M.J., 1977, Calcification of exposed filaments of endolithic algae, micrite envelope formation and sediment production; Jour. Sed. Petr. 47, 517–528

Koch, R. and Schorr, M., 1986, Diagenesis of Upper Jurassic sponge-algal reefs in SW Germany; pp. 224–244 in Schroeder, J.H. and Purser, B.H., Reef Diagenesis; Springer-Verlag, Berlin-Heidelberg-New York; 455 p.

Kocurek, G., 1981, Significance of interdune deposits and bounding surfaces in aeolian dune sands; Sedimentology 28, 753–780

Kocurek, G. and Dott, R.H., Jr., 1981, Distinctions and uses of stratification types in the interpretation of eolian sand; Jour. Sed. Petr. 51, 579–595

Kocurek, G. and Fielder, G., 1982, Adhesion structures; Jour. Sed. Petr. 52, 1229–1241

Komar, P.D., 1971, The mechanics of sand transport on beaches; Jour. Geophys. Res. 76, 713–721

Komar, P.D., 1974, Oscillatory ripple marks and the evaluation of ancient wave conditions and environments; Jour. Sed. Petr. 44, 169–180

Komar, P.D., 1975, Supercritical flow in density currents: a discussion; Jour. Sed. Petr. 45, 747–749

Komar, P.D., 1976, Nearshore currents and sediment transport, and the resulting beach configuration; pp. 241–254 in Stanley, D.J. and Swift, D.J.P., eds., Marine Sediment Transport and Environmental Management; Wiley, New York; 602 p.

Komar, P.D. and Inman, D.L., 1970, Longshore sand transport on beaches; Jour. Geophys. Res. 75, 5914–5927

Komar, P.D. and Miller, M.C., 1973, The threshold of sediment movement under oscillatory water waves; Jour. Sed. Petr. 43, 1101–1110

Komar, P.D. and Miller, M.C., 1975, On the comparison between the threshold of sediment motion under waves and unidirectional currents with a discussion of the practical evaluation of the threshold; Jour. Sed. Petr. 51, 362–367

Komar, P.D. and Reimers, C.E., 1978, Grain shape effects on settling rates; Jour. Geol. 86, 193–209

Kraft, J.C., 1971, Sedimentary facies patterns and geologic history of a Holocene marine transgression; GSA Bull. 82, 2131–2158

Kraft, J.C., Chrzastowski, M.J., Belknap, D.F., Toscano, M.A., and Fletcher, C.H., III, 1987, The transgressive barrier-lagoon coast of Delaware: morphostratigraphy, sedimentary sequences and responses to relative rise in sea level; pp. 129–143 in Nummedal, D., Pilkey, O.H., and Howard, J.D., eds., Sea-Level Fluctuation and Coastal Evolution; SEPM spec. publ. 41; 267 p.

Kramer, J.R., 1965, History of sea water, constant temperature-pressure equilibrium models compared to liquid inclusion analyses; Geochim. Cosmochim. Acta 29, 921–945

Kraus, M.J., 1988, Nodular remains of early Tertiary forests, Bighorn basin, Wyoming; Jour. Sed. Petr. 58, 888–893

Krauskopf, K.B., 1979, Introduction to Geochemistry; 2nd ed., McGraw-Hill, New York; 617 p.

Kreisa, R.D. and Moiola, R.J., 1986, Sigmoidal tidal bundles and other tide-generated sedimentary structures of the Curtis Formation, Utah; GSA Bull. 97, 381–387

Krinsley, D.H. and Doornkamp, J.C., 1973, Atlas of Quartz Sand Surface Textures; Cambridge Univ. Press, Cambridge; 91 p.

Krinsley, D.H. and Smalley, I.J., 1973, Shape and nature of small sedimentary quartz particles; Sci. 180, 1177–1179

Krinsley, D.H. and Trusty, P., 1986, Sand grain surface textures; pp. 201–207 in Sieveking, G. deG. and Hart, M.B., eds., The Scientific Study of Flint and Chert, Cambridge Univ. Press; Cambridge; 290 p.

Kruger, J., 1979, Structures and textures in till indicating subglacial deposition; Boreas 8, 323–340

Krumbein, W.C., 1934, Size frequency distributions of sediments; Jour. Sed. Petr. 4, 65–77

Krumbein, W.C. and Monk, G.D., 1942, Permeability as a function of the size parameters of unconsolidated sands; Am. Inst. Mining Metallurg Engineers tech. publ. 1492; 11 p.

Krumbein, W.C. and Pettijohn, F.J., 1938, Manual of Sedimentary Petrography; Appleton-Century-Crofts, New York; 549 p.

Krynine, P.D., 1940, Petrology and genesis of the Third Bradford sand; Penn. State College Mineral Industry Expt. Sta. Bull. 29; 134 p.

Krynine, P.D., 1941, Graywackes and the petrology of Bradford oil field, Pennsylvania; AAPG Bull. 25, 2071–2074

Krynine, P.D., 1942, Differential sedimentation and its products during one complete geosynclinal cycle; Santiago, Pan Am. Ing. Minas Geol. Ann. Cong., 536–561

Krynine, P.D., 1946, The tourmaline group in sediments; Jour. Geol. 54, 65–87

Kumar, N. and Sanders, J.E., 1974, Inlet sequence: a vertical succession of sedimentary structures and textures created by the lateral migration of tidal inlets; Sedimentology 21, 491–532

Kumar, N. and Sanders, J.E., 1976, Characteristics of shoreface storm deposits; Jour. Sed. Petr. 46, 145–162

Kushnir, J., 1981, Formation and early diagenesis of varved evaporite sediments in a coastal hypersaline pool; Jour. Sed. Petr. 51, 1193–1203

Kvale, E.P., Archer, A.W., and Johnson, H.R., 1989, Daily, monthly, and yearly tidal cycles within laminated siltstones in the Mansfield Formation (Pennsylvanian) of Indiana; Geol. 17, 365–368

Kyle, J.R., 1983, Economic aspects of subaerial carbonates; pp. 73–92 in Scholle, P.A., Bebout, D.G., and Moore, C.L., eds., Carbonate Depositional Environments; AAPG Mem. 33; 708 p.

Lamb, H., 1932, Hydrodynamics; 6th ed.; Cambridge Univ. Press, Cambridge; (reprinted 1945 by Dover, New York; 738 p.)

Laming, D.J.C., 1966, Imbrication, paleocurrents and other sedimentary features in the lower Red Sandstone, Devonshire, England; Jour. Sed. Petr. 36, 940–959

Lancaster, N., 1988, Controls of eolian dune size and spacing; Geol. 16, 972–975

Land, L.S., 1984, Frio Sandstone diagenesis, Texas Gulf coast: a regional isotopic study; p. 47–62 in McDonald, D.A. and Surdam, R.C., eds., Clastic Diagenesis; AAPG Mem. 37; 434 p.

Land, L.S., 1986, Environments of limestone and dolomite diagenesis: some geochemical considerations; pp. 26–41 in Bathurst, R.G.C. and Land, L., eds., Carbonate Depositional Environments Modern and Ancient Part 5: Diagenesis; Colo. Sch. Mines Quart. 81; 41 p.

Land, L.S., Behrens, E.W., and Frishman, S.A., 1979, The ooids of Baffin Bay, Texas; Jour. Sed. Petr. 49, 1269–1278

Land, L.S. and Dutton, S.P., 1978, Cementation of a Pennsylvanian deltaic sandstone: isotope data; Jour. Sed. Petr. 48, 1167–1176

Land, L.S. and Goreau, T.F., 1970, Submarine lithification of Jamaican reefs; Jour. Sed. Petr. 40, 457–462

Land, L.S. and Moore, C.H., 1980, Lithification, micritization and syndepositional diagenesis of biolithites on the Jamaican island slope; Jour. Sed. Petr. 50, 357–370

Landim, P.M.B. and Frakes, L.A., 1968, Distinction between tills and other diamictons based on textural characteristics; Jour. Sed. Petr. 38, 1213–1223

Landing, W.M. and Bruland, K.W., 1980, Manganese in the North Pacific; Earth Planet. Sci. Let. 49, 45–56

Lane, E.W. and Kalinske, A.A., 1939, The relation of suspended to bed material in rivers; Trans. Am. Geophys. Union 20, 637–641

Langford, R. and Bracken, B., 1987, Medano Creek, Colorado, a model for upper-flow-regime fluvial deposition; Jour. Sed. Petr. 57, 863–870

Laporte, L.F., 1967, Carbonate deposition near mean sea level and resultant facies mosaic; Manlius Formation (Lower Devonian) of New York State; AAPG Bull. 51, 73–101

Lecompte, M., 1937, Contribution a la connaissance des recifs de l'Ardennes; sur la presence de structures conservees dans des efflorescences cristallines du type "stromatactis"; Musee Royal Histoire Naturelle Belgique Bull. 13, 1–14

Lecompte, M., 1970, Die Riffe im Devon der Ardeene und irhe Bildungs Bedingungen; Geol. Palaeont. 4, 25–71

Leeder, M.R., 1973, Fluviatile fining-upward cycles and the magnitude of paleochannels; Ged. Mag. 110, 265–276

Lees, A., 1961, The Waulsortian "reefs" of Eire: a carbonate mudbank complex of Lower Carboniferous age; Jour. Geol. 69, 101–109

Lees, A., 1964, The structure and origin of the Waulsortian (Lower Carboniferous) "reefs" of west-central Eire; Phil. Trans. Roy. Soc. (London), ser. B, 247, 483–531

Leggett, J.K., 1980, British lower Paleozoic black shales and their palaeo-oceanographic significance; Jour. Geol. Soc. (London) 137, 139–156

Leighton, W.M. and Pendexter, C., 1962, Carbonate rock types; pp. 33–61 in Ham, W.E., ed., Classification of Carbonate Rocks; AAPG Mem. 1; 279 p.

Leliavski, S., 1955, An Introduction to Fluvial Hydraulics; Constable; London (reprinted 1959 by Dover, New York; 257 p.)

Le Mehaute, B., 1976, An Introduction to Hydromechanics and Water Waves; Springer-Verlag; 315 p.

Leopold, L.B., Wolman, M.G., and Miller, J.P., 1964, Fluvial Processes in Geomorphology; Freeman, San Francisco-New York; 522 p.

Leroy, S.D., 1981, Grain size and moment measures: a new look at Karl Pearson's ideas on distribution; Jour. Sed. Petr. 51, 625–630

Lindsay, R.F. and Kendall, C.G.St.C., 1985, Depositional facies and reservoir character of Mississippian cyclic carbonates in the Mission Canyon Formation, Little Knife field, Williston basin, North Dakota; pp. 177–190 in Roehl, P.O. and Choquette, P.W., eds., Carbonate Petroleum Reservoirs; Springer-Verlag, New York; 622 p.

Lindsay, R.F. and Roth, M.S., 1982, Carbonate and evaporite facies, dolomitization and reservoir distribution of the Mission Canyon Formation, Little Knife field, North Dakota; 4th Internat. Williston Basin Symposium; 153–179

Lineback, J.A., 1966, Deep-water sediments adjacent to the Borden Siltstone (Mississippian) delta in southern Illinois; Ill. Geol. Surv. circ. 401; 48 p.

Lineback, J.A., 1971, Pebble orientation and ice movement in south-central Illinois, pp. 328–344 in Goldthwait, R.P., ed., Till: a symposium; Ohio State Univ. Press, Columbus, OH; 402 p.

Little, R.D., 1982, Lithified armored mud balls of the Lower Jurassic Turner Falls Sandstone, north-central Massachusetts; Jour. Geol. 90, 203–207

Livingstone, D.A., 1963, The sodium cycle and the age of the ocean; Geochim. Cosmochim. Acta 27, 1055–1069

Loeblich, A.R. and Tappan, H., 1961, The genera Microaulopora Kuntz, 1895, and Guembelina Kuntz, 1895, and the status of Guembilina Egger, 1899; Jour. Paleont. 35, 625–627

Logan, B.W., Rezak, R., and Ginsburg, R.N., 1964, Classification and environmental significance of algal stromatolites; Jour. Geol. 72, 68–83

Lohmann, K.C. and Meyers, W.J., 1977, Microdolomite inclusions in cloudy prismatic calcites: a proposed criterion for former high magnesium calcites; Jour. Sed. Petr. 47, 1078–1088

Longiaru, S., 1987, Visual comparators for estimating the degree of sorting from plane and thin section; Jour. Sed. Petr. 57, 791–794

Longman, M.W., 1980, Carbonate diagenetic textures from near surface diagenetic environments; AAPG Bull. 64, 461–487

Longuet-Higgins, M.S., 1970, Longshore currents generated by obliquely incident sea waves; Jour. Geophys. Res. 75, 6778–6801

Lonsdale, P. and Malfait, B.T., 1974, Abyssal dunes of foraminiferal sands on the Carnegie Ridge; GSA Bull. 85, 1697–1712

Lonsdale, P., Malfait, B.T., and Spiess, F.N., 1972, Abyssal sand waves on the Carnegie Ridge; Geol. 4, 579–580

Lonsdale, P. and Smith, S.M., 1980, "Lower insular rise hills" shaped by a bottom boundary current in the mid-Pacific; Mar. Geol. 34, M19–M25

Loreau, J.P. and Purser, B.H., 1973, Distribution and ultrastructure of Holocene ooids in the Persian Gulf; pp. 279–328 in Purser, B.H., ed., The Persian Gulf. Holocene Carbonate Sedimentation in a Shallow Epicontinental Sea; Springer-Verlag, New York-Heidelberg-Berlin; 471 p.

Loughman, D.L., 1984, Phosphate authigenesis in the Aramachay Formation (Lower Jurassic) of Peru; Jour. Sed. Petr. 54, 1147–1156

Louks, R.G., Dodge, M.M., and Galloway, W.E., 1984; Regional controls on diagenesis and reservoir quality in lower Tertiary sandstones along the Texas Gulf coast; pp. 15–45 in McDonald, D.A. and Surdam, R.C., eds., Clastic Diagenesis; AAPG Mem. 37; 434 p.

Love, L.G., 1971, Early diagenetic polyframboidal pyrite, primary and redeposited, from the Wenlockian Dengihn Grit Group, Conway, North Wales, U.K., Jour. Sed. Petr. 41, 1038–1044

Lovering, T.S., 1959, Significance of accumulator plants in rock weathering; GSA Bull. 70, 781–800

Lowe, D.R., 1976a; Grain flow and grain flow deposits; Jour. Sed. Petr. 46, 188–199

Lowe, D.R., 1976b, Subaqueous liquefied and fluidized sediment flows and their deposits; Sedimentology 23, 285–308

Lowe, D.R., 1979, Sediment gravity flows: their classification and some problems of application to natural flows and deposits; pp. 75–82 in Doyle, L.J. and Pilkey, O.H., eds., Geology of Continental Slopes; SEPM spec. publ. 27, 374 p.

Lowe, D.R., 1982, Sediment gravity flows: II Depositional models with special reference to the deposits of high-density turbidity currents· Jour. Sed. Petr. 52, 279–297

Lowe, D.R. and LoPiccolo, R.D., 1974, The characteristics and origins of dish and pillar structures; Jour. Sed. Petr. 44, 484–501

Lowenstam, H.A., 1950, Niagaran reefs of the Great Lakes area; Jour. Geol. 58, 430–487

Lucia, F.J., 1962, Diagenesis of a crinoidal sediment; Jour. Sed. Petr. 32, 848–865

Lumsden, D.N., 1971, Markov chain analysis of carbonate rocks: applications, limitations, and implications as exemplified by the Pennsylvanian System in southern Nevada; GSA Bull. 82, 447–462

Lundqvist, J., 1969, Problems of the so-called Rogen moraine; Sveriges Geol. Undersokning; ser. C 648; Arsbok 64, 1–32

MacDonald, D.I.M., 1986, Proximal to distal sedimentological variation in a linear turbidite trough: implications for the fan model; Sedimentology 33, 243–259

Machel, H.-G., 1986, Early lithification, dolomitization, and anhydritization of Upper Devonian Nisku buildups, subsurface of Alberta, Canada; pp. 336–356 in Schroeder, J.H. and Purser, B.H., Reef Diagenesis; Springer-Verlag, Berlin-Heidelberg-New York; 455 p.

Machel, H.-G. and Anderson, J.H., 1989, Pervasive subsurface dolomitization of the Nisku Formation in central Alberta; Jour. Sed. Petr. 59, 891–911

Mackenzie, F.T. and Garrels, R.M., 1965, Silicates: reactivity with sea water; Sci. 150, 57–58

Mackenzie, F.T. and Pigott, J.D., 1981, Tectonic controls of Phanerozoic sedimentary rock cycling; Jour. Geol. Soc. (London) 138, 183–196

MacQuown, W.C. and Perkins, J.H., 1982, Stratigraphy and petrology of petroleum-producing Waulsortian-type carbonate mounds in Fort Payne Formation (Lower Mississippian) of north-central Tennessee; AAPG Bull. 66, 1055–1075

Madsen, O.S. and Grant, W.D., 1975, The threshold of sediment movement under oscillatory waves: a discussion; Jour. Sed. Petr. 45, 360–361

Magara, K., 1980, Comparison of porosity-depth relationships of shale and sandstone; Jour. Petroleum Geol. 3, 175–185

Maiklem, W.R., 1968, Some hydraulic properties of bioclastic carbonate grains; Sedimentology 10, 101–109

Maiklem, W.R., Bebout, D.G., and Glaister, R.P, 1969, Classification of anhydrite—a practical approach; Bull. Canadian Petroleum Geol. 17, 194–233

Malde, H.E., 1968, The catastrophic late Pleistocene Bonneville flood in the Snake River plain, Idaho; U.S. Geol. Surv. prof. paper 596; 52 p.

Maliva, R.G., 1987, Quartz geodes: early diagenetic silicified anhydrite nodules related to dolomitization; Jour. Sed. Petr. 57, 1054–1059

Maliva, R.G., Knoll, A.H., and Siever, R., 1989, Secular change in chert distribution: a reflection of evolving biological participation in the silica cycle; Palaios 4, 519–532

Maliva, R.G. and Siever, R., 1988, Diagenetic replacement controlled by force of crystallization; Geol. 16, 688–691

Mandelbrot, B.B., 1982, The Fractal Geometry of Nature; Freeman, San Francisco; 460 p.

Mandelbrot, B.B. and Wallis, J.R., 1969, Varve thicknesses in Timiskaming, Canada; some long-run properties of geophysical records; Water Resources Res. 5, 321–340

Manten, A.A., 1966, Some current trends in palynology; Earth Sci. Rev. 2, 317–343

Manus, R.W. and Coogan, A.H., 1974, Bulk volume reduction and pressure-solution derived cement; Jour. Sed. Petr. 44, 466–471

Margolis, S. and Rex, R., 1971, Endolithic algae and micrite envelope formation in Bahamian oolites as revealed by scanning electron microscopy; GSA Bull. 82, 843–852

Markert, J.C. and Al-Shaieb, Z., 1984, Diagenesis and evolution of secondary porosity in upper Minnelusa sandstones, Powder River basin, Wyoming; pp. 367–389 in McDonald, D.A. and Surdam, R.C., eds., Clastic Diagenesis; AAPG Mem. 37; 434 p.

Marshall, J.F., 1983, Submarine cementation in a high-energy platform reef: One Tree reef, southern Great Barrier reef, Jour. Sed. Petr. 53, 1133–1149

Marshall, J.F., 1986, Regional distribution of submarine cements within an epicontinental reef system: central Great Barrier Reef, Australia; pp. 8–26 in Schroeder, J.M. and Purser, B.H., eds., Reef Diagenesis; Springer-Verlag, Berlin-Heidelberg-New York; 455 p.

Martens, C.S. and Harriss, R.C., 1970, Inhibition of apatite precipitation in the marine environment by magnesium ions; Geochim. Cosmochim. Acta 34, 621–625

Mason, C.C. and Folk, R.L., 1958, Differentiation of beach, dune, and eolian flat environments by size analysis, Mustang Island, Texas; Jour. Sed. Petr. 28, 211–226

Mast, R.F. and Potter, P.E., 1963, Sedimentary structures, sand shape fabrics, and permeability II; Jour. Geol. 71, 548–565

Mattes, B.W. and Mountjoy, E.W., 1980, Burial dolomitization of the Upper Devonian Miette buildup, Jasper National Park, Alberta; pp. 259–297 in Zenger, D.H., Dunham, J.B., and Ethington, R.L., eds., Concepts and Models of Dolomitization; SEPM spec. publ. 28; 320 p.

Mattson, P.H. and Alvarez, W., 1973, Base surge deposits in Pleistocene volcanic ash near Rome; Bull. Volcanol. 37, 553–572

Maxwell, D.T. and Hower, J., 1967; High-grade diagenesis

and low-grade metamorphism of illite in the Precambrian Belt Series; Am. Mineralogist 52, 843–857

Maynard, J.B., 1983, Geochemistry of Sedimentary Ore Deposits; Springer-Verlag, New York; 305 p.

Maynard, J.B., 1991, Iron: syngenetic deposition controlled by the evolving ocean-atmosphere system; Chapter 10 in Force, E.R., Eidel, J.J., and Maynard, J.B., eds., Sedimentary and Diagenetic Mineral Deposits: A Basin Analysis Approach to Exploration; Rev. Econ. Geol. 5.

Mazzullo, S.J., 1978, Early Ordovician tidal flat sedimentation, western margin of proto-Atlantic Ocean; Jour. Sed. Petr. 48, 49–62

Mazzullo, S.J., 1980, Calcite pseudospar replacive of marine acicular aragonite and implications for aragonite cement diagenesis; Jour. Sed. Petr. 50, 409–422

Mazzullo, S.J. and Harris, P.M., 1992, Mesogenetic dissolution: its role in porosity development in carbonate reservoirs; AAPG Bull. 76, 607–620

McArthur, J.M., Coleman, M.L., and Bremner, J.M., 1980, Carbon and oxygen isotopic composition of structural carbonate in sedimentary francolite; Jour. Geol. Soc. London 137, 669–673

McBride, E.F., 1962, Flysch and associated beds of the Martinsburg Formation (Ordovician), central Appalachians; Jour. Sed. Petr. 32, 39–91

McBride, E.F., 1988, Contrasting diagenetic histories of concretions and host rock, Lion Mountain Sandstone (Cambrian), Texas; GSA Bull. 100, 1803–1810

McBride, E.F. and Thomson, A., 1970, The Caballos Novaculite, Marathon Region, Texas; GSA spec. paper 122; 129 p.

McCave, I.N., 1971, Sand waves in the North Sea off the coast of Holland; Mar. Geol. 10, 199–225

McCrossan, R.G., 1958, Sedimentary "boudinage" structures in the Upper Devonian Ireton Formation of Alberta; Jour. Sed. Petr. 28, 316–320

McCubbin, D.G., 1982, Barrier island and strand plain facies; pp. 247–279 in Scholle, P.A. and Spearing, D., eds., Sandstone Depositional Environments; AAPG Mem. 31; 410 p.

McGowan, J.H. and Garner, L.E., 1970, Physiographic features and stratification types of coarse-grained point bars: modern and ancient examples; Sedimentology 14, 77–111

McGowan, J.H. and Groat, C.G., 1971, Van Horn Sandstone, West Texas; an alluvial fan model for mineral exploration; Texas Univ. at Austin; Bur. Econ. Geol. Rept. Inv. 72; 57 p.

McHargue, T.R. and Price, R.C., 1982, Dolomite from clay in argillaceous shale- associated marine carbonates; Jour. Sed. Petr. 52, 873–886

McKee, E.D., 1933, The Coconino Sandstone—its history and origin; Carnegie Inst. publ. 440, 77–115

McKee, E.D., 1945, Small-scale structures in the Coconino Sandstone of Northern Arizona; Jour. Geol. 53, 313–325

McKee, E.D., 1966, Structures of dunes at White Sands National Monument, Mexico (and a comparison with structures of dunes from other selected areas); Sedimentology 7, 1–69

McKee, E.D. and Bigarella, J.J., 1979a, Ancient sandstones considered to be aeolian; pp. 187–238 in McKee, E.D., ed., A Study of Global Sand Seas; U.S. Geol. Surv. prof. paper 1052; 429 p.

McKee, E.D. and Bigarella, J.J., 1979b, Sedimentary structures in dunes; pp. 83–134 in McKee, E.D., ed., A study of global sand seas; U.S. Geol. Surv. prof. paper 1052; 429 p.

McKee, E.D. and Gutschick, R.C., 1969, History of the Redwall Limestone of northern Arizona; GSA Mem. 114, 726 p.

McKee, E.D. and Wier, G.W., 1953, Terminology for stratification and cross-stratification in sedimentary rocks; GSA Bull. 64, 381–389

McKelvey, V.E., Williams, J.S., Sheldon, R.P., Cressman, E.R., Cheney, T.M., and Swanson, R.W., 1959, The Phosphoria, Park City, and Shedhorn Formations in the western phosphate field, U.S. Geol. Surv. prof. paper 313–A, 47 p.

McLane, M.J., 1972, Sandstone: secular trends in lithology in southwestern Montana; Sci. 179, 502–504

McLane, M.J., 1982, Upper Cretaceous coastal deposits in south-central Colorado—Codell and Juana Lopez Members of Carlile Shale; AAPG Bull. 66, 71–90

McLane, M.J., 1983, Codell and Juana Lopez in south-central Colorado; pp. 49–66 in Merewether, E.A., ed., Mid-Cretaceous Codell Sandstone Member of Carlile Shale, eastern Colorado; SEPM Fld. Trip Gdbk.; 100 p.

McLaren, D.J. and Goodfellow, W.D., 1990, Geological and biological consequences of giant impacts; Ann. Rev. Earth Planet. Sci. 18, 123–171

McLaren, P., 1981, An interpretation of trends in grain-size measures; Jour. Sed. Petr. 51, 616–624

McLennan, S.M., 1982, On the geochemical evolution of sedimentary rocks; Chem. Geol. 37, 335–350

McPherson, J.G., Shanmugam, G., and Moiola, R.J., 1987, Fan deltas and braid deltas: varieties of coarse-grained deltas; GSA Bull. 99, 331–340

Meade, R.H., 1966, Factors influencing the early stages of the compaction of clays and sands—review; Jour. Sed. Petr. 36, 1085–1101

Meckel, L.D., 1967, Origin of Pottsville conglomerates (Pennsylvanian) in the central Appalachians; GSA Bull. 78, 223–258

Meissner, F.F., 1972, Cyclic sedimentation in middle Permian strata of the Permian basin, West Texas and Mexico; pp. 203–232 in Elam, J.G. and Chuber, S. eds., Cyclic Sedimentation in the Permian Basin; 2nd ed., West Texas Geol. Soc.; Midland; 232 p.

Meyers, J.H., 1987, Marine vadose beachrock cementation by cryptocrystalline magnesian calcite—Maui, Hawaii; Jour. Sed. Petr. 57, 558–570

Meyers, W.J., 1974, Carbonate cement stratigraphy of the

Lake Valley Formation (Mississippian), Sacramento Mountains, Mexico; Jour. Sed. Petr. 44, 837–861

Meyers, W.J., 1977, Chertification of the Mississippian Lake Valley Formation, Sacramento Mountains, New Mexico; Sedimentology 24, 75–105

Meyers, W.J., 1978, Carbonate cements: their regional distribution and interpretation in Mississippian limestones of southwestern Mexico; Sedimentology 25, 371–400

Meyers, W.J., 1988, Paleokarstic features in Mississippian limestones, Mexico; pp. 306–328 in James, N.P. and Choquette, P.W., eds., Paleokarst; Springer-Verlag, New York-Berlin-Heidelberg; 416 p.

Meyers, W.J. and Lohmann, K.C., 1978, Microdolomite-rich syntaxial cements: proposed meteoric-marine mixing zone phreatic cements from Mississippian limestones, Mexico; Jour. Sed. Petr. 48, 475–488

Miall, A.D., 1974, Paleocurrent analysis of alluvial sediments: a discussion of directional variance and vector magnitude; Jour. Sed. Petr. 44, 1174–1185

Miall, A.D., 1986, Eustatic sea level changes interpreted from seismic stratigraphy: a critique of the methodology with particular reference to the North Sea Jurassic record; AAPG Bull. 70, 131–137

Miall, A.D., 1988, Facies architecture in clastic sedimentary basins; pp. 67–81 in Kleispehn, K.L. and Paolo, C., eds, New Perspectives in Basin Analysis; Springer-Verlag, New York-Berlin-Heidelberg; 453 p.

Middleton, G.V. and Hampton, M.A., 1976, Subaqueous sediment transport and deposition by sediment gravity flows; pp. 197–218 in Stanley, D.J. and Swift, D.J.P., Marine Sediment Transport and Environmental Management; Wiley, New York; 602 p.

Migniot, C., 1968, Etude des proprietes physiques de differents sediments tres fin et leur comportement sous des actions hydrodynamiques; La Houille Blanche 23, 591–620

Milankovitch, M., 1941, Kanon de Erdbestrahlung und seine Anwendung auf das Eiszeitproblem; Acad. Roy. Serbe, 133; 633 p.

Miller, J., 1986, Facies relationships and diagenesis in Waulsortian mudmounds from the Lower Carboniferous of Ireland and N. England; pp. 311–335 in Schroeder, J.H. and Purser, B.H., Reef Diagenesis; Springer-Verlag, Berlin-Heidelberg-New York; 455 p.

Miller, M.A., McCave, I.N., and Komar, P.D., 1977, Threshold of sediment motion under unidirectional currents; Sedimentology 24, 507–528

Miller, M.C. and Komar, P.D., 1980, Oscillation sand ripples generated by laboratory apparatus; Jour. Sed. Petr. 50, 173–182

Milliken, K.L., 1979, The silicified evaporite syndrome—two aspects of silicification history of former evaporite nodules from southern Kentucky and northern Tennessee; Jour. Sed. Petr. 49, 245–256

Millot, G., 1970, Geology of Clays; Springer-Verlag, New York-Heidelberg-Berlin; 429 p.

Mitchum, R.M., Jr., Vail, P.R., and Sangree, J.B., 1977, Seismic stratigraphy and global changes of sea level, Pt. 6: Stratigraphic interpretation of seismic reflection patterns in depositional sequences; pp. 117–133 in Payton, C.E., ed., Seismic Stratigraphy—Applications to Hydrocarbon Exploration; AAPG Mem. 26; 516 p.

Mitra, S. and Beard, W.C., 1980, Theoretical models of porosity reduction by pressure solution for well-sorted sandstones; Jour. Sed. Petr. 50, 1347–1360

Mizutani, S., 1970, Silica minerals in the early stage of diagenesis; Sedimentology 15, 419–436

Mizutani, S., 1977, Progressive ordering of cristobalitic silica in the eraly stage of diagenesis; Contrib. Miner. Petr. 61, 129–140

Moiola, R.J. and Weiser, D., 1968, Textural parameters: an evaluation; Jour. Sed. Petr. 38, 45–53

Moore, C.H., 1989, Carbonate Diagenesis and Porosity; Developments in Sedimentology 46; Elsevier, Amsterdam; 338 p.

Moore, C.H. and Druckman, Y., 1981, Burial diagenesis and porosity evolution, Upper Jurassic Smackover, Arkansas and Louisiana; AAPG Bull. 65, 597–628

Moore, D.M. and Reynolds, R.C., Jr., 1989, X-ray Diffraction and the Identification and Analysis of Clay Minerals; Oxford Univ Press, New York; 332 p.

Moore, G.W. and Moore, J.G., 1988, Large-scale bedforms in boulder gravel produced by giant waves in Hawaii; pp. 101–109 in Clifton, H.E., ed., Sedimentologic Consequences of Convulsive Geologic Events; GSA spec. paper 229; 157 p.

Moore, J.G., 1967, Base surge in recent volcanic eruptions; Bull. Volcanol. 30, 337–363

Morgan, J.P., Coleman, J.M., and Gagliano, S.M., 1968, Mud lumps: diapiric structures in Mississippi delta sediments; pp. 145–161 in Braustein, J. and O'Brien, G.D., eds., Diapirism and Diapirs, a Symposium; AAPG Mem. 8; 444 p.

Morrow, D.W., 1982, Descriptive field classification of sedimentary and diagenetic breccia fabrics in carbonate rocks; Bull. Canadian Petroleum Geol. 30, 227–229

Morrow, D.W. and Ricketts, B.D., 1988, Experimental investigation of sulphate inhibition of dolomite and its mineral analogues; pp. 25–38 in Shukla, V. and Baker, P.A., eds., Sedimentology and Geochemistry of Dolostones; SEPM spec. publ. 43; 266 p.

Mosier, S.O., 1989, Microporosity in micritic limestones: a review; Sed. Geol. 63, 191–213

Moss, A.J., 1966, Origin, shaping and significance of quartz sand grains; Geol. Soc. Australia Jour. 13, 97–136

Moss, A.J., Walker, P.H., and Hutka, J., 1973, Fragmentation of granitic quartz in water; Sedimentology 20, 489–511

Mount, J.F. and Ward, P., 1986, Origin of limestone/marl alternations in the upper Maastrichtian of Zumaya, Spain; Jour. Sed. Petr. 228–236

Mudge, M.R. and Sheppard, R.A., 1968, Provenance of igneous rocks in Cretaceous conglomerates in northwestern Montana; U.S. Geol. Surv. prof. paper 600–D, 137–146

Muir, M., Lock, D., and von der Borch, C., 1980, The Coorong model for penecontemporaneous dolomite formation in the middle Proterozoic McArthur Group, Northern Territory, Australia; pp. 51–67 *in* Zenger, D.H., Dunham, J.B., and Ethington, R.L., eds., Concepts and Models of Dolomitization; Soc. Econ. Mineralogists and Paleontologists spec. publ. 28; 320 p.

Muller, R.A. and Morris, D.C., 1986, Geomagnetic reversals from impacts on the earth; Geophys. Res. Letters 13, 1177–1180

Muller-Feuga, R., 1952, Contribution a l'etude de la geologie, de la petrographie et des ressources hydrauliques et minerales du Fezzan; These Sci. Nancy et Mem. 12 des Ann. Min. et Geol. Tunisie (1954)

Mullins, H.T. and Cook, H.E., 1986, Carbonate apron models: alternatives to the submarine fan model for paleoenvironmental analysis and hydrocarbon exploration; Sed. Geol. 48, 37–79

Mullins, H.T., Neumann, A.C., Wilber, R.J., and Boardman, M.R., 1980, Nodular carbonate sediment in Bahamian slopes: possible precursors to nodular limestones; Jour. Sed. Petr. 50, 117–131

Munk, W.H. and Traylor, M.A., 1947, Refraction of ocean waves: a process linking underwater topography to beach erosion; Jour. Geol. 55, p. 1–34

Mussman, W.J., Montanez, I.P, and Read, J.F., 1988, Ordovician Knox paleokarst unconformity, Appalachians; pp. 211–228 *in* James, N.P. and Choquette, P.W., Paleokarst; Springer-Verlag, New York-Berlin-Heidelberg; 416 p.

Mutti, E., 1974, Examples of ancient deep-sea fan deposits from circum-Mediterranean geosynclines; pp. 92–105 *in* Dott, R.H., Jr. and Shaver, R.H., eds., Modern and Ancient Geosynclinal Sedimentation; SEPM spec. publ. 19; 380 p.

NACSN, 1983; North American stratigraphic code; Am. Assoc. Petroleum Geologists Bull. 67, p. 841–875

Naeser, N.D., Naeser, C.W., and McCulloh, T.H., 1989, The application of fission-track dating to the depositional and thermal history of rocks in sedimentary basins; pp. 157–180 *in* Naeser, N.D. and McCulloh, T.H., eds., Thermal History of Sedimentary Basins, Methods and Case Histories; Springer-Verlag, New York; 319 p.

Nagle, J.S., 1967, Wave and current orientation of shells; Jour. Sed. Petr. 37, 1124–1138

Nahon, D.B., 1991, Introduction to the Petrology of Soils and Chemical Weathering; Wiley, New York; 313 p.

Nandi, K., 1967, Garnets as indicators of progressive regional metamorphism; Mineral. Mag. 36, 89–93

Naylor, M.A., 1980, The origin of inverse grading in muddy debris flow deposits—a review; Jour. Sed. Petr. 50, 1111–1116

Neal, J.T., Langer, A.M., and Kerr, P.F., 1968, Giant desiccation polygons of Great Basin playas; GSA Bull. 79, 69–90

Nemec, W. and Steel, R.J., 1984, Alluvial and coastal conglomerates: their significant features and some comments on gravelly mass-flow deposits; pp. 1–31 *in* Koster, E.H. and Steel, R.J., eds., Sedimentology of Gravels and Conglomerates; Canadian Soc. Petroleum Geologists Mem. 10; Calgary; 441 p.

Neumann, A.C., Kofoed, J.W., and Keller, G.H., 1977, Lithoherms in the Straits of Florida; Geol. 5, 4–11

Newell, N.D., 1967, Paraconformities; p. 349–367 *in* Essays in Paleontology and Stratigraphy; Univ. Kansas Dept. Geol spec. publ. 2; 626 p.

Newell, N.D., 1982, Mass extinctions—illusions or realities; pp. 257–263 *in* Silver, L.T. and Schulz, P.H., eds., Geological implications of impacts of large asteroids and comets on the earth; GSA spec. paper 190; Boulder, CO; 528 p.

Newell, N.D., Purdy, E.G., and Imbrie, J., 1960, Bahamian oolitic sand; Jour. Geol. 68, 481–497

Newell, N.D., Rigby, J.K., Fischer, A.G., Whiteman, A.J., Hickox, J.E., and Bradley, J.S., 1953, The Permian Reef Complex of the Guadalupe Mountains Region, Texas and Mexico: a Study in Paleoecology; Freeman, San Francisco; 236 p.

Nilsen, T.H., 1982, Alluvial fan deposits; pp. 49–86 *in* Scholle, P.A. and Spearing, D., eds., Sandstone Depositional Environments; AAPG Mem. 31; 410 p.

Nishiwaki, 1979, Simulation of bed-thickness distribution based on waiting time in the Poisson process; pp. 17–32 *in* Gill, D. and Merriam, D.F., eds., Geomathematical and Petrophysical Studies in Sedimentology; Pergamon Press, Oxford; 267 p.

Normark, W.R., 1970, Growth patterns of deep-sea fans; AAPG Bull. 54, 2170–2195

Nummedal, D. and Swift, D.J.P., 1987, Transgressive stratigraphy at sequence-boundary unconformities: some principles derived from Holocene and Cretaceous examples; pp. 241–260 *in* Nummedal, D., Pilkey, O.H., and Howard, J.D., eds., Sea-Level Fluctuation and Coastal Evolution; SEPM spec. publ. 41; 267 p.

O'Brien, G.W., Harris, J.R., Milnes, H.R., and Veeh, H.H., 1981, Bacterial origin of East Australian continental margin phosphorites; Nature 294, 442–444

O'Brien, N.R., 1970, The fabric of shale—an electron- microscope study; Sedimentology 15, 229–246

O'Brien, N.R., 1987, The effects of bioturbation on the fabric of shale; Jour. Sed. Petr. 57, 449–455

Odin, G.S. and Matter, A., 1981, De glauconiarum irigine; Sedimentology 28, 611–641

Odom, I.E., 1967, Clay fabric and its relation to structural properties in mid-continent Pennsylvanian sediments; Jour. Sed. Petr. 37, 610–623

Odom, I.E., 1975, Feldspar grain size relations in Cambrian arenites, upper Mississippi valley; Jour. Sed. Petr. 45, 636–650

Odom, I.E., Doe, T.W., and Dott, R.H., Jr., 1976, Nature of feldspar-grain size relations in some quartz-rich sandstones; Jour. Sed. Petr. 46, 862–870

Odum, H.T., 1957, Biochemical deposition of strontium; Inst. Mar. Sci. publ. 4, 38–114

Oldershaw, A.E., and Scoffin, T.P., 1967, The source of ferroan and non-ferroan calcite cements in the Halkin and Wenlock Limestones; Geol. Jour. 5, 309–320

Osleger, D. and Read, J.F., 1991, Relation of eustacy to stacking patterns of meter-scale carbonate cycles, Late Cambrian, U.S.A.; Jour. Sed. Petr. 61, 1225–1252

Pannella, G., MacClintock, C., and Thompson, M.N., 1968, Paleontologic evidence of variations in length of synodic month since Late Cambrian; Sci. 162, 792–796

Park, R.K., 1977, The preservation potential of some recent stromatolites; Sedimentology 24, 485–506

Parkinson, D., 1957, Lower Carboniferous reefs of northern England; AAPG Bull. 41, 511–537

Parrish, J.T., 1982, Upwelling and petroleum source beds, with reference to Paleozoic; AAPG Bull. 66, 750–774

Passega, R., 1957, Texture as characteristic of clastic deposition; AAPG Bull. 41, 1952–1984

Passega, R., 1964, Grain size representation by CM patterns as a geologic tool; Jour. Sed. Petr. 34, 830–847

Patterson, R.J. and Kinsman, D.J.J., 1982, Formation of diagenetic dolomite in coastal sabkha along Arabian (Persian) Gulf; AAPG Bull. 66, 28–43

Pedersen, T.F. and Calvert, S.E., 1990, Anoxia vs. productivity: what controls the formation of organic-carbon-rich sediments and sedimentary rocks?; AAPG Bull. 74, 454–466

Pelletier, B.R., 1958, Pocono paleocurrents in Pennsylvania and Maryland; GSA Bull. 69, 1033–1064

Perrier, R. and Quiblier, J., 1974, Thickness changes in sedimentary layers during compaction history; methods for quantitative evaluation; AAPG Bull. 58, 507–520

Perry, E.A. and Hower, J., 1970, Burial diagenesis in Gulf Coast pelitic sediments; Clays and Clay Minerals 18, 165–177

Perry, E.A. and Hower, J., 1972, Late-stage dehydration in deeply buried pellitic sediments; AAPG Bull. 56, 2013–2021

Peterson, M.N.A. and von der Borch, C.C., 1965, Chert: modern inorganic deposition in a carbonate-precipitating locality; Sci. 149, 1501–1503

Pettijohn, F.J., 1941, Persistence of minerals and geologic age; Jour. Geol. 49, 610–625

Pettijohn, F.J., 1962, Paleocurrents and paleogeography; AAPG Bull. 46, 1468–1493

Pettijohn, F.J., 1975, Sedimentary Rocks; 2nd ed., Harper, New York; 718 p.

Pettijohn, F.J. and Potter, P.E., 1964, Atlas and Glossary of Primary Sedimentary Structures; Springer-Verlag, Heidelberg-Berlin-New York; 370 p.

Pettijohn, F.J., Potter, P.E., and Sevier, R., 1987, Sand and Sandstone; Springer-Verlag, New York-Heidelberg-Berlin; 553 p.

Phleger, F.B., 1969, A modern evaporite deposit in Mexico; AAPG Bull. 53, 824–829

Picard, M.D., 1971, Classification of fine grained sedimentary rocks; Jour. Sed. Petr. 41, 179–195

Picard, M.D. and High, L.D., Jr., 1981, Physical stratigraphy of ancient lacustrine deposits; pp. 233–259 in Ethridge,

F.G. and Flores, R.M., eds., Recent and Ancient Nonmarine Depositional Environments: Models for Exploration; SEPM spec. publ. 31; 349 p.

Pierre, C., Luc, O., and Person, A., 1984, Supratidal evaporitic dolomite at Ojo de Liebre lagoon: mineralogical and isotopic arguments for primary crystallization; Jour. Sed. Petr. 54, 1049–1061

Pilkey, O.H., 1988, Basin plains; giant sedimentation events; pp. 93–99 in Clifton, H.E., ed., Sedimentologic Consequences of Convulsive Geologic Events; GSA spec. paper 229; 157 p.

Pisciotto, K.A. and Garrison, R.E., 1981, Lithofacies and depositional environments of the Monterey Formation, California; pp. 97–122 in Garrison, R.E., Douglas, R.G., Pisciotto, K.A., Isaacs, C.M., and Ingle, J.C., eds., The Monterey Formation and Related Siliceous Rocks of California; SEPM Pacific Sec.; 327 p.

Pittman, E.D. and Duschatko, R.W., 1970, Use of pore casts and scanning-electron microscope to study pore geometry; Jour. Sed. Petr. 40, 1153–1157

Pittman, E.D. and Larese, R.E., 1991, Compaction of lithic sands: experimental results and applications; AAPG Bull. 75, 1279–1299

Playford, P.E. and Cockbain, A.E., 1969, Algal stromatolites: deepwater forms in the Devonian of Western Australia; Science 165, 1008–1010

Playford, P.E. and Lowry, D.C., 1966, Devonian reef complexes of the Canning basin, Western Australia; Bull. Geol. Surv. Western Australia 118; 150 p.

Plummer, L.N., 1975, Mixing of sea water with calcium carbonate ground water; pp. 219–238 in Whitten, E.H.T., ed., Quantitative Studies in the Geological Sciences; GSA Mem. 142; 406 p.

Poldervaart, A., 1956, Zircons in rocks, Pt. 2, Igneous rocks; Am. Jour. Sci. 254, 521–554

Porrenga, D.H., 1965, Chamosite in Recent sediments of the Niger and Orinoco deltas; Geol. Mijnb. 44, 400–403

Porrenga, D.H., 1967, Glauconite and chamosite as depth indicators in the marine environment; Mar. Geol. 5, 495–501

Posamentier, H.W., Allen, G.P., James, D.P., and Tesson, M, 1992, Forced regressions in a sequence stratigraphic framework: concepts, examples, and exploration significance; AAPG Bull. 76, 1687–1709

Potter, P.E., 1978, Petrology and composition of modern big-river sands; Jour. Geol. 86, 423–449

Potter, P.E. and Blakely, R.F., 1968, Random processes and lithologic transitions; Jour. Geol. 76, 154–170

Potter, P.E. and Glass, H.D., 1958, Petrology and sedimentation of the Pennsylvanian sediments in southern Illinois: a vertical profile; Ill. Geol. Surv. Rept. of Inv. 204; 60 p.

Potter, P.E. and Mast, R.F., 1963, Sedimentary structures, sand shape fabrics and permeability, Pt. I; Jour. Geol. 71, 441–471

Potter, P.E., Maynard, J.B., and Pryor, W.A., 1980, Sedimentology of Shale; Springer-Verlag, New York-Heidelberg-Berlin; 306 p.

Potter, P.E. and Pettijohn, F.J., 1977, Paleocurrents and Basin Analysis; 2nd ed.; Springer-Verlag, Berlin-Heidelberg-New York; 425 p.

Potter, P.E. and Pryor, W.A., 1961, Dispersal centers of Paleozoic and later clastics of the upper Mississippi valley and adjacent areas; GSA Bull. 72, 1195–1250

Potter, P.E. and Scheidegger, A.E., 1966, Bed thickness and grain size: graded beds; Sedimentology 7, 233–240

Potter, P.E. and Siever, R. 1955, A comparative study of upper Chester and Lower Pennsylvanian stratigraphic variability; Jour. Geol. 63, 429–451

Power, G.M., 1968, Chemical variation in tourmalines from southwest England; Mineralogical Mag. 36, 1078–1089

Powers, M.C., 1953, A new roundness scale for sedimentary particles; Jour. Sed. Petr. 23, 117–119

Powers, M.C., 1967, Fluid-release mechanisms in compacting marine mudrocks and their importance in oil exploration; AAPG Bull. 51, 1240–1254

Pratt, B.R., 1982a, Stromatolitic framework of carbonate mud-mounds; Jour. Sed. Petr. 52, 1203–1227

Pratt, B.R., 1982b, Stromatolite decline—a reconsideration; Geol. 10, 512–515

Pratt, B.R., 1984, *Epiphyton* and *Renalcis*—diagenetic microfissils from calcification of coccoid blue-green algae; Jour. Sed. Petr. 54, 948–971

Pratt, B.R. and James, N.P., 1982, Cryptalgal-metozoan bioherms of Early Ordovician age in the St. George Group, western Newfoundland; Sedimentology 29, 543–569

Pratt, L.M., 1984, Influence of paleoenvironmental factors on preservation of organic matter in middle Cretaceous Greenhorn Formation, Pueblo, Colorado; AAPG Bull. 68, 1146–1159

Pray, L.C., 1958, Fenestrate bryozoan core facies, Mississippian bioherms, southwestern United States; Jour. Sed. Petr. 28, 261–273

Pryor, W.A., 1960, Cretaceous sedimentation in upper Mississippi embayment; AAPG Bull. 44, 1473–1504

Pryor, W.A., 1975, Biogenic sedimentation and alteration of argillaceous sediments in shallow marine environments; GSA Bull. 86, 1244–1254

Pryor, W.A. and Glass, H.D., 1961, Cretaceous-Tertiary clay mineralogy of the upper Mississippi embayment; Jour. Sed. Petr. 31, 38–51

Pryor, W.A. and van Wie, W.A., 1971, The "sawdust sand"—an Eocene sediment of floccule origin; Jour. Sed. Petr. 41, 763–769

Puigdefabregas, C. and van Vliet, A., 1978, Meandering stream deposits from the Tertiary of the southern Pyrenees; pp. 469–485 *in* Miall, A.D., ed., Fluvial Sedimentology; Canadian Soc. Petroleum Geol. Mem. 5; 859 p.

Purser, B.H. and Aissaoui, D.M., 1985, Reef diagenesis: dolomitization and dedolomitization at Mururoa atoll, French Polynesia; Proc. 5th Internat. Coral Reef Congr., Tahiti, 3, 263–269

Pytte, A.M. and Reynolds, R.C., 1989, The thermal transformation of smectite to illite; pp. 133–140 *in* Naeser, N.D. and McCulloh, T.H., eds., Thermal History of Sedimentary Basins; Methods and Case Histories; Springer-Verlag, New York; 320 p.

Radke, B.M. and Mathis, R.L., 1980, On the formation and occurrence of saddle dolomite; Jour. Sed. Petr. 50, 1149–1168

Raiswell, R., 1971, The growth of Cambrian and Liassic concretions; Sedimentology 17, 147–171

Raiswell, R., 1976, The microbiological formation of carbonate concretions in the upper Lias of NE England; Chem. Geol. 18, 227–244

Rampino, M.R. and Stothers, R.B., 1984, Terrestrial mass extinctions, cometary impacts and the sun's motion perpendicular to the galactic plane; Nature 308, 709–712

Ramseyer, K and Boles, J.R., 1986, Mixed-layer illite/smectite minerals in Tertiary sandstones and shales, San Joachin basin, California; Clays Clay Miner. 34, 115–124

Rau, R.C. and Amaral, E.J., 1969, Electron microscopy of precious opal; Metallography 2, 323–328

Raup, D.M. and Seilacher, A., 1969, Computer simulation of fossil foraging behavior; Sci. 166, 994–995

Raup, D.M. and Sepkoski, J.J., Jr., 1982, Mass extinctions in marine fossil record; Sci. 215, 1501–1503

Raup, D.M. and Sepkoski, J.J., Jr., 1984, Periodicity of extinctions in the geologic past; Natl. Acad. Sci. U.S.A. Proc. 81; 801–805

Raup, D.M. and Sepkoski, J.J., Jr., 1986, Periodic extinctions of families and genera; Sci. 231, 833–836

Read, J.F., 1973a, Carbonate cycles, Pillara Formation (Devonian), Canning basin, Western Australia; Canadian Petroleum Geol. 21, 38–51

Read, J.F., 1973b, Paleoenvironments and paleography, Pillara Formation (Devonian), Western Australia; Bull. Canadian Petroleum Geol. 21, 344–394

Read, J.F., 1980, Carbonate ramp-to-basin transitions and foreland basin evolution, Middle Ordovician, Virginia Appalachians; AAPG Bull. 64, 1575–1612

Read, J.F., 1985, Carbonate platform facies models; AAPG Bull. 69, 1–21

Reading, H.G., 1987, Fashions and models in sedimentology: a personal perspective; Sedimentology 34, 3–9

Rees, A.J., 1983, Experiments on the production of transverse grain alignment in a sheared dispersion; Sedimentology 30, 437–448

Reid, R.P. and Browne, K.M., 1991, Intertidal stromatolites in a fringing Holocene reef complex, Bahamas; Geol. 19, 15–18

Reineck, H.-E. and Singh, I.B., 1980, Depositional Sedimentary Environments; 2nd. ed., Springer-Verlag, Berlin-Heidelberg-New York; 549 p.

Reineck, H.-E. and Wunderlich, F., 1968, Classification and origin of flaser and lenticular bedding; Sedimentology 11, 99–104

Renton, J.J., Heald, M.T., and Cecil, C.B., 1969, Experimental investigation of pressure solution of quartz; Jour. Sed. Petr. 39, 1107–1117

Retallack, G.J., 1984, Trace fossils of burrowing beetles and bees in an Oligocene paleosol, Badlands National Park, South Dakota; Jour. Paleont. 58, 571–592

Reynolds, S. and Gorsline, D.S., 1992, Clay microfabric of deep-sea, detrital mud(stone)s, California continental borderland; Jour. Sed. Petr. 62, 41–53

Rhodes, E.G., 1982, Depositional model for a chenier plain, Gulf of Carpentaria, Australia; Sedimentology 29, 201–221

Ricci-Lucchi, F., 1969, Channelized deposits in the middle Miocene flysch of Romagna (Italy); Gior. Geologica, ser. 2a, 36, 203–282

Ricci-Lucchi, F., 1975, Depositional cycles in two turbidite formations of northern Appenines (Italy); Jour. Sed. Petr. 45, 3–43

Ricci-Lucchi, F. and Valmori, E., 1980, Basin-wide turbiditesin a Miocene over-supplied deep-sea plain: a geometrical analysis; Sedimentology 27, 241–270

Rich, M., 1965, Calcispheres from the Duperow Formation (Upper Devonian) in western North Dakota; Jour. Paleont. 39, 143–145

Richter, D.K. and Fuchtbauer, H., 1981, Merkmale und genese von Breccien und irhe Bedeutung im Mesozoikum von Hydra (Griechenland); Zeitschr. Deutchen Geol. Gesellsch. 132; 451–501

Richter, D.K., 1972, Authigenic quartz preserving skeletal material; Sedimentology 19, 211–218

Riggs, S.R., 1980, Intraclast and pellet phosphorite sedimentation in the Miocene of Florida; Jour. Geol. Soc. (London) 137, 741–748

Rimsaite, J., 1964, On micas from magmatic and metamorphic rocks; Beitrag Mineralogie Petrographie 10, 152–183

Rimsaite, J., 1967, Optical heterogeneity of feldspars observed in diverse Canadian rocks; Schweiz. Min. Petr. Mitt. 47, 61–76

Rink, 1976, Jour. Microscopy 107, pt. 3, 267–386

Roberts, A.E., 1966, Stratigraphy of Madison Group near Livingston, Montana, and discussion of karst and solution-breccia features; U.S. Geol. Surv. prof. paper 526–B; 23 p.

Roberts, H.H., Aharon, P., Carney, R., Larkin, J., and Sassen, R., 1990, Sea floor response to hydrocarbon seeps, Louisiana continental slope; Geo-Marine Letters 10, 232– 241

Roedder, E., 1972, Composition of fluid inclusions; U.S. Geol. Surv. prof. paper 440–JJ; 164 p.

Roedder, E., 1979, Fluid inclusion evidence on the environments of sedimentary diagenesis, a review; pp. 89–107 in Scholle, P.A. and Schluger, P.R., eds., Aspects of Diagenesis; SEPM spec. publ. 26; 443 p.

Rose, P.R., 1976, Mississippian carbonate shelf margins, western United States; U.S. Geol. Surv. Jour. Res. 4, 449–466

Rosin, P. and Rammler, E., 1933, Laws governing the fineness of powdered coal; Jour. Inst. Fuel 7, 29–36

Ross, R.J., Jr., Jaanusson, V., and Friedman, I., 1975, Lithology and origin of Middle Ordovician calcareous mudmound at Meiklejohn Peak, southern Nevada; U.S. Geol. Surv. prof. paper 871; 48 p.

Rubin, D.M. and Hunter, R.E., 1982, Bedform climbing in theory and nature; Sedimentology 29, 121–138

Rubin, D.M. and Hunter, R.E., 1984, Sedimentary structures formed in sand by surface tension on melting hailstones; Jour. Sed. Petr. 54, 581–582

Runnells, D.D., 1969, Diagenesis, chemical sediments, and the mixing of natural waters; Jour. Sed. Petr. 39, 1188–1201

Russell, R.D. and Taylor, R.E., 1937, Roundness and shape of Mississippi River sands; Jour. Geology 45, 225–267

Rust, B.R., 1981, A classification of alluvial channel systems; pp. 187–198 in Miall, A.D., ed., Fluvial Sedimentology; Canadian Soc. Petroleum Geologists Mem. 5; 859 p.

Sahu, B.K., 1964, Depositional mechanisms from the size analysis of clastic sediments; Jour. Sed. Petr. 34, 73–84

Sahu, B.K., 1965, Transformation of weight- and number-frequencies for phi-normal size distributions; Jour. Sed. Petr. 35, 973–975

Sallenger, A.H., 1979, Inverse grading and hydraulic equivalence in grain-flow deposits; Jour. Sed. Petr. 49, 553–562

Saller, A.H., 1984, Petrologic and geochemical constraints on the origin of subsurface dolomite, Enewetak Atoll: an example of dolomitization by normal seawater; Geology 12, 217–220

Saller, A.H., 1986, Radiaxial calcite in lower Miocene strata, subsurface Enewetak Atoll; Jour. Sed. Petr. 56, 743–762

Sandberg, C.A. and Gutschick, R.C., 1984, Distribution, microfauna, and source-rock potential of Mississippian Delle Phosphatic Member of Woodman Formation and equivalents, Utah and adjacent states; pp. 135–178 in Woodward, J., Meissner, F.F., and Clayton, J.L., eds., Hydrocarbon Source Rocks in the Greater Rocky Mountain Region; Rocky Mountain Assoc. Geologists; 557 p.

Sandberg, C.A., Ziegler, W., Dresen, R., and Butler, J.L., 1992, Conodont biochronology, biofacies, taxonomy, and event stratigraphy around the middle Frasnian Lion Mudmound (F2h), Frasnes, Belgium: Cour. Forsch.-Inst. Senkenberg 150; 87 p.

Sandberg, P.A., 1975, New interpretation of Great Salt Lake ooids and of ancient non-skeletal carbonate minerology; Sedimentology 22, 497–537

Sandberg, P.A., 1983, An oscillating trend in Phanerozoic non-skeletal carbonate mineralogy; Nature 305, 19–22

Sandberg, P.A., 1985, Aragonite cements and their occurrence in ancient limestones; pp. 33–57 in Schneidermann, N. and Harris, P.M., eds., Carbonate Cements; SEPM spec. publ. 36, 379 p.

Sanders, J.E., 1965, Primary sedimentary structures formed by turbidity currents and related resedimentation mechanisms; pp. 192–219 in Middleton, G.V., ed., Primary Sedimentary Structures and Their Hydrodynamic Interpretation; SEPM spec. publ. 12; Tulsa, OK; 265 p.

Sanderson, I.D., 1974, Sedimentary structures and their environmental significance in the Navajo Sandstone, San Raphael swell, Utah; Brigham Young Univ. Geol. Studies 21, 215–246

Sando, W.J., 1988, Madison Limestone (Mississippian) paleokarst: a geologic synthesis; pp. 256–277 in James, N.P. and Choquette, P.W., Paleokarst; Springer-Verlag, New York-Berlin-Heidelberg; 416 p.

Sass, E. and Kolodny, Y., 1972, Stable isotopes, chemistry and petrology of carbonate concretions (Mishash Formation, Israel); Cem. Geol. 10, 261–286

Savrda, C.E. and Bottjer, D.J., 1986, Trace-fossil model for reconstruction of paleo-oxygenation in bottom waters; Geol. 14, 3–6

Schafer, A. and Stapf, K.R.G., 1978, Permian Saar-Nahe basin and Recent Lake Constance (Germany): two environments of lacustrine algal carbonates; pp. 81–107 in Matter, A. and Tucker, M.E., eds., Modern and Ancient Lake Sediments; Internat. Assoc. Sedimentologists spec. publ. 2; Blackwell, Oxford; 290 p.

Scheidegger, A.E., 1957, The Physics of Flow through Porous Media; Macmillan, New York; 236 p.

Schenk, C.J. and Richardson, R.W., 1985, Recognition of interstitial anhydrite dissolution: a cause of secondary porosity, San Andres Limestone, Mexico, and upper Minnelusa Formation, Wyoming; AAPG Bull. 69, 1064–1076

Scherer, M., 1987, Parameters influencing porosity in sandstones: a model for sandstone porosity prediction; AAPG Bull. 71, 485–491

Schlager, W. and Bolz, H., 1977, Clastic accumulation of sulphate evaporites in deep water; Jour. Sed. Petr. 47, 600–609

Schlager, W., 1992, Sedimentology and Sequence Stratigraphy of Reefs and Carbonate Platforms; Continuing Education Course Note Ser. 34; AAPG; 71 p.

Schlanger, S.O. and Jenkyns, H.C., 1976, Cretaceous oceanic anoxic events: causes and consequences; Geol. Mijnb. 55, 179–184

Schlee, J., 1957, Upland gravels of southern Maryland; GSA Bull. 68, 1371–1410

Schmalz, R.F., 1969, Deep-water evaporite deposition: a genetic model; AAPG Bull. 53, 798–823

Schmidt, R., 1981, Descriptive nomenclature and classification of pyroclastic deposits and fragments: recommendations of the IUGS Subcommission on the Systematics of Igneous Rocks; Geology 9, 41–43

Schmidt, V. and McDonald, D.A., 1979, The role of secondary porosity in the course of sandstone diagenesis;

pp. 175–207 in Scholle, P.A. and Schluger, P.R., eds., Aspects of Diagenesis; SEPM spec. publ. 26; 443 p.

Schnitzer, W.A., 1957, Die Quarzkornfarbe als Hilfsmittel fuer die stratigraphische und palaeogeographische Erforschung sandiger Sedimente (aufgezeigt an Beispielen auf Ostbayern); Erlanger Geol. Abh. 23, p.

Scholle, P.A., 1971, Diagenesis of deep-water carbonate turbidites, Upper Cretaceous Monte Antola flysch, northern Appenines, Italy; Jour. Sed. Petr. 41, 233–250

Scholle, P.A., 1978, A Color Illustrated Guide to Carbonate Rock Constituents, Textures, Cements and Porosities; AAPG Mem. 27; 241 p.

Schopf, J.M., 1956, A definition of coal; Econ. Geol. 51, 521–527

Schrader, H.-J., 1971, Fecal pellets: role in sedimentation of pelagic diatoms; Sci. 174, 55–57

Schreiber, B.C., 1986, Arid shorelines and evaporites; pp. 189–228 in Reading, H.G., ed., Sedimentary Environments and Facies; 2nd ed.; Blackwell, Oxford-London; 615 p.

Schreiber, B.C., Friedman, G.M., Decima, A., and Schreiber, E., 1976, Depositional environments of upper Miocene (Messinian) evaporite deposits of the Sicilian basin; Sedimentology 23, 729–760

Schroeder, J.H., 1973, Submarine and vadose cements in Pleistocene Bermuda reef rock; Sed. Geol. 10, 179–204

Schumm, S.A., 1960a, The shape of alluvial channels in relation to sediment type; U.S. Geol. Surv. prof. paper 352-B, 17–30

Schumm, S.A., 1960b, The effect of sediment type on the shape and stratification of some modern fluvial deposits; Am. Jour. Sci. 258, 177–184

Schumm, S.A., 1963, Sinuosity of alluvial channels of the Great Plains; GSA Bull. 74, 1089–1100

Schumm, S.A., 1968, Speculations concerning paleohydrologic controls of terrestrial sedimentation; GSA Bull. 79, 1573–1588

Schumm, S.A., 1972, Fluvial paleochannels, pp. 98–107 in Rigby, J.K. and Hamblin, W.K., eds., Recognition of Ancient Sedimentary Environments; SEPM spec. publ. 16; 340 p.

Schwab, F.L., 1969a, Geosynclines: what contribution to the crust?; Jour. Sed. Petr. 39, 150–158

Schwab, F.L., 1969b, Cyclic geosynclinal sedimentation: a petrographic evaluation: Jour. Sed. Petr. 39, 1325–1343

Schwarcz, H.P. and Shane, K.C., 1970, Measurement of particle shape by Fourier analysis; Sedimentology 13, 213–231

Schwartz, D.E., 1978, Hydrology and current orientation analysis of a braided-to-meandering transition: the Red River in Oklahoma and Texas, U.S.A.; pp. 231–255 in Miall, A.D., ed., Fluvial Sedimentology; Canadian Soc. Petroleum Geol. Mem. 5; 859 p.

Schwarzacher, W., 1953, Cross-bedding and grain size in Lower Cretaceous sands of East Anglia; Geol. Mag. 90, 322–330

Schwarzacher, W., 1961, Petrology and structure of some

Lower Carboniferous reefs in northwestern Ireland; AAPG Bull. 45, 1481–1503

Scrutton, C.T., 1965, Periodicity in Devonian coral growth; Paleontology 7, 552–558

Seilacher, A., 1953, Studien zur Palichnologie. I. Uber die Methoden den Palichnologie; Neues Jahrb. Geol. Palaont. Abh. 96, 421–452

Seilacher, A., 1959, Fossilien als Stromungsanzeiger; Aus der Heimat 67, 171–177

Seilacher, A., 1960, Stromungsanzeichnen im Hunsruckschiefer; Notizbl. Hess. Landesamtes Bodenforsch. 88, 88–106

Seilacher, A., 1967a, Bathymetry of trace fossils; Mar. Geol. 5, 413–428

Seilacher, A., 1967b, Fossil behavior; Sci. Am. 217, 72–80

Sellwood, B.W., 1986, Shallow-marine carbonate environments; p 283–342 in Reading, H.G., ed., Sedimentary Environments and Facies; Blackwell; 615 p.

Sepkoski, J.J., Jr., 1982a, Flat pebble conglomerates, storm deposits, and the Cambrian bottom fauna; pp. 371–388 in Einsele, G. and Seilacher, A., eds., Cyclic and Event Stratigraphy; Springer-Verlag, Berlin-Heidelberg-New York; 536 p.

Sepkoski, J.J., Jr., 1982b, Mass extinctions in the Phanerozoic oceans: a review; pp. 283–289 in Silver, L.T. and Schulz, P.H., eds., Geological Implications of Impacts of Large Asteroids and Comets on the Earth; GSA spec. paper 190; 528 p.

Shackleton, J.S., 1962, Cross strata of the Rough Rock (Millstone Grit Series) in the Pennines; Liverpool and Manchester Geol. Jour. 3, 109–120

Shaw, D.B. and Weaver, C.E., 1965, The mineralogical composition of shale; Jour. Sed. Petr. 35, 213–222

Shaw, D.M. and Bugry, R., 1966, A review of boron sedimentary geochemistry in relation to new analyses of some North American shales; Canadian Jour. Earth Sciences 3, 49–63

Shaw, J., and Kellerhalls, R., 1977, Paleohydraulic interpretation of antidune bedforms with applications to antidunes in gravel; Jour. Sed. Petr. 47, 257–266

Shearman, D.J., 1966, Origin of marine evaporites by diagenesis; Inst. Mining Metallurgy, Trans. 75, B208–B215

Shearman, D.J., 1970, Recent halite rock, Baja California, Mexico; Inst. Mining Metallurg. Trans. 79, p. 155–162

Shearman, D.J. and Fuller, J.G., 1969, Anhydrite diagenesis, calcitization, and organic laminites, Winnipegosis Formation, Middle Devonian, Saskatchewan; Bull. Canadian Petroleum Geol. 17, 496–525

Shearman, D.J., Khouri, J., and Taha, S., 1961, On the replacement of dolomite by calcite in some Mesozoic limestones from the French Jura; Proc. Geologists' Assoc. 72, 1–12

Sheldon, R.P., 1981, Ancient marine phosphorites; Ann. Rev. Earth Planet. Sci. 9, 251–284

Shepard, F.P. and Dill, R.F., 1966, Submarine Canyons and Other Sea Valleys; Rand-McNally, Chicago; 381 p.

Sherer, M., 1987, Parameters influencing porosity in sandstones: a model for sandstone porosity prediction; AAPG Bull. 71, 485–491

Sheriff, R.E., 1976, Inferring stratigraphy from seismic data; AAPG Bull. 60, 528–542

Sheriff, R.E., 1980, Seismic Stratigraphy; Internat. Human Res. Development Corp; 227 p.

Shinn, E.A. and Lidz, B.H., 1988, Blackened limestone pebbles: fire at subaerial unconformities; pp. 117–131 in James, N.P. and Choquette, P.W., eds., Paleokarst; Springer-Verlag, New York-Berlin-Heidelberg; 416 p.

Shinn, E.A. and Robbin, D.A., 1983, Mechanical and chemical compaction in fine-grained shallow-water limestones; Jour. Sed. Petr. 53, 595–618

Shoemaker, E.M., 1977, Why study impact craters?; pp. 1–10 in Roddy, D.J., Pepin, R.O., and Merrill, R.B., eds., Impact and Explosion Cratering; Pergamon; 1301 p.

Sibley, D.F., 1982, The origin of common dolomite fabrics: clues from the Pliocene; Jour. Sed. Petr. 52, 1087–1100

Sibley, D.F. and Blatt, H., 1976, Intergranular pressure solution and cementation of the Tuscarora orthoquartzite; Jour. Sed. Petr. 46, 881–896

Siebert, R.M., Moncure, G.K., and Lahann, R.W., 1984, A theory of framework grain dissolution in sandstones; pp. 163–175 in McDonald, D.A. and Surdam, R.C., eds., Clastic Diagenesis; AAPG Mem. 37; 434 p.

Siever, R., 1959, Petrology and geochemistry of silica cementation in some Pennsylvanian sandstones; pp. 55–79 in Ireland, H.A., ed., Silica in Sediments; SEPM spec. publ. 7; 185 p.

Siever, R., 1983, Evolution of chert at active and passive continental margins; pp. 7–24 in Iijima, A., Hein, J.R., and Siever, R., eds., Siliceous Deposits in the Pacific Region; Elsevier; 472 p.

Sillen, L.G., 1967, The ocean as a chemical system; Sci. 156, 1189–1197

Simon, H.A., 1978, The sizes of things; pp. 195–202 in Tanur, J.M., et al., eds., Statistics: a Guide to the Unknown; Holden-Day, San Francisco; 430 p.

Simons, D.B, Richardson, E.V., and Nordin, C.F., Jr., 1965, Sedimentary structures generated by flow in alluvial channels; pp. 34–52 in Middleton, G.V., ed., Primary Sedimentary Structures and their Hydrodynamic Interpretation—a Symposium; SEPM spec. publ. 12; 265 p.

Simpson, S., 1975, Classification of trace fossils; pp. 39–54 in Frey, R.W., ed., The Study of Trace Fossils; Springer-Verlag, New York-Heidelberg-Berlin; 562 p.

Sippel, R.F., 1968, Sandstone petrology, evidence from luminescence petrography; Jour. Sed. Petr. 38, 530–554

Skipper, K., 1971, Antidune cross-stratification in a turbidite sequence, Cloridorme Formation, Gaspe, Quebec; Sedimentology 17, 51–68

Sleep, N.H. and Windley, B.F., 1982, Archen plate tectonics: constraints and inferences; Jour. Geol. 90, 363–379

Sloss, L.L., 1963, Sequence in the cratonic interior of North America; GSA Bull. 74, 93–114

Sloss, L.L., 1972, Synchrony of Phanerozoic sedimentary-tectonic events of the North American craton and the Russian platform; 24th Internat. Geol. Congr., Montreal, sect. 6; 24–32

Sloss, L.L., 1984, Comparative anatomy of cratonic unconformities; pp.1–6 in Schlee, J.S., ed., Interregional Unconformities and Hydrocarbon Accumulation; AAPG Mem. 36; 184 p.

Smith, C.S., 1981, A Search for Structure; MIT Press, Cambridge, MA; 410 p.

Smith, J.V., 1974, Feldspar Minerals, 2. Chemical and Textural Properties; Springer-Verlag, New York; 690 p.

Smith, J.V. and Stenstrom, R.C., 1965, Electron excited luminescence as a petrologic tool; Jour. Geol. 73, 627–635

Smith, N.D., 1972, Flume experiments on the durability of mud clasts; Jour. Sed. Petr. 42, 378–383

Smith, S.V., Buddemeier, R.W., Redalje, R.C., and Houck, J.E., 1979, Strontium-calcium thermometry in coral skeletons; Sci. 204, 404–407

Snarsky, A.N., 1962, Die primare Migration des Erdols; Freiberger Forsch. 123, 63–73

Sneed, E.P. and Folk, R.L., 1958, Pebbles in the lower Colorado River, Texas, a study in particle morphogenesis; Jour. Geology 66, 114–150

Soeder, D.J., 1990, Applications of fluorescence microscopy to study of pores in tight rocks; AAPG Bull. 74, 30–40

Sonnenfeld, P., 1991, Evaporite basin analysis; Chapter 12 in Force, E.R., Eidel, J.J., and Maynard, J.B., eds., Sedimentary and Diagenetic Mineral Deposits: A Basin Analysis Approach to Exploration; Rev. Econ. Geol. 5.

Southard, J.B., 1971, Representation of bed configurations in depth-velocity-size diagrams; Jour. Sed. Petr. 41, 903–915

Southard, J.B. and Boguchwal, L.A., 1990, Bed configurations in steady unidirectional water flows. Pt. 2. synthesis of flume data; Jour. Sed. Petr. 60, 658–678

Sparks, R.S.J., 1976, Grain size variations in ignimbrites and implications for the transport of pyroclastic flows; Sedimentology 23, 147–188

Spearing, D.R., 1974, Model of an offshore, shallow marine sand body: Shannon sandstone, Wyoming; (abs); AAPG ann. mtg. abs. 1974, v. 1; 85

Spears, D.A., 1989, Aspects of iron incorporation into sediments with special reference to the Yorkshire ironstones; pp. 19–30 in Young, T.P. and Taylor, W.E.G., eds., Phanerozoic Ironstones; G.S.A. spec. publ. 46; 251 p.

Spencer, D.W., 1963, The interpretation of grain size distribution curves of clastic sediments; Jour. Sed. Petr. 33, 180–190

Sprinkle, J. and Gutschick, R.C., 1967, Costatoblastus, a channel fill blastoid from the Sappington Formation of Montana; Jour. Paleont. 41, 385–403

Stach, E., Mackowsky, M.-Th., Teichmuller, M., Taylor, G.H., Chandra, D., and Teichmuller, R., 1982, Stach's Textbook of Coal Petrology; 3rd. ed.; Gebruder Borntraeger, Berlin-Stuttgart; 535 p.

Stauffer, D. and Aharony, A., 1992, Indtroduction to Percolation Theory; 2nd ed.; Taylor and Francis, London; 200 p.

Stehli, F.G. and Hower, J., 1961, Mineralogy and early diagenesis of carbonate sediments; Jour. Sed. Petr. 31, 358–371

Stein, C.L., 1982, Silica recrystallization in petrified wood; Jour. Sed. Petr. 52, 1277–1282

Steiner, J., 1967, The sequence of geological events and the dynamics of the Milky Way Galaxy; Geol. Soc. Australia Jour. 14, 99–131

Steiner, J., 1973, Possible galactic causes for synchronous sedimentation sequences of the North American and eastern European cratons; Geology 1, 89–92

Steiner, J. and Grillmair, E., 1973, Possible galactic causes for periodic and episodic glaciations; GSA Bull. 84, 1003–1018

Stoessell, R.K. and Pittman, E.D., 1990, Secondary porosity revisited: the chemistry of feldspar dissolution by carboxylic acids and anions; AAPG Bull. 74, 1795–1805

Stokes, W.L., 1968, Multiple parallel-truncation bedding planes—a feature of wind-deposited sandstone formations; Jour. Sed. Petr. 38, 510–515

Stowe, D.A.V. and Lovell, J.P.B., 1979, Contourites; their recognition in modern and ancient sediments; Earth Sci. Rev. 14, 251–291

Strasser, A., 1984, Black-pebble occurrence and genesis in Holocene carbonate sediments (Florida keyes, Bahamas, and Tunisia); Jour. Sed. Petr. 54, 1097–1109

Sugden, D.E. and John, B.S., 1976; Glaciers and Landscape—a Geomorphological Approach; Wiley, New York; 376 p.

Sundborg, A., 1956, The River Klaralven, a study in fluvial processes; Geol. Annlr. 38, 127–316

Surdam, R.C., Boese, S.W., and Crossey, L.J., 1984, The chemistry of secondary porosity; pp. 127–161 in McDonald, D.A. and Surdam, R.C., eds., Clastic Diagenesis; AAPG Mem. 37; 434 p.

Surdam, R.C. and Boles, J.R., 1979, Diagenesis of volcanic sandstones; pp. 227–242 in Scholle, P.A. and Schluger, P.R., eds., Aspects of Diagenesis; SEPM spec. publ. 26; 443 p.

Suttner, L.J., 1974, Sedimentary petrographic provinces: an evaluation; pp. 75–84 in Ross, C.A., ed., Paleogeographic Provinces and Provinciality; SEPM spec. publ. 21, 233 p.

Suttner, L.J., Basu, A., and Mack, G.H., 1981, Climate and the origin of quartzarenites; Jour. Sed. Petr. 51, 1235–1246

Suttner, L.J. and Leininger, R.K., 1972, Comparison of the trace element content of plutonic, volcanic, and metamorphic quartz from southwestern Montana; GSA Bull. 83, 1855–1862

Swann, D.H., Lineback, J.A., and Frund, E., 1965, The Borden siltstone (Mississippian) delta in southwestern Illinois; Ill. Geol. Surv. circ. 386, 20 p.

Sweeney, R.E. and Kaplan, I.R., 1978, Pyrite framboid formation: laboratory synthesis and marine sediments; Econ. Geol. 68, 618–634

Swett, K., 1968, Authigenic feldspars and cherts resulting from dolomitization of illitic limestones: a hypothesis; Jour. Sed. Petr. 38, 128–135

Swett, K., Klein, G. deV., and Smit, D., 1971, A Cambrian tidal sand body; the Eriboll Sandstone of northwest Scotland, an ancient-recent analog; Jour. Geol. 79, 400–415

Swift, D.J.P., 1968, Coastal erosion and transgressive stratigraphy; Jour. Geol. 76, 444–456

Swift, D.J.P., Sanford, R.B., Dill, C.E., Jr., and Avignone, N.F., 1971, Textural differentiation on the shoreface during erosional retreat of an unconsolidated coast, Cape Henry to Cape Hatteras, western North Atlantic shelf; Sedimentology 16, 221–250

Tada, R., 1991, Origin of rhythmical bedding in middle Miocene siliceous rocks of the Onnagama Formation, northern Japan; Jour. Sed. Petr. 61, 1123–1145

Taira, A. and Scholle, P.A., 1979, Discrimination of depositional environments using settling tube data; Jour. Sed. Petr. 49, 787–800

Tanner, W.F., 1958, The zig-zag nature of type I and type IV curves; Jour. Sed. Petr. 28, 372–375

Tanner, W.F., 1963, Crushed pebble conglomerate of southwestern Montana; Jour. Geol. 71, 637–641

Tanner, W.F., 1967, Ripple-mark indices and their uses; Sedimentology 9, 89–104

Taylor, J.C.M. and Illing, L.V., 1969, Holocene intertidal calcium carbonate cementation, Qatar, Persian Gulf; Sedimentology 12, 69–109

Taylor, J.M., 1950, Pore-space reduction in sandstone; AAPG Bull. 34, 701–716

Thiry, M. and Millot, G., 1987, Mineralogical forms of silica and their sequence of formation in silcretes; Jour. Sed. Petr. 57, 343–352

Thomas, C.M., 1965, Origin of pisolites (abs.); AAPG Bull. 49, 360

Thomas, G.S.P. and Connell, R.J., 1985, Iceberg drop, dump, and grounding structures from Pleistocene glacio-lacustrine sediments, Scotland; Jour. Sed. Petr. 55, 243–249

Thomson, A., 1959, Pressure solution and porosity; p. 92–110 in Ireland, H.A., ed., Silica in Sediments—a Symposium; SEPM spec. publ. 7; 185 p.

Tillman, R.W. and Martinsen, R.S., 1984, The Shannon shelf-ridge sandstone complex, Salt Creek anticline area, Powder River basin, Wyoming; pp. 85–142 in Tillman, R.W. and Siemers, C.T., eds., Siliciclastic Shelf Sediments; SEPM spec. publ. 34; 268 p.

Tissot, B.P. and Pelet, R., 1971, Nouvelles donnees sur les mecanismes de genese et de migration du petrole, simulation mathematique et application a la prospection; Proc. 8th World Petroleum Cong., 35–46

Tissot, B.P. and Welte, D.H., 1984, Petroleum Formation and Occurrence; Springer-Verlag, Berlin-Heidelberg-New York; 699 p.

Tomita, T., 1954, Geologic significance of the color of granite zircon, and the discovery of the Pre-Cambrian in Japan; Mem. Fac. Sci. Kyushu Univ., Ser. D, Geology, 4, 135–161

Tortosa, A., Palomares, M., and Arribas, J., 1991, Quartz grain types from the Spanish Central System: some problems in provenance analysis; pp. 47–54 in Morton, A.C., Todd., S.P., and Haughton, P.D.W., eds., Developments in Sedimentary Provenance Analysis; Geol. Sec. London spec. publ. 57; 370 p.

Trendall, A.F., 1968, Three great basins of Precambrian banded iron formation deposition: a systematic comparison; GSA Bull. 79, 1527–1544

Tucholke, B.E. and Embley, R.W., 1984, Cenozoic regional erosion of the abyssal sea floor off South Africa; pp. 145–164 in Schlee, J.S., ed., Interregional Unconformities and Hydrocarbon Accumulation; AAPG Mem. 36; 184 p.

Tucker, M.E., 1983, Diagenesis, geochemistry, and origin of a Precambrian dolomite: the Beck Spring Dolomite of eastern California; Jour. Sed. Petr. 53, 1097–1119

Tucker, R.W. and Vacher, H.L., 1980, Effectiveness of discriminating beach, dune, and river sands by moments and the cumulative weight percentage; Jour. Sed. Petr. 50, 165–172

Underwood, E.E., 1970, Quantitative Stereology; Addison-Wesley, Reading, MA; 247 p.

Vail, P.R., Hardenbol, J., and Todd, R.G., 1984, Jurassic unconformities, chronostratigraphy, and sea-level changes from seismic stratigraphy and biostratigraphy; pp. 129–144 in Schlee, J.S., ed., Interregional Unconformities and Hydrocarbon Accumulation; AAPG Mem. 36; 184 p.

Vail, P.R., Mitchum, R.M., and Thompson, S., III, 1977, Seismic stratigraphy and global changes of sea level, Part 4; Global cycles of relative changes of sea level; pp. 83–98 in Payton, C.E., ed., Seismic Stratigraphy—Applications to Hydrocarbon Exploration; AAPG Mem. 26; 516 p.

Valentine, J.W. and Moores, E., 1970, Plate tectonic regulation of faunal diversity and sea level; Nature 228, 657–669

Valloni, R. and Maynard, J.B., 1981, Detrital modes of recent deep-sea sands and their relation to tectonic setting: a first approximation; Sedimentology 28, 75–83

Valloni, R. and Mezzadri, G., 1984, Compositional suites of terrigenous deep-sea sands of the present continental margins; Sedimentology 31, 353–364

van Andel, Tj. H. and Komar, P.D., 1969, Ponded sediments of the Mid-Atlantic Ridge between 22° and 23° north latitude; GSA Bull. 80, 1163–1190

van de Graaff, F.R., 1970, Fluvial-deltaic facies of the Castlegate Sandstone (Cretaceous), east-central Utah; Jour. Sed. Petr. 42, 558–571 van Houten, F.B., 1964, Cyclic lacustrine sedimentation, Upper Triassic Lockatong Formation, central New Jersey and adjacent Pennsylvania; Kansas Geol. Soc. Bull. 169, 497–531

van Loon, A.J., 1992, The recognition of soft-sediment deformations as early-diagenetic features—a literature review; pp. 135–189 in Wolf, K.H. and Chilingarian, G.V., eds., Diagenesis III; Developments in Sedimentology 47; Elsevier, Amsterdam; 674 p.

van Moort, J.C., 1972, The K_2O, CaO, MgO and CO_2 contents of shales and related rocks and their implications for sedimentary evolution since the Proterozoic; 24th Internat. Geol. Congr., Montreal, sect. 10, 427–439

Vanoni, V.A., 1946, Transportation of suspended sediment by water; Trans. Am. Soc. Civil Engineers 111, 67–102

Veevers, J.J., 1990, Tectono-climatic supercycle in the billion-year plate-tectonic eon: Permian Pangean icehouse alternates with Cretaceous dispersed-continent greenhouse; Sed. Geol. 68, 1–16

Veizer, J., 1988, The evolving exogenic cycle, p 175–220 in Gregor, C.B., Garrels, R.M., Mackenzie, F.T., and Maynard, J.B., eds., Chemical Cycles in the Evolution of the Earth; Wiley, New York

Velde, B., 1989, Phyllosilicate formation in berthierine peloids and iron oolites; pp. 3–8 in Toung, T.P. and Taylor, W.E.G., eds., Phanerozoic Ironstones; Geol. Soc. London spec. publ. 46, 251 p.

Velde, B. and Hower, J., 1963, Petrological significance of illite polymorphism in Paleozoic sedimentary rocks; Am. Mineralogist 48, 1239–1254

Vinogradov, A.P. and Ronov, A.B., 1956, Evolution of chemical composition of clays of the Russian platform; Geochem. 2, 123–139

Visher, G.S., 1965a, Fluvial processes as interpreted from ancient and recent fluvial deposits; pp. 116–132 in Middleton, G.V., ed., Primary Sedimentary Structures and their Hydrodynamic Interpretation, SEPM spec. publ. 12, 265 p.

Visher, G.S., 1965b, Use of vertical profile in environmental reconstruction; AAPG Bull. 49, 41–61

Visher, G.S., 1969, Grain size distributions and depositional processes; Jour. Sed. Petr. 39, 1074–1106

Vitaliano, D., 1973, Legends of the Earth; Indiana Univ. Press, Bloomington, IN; 305 p.

von Bitter, P.H., Scott, S.D., and Schenk, P.E., 1992, Chemosynthesis: an alternative hypothesis for Carboniferous biotas in bryozoan/microbial mounds, Newfoundland, Canada; Palaios 7, 466–484

von Bubnoff, S., 1963, Fundamentals of Geology; Oliver and Boyd; London 289 p.

von Burkalow, A., 1945, Angle of repose and angle of sliding friction: an experimental study; GSA Bull. 56, 669–708

von der Borch, C.C., 1976, Stratigraphy and formation of Holocene dolomitic carbonate deposits of the Coorong area, South Australia; Jour. Sed. Petr. 46, 952–966

von der Borch, C.C. and Lock, D.E., 1979, Geological significance of Coorong dolomites; Sedimentology 26, 813–824

von der Borch, C.C., Lock, D.E., and Schwebel, D., 1975, Ground-water formation of dolomite in the Coorong region of South Australia; Geol. 3, 283–285

Waag, K.M. and Ogren, D.E., 1984, Shape evolution and fabric in a boulder beach, Monument Cove, Maine; Jour. Sed. Petr. 54, 98–102

Wadell, H., 1932, Volume, shape, and roundness of rock particles; Jour. Geol. 40, 443–451

Wager, L.R., and Brown, G.M., 1967, Layered Igneous Rocks; Freeman, San Francisco; 588 p.

Walker, R.G., 1967, Turbidite sedimentary structures and their relationship to proximal and distal depositional environments; Jour. Sed. Petr. 37, 25–43

Walker, R.G., 1975, Generalized facies models for resedimented conglomerates of turbidite association; GSA Bull. 86, 737–748

Walker, R.G., 1978, Deep-water limestone facies and ancient submarine fans: models for exploration for stratigraphic traps; AAPG Bull. 62, 932–966

Walker, R.G. and Mutti, E., 1973, Turbidite facies and facies associations; pp. 119–157 in Middleton, G.V. and Bouma, A.H., eds., Turbidites and Deep Sea Sedimentation; SEPM, Pacific Coast sect.; 157 p.

Walker, T.R., 1960, Carbonate replacement of detrital crystalline silicate minerals as a source of authigenic silica in sedimentary rocks; GSA Bull. 71, 145–152

Walker, T.R., 1962, Reversible nature of chert-carbonate replacement in sedimentary rocks; GSA Bull. 73, 1237–1242

Walker, T.R., 1967, Formation of red beds in modern and ancient deserts; GSA Bull. 78, 353–368

Walker, T.R. and Harms, J.C., 1972, Eolian origin of flagstone beds, Lyons Sandstone (Permian), type area, Boulder County, Colorado; Mtn. Geologist 9, 279–288

Walker, T.R. and Honea, R.M., 1969, Iron content of modern deposits in the Sonoran Desert: a contribution to the origin of red beds; GSA Bull. 80, 535–543

Wallace, M.W., 1987, The role of internal erosion and sedimentation in the formation of stromatactis mudmounds and associated lithologies; Jour. Sed. Petr. 57, 695–700

Walter, L., 1985, Relative reactivity of skeletal carbonates during dissolution: implications for diagenesis; pp. 3–16 in Schneidermann, N. and Harris, P.M., eds., Carbonate Cements; SEPM spec. publ. 36; 379 p.

Waples, D.W., 1980, Time and temperature in petroleum formation: application of Lopatin's method to petroleum exploration; AAPG Bull. 64, 916–926

Waples, D.W., 1983, Reappraisal of anoxia and organic richness, with emphasis on Cretaceous in North Atlantic; AAPG Bull. 67, 963–978

Ward, W.C., Folk, R.L., and Wilson, J.L., 1970, Blackening of eolianite and caliche adjacent to saline lakes, Isla Mujeres, Quintana Roo, Mexico; Jour. Sed. Petr. 40, 548–555

Ward, W.C. and Halley, R.B., 1985, Dolomitization in a

mixing zone of near-seawater composition, late Pleistocene, northeastern Yucatan peninsula; Jour. Sed. Petr. 55, 407–420

Wardlaw, N.C., 1976, Pore geometry of carbonate rocks as revealed by pore casts and capillary pressure; AAPG Bull. 60, 245–257

Wardlaw, N.C., 1979, Pore systems in carbonate rocks and their influence on hydrocarbon recovery efficiency; pp. E1–E23 in Moore, C.H., ed., Geology of Carbonate Porosity; AAPG cont. ed. course note ser. 11

Wardlaw, N.C. and Reinson, G.E., 1971, Carbonate and evaporite deposition and diagenesis, Middle Devonian Winnipegosis and Prairie Evaporite Formations of south-central Saskatchewan; AAPG Bull. 55, 1759–1786

Warrak, M., 1974, The petrography and origin of dedolomitized, veined or brecciated carbonate rocks, the ''cornieules'', in the Frejus region, French Alps; Jour. Geol. Soc. (London) 130, 229–247

Warren, J.K. and Kendall, C.G.St.C., 1985, Comparison of sequences formed in marine sabkha (subaerial) and salina (subaqueous) settings—modern and ancient; AAPG Bull. 69, 1013–1023

Weaver, C.E., 1967, Potassium, illite, and the ocean; Geochim. Cosmochim. Acta. 31, 2181–2196

Weaver, C.E., 1989, Clays, Muds, and Shales; Developments in Sedimentology 44; Elsevier, New York-Amsterdam; 819 p.

Weber, J.N., 1958, Recent grooving in lake bottom sediments at Great Slave Lake, Northwest Territories; Jour. Sed. Petr. 28, 333–341

Weber, J.N., 1973, Incorporation of strontium into reef coral skeletal carbonate; Geochim. Cosmochim. Acta 37, 2170–2190

Weeks, L.G., 1953, Environment and mode of origin and facies relationships of carbonate concretions in shales; Jour. Sedimentary Petrology 23, 162–173

Weimer, P., 1990, Sequence stratigraphy, facies geometries, and depositional history of the Mississippi fan, Gulf of Mexico; AAPG Bull. 74, 425–453

Weimer, R.J. and Hoyt, J.H., 1964, Burrows of Callianassa major Say, geologic indicators of littoral and shallow neritic environments; Jour. Paleont. 38, 761–767

Weissman, P., 1990, Are periodic bombardments real?; Sky and Telescope 79, 266–270

Wells, J.W., 1963, Coral growth and geochronometry; Nature 197, 948–950

Welton, J.E., 1984, SEM Petrology Atlas; AAPG Methods in Exploration Series; 237 p.

Wengerd, S.A. and Strickland, J.W., 1954, Pennsylvanian stratigraphy of Paradox salt basin, Four Corners region, Colorado and Utah; AAPG Bull. 38, 2157–2199

Weyl, P.K., 1958, The solution kinetics of calcite; Jour. Geol. 66, 163–176

Weyl, P.K., 1959, Pressure solution and the force of crystallization—a phenomenological theory; Jour. Geophys. Res. 64, 2001–2025

Wheeler, H.E., 1958, Time stratigraphy; AAPG Bull. 42, 1047–1063

White, L.D., Garrison, R.E., and Barron, J.A., 1992, Miocene intensification of upwelling along the California margin as recorded in siliceous facies of the Monterey Formation and offshore DSDP sites; pp. 429–442 in Summerhayes, C.P., Prell, W.L., and Eneis, K.C., eds., Upwelling Systems: Evolution Since Early Miocene; Geol. Soc. London spec. publ. 64; 519 p.

Whittaker, R.H. and Margulis, L., 1978, Protist classification and the kingdoms of organisms; BioSystems 10, 3–18

Whyte, M.A., 1977, Turning points in Phanerozoic history; Nature 267, 679–682

Wicksell, S.D., 1925, The corpuscle problem: a mathematical study of a biometric problem; Biometrika 17, 84–99

Wilkinson, B.H., 1979, Biomineralization, paleoceanography, and the evolution of calcareous marine organisms; Geol. 7, 524–527

Wilkinson, B.H., Owen, R.M., and Carroll, A.R., 1985, Submarine hydrothermal weathering, global eustacy, and carbonate polymorphism in Phanerozoic marine oolites; Jour. Sed. Petr. 55, 171–183

Williams, G.E., 1989, Late Precambrian tidal rhythmites in South Australia and the history of the Earth's rotation; Jour. Geol. Soc. (London) 146, 97–111

Williams, P.F. and Rust, B.R., 1969, The sedimentology of a braided river; Jour. Sed. Petr. 39, 649–679

Wilson, I.G., 1972, Aeolian bedforms—their development and origins; Sedimentology 19, 173–210

Wilson, J.C. and McBride, E.F., 1988, Compaction and porosity evolution of Pliocene sandstones, Ventura basin, California; AAPB Bull. 72, 664–681

Wilson, J.L., 1967, Carbonate-evaporite cycles in lower Duperow Formation of Williston basin; Canadian Petroleum Geol. 15, 230–312

Wilson, J.L., 1974, Characteristics of carbonate platform margins; AAPG Bull. 58, 810–824

Wilson, J.L., 1975, Carbonate Facies in Geologic History; Springer-Verlag, New York-Heidelberg-Berlin; 471 p.

Wilson, M.D., 1970, Upper Cretaceous-Paleocene synorogenic conglomerates of southwestern Montana; AAPG Bull. 54, 1843–1867

Wilson, M.D., 1992, Inherited grain-rimming clays in sandstones from aeolian and shelf environments: their origin and control on reservoir properties; p. 209–225 in Houseknecht, D.W. and Pittman, E.D., eds., Origin, Diagenesis, and Petrophysics of Clay Minerals in Sandstones; SEPM spec. publ. 47; 282 p.

Wilson, M.D. and Pittman, E.D., 1977, Authigenic clays in sandstones: recognition and influence on reservoir properties and paleoenvironmental analysis; Jour. Sed. Petr. 47, 3–31

Winkler, H.G.F., 1979, Petrogenesis of Metamorphic Rocks; 5th ed.; Springer-Verlag; 348 p.

Winkler, E.M. and Singer, P.C., 1972, Crystallization pressure of salts in stone and concrete; GSA Bull. 83, 3509–3514

Winland, H.D. and Matthews, R.K., 1974, Origin and significance of grapestone, Bahama Islands; Jour. Sed. Petr. 44, 921–927

Winter, J., 1982, Habits of zircon as a tool for precise tephrostratigraphic correlation; p. 423–428 in Einsele, G. and Seilacher, A., eds., Cyclic and Event Stratigraphy; Springer-Verlag; 536 p.

Wobber, F.J., 1967, Post-depositional structures in the Lias, South Wales; Jour. Sed. Petr. 37, 166–174

Wolfe, M.J., 1967, An electron microscope study of the surface texture of sand grains from a basal conglomerate; Sedimentology 8, 239–247

Wolfe, M.J., 1970, Dolomitization and dedolomitization in the Senonian chalk of northern Ireland; Geol. Mag. 107, 39–49

Wong, P.K. and Oldershaw, A., 1981, Burial cementation in the Devonian Kaybob reef complex, Alberta, Canada; Jour. Sed. Petr. 51, 507–520

Woodworth-Lynas, C.M.T. and Guigue, J.Y., 1990, Iceberg scours in the geological record: examples from glacial Lake Agassiz; pp. 217–223 in Dowdeswell, J.A. and Scourse, J.D., eds., Glacimarine Environments: Processes and Sediments; Geol. Soc. London spec. publ. 53, 423 p.

Woronick, R.E. and Land, L.S., 1985, Late burial diagenesis, Lower Cretaceous Pearsall and lower Glen Rose Formations, south Texas; pp. 265–275 in Schneidermann, N. and Harris, P.M., eds., Carbonate Cements; SEPM spec. publ. 36; 379 p.

Wray, J.L., 1977, Calcareous Algae; Elsevier, Amsterdam; 186 p.

Wright, L.D., 1977, Sediment transport and deposition at river mouths: a synthesis; GSA Bull. 88, 857–868

Wright, W.I., 1938, The composition and occurrence of garnets; Am. Mineralogist 23, 436–449

Wyrwoll, K.-H. and Smyth, G.K., 1985, On using the log-hyperbolic distribution to describe the textural characteristics of aeolian sediments; Jour. Sed. Petr. 55, 471–478

Yalin, M.S., 1964, Geometrical properties of sand waves; Am. Soc. Civil Engineers Proc. 90; HY5, 105–119

Yamamoto, S., 1992, Diagenetic enrichment of manganese and other heavy metals in hemipelagic brown clay of the Palau trench floor; Jour. Sed. Petr. 62, 706–717

Yeakel, L.S., 1959, Tuscarora, Juniata, and Bald Eagle paleocurrents and paleogeography in the central Appalachians; GSA Bull. 73, 1515–1540

Zahner, R., 1968, Biologische Abbauvorgange im Bodensediment von Seen; Wasser- und Abwasser-Forsch. 4/68, 1–5

Zenger, D.H., 1983, Burial dolomitization in the Lost Burrow Formation (Devonian), east-central California, and the significance of late diagenetic dolomitization; Geol. 11, 519–522

Zingg, Th., 1935, Beitrage zur Schotteranalyse; Schweiz. Mineralog. Petrog. Mitt. 15, 39–140

Zinkernagel, U., 1978, Cathodoluminescence of quartz and its application to sandstone petrology; Contributions to Sedimentology 8; 69 p.

Name Index

Subject Index